РОСТ КРИСТАЛЛОВ
Rost Kristallov
GROWTH OF CRYSTALS

VOLUME 13

Growth of Crystals

Volume 13

Edited by

E. I. Givargizov

Institute of Crystallography
Academy of Sciences of the USSR, Moscow

Translated by

V. I. Kisin

Moscow, USSR

Translation edited by

E. A. D. White

British Telecom Research Laboratories

CONSULTANTS BUREAU · NEW YORK AND LONDON

The Library of Congress cataloged the first volume of this title as follows:

Growth of crystals. v. [1]
   New York, Consultants Bureau, 1958–

    v. illus., diagrs. 28 cm.
   Vols. 1, 3–   constitute reports of 1st–   Conference on Crystal Growth, 1956–   v. 2 contains interim reports between the 1st and 2nd Conference on Crystal Growth, Institute of Crystallography, Academy of Sciences, USSR.

   "Authorized translation from the Russian" (varies slightly)
   Editors: 1958–   A. V. Shubnikov and N. N. Sheftal.

   1. Crystal–Growth. I. Shubnikov, Aleksei Vasil'evich, ed. II. Sheftal', N. N., ed. III. Consultants Bureau Enterprises, inc., New York, IV. Soveshchanie po rostu kristallov. V. Akademiia nauk SSSR. Institut kristallografii.
QD921.R633   548.5   58-1212

ISBN 978-1-4615-7121-6   ISBN 978-1-4615-7119-3 (eBook)
DOI 10.1007/978-1-4615-7119-3

The original Russian text was published for the Institute of Crystallography of the Academy of Sciences of the USSR by Nauka Press in Moscow in 1980.

© 1986 Consultants Bureau, New York
Softcover reprint of the hardcover 1st edition 1986
A Division of Plenum Publishing Corporation
233 Spring Street, New York, N.Y. 10013

All rights reserved

No part of this book may be reproduced, stored in a retrieval system, or transmitted in any form or by any means, electronic, mechanical, photocopying, microfilming, recording, or otherwise, without written permission from the Publisher

PREFACE

The present volume continues the tradition of the preceding volumes, covering a wide range of crystal growth problems and treating aspects of critical importance for crystallization. Changes in this field of knowledge have, however, changed the criteria for selection of papers for inclusion in this series.

The increasing role of crystals in science and technology is even more apparent today. The study and utilization of these highly perfect objects of nature considerably facilitates progress in the physics and chemistry of solids, quantum electronics, optics, microelectronics, and other sciences. The demand for crystals and crystal devices has grown steadily and has led to the emergence and rapid growth of the single crystal industry (we can safely say that the state of the art in this industry is indicative of the overall scientific and technological potential of a country). At the same time, the introduction of crystallization techniques into other industries is gaining ever-increasing importance. To illustrate this last statement, we can mention the fabrication of textured structural materials and direct methods of metal reduction in ores by using chemical vapor transport techniques. Crystallization techniques progress both in "width" and in "depth": traditional methods are modernized, and novel techniques appear, some of them at the junction of the already existing technologies (for example, flux growth of crystals, growth from vapor with participation of the liquid phase, etc.). As a consequence (and to some extent a prerequisite) of this process, research effort has increased and the number of publications on crystal growth, especially on growth techniques, is mounting. Numerous, and increasingly narrowed and specialized, conferences are organized, directed at specific types of growth process and sometimes to specific materials (silicon, gallium arsenide, etc.).

This makes fundamental research on crystal growth more important than ever before. The demands on size and quality (and, in particular, of structural perfection) of crystals have become more stringent, and this can be met only via a detailed analysis of atomic-scale crystallization events and the mechanisms of crystal lattice growth. This calls for highly sophisticated experiments and broad theoretical generalizations, which were the requirements demanded of the invited papers at the 5th USSR conference on crystal growth (Tbilisi, September 1977). The present volume is based on reports delivered at this conference and the discussions that followed.

This volume reflects all fundamental approaches to crystal growth research: theory, experimental crystal growth, and production-scale growth. Some of the contributions present generalizations and some are review papers, but most of them contain new, original results. The papers are grouped into topically defined parts.

In the first paper, B. K. Vainshtein discusses the principles of self-organization and crystallization of protein molecules, reviews the methods of protein crystal growth, and reports new results.

Parts II and III (12 papers) are devoted to the mechanisms and kinetics of crystal growth by vapor deposition and to epitaxial growth. This aspect is of considerable practical importance, especially in microelectronics. One unquestionable advantage of gas or vapor-phase crystallization is the fact that, from the standpoint of crystal growth, this technique embodies, more than any other technique, the atomic (or molecular) nature of lattice growth; this feature facilitates both crystallization experiments (including direct in situ observation of elementary attachment and detachment atomic events) and various theoretical analyses, both qualitative and quantitative, including computer simulations. The papers in Parts II and III analyze the role of adsorbed layers in chemical vapor deposition, micromorphology of homoepitaxial layers of semiconductor compounds, and the anisotropy of layered growth in the process of formation of epitaxial structures. Another important aspect is artificial epitaxial growth, the so-called graphoepitaxy or diataxy, regarded as a way of controlling the structure of crystal layers on amorphous substrates. One of the papers discusses vapor condensation in strong electric fields. Parts II and III reflect to some extent certain experimental techniques and the relevant results.

Part IV is devoted to crystallization from the melt and to flux growth. In such processes interfaces can be observed directly only in exceptional situations, so that the potentials of experimental studies are necessarily limited; it is therefore not surprising that three out of five papers in Part IV are based on theoretical calculations, and in particular on computer simulation of crystal growth. On the whole Part IV presents a reasonably complete picture of the mechanisms and kinetics of crystal growth from the liquid phase.

The remaining parts of this volume analyze crystal growth techniques.

In Part V, devoted to crystallization from the melt, I would single out three papers which discuss the Stepanov technique of edge-defined growth of single crystals from the melt. The special interest in growing ribbon crystals, for example, silicon ribbons, stems from the energy crisis and hence from the requirement for minimum-cost solar batteries. This field still faces a number of technological problems which are outlined fairly extensively in the papers of this part.

Part VI reviews crystal growth from high-temperature solutions: growth from fluxes and from hydrothermal media. One of the papers is devoted to the growth of emerald by various methods, including gas transport techniques; substantial success has been achieved in the synthesis of this rare mineral. The remaining papers of this part discuss various methods of flux crystallization and the methodology of hydrothermal crystallization and its physicochemical features.

Part VII discusses the defect structure of crystals in relation to growth conditions. The generation of dislocations in semiconductors grown from melts is treated in two papers which use different approaches to the problem: one derives the basic mechanism of defect formation from trapping of the melt in the course of growth, and the other deals with thermally induced stress. One of the papers discusses defect generation in heteroepitaxial compositions based on ternary and quaternary solutions of $A^{III}B^V$ compounds. The last two papers of the part are devoted to the formation of spherulites in thin films.

The concluding part (Part VIII) deals with new materials and crystallization equipment. Among the new materials I would single out (again in connection with the energy crisis) crystals with enhanced ionic conductivity whose growth is discussed in one of the papers. The last two papers in Part VIII deal with equipment for crystal growth: A rather complete survey is given of the methods of controlling complex parameters of the crystallization process, followed by a short article on crystallization by laser heating.

I am grateful to N. N. Sheftal', A. A. Chernov, K. S. Bagdasarov, and V. A. Kuznetsov at the Institute of Crystallography of the USSR Academy of Science for helpful advice. Considerable work was done by S. A. Grinberg and L. A. Zadorozhnaya in compiling the manuscript, and by V. I. Muratova, M. A. Golosova, and L. N. Obolenskaya in bringing them to the final acceptable form.

E. I. Givargizov

CONTENTS

	PAGE	RUSS. PAGE

I. CRYSTALLIZATION OF PROTEINS

Protein Molecules and Crystals. B. K. Vainshtein — 3, 9

II. MECHANISMS AND KINETICS OF CRYSTAL GROWTH BY VAPOR DEPOSITION

Equilibrium Adsorption Layers on GaAs (111) and Si (111) Surfaces in CVD Growth. A. A. Chernov and M. P. Ruzaikin — 19, 20

Role of Adsorption Layer in Chemical Vapor Deposition. E. I. Givargizov — 27, 27

Kinetics and Mechanism of Gallium Arsenide Growth in Gas-Transport Systems. L. G. Lavrentyeva, I. V. Ivonin, and L. P. Porokhovnichenko — 34, 33

Local Epitaxy Under Conditions of Strong Growth Rate Anisotropy. P. B. Pashchenko, G. A. Aleksandrova, and I. M. Skvortsov — 46, 45

Electron Microscopic Observation and Computer Simulation of Step Redistribution in Step Trains due to Changes in Step Density. K. W. Keller, D. Katzer, and H. Höche — 54, 52

Crystal Growth Under Thermodynamically Metastable Conditions. B. V. Spitsyn — 58, 55

Role of Defects in the Nucleation of Whiskers Growing from Vapor. S. A. Ammer and A. F. Tatarenkov — 67, 63

Field Emission Microscopic Study of Thermal Field- and Condensation-Induced Growth Forms of Crystal Tips. V. N. Shrednik — 74, 68

III. EPITAXY

Peculiarities and Mechanism of Graphoepitaxy. N. N. Sheftal' and V. I. Klykov — 89, 80

Application of Electron Microscopy to a Study of Kinetics and Mechanism of Crystallization. Gl. S. Zhdanov — 95, 85

The Effects of Substrate-Mediated Interaction Between Adsorbed Atoms on the Structure of Two-Dimensional Crystals Formed from these Atoms. V. K. Medvedev, A. G. Naumovets, and A. G. Fedorus — 108, 96

Crystal Growth and Polytypism in Silicon Carbide. Yu. M. Tairov and V. F. Tsvetkov — 117, 104

IV. MECHANISMS AND KINETICS OF CRYSTAL GROWTH FROM THE MELT AND FROM HIGH-TEMPERATURE SOLUTIONS

Motion of Low-Angle Macrosteps. V. V. Voronkov	127	112
Peculiarities of Melt Growth of Crystals with Different Entropies of Melting. G. A. Alfintsev and D. E. Ovsiyenko	137	121
Kinetic Conditions at the Growth Interface of a Mixed Crystal. D. E. Temkin	152	134
General Approach to Monte Carlo Simulation of Crystal Growth. T. A. Cherepanova	162	143
Parameters Characterizing the Kinetics of Dissolution of Crystalline Germanium in Liquid Ge-Au. S. A. Grinberg	174	153

V. GROWTH OF CRYSTALS FROM THE MELT

Stability of Crystallization in Edge-Defined Film-Fed Growth from the Melt. V. A. Tatarchenko	185	160
Shape and Properties of Crystals Grown from the Melt by the Stepanov Techniques. P. I. Antonov	198	171
Crystal Shape Stability in Meniscus-Controlled Growth Processes. T. Surek, S. R. Coriell, and B. Chalmers	208	180
Numerical Analysis of Heat and Mass Transfer in the Growth of Large Single Crystals from the Melt. N. A. Avdonin and V. A. Smirnov	221	191
Phase Diagrams of Binary Systems Formed by Rare Earth Trifluorides. B. P. Sobolev, P. P. Fedorov, A. K. Galkin, V. S. Sidorov, and D. D. Ikrami	229	198

VI. GROWTH OF CRYSTALS FROM SOLUTIONS

Progress in Flux Growth of Large Crystals. V. A. Timofeeva	239	205
Growth of Emerald Single Crystals. G. V. Bukin, A. A. Godovikov, V. A. Klyakhin, and V. S. Sobolev	251	215
Some Technological Procedures and Equipment for Hydrothermal Growth of Single Crystals. V. I. Popolitov, A. N. Lobachev, and A. Ya. Shapiro	261	223
The Role Played by Me^{2+} in Hydrothermal Crystallization of Germanates of Divalent Metals. L. N. Demianets, N. G. Duderov V. A. Kuznetsov, and T. N. Nadezhina	269	231

VII. DEFECT STRUCTURE IN CRYSTALS: RELATION TO GROWTH CONDITIONS

Theoretical and Experimental Studies of Generation of Stress and Dislocations in Growing Crystals. V. L. Indenbom and V. B. Osvensky	279	240
Growth Defects in Semiconductor Crystals. A. N. Buzynin, N. I. Bletskan, Yu. N. Kuznetsov, and N. N. Sheftal'	291	251
Formation of Defects in Epitaxial Heterostructures and Multicomponent Solid Solutions of Semiconductor Compounds. M. G. Mil'vidsky and L. M. Dolginov	301	260

Growth and Structure of Synthetic Amethyst Crystals. L. I. Tsinober
V. E. Khadzhi, E. M. Tsyganov, M. I. Samoilovich, and
A. A. Shaposhnikov   312   271

Defect Structure of Sb_2S_3 Crystals Revealed by Electron Microscope
Crystal Lattice Imaging Techniques. A. A. Sokol, V. M. Kosevich
and A. G. Bagmut   322   280

On the Structure of Crystalline Graphite Intergrowths. V. M. Kosevich
A. P. Lyubchenko, S. N. Grigorov, and G. P. Umansky   332   288

VIII. NEW MATERIALS: EQUIPMENT FOR CRYSTAL GROWTH

Crystal Growth and Properties of Some New Ionic Conductors.
A. Rabenau   343   296

Modern Methods of Monitoring and Control in Crystal Growth.
É. L. Lube   353   304

Application of Laser Heating to Crystal Growth. Kh. S. Bagdasarov,
V. V. Dyachenko, A. M. Kevorkov, and A. Kholov   364   314

SUBJECT INDEX   371

Part I
Crystallization of Proteins

PROTEIN MOLECULES AND CRYSTALS

B. K. Vainshtein

The Institute of Crystallography of the USSR Academy of Sciences, Moscow

Introduction

Protein molecules and crystals formed of them are unique objects which differ in many respects from ordinary organic molecules and crystals of organic and inorganic substances. A globular protein molecule is a system consisting of 10^3-10^4 atoms, built according to the genetic information inherited by a given organism; assembly of the molecules proceeds at the self-organization level. Globular protein molecules may be crystallized, and the crystals thus obtained have a number of striking features which will be discussed below. One has to mention immediately that the protein molecules may associate not only into three-dimensional, triply periodic crystals but also into two-dimensional and tubular crystals, as well as into pseudospherical shells with point symmetry.

The formation of protein molecules is thus governed by specific biological and physicochemical factors which operate within the framework of the laws of physics. The specificity of existence of living systems, which represent the most complex form of organization of matter, determines the appearance of new relationships not observed in non-living matter.

One of the main features of crystals built of biological macromolecules (comprising, in addition to protein crystals, crystals of nucleic acids and viruses) is the very large unit-cell dimensions dictated by the molecular dimensions (see Table 1).

TABLE 1

	Molecular mass, Daltons	Unit cell dimensions, Å	Unit cell volume, Å^3	Number of atoms per unit cell
Typical inorganic and organic crystals	10-10^2	5-20	10^2-10^4	$\lesssim 10^3$
Globular proteins	10^4-10^5	50-250	10^5-10^7	10^4-10^5
Viruses	10^6-10^7	200-1000	10^8-10^{10}	10^7-10^9

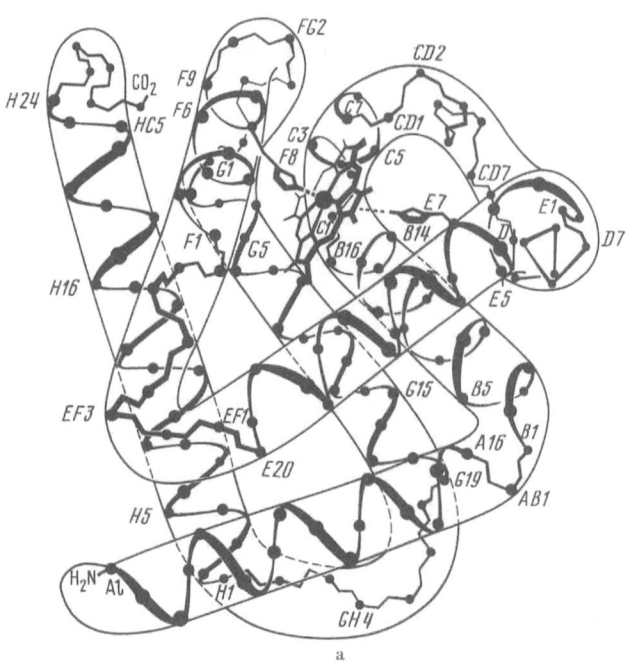

Fig. 1. Structure of myoglobin-type protein molecules. (a) a schematic of the folding of the polypeptide chain in the myoglobin molecule (A—H and the attached numbers mark the α-helical segments and the amino residues); (b) a model of the atomic arrangement in the leghemoglobin molecule (the chain folding in this molecule is similar to that in the myoglobin molecules).

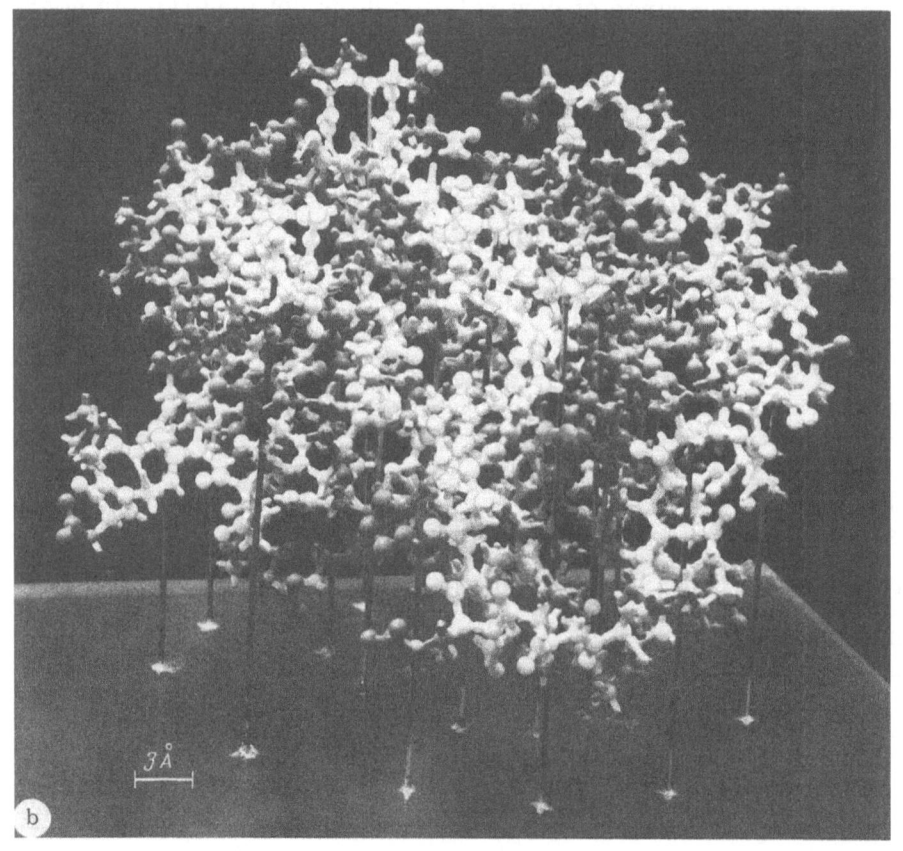

Another important feature of the structure of such crystals is that they not only contain biological macromolecules but also carry mother solution in the intermolecular spaces, mostly water with different ions.

Molecular Structure of Globular Proteins

Proteins are polymer molecules with large molecular mass, built of amino acid residues [1, 2]:

$$H_2N-C_\alpha(H)(R)-COOH$$

Amino acid residues link together and form polypeptide chains

from which the protein molecule is built. The number of amino acid residues in a chain varies from approximately 100 to 500. Proteins are predominantly built of twenty "basic" amino acids. Individual properties of a given amino acid are determined by the radical R. Radicals of one group of amino acids (for example, glycine, phenylalanine, etc.) are not charged; they repel water molecules, that is, they are hydrophobic (nonpolar). Radicals of other amino acids have active polar groups OH, COO^-, NH_2^+, capable of forming hydrogen or ionic bonds. These are hydrophilic, that is, they easily bond to water molecules.

The structural hierarchy of proteins comprises the primary, secondary, tertiary, and quaternary structure levels. The primary structure is the chemical formula of the chain reflecting the sequence of amino residues. The secondary, tertiary, and quaternary structures are geometric in nature and are determined by the chain structure and chain folding and packing.

The secondary structure is a typical stable conformation of a segment of a polypeptide chain determined by the minimum of interaction energy between chain links and by a convenient method of locking hydrogen bonds within the chain and between adjacent chains. The most frequently encountered conformations are the α-helix of Pauling and Corey and the β-structure characterized by a parallel or antiparallel packing of stretched chains [3].

A globular protein molecule is a polypeptide chain folded into a compact globule and manifesting a secondary structure in some of its segments (Fig. 1a, b) [4, 5]; the conformation of the chain between these segments is irregular but this irregular conformation is definitive and stable for a given protein.

The three-dimensional spatial structure of a protein molecule with a specific sequence of amino residues and known locations of all radicals and atoms is precisely the tertiary structure of the protein.

The tertiary spatial structure of a protein molecule (specific for a given protein) follows from the chemically determined primary structure, that is, by the amino acid sequence of the protein dictated by the genetic information encoded in the genes of a given organism. After a protein chain synthesized in a ribosome has been moved to the ambient aqueous medium, the process of folding into a globule proceeds spontaneously and is controlled exclusively by physio-chemical factors: interactions between chain links carrying various amino acid residues and interaction of these residues with the ambient aqueous medium.

Water is not merely a passive medium with protein molecules swimming in it: this ambient helps to form the globular structure. Roughly speaking, a protein molecule having both hydrophilic (polar) and hydrophobic (nonpolar) residues behaves in solution similarly to micella of a fatty acid whose hydrophilic ends readily contact water molecules and whose "nonwetted" hydrophobic ends repel water molecules and are inside the micella in contact with one another [6, 7]. Water molecules attached by hydrogen bonds to polar radicals form around a globule in an ordered layer. At the same time, it is energetically advantageous for hydrophobic groups to avoid contact with water molecules. All this leads to minimizing the free energy and maximizing the entropy of the globule + water system, that is, to the stabilization of this system. Furthermore, van der Waals interactions between nonpolar residues are sufficient to explain the attraction and the resulting compact packing of chains inside the globule. It is significant that a single configuration, reproduced in an enormous number of copies of the molecule, appears in given thermodynamic conditions corresponding to the internal conditions of a living organism, despite the tremendous complexity of the structure.

X-ray structure analyses of protein molecules revealed the basic regularities in their structure. This should be regarded as one of the most important advances in modern molecular biology. By fixing the molecule not only in its free state but also introducing into it inhibitors and pseudosubstrates and analyzing the corresponding crystals by X-ray techniques, it was possible to determine conformational changes in a number of molecules and to find the mechanisms by which their biological functions are implemented. At present about 1,000 various protein species have been crystallized and more than a hundred structures have been completely deciphered. The work on the study of crystals of nucleic acids and viruses is expanding.

Numerous attempts have been made to theoretically predict the spatial structure of protein molecules on the basis of the chemical sequence of amino acid residues. So far this problem remains extremely difficult. Nevertheless, it is already possible to predict with a probability of about 80% the presence of α- or β-segments in the chain [8].

Many proteins are not represented by an individual globule but consist of several symmetrically arranged individual globular subunits. Such aggregation is called a quaternary structure (Fig. 2a) [9]. As a rule, the quaternary structure is not merely a passive "coagulation" of protein subunits, but is essential for the functioning of a protein molecule as a whole (Fig. 2a) [10]. These are the basic factors which determine the formation of the spatial structure of protein molecules.

It must be emphasized that substantial conformational changes may occur in the course of functioning of the protein molecules of enzymes, and displacements of some groups may reach several Ångstroms.

Denaturation (degradation of the native structure) of a protein molecule by various means (heating, chemical factors) can be described as a first-order phase transition.

The interaction between protein molecules attached to a growing crystal is mostly electrostatic [11-13] owing to the presence of charged groups on the crystal surface. Screening by water molecules weakens this interaction. The approximately spherical shape of globules contacting one another in a limited number of points leaves considerable space

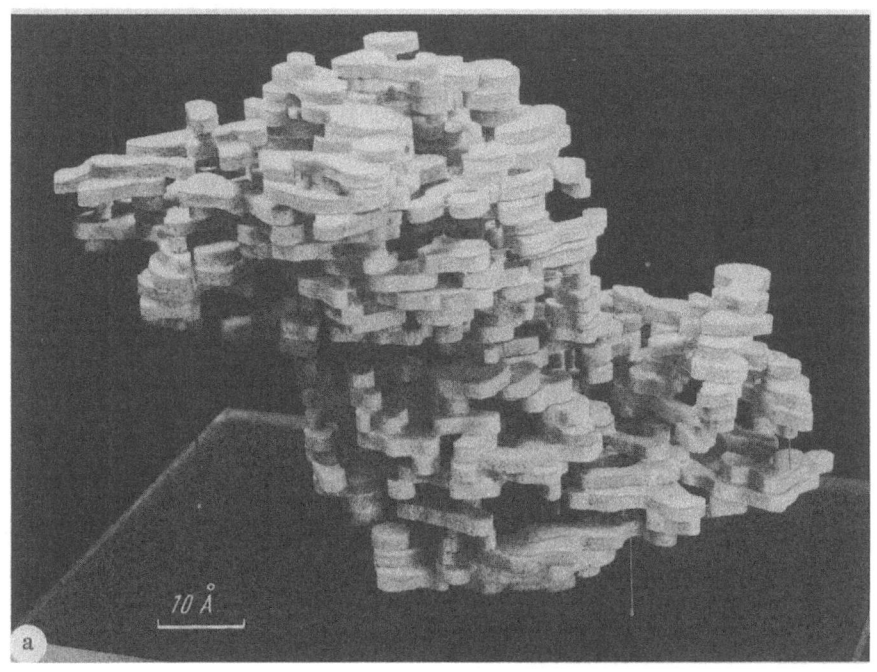

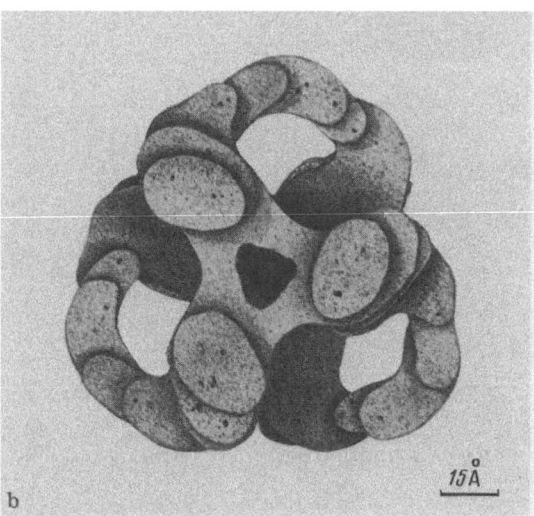

Fig. 2. Quaternary structure of (a) aspartate transaminase (dimer) and (b) leucincaminopeptidase (hexamer).

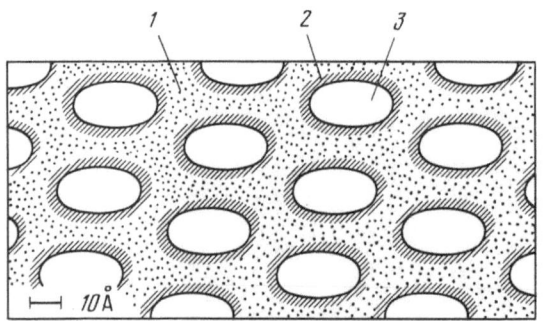

Fig. 3. Scheme of the protein crystal structure. (1) water; (2) bound water; (3) protein.

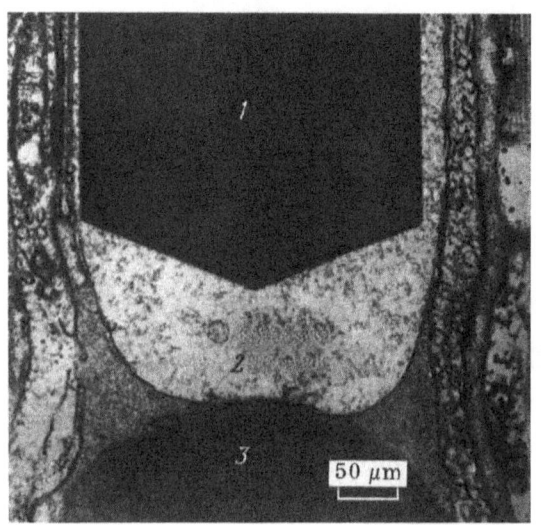

Fig. 4. Hemoglobin crystal in the myocardium of a cat. (1) hemoglobin crystal; (2) erythrocyte membrane; (3) normal erythrocyte.

Fig. 5. Crystals of bacterial catalase.

filled with the mother solution, namely water with inorganic ions, such as sulphate SO_4^{2-}, phosphate PO_4^{3-}, halides F^-, Cl^-, I^-, and others (Fig. 3). From 30 to 70% of the total volume of the crystal is occupied by the solvent. X-ray diffraction data show that only those water molecules directly adjacent to the surface are ordered while the remaining part of the solvent is disordered, although the system as a whole is of course a crystal because of the three-dimensional ordering of the molecular arrangement. In this respect protein crystals are unique: they enclose the same disordered solvent as that surrounding the crystal as a whole.

Naturally, no "melting" can occur in protein crystals, and temperature increase results in "desiccation" and destruction of the crystals and denaturation of the molecules.

Principles and Techniques of Protein Crystallization

Natural three-dimensional protein crystals are found in organisms only rarely since the fundamental function of protein molecules consists in implementing enzymatic reactions in solutions or cell plasma. Such crystals are nevertheless observed sometimes (Fig. 4) [11]. The artificial growth of protein crystals is a very peculiar process, very different from crystallization of conventional organic and inorganic compounds [13]. Crystallization of proteins is preceded by complex biochemical methods of their extraction and purification which will not be dealt with here. It should be emphasized that a researcher has to work with minute amounts of a protein (tens or hundreds of milligrams) because of the difficulties of extraction. All basic thermodynamic relationships, such as dependence on temperature, pressure, concentration (supersaturation), remain valid for protein crystallization. However, the possibility of using minute admixtures of chemical additives in order to affect the surface of protein molecules ("closing" or "opening" specific groups on this surface, thus modifying the interaction of the molecules with the internal medium of the crystal (solvent)) provides additional parameters of the crystallization processes.

The main purpose of the current work on protein crystallization is to grow 0.3-0.5 mm crystals suitable for X-ray structure investigation. Typical shapes of as-grown protein crystals are given in Fig. 5.

The most widespread technique of preparing supersaturated protein solutions are based on lowering the solubility of proteins by adding to the solutions salts or organic solvents. Other methods employ the change in relative concentrations of components of a solution caused by evaporation of the solvent, as well as the temperature and concentration dependence of solubility of hydrogen ions in the solution, that is, the dependence on pH of the solution. The factors affecting protein crystallization are shown in Fig. 6.

Protein solubility curves at different values of pH usually have a minimum in the neighborhood of the isoelectric point. Several minima may point to the existence of several phases with different charge distributions on the molecules. Buffer solutions are used to monitor and control pH. Temperature effect on protein solubility depends on the ionic strength of the solution. In solutions with low ionic strength an increase in temperature enhances the solubility of most proteins. Contrary to this, protein solubility in high-concentration salt solutions usually diminishes with increasing temperature [13]. Organic solvents also diminish protein solubility and are often used in their crystallization.

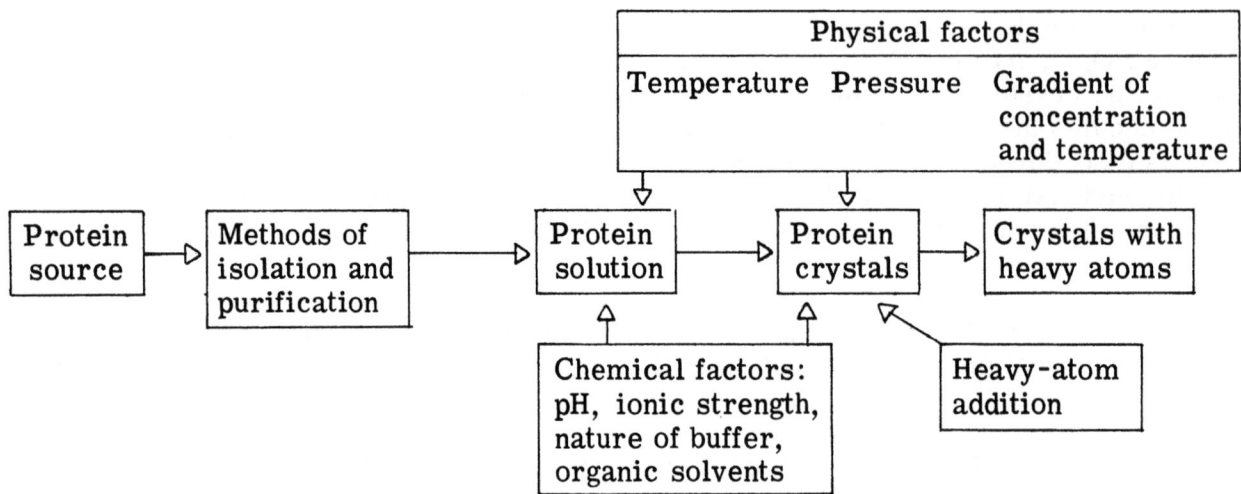

Fig. 6. Schematic diagram of formation of protein crystals.

As a rule the growth of perfect protein crystals takes from several weeks to several months. Protein crystallization techniques can be provisionally classified into crystallization in test tubes, crystallization in dialysis cells, and crystallization based on solvent diffusion through a gas. One ingenious method of crystal growth is based on producing protein supersaturation in a centrifuge at the bottom of centrifuge tubes [14].

The techniques based on the equilibrium dialysis are the most convenient to determine optimum conditions of crystallization. These techniques are based on the impenetrability of cellophane membranes by proteins, and their high penetrability for low-molecular weight compounds such as water, salts, and organic solvents. A schematic of such a cell is shown in Fig. 7 [15].

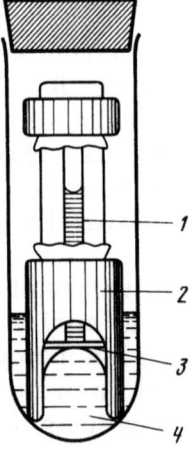

Fig. 7. Microdiffusion (dialysis) cell. (1) protein solution; (2) rubber coupling; (3) membrane; (4) buffer solution containing the precipitating agent.

The above data on protein crystallization can be summarized as follows:

I. Composition of Solutions

1. Salt solutions with high ionic strength (ammonium sulphate, phosphates).
2. Solutions in organic solvents (ethanol, MPD, etc.) with low ionic strength.

II. Methods of Producing Supersaturation in Solutions

1. Changing the solution composition:
   (a) addition of a precipitating agent (dialysis, gaseous diffusion);
   (b) reduction of ionic strength by dialysis;
   (c) alteration of pH;
   (d) evaporation of the solvent.
2. Temperature variation (heating or cooling).
3. Centrifugation.

III. Crystallization Techniques

1. In test tubes.
2. In dialysis cells.
3. In liquid drops.

X-ray structural analysis requires that we obtain not only the crystals of pure native protein but also isomorphs with added heavy atoms, which are required as markers to determine the phase of X-ray reflections. They can be introduced from inorganic or organic ions, such as $AuCl_4^{2+}$, PtI_4^{2+}, $UO_2(OH)_3$, $ClHg-\langle\bigcirc\rangle-SO_3H$, or others [1]. Such ions are adsorbed at specific sites on the protein molecule surface.

Heavy atoms are introduced either by growing crystals in solutions containing these atoms or (more frequently) by saturating as-grown crystals by soaking them in an appropriate ion-containing solution. The ions containing heavy atoms freely diffuse to protein molecules through the internal liquid of the crystal.

Polymorphism of Protein Crystals

Because of the high lability of bonds between protein molecules, polymorphism of protein crystals is a very frequent phenomenon. Polymorphism due to natural configuration changes in protein molecules, related to their biological functions, is also common. In some proteins polymorphism takes another form: in addition to three-dimensional crystals these proteins also form ordered symmetrical associations without a 3-D periodicity. If the number of dimensions in space is n and the number of dimensions along which a periodicity is observed is m, then for $n \geq m$ the possible symmetry groups are G_m^n. In three-dimensional space groups, G_3^3 are space groups, G_2^3 are layer groups, G_1^3 are cylindrical (helical) groups, and G_0^3 are symmetry point groups.

The formation of three-dimentional crystals follows from an approximate equality of the strengths of chemical bonds between atoms or molecules along all directions. In this case the binding elements of the structure promote crystalline order in all three dimensions. But in the general case particles with different symmetries may be formed. In appropriate conditions the structures of biological origin are observed to have G_2^3 and G_1^3 symmetries: layers and tubes [16].

Systems with point symmetry G_0^3 also exist. The formation of assemblies of biological molecules is dictated not only by physico-chemical laws but also by the principle of biological expediency enforcing the behavior of the molecule in accordance with its function in a living system. For example, consider the structure of the tobacco mosaic virus [17]; molecules in this virus are stacked in helical symmetry (Fig. 8). The shape of these molecules and the interaction between contacting segments of their surfaces stimulates this packing whose function is a "sheath" for the virus RNA.

Among structures with cylindrical symmetry are tubular crystals of enzymes discovered several years ago. They are formed by monomolecular layers rolled into a tube, with inner channel of considerable size. Such tubes were first observed in catalase protein which has a number of conventional crystalline modifications [18]. Some information on

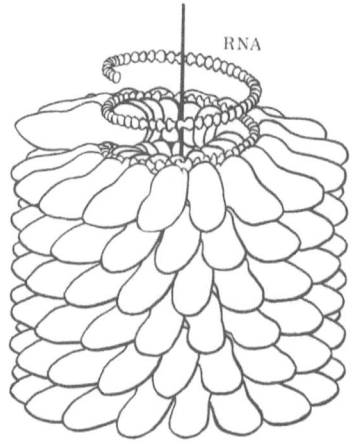

Fig. 8. Packing of protein molecules in tobacco mosaic virus.

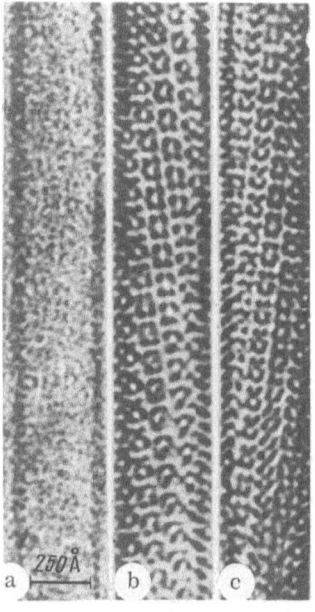

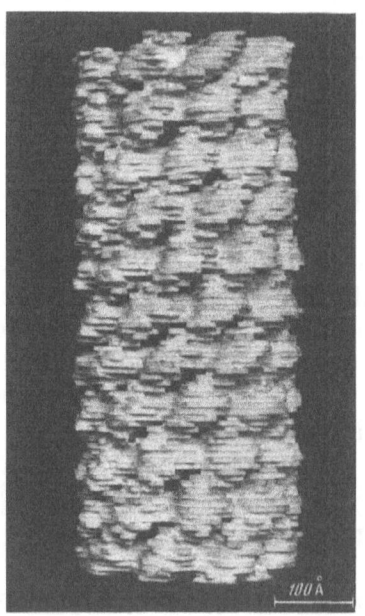

Fig. 9. Tubular crystals of phosphorylase (a) electron micrograph; (b, c) optical filtrated images ("near" side (b) and "far" side (c) of the crystal).

Fig. 10. Model of the tubular catalase crystal.

tubular structures can be obtained by using optical diffraction and filtration of the electron microscope images [19]. This technique allows refinement of the geometric parameters of the packing and its symmetry $S_{p/q}$, to give separately the images of the "near" and "far" walls of the tube, and to provide information on the size and shape of the molecule (Fig. 9) [20]. The final reconstruction of the tubular crystal model (Fig. 10) is achieved by applying the mathematical method of three-dimensional reconstruction [21, 22].

A similar cylindrical helical packing is found in protein molecules forming bacteriophage tails. Figure 11 shows this arrangement for the Phi-5 phage obtained by the 3-D reconstruction technique [23].

A realistic mechanism of buildup (crystallization) of tubes is the addition of strands, coils or discs, and not the folding of a completed monomolecular layer.

Fig. 11. Model of packing in Phi-5 bacteriophage tail.

Fig. 12. Thin crystal of leucine aminopeptidase. (a) electron micrograph; (b) optical filtration pattern.

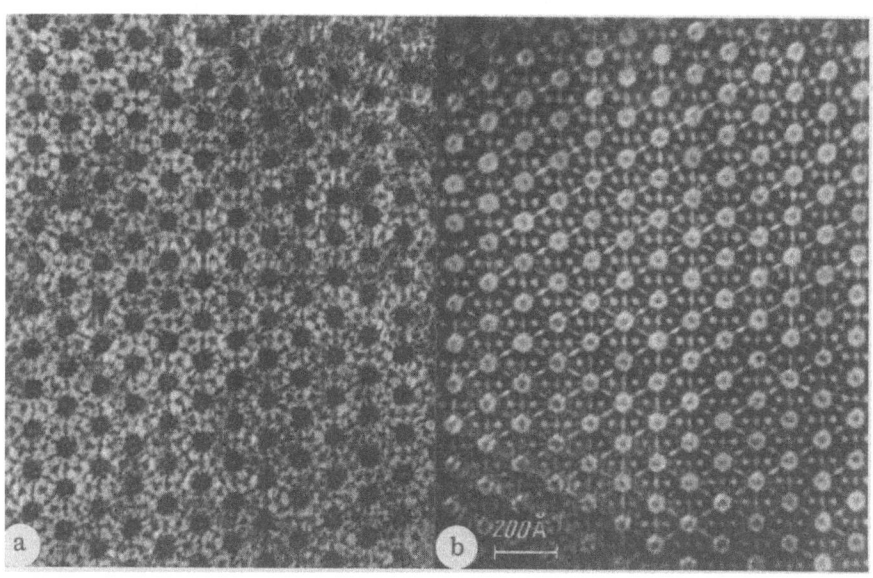

Formation of planar monomolecular two-dimensional crystals is seen in numerous membrane proteins, proteins of the bacteriophage capsules, and in a number of enzymes, such as catalase, or leucine aminopeptidase [10] (Fig. 12). Capsules of spherical viruses [21] are examples of closed monomolecular shells with point symmetry G_0^3. Crystallization in two spherical layers reveals icosahedral symmetry.

Conclusion

The preparation of protein crystals is important, *inter alia*, for protein X-ray structural analysis. The study of these extremely interesting and peculiar objects is still in the infancy stage. Some of their physical properties (such as electric and optical properties) are still almost unknown. There is no doubt that this field has much interest in store for physicists and biologists.

Literature Cited

1. B. K. Vainshtein and V. V. Borisov. Structure and functions of globular proteins in the light of X-ray structural data. Uspekhi Biol. Nauk, 14, 91-145 (in Russian) (1973).
2. R. E. Dickerson and I. Geis. The Structure and Action of Proteins. New York: Harper and Row, p. 120 (1969).
3. L. Pauling and R. B. Correy. Fundamental dimensions of polypeptide chains Proc. Roy. Soc. London, B141, 10-20 (1953).
4. J. C. Kendrew, Myoglobin and the structure of proteins: crystallographic analysis and data-processing techniques reveal the molecular architecture. Science, 139, 1259-1266 (1963).
5. B. K. Vainshtein, E. G. Arutyunyan, I. P. Kuranov, V. V. Borisov, N. I. Sosfenov, A. G. Pavlovsky, A. I. Grebenko, N. V. Konarev, and Yu. V. Nekrasov. Spatial structure of lupin leghemoglobin at 2.8 Å resolution. Dokl. AN SSSR, 233, 238-241 (in Russian) (1977).
6. S. E. Bresler and D. L. Talmud. On the nature of globular proteins. Dokl. AN SSSR, 43, 326-330 (in Russian) (1944).
7. H. Fisher. A limiting law relating the size and shape of protein molecules to their composition. Proc. Nat. Acad. Sci. USA, 51, 1285-1291 (1964).
8. O. B. Ptitsin. Self-organization in protein molecules. Vestnik AN SSSR, No. 5, 57-68 (in Russian) (1973).
9. V. V. Borisov, S. N. Borisova, N. I. Sosfenov, G. S. Kachalova, A. A. Voronova, B. K. Vainshtein, Yu. M. Torchinsky, G. A. Volkova, and A. E. Braunshtein. X-ray structure of asparate transaminase at 5Å resolution. Dokl. AN SSSR, 235, 212-215 (in Russian) (1977).
10. N. A. Kiselev, V. Ya. Stel'mashchuk, V. L. Tsuprun, M. Ludewig, and H. Hanson. Electron microscopy of leucine aminopeptidase. J. Mol. Biol., 115, 33-43 (1977).
11. D. C. Hodgkin. Problems in the X-ray analysis of proteins. Chem. Scr., 6, 145-157 (1974).
12. J. H. Northrop. In: Crystalline Enzymes. Columbia Univ. Press, 16, 176 (1939).
13. V. R. Melik-Adamyan. Crystallization of proteins. Kristallografiya, 20, 687-696 (in Russian) (1975).
14. S. Ya. Karpukhina, V. V. Barynin, and G. M. Lobanova. Catalase crystallization in a centrifuge. Kristallografiya, 20, 680-681 (in Russian) (1975).
15. M. Zeppezauer, H. Elkund, and E. S. Zeppezauer. Microdiffusion cells for the growth of single protein crystals by means of equilibrium dialyses. Arch. Biochem. and Biophys., 125, 564 (1968).

16. B. K. Vainshtein. Tubular crystals of globular proteins. Mat. Res. Bull., 7, 1347-1356 (1972).
17. J. T. Finch. Discovery, 28, 10-17 (1966).
18. N. A. Kiselev, C. L. Shpitzberg, and B. K. Vainshtein. Crystallization of catalase in the form of tubes with monomolecular walls. J. Mol. Biol., 25, 433-446 (1967).
19. A. Klug and J. E. Berger. An optical method for the analysis of periodicities in electron micrographs and some observations on the mechanism of negative staining. J. Mol. Biol., 10, 565-569 (1964).
20. N. A. Kiselev, F. Ya. Lerner, and N. B. Livanova. Electron microscopy of muscle phosphorylase b. J. Mol. Biol., 62, 537-549 (1971).
21. B. K. Vainshtein. Three-dimensional electron microscopy of biological macromolecules. Uspekhi Fiz. Nauk, 109, 455-497 (in Russian) (1973).
22. V. V. Barynin, B. K. Vainshtein, O. N. Zograf, and S. Ya. Karpukhina. Three-dimensional reconstruction of tubular catalase crystals. Molekulyar. Biol., 13, 1189-1197 (in Russian) (1979).
23. B. K. Vainshtein, A. M. Mikhailov, I. A. Andriashvili, A. S. Kaftanova, and G. V. Petrovsky. Electron-microscopic study of the 3-D structure of the Phi-S phage tail. Dokl. AN SSSR, 234, 699-702 (in Russian) (1977).

Part II
Mechanisms and Kinetics of Crystal Growth by Vapor Deposition

EQUILIBRIUM ADSORPTION LAYERS ON GaAs (111) and Si (111) SURFACES IN CVD GROWTH

A. A. Chernov

The Institute of Crystallography of the USSR Academy of Sciences, Moscow

M. P. Ruzaikin

The Siberian Physico-Technical Institute, Tomsk

Introduction

Estimates based on the simplest kinetic equation of adsorption show that the Si (111) face which is in equilibrium with a gas system Si-H-Cl at T=1500 K must be covered almost completely by chlorine and hydrogen atoms and by $SiCl_2$ molecules [1] (later Chernov and Popkov studied the silicon system in more detail). A similar conclusion was drawn for the Si-H-Br system on the basis of similar but even cruder estimates [2] and for the system Si-H-Cl on the basis of a quantitative analysis of experimental data [3]. A more consistent analysis of adsorption and normal growth was made for the GaAs (100) face in the system Ga-As-H-Cl [4]; however, it seems that in this case not all of the principal species present were taken into account, and the method of calculation has not been published. Adsorption on the GaAs (111) face has not yet been calculated. The present paper presents a fundamental approach to calculations dealing with adsorbed layers in growth systems involving chemical reactions, using the faces GaAs (111) and Si (111) in the systems Ga-As-H-Cl and Si-H-Cl, respectively, as specific illustrations.

The Model

Let us identify adsorption sites as the sites in which unsaturated bonds are located on the surface, and let us define the coverage to be found, θ_i, as a fraction of sites occupied by atoms or molecules of species i. The surface under the adsorption layer will be assumed unreconstructed (Fig. 1). This assumption will be valid if the energy gain in the crystal-vapor system for adsorption exceeds the gain for the case when dangling bonds link to one another, and if the adsorption layer is sufficiently dense. Estimates show that for the systems discussed in the present paper these conditions are satisfied to different extents. The data of ultraviolet spectroscopy [5, 6] indeed show that Si (111) faces covered with a chlorine monolayer are not reconstructed. The GaAs (111) interface with vacuum are reconstructed [7, 8], as also are Si (111) faces; LEED data have revealed a superstructure on these faces after they were brought into contact with a Ga-As-H-Cl system [4]. One of the reasons for the appearance of this superstructure may be an ordering of the adsorption layer; the structure of the surface under this layer remains unknown. This constitutes one of the difficulties in the model under discussion, so that the model is only meant to represent the principal characteristics of the adsorption layer.

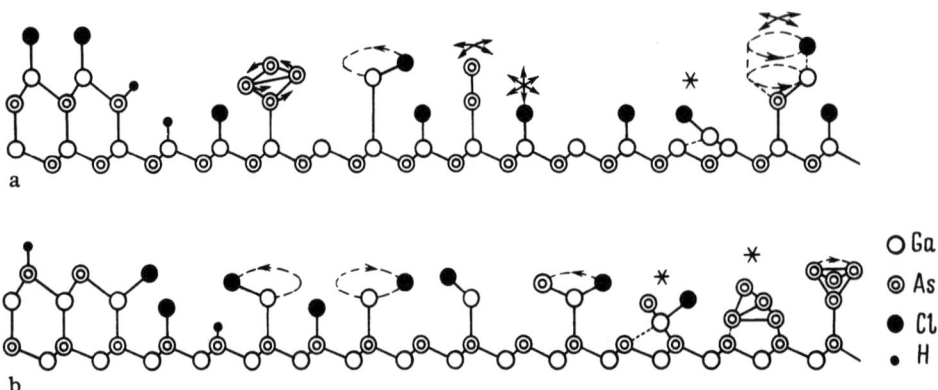

Fig. 1. A schematic representation of gallium arsenide (111) Ga (a) and (111) As (b) surfaces containing a step (on the left) and adsorbed atoms and molecules. Arrows indicate directions of oscillations and rotations of admolecules, dashed curves are rotation orbitals, and solid and dotted lines denote bonds in the ground and activated states; asterisks mark GaCl, AsGaCl, and As_4 molecules activated for surface diffusion.

The coverage characteristics θ satisfy the following system of equations:

$$\theta_i = K_i P_i (1 + \sum_j K_j P_j)^{-1}, \tag{1}$$

$$K_i = (Z_{si}/Z_{gi})(h^2/2\pi m_i kT)^{3/2} \exp(\varepsilon_{si}/kT),$$

$$\varepsilon_{si} = \varepsilon_{si}^0 + \sum_j E_{ij}\theta_j, \tag{2}$$

where P_i is the partial pressure of vapor of the ith species; Z_{si} and Z_{gi} are statistical sums of the ith species of molecules on the surface and in the vapor phase, respectively; $h = 6.63 \cdot 10^{-27}$ erg·s; m_i is the molecular mass; k is Boltzmann's constant; ε_{si} is the total adsorption energy of an ith particle; ε_{si}^0 is the adsorption energy of an isolated ith particle; and E_{ij} are the coefficients of electrostatic interaction between adsorbed particles. The product $K_i P_i$ in the set of equations (1) for the complex i = AsGaCl is replaced by

$$(Z_{si}/Z_{gl}Z_{gn})(h^2/2\pi kT)^3 (m_l m_n)^{-3/2} \exp[(\varepsilon_{s} + \varepsilon_{ln})/kT] P_l P_n,$$

where n = As, l = GaCl on Ga (111) face and n = GaCl, l = As on As (111) face; ε_{ln} is equal to the Ga–As bonding energy.

Calculated thermodynamic equilibrium partial pressures of vapor above GaAs solid phases were used for P_i [9–11]. A discrepancy below one order of magnitude between the calculated and measured values of pressure has been demonstrated for the Si–H–Cl system in ref. 12.

The statistical sums reflect possible rotations of molecules in the gas phase as well as rotations and vibrations of the adsorbed molecules (see Fig. 1). Two types of vibrations parallel to the surface (frequency $\nu_{\parallel}$) and a vibration normal to this surface ($\nu_{\perp}$) were taken into account for the center-of-mass of an adsorbed particle. An estimate of the statistical sums for rotational vibration around axes parallel to the surface and perpendicular to it shows that these sums are nearly equal to unity. Frequencies $\nu_{\parallel}$ and $\nu_{\perp}$ were found from the value of the pre-exponential factor for the lifetime of gallium atoms on faces (111) Ga and (111) As [13] and from the empirical estimate $\nu_{\parallel} \approx \nu_{\perp}/4$. The applicability of this estimate is justified by comparatively equal values of rigidity constants for deformational ($\nu_{\parallel}$) and valence ($\nu_{\perp}$)

TABLE 1. Coefficients of electrostatic interaction between adsorbed particles, E_{ij} (kcal mol^{-1}), on Si(111) face for the Si-H-Cl system

Component	Cl	H	Si	SiCl	SiCl$_2$
Cl	38.3	−2.1	0	10.1	20.2
H	−2.1	0.1	0	−0.5	−0.9
Si	0	0	0	0	0
SiCl	10.1	−0.5	0	−9.4	−4.6
SiCl$_2$	20.2	−0.9	0	−4.6	−9.2

TABLE 2. Coefficients of electrostatic interaction between adsorbed particles, E_{ij} (kcal mol^{-1}), on Ga(111) face of gallium arsenide for the Ga-As-Cl system.

Component	Cl	H	As	As$_2$	As$_4$	GaCl	AsGaCl
Cl	53.41	6.13	5.83	5.83	5.83	1.50	3.66
H	6.13	0.64	0.58	0.58	0.58	0.06	0.56
As	5.83	0.58	0.54	0.54	0.54	0.20	0.37
As$_2$	5.83	0.58	0.54	0.54	0.54	0.20	0.37
As$_4$	5.83	0.58	0.54	0.54	0.54	0.20	0.37
GaCl	1.50	0.06	0.2	0.2	0.2	3.71	4.34
AsGaCl	3.66	0.56	0.37	0.37	0.37	4.34	32.37

TABLE 3. Coefficients of electrostatic interaction between adsorbed particles, E_{ij} (kcal mol^{-1}), on As(111) face of gallium arsenide for the Ga-As-Cl system

Component	Cl	H	Ga	GaCl	AsGaCl
Cl	16.72	0	−2.85	2.33	2.60
H	0	0	0	0	0
Ga	−2.85	0	−0.56	−0.02	−0.28
GaCl	2.33	0	−0.02	5.05	8.32
AsGaCl	2.6	0	−0.28	8.32	8.47

vibrations of atoms of various elements in hydrocarbons and some other compounds [14]. Vibration frequencies for atoms or molecules of species i, with the exception of Ga, were calculated by the formula

$$\nu_i = \nu_{Ga}(m_{Ga}/m_i)^{1/2}.$$

Frequencies for Si (111) were calculated in a similar manner. Internal vibrations in an adsorbed molecule were assumed to be identical to its vibrations in the gas phase.

Each adsorbed atom is charged in accordance with Pauling's electronegativity. As a result, adsorbed particles interact with their own mirror images and with one another. This last interaction results in the dependence (2) of the total adsorption energy (ε_{si}) of the ith particle on the coverage of the surface by the particles of the jth species, and consequently, on temperature as well. In this respect ε_{si}^o is the sum of the valence bond energy and the energy of attraction of the particle to the mirror image. The calculated values of coefficients E_{ij} (kcal/mole) are listed in Tables 1-3. Positive energies E_{ij} correspond to attraction of particles of species i by its neighbors of species j, and negative E_{ij} correspond to repulsion. Valence components of energy ε_{0i}^o (Table 4) were chosen on the basis of tabulated values of bonding energy in diatomic molecules [15] (upper figures) and measured life times [13] (lower figures). Correspondingly, two series of calculations have been carried out.

TABLE 4. Valence bond energies (kcal mol^{-1}) in the Ga-As-H-Cl system

Component	Ga	As	Cl	H	Component	Ga	As	Cl	H
Ga	32	36	114.5	68	As	36	32	68.8	65
	57	60	114.5	68		60	32	68.8	65

Similar calculations have been carried out for Si (111) surface in contact with the Si-H-Cl ambient for the following values of bonding energies: ε_{Si-Si} = 55 kcal mol^{-1}, ε_{Si-H} = 74 kcal mol^{-1}, ε_{Si-Cl} = 96 and 108 kcal mol^{-1}.

Adsorption Layer on GaAs (111) Face

The values of coverage θ_i were calculated for the concentrations of the gas mixture $Q = P_{AsCl_3}/P_{H_2} = \alpha \cdot 10^{-3}$ (α = 1, 2, 3, 4, 5, 36.6), at T = 973, 1023, 1073, 1123, 1173, 1223 K (vapor pressure was calculated in the Institute of Inorganic Chemistry of the Siberian Branch of the USSR Academy of Sciences [9]). For $Q = 36.6 \cdot 10^{-3}$ these values are close to those calculated in ref. 16. After preliminary estimates seven components were selected for computer solution of a transcendental system of equations (1)-(2) with respect to θ_i: i = Cl, H, As, As$_2$, As$_4$, GaCl, AsGaCl, as well as vacancies. Functions $\theta_i(Q)$ corresponding to bonding energies in Table 4 obtained on the basis of values tabulated in ref. 15, and to T = 1073 K, are given in Fig. 2. It is clear from Fig. 2a, that the (111) Ga surface of gallium arsenide must be mostly covered by atomic chlorine and hydrogen and must contain 10-20% of dangling bonds. Among the gallium- and arsenic-containing components the most probable are GaCl and As$_4$ (up to several per cent for $Q = 36.6 \cdot 10^{-3}$). The fraction of vacancies θ_{vac} diminishes as the concentration Q of the gas mixture increases, and increases as temperature rises; an increase in temperature reduces the number of all particles on the surface but within an order of magnitude the values of θ_i in the temperature range investigated are retained. For example, for $Q = 36.6 \cdot 10^{-3}$ the coverage in the range 973-1173 K decreases from 85 to 77% for chlorine, from 3.5 to 1.5% for As$_4$, and from 3.7 to 2.7% for GaCl. In contrast, hydrogen coverage θ_H rises from 3.3 to 5.4% because increasing temperature raises the atomic hydrogen pressure in the vapor.

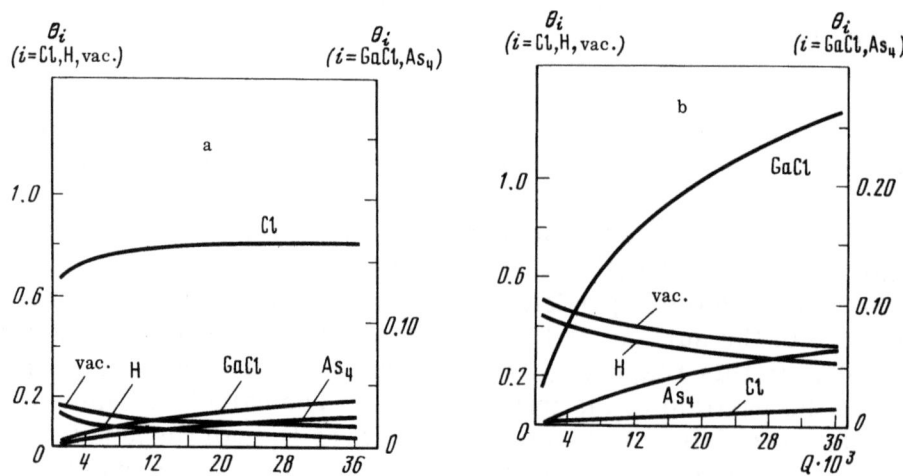

Fig. 2. Calculated coverage of gallium arsenide (111) Ga (a) and (111) As (b) surfaces by various components at T = 1073 K as a function of the gas mixture concentration.

The coverage of a (111) As face (Fig. 2b) differs from that of a (111) Ga face first of all in a lower density: 25-50% of vacancies and a minute amount of chlorine. The reason lies in weaker Cl-As bonding as compared to Cl-Ga bonding (see Table 4). On the other hand, surface concentrations of GaCl and As_4 on (111) As faces are slightly higher. Functions $\theta_i(T)$ for this surface are of the same nature as for (111) Ga.

The AsGaCl complexes which, for stereochemical reasons, are assumed to play the principal role in the growth of (111) faces [4], are present on these faces in much smaller amounts than GaCl and As_4. For $Q = 36.6 \cdot 10^{-3}$ and $T = 1073$ K the values of θ_{AsGaCl} on (111) Ga and (111) As faces are thus equal to $0.34 \cdot 10^{-5}$ and $0.72 \cdot 10^{-5}$, respectively.

If we use bonding energies obtained from measured lifetimes (see Fig. 4), the values of coverage are higher. Thus, for $Q = 36.6 \cdot 10^{-3}$ and $T = 1073$ K we have: $\theta_{vac} = 1.3 \cdot 10^{-4}$, $\theta_{Cl} = 0.47$, $\theta_{GaCl} = 0.26$, $\theta_{AsGaCl} = 0.17$, $\theta_{As_4} = 0.09$, $\theta_{As_2} = 8 \cdot 10^{-3}$, $\theta_H = 2.8 \cdot 10^{-4}$. However, gallium desorption heat values measured in ref. 13 (57 and 60 kcal mol^{-1} for (111) Ga and (111) As faces, respectively) presumably reflect mostly the double Ga-Ga and Ga-As bonds on reconstructed surfaces; hence, it would be unjustified to use these values for single bonds in further discussion.

Adsorption Layer on Si (111) Face

The values of coverage θ_i for Cl/H = 0.1 and 0.01 satisfying the system of equations (1)-(2), for values of P_i taken from [10, 11] and for Si-Cl bonding energies 96 and 108 kcal mol^{-1}, are listed in Table 5. The results obtained for these energies are close: the surface is mostly covered with chlorine and hydrogen atoms whose fractions are almost equal; $\theta_{Cl} \approx \theta_H \approx 0.5$.

TABLE 5. Coverage of Si(111) Face with Different Components

Component	1425 K	1475 K	1525 K	Component	1425 K	1475 K	1525 K
$\varepsilon_{Si-Cl} = 96$ kcal mol^{-1}				$\varepsilon_{Si-Cl} = 108$ kcal mol^{-1}			
Cl	0.36 / 0.47	0.36 / 0.47	0.34 / 0.45	Cl	0.62 / 0.73	0.62 / 0.73	0.60 / 0.71
H	0.63 / 0.52	0.62 / 0.51	0.65 / 0.54	H	0.37 / 0.26	0.36 / 0.26	0.39 / 0.29
Si	$3.7 \cdot 10^{-7}$ / $3.3 \cdot 10^{-7}$	$1 \cdot 10^{-6}$ / $9 \cdot 10^{-7}$	$2.5 \cdot 10^{-6}$ / $2.2 \cdot 10^{-6}$	Si	$2.6 \cdot 10^{-7}$ / $2.0 \cdot 10^{-7}$	$7.1 \cdot 10^{-7}$ / $5.5 \cdot 10^{-7}$	$1.8 \cdot 10^{-6}$ / $1.4 \cdot 10^{-6}$
SiCl	$3.49 \cdot 10^{-7}$ / $1.42 \cdot 10^{-6}$	$5.6 \cdot 10^{-7}$ / $1.95 \cdot 10^{-6}$	$5.31 \cdot 10^{-7}$ / $2.48 \cdot 10^{-6}$	SiCl	$1.02 \cdot 10^{-7}$ / $3.54 \cdot 10^{-7}$	$1.68 \cdot 10^{-7}$ / $4.96 \cdot 10^{-7}$	$1.68 \cdot 10^{-7}$ / $7.09 \cdot 10^{-7}$
SiCl$_2$	$8.3 \cdot 10^{-6}$ / $9.6 \cdot 10^{-5}$	$3.4 \cdot 10^{-6}$ / $9.2 \cdot 10^{-5}$	$4.6 \cdot 10^{-6}$ / $7.4 \cdot 10^{-5}$	SiCl$_2$	$9.6 \cdot 10^{-7}$ / $9.6 \cdot 10^{-6}$	$8.2 \cdot 10^{-7}$ / $9.9 \cdot 10^{-6}$	$6.3 \cdot 10^{-7}$ / $8.9 \cdot 10^{-6}$
Vacancies	$1.0 \cdot 10^{-2}$ / $1.0 \cdot 10^{-2}$	$1.5 \cdot 10^{-2}$ / $1.3 \cdot 10^{-2}$	$1.4 \cdot 10^{-2}$ / $1.3 \cdot 10^{-2}$	Vacancies	$0.8 \cdot 10^{-2}$ / $0.6 \cdot 10^{-2}$	$1.0 \cdot 10^{-2}$ / $0.8 \cdot 10^{-2}$	$1 \cdot 10^{-2}$ / $0.8 \cdot 10^{-2}$

Note: The upper value of coverage for each component corresponds to the ratio Cl/H = 0.01, and the lower one to Cl/H = 0.1.

Vacancies are rare: $\theta_{vac} = 0.015$. The main silicon-containing component is SiCl$_2$, for which $\theta_{SiCl_2} \approx 10^{-4} - 10^{-6}$. The next two components, in order of importance, are Si and SiCl with coverages of about $10^{-6} - 10^{-7}$. As the temperature increases (1425 $\leq$ T $\leq$ 1525 K) θ_{SiCl_2} increases and θ_{SiCl} diminishes, reflecting the changes in SiCl$_2$ and SiCl partial pressures. An increase in Cl/H ratio leads to a rapid enhancement of the surface concentration of $SiCl_2$, a less pronounced increase in Θ_{SiCl}, and only a slight increase in Θ_{Si}.

The values of Θ_{SiCl_2} given in Table 5 are by two to three orders of magnitude smaller than those obtained in ref. 1. This is due to the fact that ref. 1 ignored the rotational and vibrational statistical sums for $SiCl_2$, which emphasizes the important role played by these processes in the behavior of molecules.

Surface Diffusion

An analysis of surface diffusion in systems involving chemical reactions must take into account two factors which are insignificant for crystal growth from atomic vapor; most of the formulas in the Burton-Cabrera-Frank theory [17] are written for this last situation. These two factors are, first, a high density of the adlayer, and second, the molecular, and not the atomic, form of the adsorbate containing the crystallizing material. Taking these factors into account, one can rewrite the surface diffusion coefficient of the ith species in the form

$$D_{si} = a^2 (kT/h)(Z^*_{si}/Z_{si}) \theta_{vac} \exp(-\varepsilon_{di}/kT),$$

where a is the diffusion jump length $a \simeq 4$ Å for Si (111) and GaCs (111)); ϵ_{di} is the diffusion activation energy; Z^*_{si} is the statistical sum of the molecules in the diffusion-activated state. Molecules in this state (for example, AsGaCl or As_4 in Fig. 1) lose their rotational degrees of freedom, so that $Z^*_{si}/Z_{si} \leq 1$ ($3 \cdot 10^{-3}$ and $3 \cdot 10^{-2}$ for AsGaCl and As_4, respectively). In the case of metals $\varepsilon_{di} \approx (0.2 - 0.3) \varepsilon_{si}$ [18]. Experimental values for silicon are $\varepsilon_d = 0.24 \cdot \varepsilon_s = 25.4$ kcal/mol^{-1} [19]. Therefore we can expect, for example, that $\varepsilon_{di} = 7$-11 kcal mol^{-1} for AsGaCl and GaCl diffusion on (111) Ga. By estimating D_{si} and the residence time (from bonding energy ε_{si}), we can estimate the free path length λ_s of the particle during its residence time on the surface. Thus, for AsGaCl on (111) Ga face at $T \approx 1000$ K we obtain $\lambda_s \approx 2 \cdot 10^{-6}$-$6 \cdot 10^{-7}$ cm, and $\lambda_s \approx 60a \approx 2.5 \cdot 10^{-6}$ cm for $SiCl_2$ on (111) Si at T = 1500 K. This is much less than the free path length of gallium and silicon atoms on pure surfaces.

Discussion of Results

Calculations show that surface concentrations of GaCl and As_4 reach rather high values of several per cent (see Fig. 2). Therefore the reaction between GaCl and hydrogen of the gas phase and (or) hydrogen of the surface layer, and the decomposition of As_4 may prove to be the main sources of gallium arsenide for the growth of the crystal. This corresponds to the important role played by GaCl and As_4, already mentioned in connection with the analysis of experimental data in ref. 22.

Equilibrium constants and surface reaction rates must depend on the adsorption layer composition. Indeed, the negatively charged chlorine atom in the adsorbed GaCl molecule on a (111) Ga face is repulsed by all the surrounding adsorbed chlorine atoms, while the positively charged gallium atom in GaCl is attracted. Calculations have demonstrated that the total Ga-Cl bonding energy in the adsorbed GaCl molecule is thereby diminished at $\Theta_{Cl} = 0.8$ to 59 kcal/mol^{-1} compared with 114.5 kcal mol^{-1} in the gas phase. The energy of electrostatic attraction of gallium atoms is 58 kcal mol^{-1}. Therefore, the local electric field produced on the surface by the adsorbed layer is highly conducive to GaAs deposition reaction on (111) Ga. On (111) As faces this effect must be practically suppressed because of the small amount of chlorine; hence, the probability of a higher growth rate of (111) Ga face as compared to (111) As, a conclusion borne out by the experiments.

A strong mutual repulsion of adsorbed chlorine atoms on silicon surfaces (111) Si and (111) Ga of GaAs (see E_{Cl-Cl} coefficients in Tables 1 and 2) must lead to their ordered arrangement on the surface, that is, to the formation of two-dimensional superlattices in the adsorbed layer, indeed revealed by the LEED data. Lattice periods of these superlattices are determined by the adsorption layer composition, that is, by the temperature and composition at the vapor phase. Adsorbed chlorine atoms on (111) As are located at sufficiently large distances from one another ($\theta_{Cl} \simeq 10^{-2} - 10^{-1}$), so that interaction between them is weak (approximately $0.4\,\text{kcal mol}^{-1}$ for $Q = 36.6 \cdot 10^{-3}$ and $T = 1073$ K), well below kT. The same is true for the GaCl-GaCl dipole-dipole interaction. Therefore at high growth temperatures the electrostatic interaction can hardly lead to the formation of superstructures on (111) As.

The conclusion on the important role played by $SiCl_2$ in the course of (111) Si growth is in agreement with the available experimental data, and especially with the observed proportionality of growth rate to the $SiCl_2$ concentration gradient in the chloride system [23]. The above arguments on local electric fields around adsorbed chlorine atoms are also applicable to the surface reaction of silicon reduction by hydrogen (from the gas phase or from the adsorbed layer).

We have mentioned above the comparatively low values of the diffusion path length and surface diffusion coefficients of adsorbed molecules in dense adsorption layers. This is true to a certain extent for adsorbed atoms as well. If the generation of new elementary growth steps is determined by the diffusion fields of the already existing steps (as in the case of 2-D nucleation [21] and in the case of nucleation on screw dislocations), the distance between neighboring steps is within an order of magnitude of $2\lambda_s$. The spacing between steps for vicinal growth hillocks with the slope 0.6-0.8° observed on (111) face of silicon [20] deposited in a chlorine system is approximately $60a$, that is, very close to λ_s. This does not contradict the model of a dense adsorption layer and of growth at the expense of $SiCl_2$. At the same time, spacings of the order of $2 \cdot 10^{-5}$ cm were observed in similar conditions [20]. Adsorbed $SiCl_2$ molecules, fixed by electrostatic forces on top of a dense adsorption layer, may have large free path lengths and high mobility. However, this possibility, suggested by Sedgwick in a discussion of the present paper, needs special checking.

The limited mobility of adsorbed molecules in the dense adsorbed layer leads one to assume that the reactions of silicon and gallium deposition on the surface involve the hydrogen of the gas phase. In this case the morphology of the macroscopic deposit must be mainly determined by the bulk diffusion, thus making it unjustifiable to identify the characteristic spacings between vicinal growth hillocks with the diffusional free path length on the surface.

The above-mentioned, as well as numerous other, problems can be solved only if we study the growth kinetics which is almost completely left out of the present paper devoted as it is to the equilibrium ground state serving as the background to the kinetic process. Among other aspects, one needs to analyze adsorption layers on growing surfaces and steps, the generation of steps, their motion in the presence of dense adsorption layers, interaction with impurities, the vapor composition directly above the growing surface, and so forth.

Literature Cited

1. A. A. Chernov and N. S. Papkov. On the mechanism of CVD crystal growth (Si-H-Cl system). Kristallografiya, 22, 35-43 (in Russian) (1977).
2. L. N. Alexandrov, E. A. Krivorotov, and Yu. G. Sidorov. Initial stage and kinetics of epitaxial film growth of A^3B^5 compounds. Krist. und Techn., 11, 591-605 (1976).
3. V. F. Dorfman. Gas-Phase Metallurgy of Semiconductors, Metallurgy, Moscow (in Russian) (1974).

4. J. B. Theeten, L. Hollan and R. Cadoret. Growth mechanisms in CVD of GaAs. In: 1976 Crystal Growth and Materials/Ed. E. Kaldis, H. J. Scheel. New York; Oxford: North Holland Publ. Co., 2, 195-236 (1977).

5. M. Schlüter, J. E. Rowe, Margaritondo, K. M. Ho, and M. L. Cohen. Chemisorption-site geometry from polarized photoemission: Si(111)Cl and Ge(111)Cl Phys. Rev. Lett., 37, 1632-1635. (1976).

6. P. K. Larsen, N. V. Smith, M. Schlüter, H. H. Farrell, K. M. Ho, and M. L. Cohen. Surface energy bands and atomic position of Cl chemisorbed on cleaved (111). Phys. Rev., B17, 2612-2619. (1978).

7. W. Ranke and K. Jacobi. Composition, structure, surface states, and O_2 sticking coefficient for differently prepared GaAs(111) As surfaces. Surface Sci., 63, 33-44 (1977).

8. B. Joyce, C. T. Foxon, and J. H. Neave. Fundamentals of molecular beam epitaxy. J. Jap. Assoc. Cryst. Growth, 5, 185-197 (1978).

9. N. A. Testova, G. A. Kokovin, and F. A. Kuznetsov. On molecular composition of the gas phase in the Ga-As-Cl-H system. In: Gallium Arsenide. Proceedings of the 4th USSR Conf. on gallium arsenide studies, Abstracts. Tomsk University Publ. House, Tomsk, p. 193 (in Russian) (1978).

10. L. P. Hunt and E. Sirtl. A thorough thermodynamic evaluation of the silicon-hydrogen-chlorine system. J. Electrochem. Soc., 119, 1741-1745 (1972).

11. E. Sirtl, L. P. Hunt, and D. H. Sawyer. High temperature reactions in the silicon-hydrogen-chlorine system. J. Electrochem. Soc. 121, 919-925 (1974).

12. V. S. Ban. Mass spectrometric studies of chemical reactions and transport phenomena in silicon epitaxy. In: Proc. 6th Intern. CVD Conf. Atlanta, Oct. (1977).

13. J. R. Arthur. Interaction of Ga and As_2 molecular beams with GaAs surface. J. Appl. Phys. 39, 4032-4034 (1968).

14. A. Gordon and R. Ford. The Chemist's Companion, A Handbook of Practical Data, Techniques, and References. Wiley-Interscience. New York, London, Toronto (1972).

15. Handbook of Chemistry and Physics. 56th ed./Ed. R. C. Weast. Cleveland (Ohio): RCR Press (1974-1975).

16. P. Klima, J. Silhavy, V. Rerabek, J. Braun, C. Cerny, P. Vonka, and R. Holub. A study of equilibrium reactions in the $Ga-PCl_3-H_2$ and $Ga-AsCl_3-H_2$ epitaxial systems. J. Cryst. Growth, 32, 279-286 (1976).

17. U. K. Burton, N. Cabrera, and F. Frank. The growth of crystals and the equilibrium structure of their surfaces. Phil. Trans. Roy. Soc., A243, 209-358 (1951).

18. A. A. Chernov. Crystallization. Ann. Rev. Mat. Sci., 3, 397-454 (1973).

19. S. F. Bedair. Activation energy for migration on silicon (111) face. Surface Sci., 42, 595-599 (1974).

20. M. Shimbo, J. Nishizawa, and T. Terasaki. Defect-free nucleation of silicon on (111) silicon surfaces. J. Cryst. Growth, 23, 267-274 (1974).

21. A. A. Chernov. Gorwth kinetics and capture of impurities during vapor crystallization. J. Cryst. Growth, 42, 55-76 (1977).

22. D. W. Shaw. Chemistry of vapor phase epitaxy. In: 4th Intern. Conf. on Vapor Growth and Epitaxy (ICVGE-4), Nagoya, Japan, 173-174 (1978).

23. T. Sedgwick and G. V. Arbads. Thermodynamic calculations of partial pressures in gas system Ga-As-H-Cl. J. Jap. Assoc. Cryst. Growth, 5, 93-102 (1978).

ROLE OF ADSORPTION LAYER IN CHEMICAL VAPOR DEPOSITION

E. I. Givargizov

The Institute of Crystallography of the USSR Academy of Sciences, Moscow

Introduction

Chemical vapor deposition is a widely used technique. This process provides the foundation for the epitaxial technology which is the key process in modern microelectronics. At the same time the understanding of many details of the mechanism of crystal growth from the gas phase involving chemical reactions remains insufficient. The Burton-Cabrera-Frank theory of crystal growth from the crystal's vapor cannot be directly applied to this case: the difficulty lies not in that the parameters of this theory (diffusional path length, residence time of atoms on the surface, etc.) can be measured only with great difficulties in the case of chemical crystallization [1], but also in that the surface is covered with a dense chemisorbed layer [2]; this means the growth conditions, and presumably, mechanisms must be essentially different from those analyzed by Burton, Cabrera, and Frank.

Features of chemical crystallization are related to quantitative characteristics of the process, and first of all the supersaturation. It has been mentioned [3, 4] that by virtue of the chemical reaction being in principle reversible, the supersaturation (in the rigorous thermodynamic meaning of this term) must be comparatively low; so it is erroneous to find the supersaturation by comparing the concentration of a material in a medium with the equilibrium pressure of the vapor of the crystallized substance as reported in [5]. In particular, the supersaturation comes to only a few per cent in the most frequently used chloride-hydrogen process of silicon crystallization at typical deposition temperatures and concentrations of the initial gas-vapor mixture [6]. According to the Volmer-Becker-Döring theory for two-dimensional nucleation, a smooth (singular) face of a perfect crystal must have a practically negligible growth rate at such supersaturation. However, the whole accumulated experience of epitaxial growth demonstrates that reasonably perfect (including dislocation-free) single crystal layers grow at moderately high rates (for example, the growth rate of silicon $V \gtrsim 1\mu m/$min at temperatures of about 1000°C). Attempts have been made to remove this contradiction by assuming, for example, a "surface reconstruction" at high temperatures [7]; however, so far no evidence has been obtained that such reconstruction takes place in chemical crystallization.

In the present paper we interpret the kinetics of chemical vapor deposition in terms of the concept of two-dimensional nucleation at low edge energy of the nucleus with an energy reduction caused by a dense adsorption layer on the surface [2] (this approach was applied earlier to explain the parabolic kinetics of whisker growth by the vapor-liquid-solid mechanism [8]; the reduction in the energy of the nucleus is caused by the growth proceeding from a solution — specifically, ref. 8 dealt with silicon growth from its solution in gold). As a result, one can also determine the surface energy of a crystal face covered with an adsorption layer. With this value of surface energy, it is possible to improve the result obtained in ref. 8 for the surface energy of the silicon/silicon-gold solution interface. Some morphological features of epitaxial silicon layers grown in a chloride-hydrogen process are also analyzed.

Kinetics of Chemical Vapor Deposition

The present paper deals with the simultaneous analysis of kinetic relationships governing the growth of epitaxial films deposited from the vapor, and of whiskers growing by the VLS mechanism. The experiments with whiskers are necessary only to measure supersaturation in the gas phase over the growing epitaxial film; it has been shown [6] that the supersaturation can be obtained from the growth rate of whiskers as a function of their diameter in the submicron range, provided the surface energy of the crystal-vapor interface is known. Such an indirect measure of supersaturation is necessary because of the great complexity involved in crystallization by vapor transport reactions; direct measurements are simply unfeasible. Indeed, the chloride-hydrogen crystallization of silicon, for example, involves at high temperatures more than ten different compounds (in addition to the initial $SiCl_4$ and H_2, the components taking part are $SiCl_2$, HCl, SiH_2Cl_2, $SiHCl_3$, SiH_3Cl, etc.) [9]. In addition, the rates of the various intermediate reactions are unknown, so the concentrations of the above compounds are also unknown. Consequently, one cannot hope to calculate supersaturations by the usual thermodynamic method from the partial pressures of the components.

The techniques of the crystallization experiments are mostly trivial. They were described for whiskers in [6, 8]. Films were either grown in the same experiment with whiskers, or in separate experiments under carefully reproduced conditions. In order to improve the reproducibility of an experiment, the single-crystal silicon wafers used as substrates were initially covered at a high temperature (1200-1250°C) with an epitaxial 5-10 μm film, with the substrate-film transition layer thus completely buried.

The crystal growth of films and whiskers was conducted in the temperature range from 930° to 1050°C (experiments have demonstrated that this is the region in which the kinetic mode of growth is still operative). The typical appearance of a freshly grown surface is shown in Fig. 1. We clearly notice the features of layered growth which are known to be among the most reliable criteria of the kinetic mode of growth [10. 11]. The growth rate of the films was found from the mass increment, the duration of the growth experiment being selected to ensure an acceptable accuracy of measurements. Typically, as-grown films were from 1 to 5 μm thick over an area of approximately 3 cm².

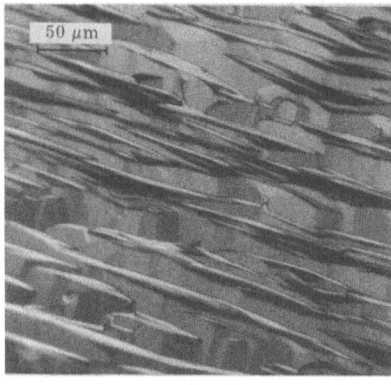

Fig. 1. Typical morphology of as-grown epitaxial silicon layers (crystallization temperature around 1000°C).

The kinetic experiments with whiskers have been described in detail in ref. 8. Similarly, in the present paper we plotted growth rate as a function of crystal diameter d, in ($V^{1/2}$, $1/d$) coordinates, and calculated the effective supersaturation in the vapor from the critical diameter. We took into account the fact that as the temperature changes, the lateral facetting of whiskers is transformed [12] and the surface energy of the vapor-crystal interface, α_{vc}, correspondingly takes on new values.

The composition of the initial gas mixture $H_2 + SiCl_4$ was kept constant in all experiments, in order to achieve a similar thickness of adsorption layer on the crystal surface; however, this was achieved only approximately, since the ratios of the various compounds vary slightly for the same average composition of the vapor phase [9].

The values of the film growth rate were analyzed assuming that the growth mechanism on a (111) face was two-dimensional nucleation of layers, and choosing the polycentric nucleation model, that is assuming kinetic roughness. It was also assumed (this is a standard procedure) that the nuclei are of monatomic height. In this case the film growth rate is [8]

$$V = V_0 \exp(-\pi\Omega\varkappa^2/3kTh\Delta\mu), \tag{1}$$

where V_0 is a constant or a slowly varying quantity; Ω is the specific volume of the crystal; $\varkappa$ is the edge energy of a nucleus; h is its thickness; $\Delta\mu$ is the effective difference between chemical potentials (in the present case, for silicon) in the gas phase and in the crystal, and k and T have a standard meaning, that is, $\Delta\mu/kT$ gives supersaturation. If we drop the assumption of polycentric nucleation, this will only remove factor 3 in (1) which obviously will not appreciably affect the final result. It will be more convenient to rewrite (1) in the form

$$V = V_0 \exp\left[-\frac{\pi\Omega\varkappa^2}{3k^2h} \frac{1}{T^2(\Delta\mu/kT)}\right], \tag{2}$$

and then, plotting ln V as a function of $1/T^2$ ($\Delta\mu/kT$), we obtain a straight line (Fig. 2) from whose slope we find the value of $\varkappa$ and with it the specific surface energy.

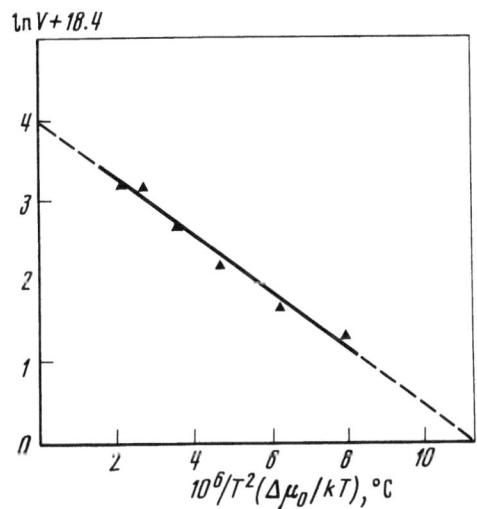

Fig. 2. Film growth rate as a function of supersaturation and temperature.

The results of experiments and calculations are shown in Table 1. The edge energy of a nucleus (i.e. the specific energy of a step {111}) at the average temperature of the range investigated, $T_{av} \approx 1250$ K, was found to be $\varkappa = 2.8 \cdot 10^{-6}$ erg cm^{-1}. Calculations by the method of [13] yielded a value for surface energy $\alpha = 235$ erg cm^{-2} which must be compared, first, with the energy of liquid-crystal interface (silicon solution in gold/silicon) $\alpha_{lq} = 210$ erg cm^{-2} [8], and second, with the value of the silicon surface energy in the chloride-hydrogen ambient (180-210 erg cm^{-2}) obtained by elementary estimates [2].

TABLE 1. Supersaturation in the gas phase and growth rate of silicon films

Parameters of the process	T, K					
	1223	1243	1263	1283	1303	1323
α_{vc}, erg/cm^2	1617	1617	1617	1734	1734	1734
$\Delta\mu/kT$	0.084	0.104	0.134	0.172	0.224	0.272
V, Å/s	3.8	5.6	9.3	17.3	24.1	30.3

The relative smallness of all these quantities supports the assumption of the two-dimensional growth mechanism for epitaxial silicon films in the chloride-hydrogen process, as well as the assumption of kinetic roughness of the growing face.

These results require that the parameters obtained earlier for whiskers and, correspondingly, the above-mentioned values of $\varkappa$ and α be somewhat corrected. The fact is that in the experiments with whisker growth the calculations of [8] were based on the value of the vapor-crystal surface energy, α_{vc} (111) 1230 erg cm^{-2}, found from experiments on silicon crystal cleavage in vacuum [14].

Obviously, instead of α_{vc} the calculations of supersaturation must use the energy value for silicon in contact with the chloride-hydrogen medium, α = 235 erg cm^{-2}. This substitution yields a corrected value of surface energy for silicon in contact with the chloride-hydrogen medium, α_{corr} = 170 erg cm^{-2}.

With this new value, we can carry out another correction, and so on, arriving by successive approximations at more and more accurate values for α. It is evident, however, that at a certain stage this becomes meaningless: the accuracy of the experiment is insufficient for further elaboration of the conclusions. One can only state that the value of the surface energy is in the range 150-200 erg cm^{-2}, which is sufficient to distinguish between the mechanisms of silicon crystal growth by chemical vapor deposition: in accordance with the results of [2], we can state that the surface of the crystal is covered with a dense adsorption layer. This factor considerably reduces the energy of nucleus formation; hence, crystal growth in these conditions is similar to the growth from solutions.

Morphology of Epitaxial Layers

The concept of crystal growth from vapor via quasiliquid layers has been discussed in the literature on numerous occasions [15-17], but usually in reference to impurity effects. In the present paper the quasiliquid layer on the crystal surface is treated as a natural and unavoidable consequence of the process. It is essential to note that quasiliquid layers are obviously inhomogeneous in thickness: condensation of the quasiliquid phase must occur along steps and in any other regions where the smoothness is disturbed (e.g., defect emergence points, as well as clusters of impurity). A model of this type was used, for example, in [15]. Several cases must be mentioned [18] where small drops (presumably of solution, so-called "protuberances") were observed to migrate along a growing face and to disappear completely when meeting the end face of a step; apparently, these drops simply spread over the step.

The concept of inhomogeneity and discontinuity of a quasiliquid layer enables one to explain a phenomenon frequently observed in epitaxial growth, namely, the formation on silicon and germanium layers of growth pyramids unrelated to any structural defects. A typical shape for such pyramids on the surface of a germanium film is shown in Fig. 3. Similar shapes were observed on silicon films [19-21]. The origin of these pyramids has been discussed in the literature but no acceptable explanation has been offered so far. The concept of self-consistent nucleation [2, 22], specially developed to explain this growth feature, is unconvincing because it opposes nucleation on a small smooth segment of a top face to nucleation on a much larger, and even stepped, area of lateral faces.

We suggest a different explanation based on an experimentally established dependence of density of these pyramids on the conditions of crystallization and on an analogy with the VLS growth of whiskers.

Fig. 3. Growth pyramids on the surface of a germanium film grown by the chloride process (GeCl$_4$+H$_2$) at T≈850°C.

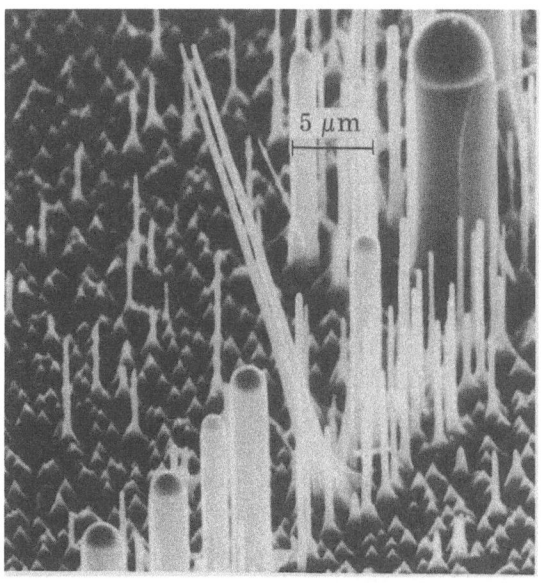

Fig. 4. Morphology of surface of a gallium arsenide wafer on which whiskers were grown by the VLS mechanism, with gold as the solvent metal, at a relatively high temperature (T≈800°C).

The density of pyramids increases with increasing concentration of the gas mixture and diminishes as the temperature rises [19-21]. Similar relationships were observed for whiskers initiated by particles of liquid phase which were formed by rupturing a thin film of high-temperature solution: the density of particles increases with increasing supersaturation, and decreases with increasing temperature [8]. In the general case, there is no principal difference between growth pyramids and columnar crystals (whiskers) growing on a substrate: this can be seen in Fig. 4 where both columnar crystals are shown (with caps of solidified solution of gallium arsenide in gold), and conical or pyramidal features; the latter developed from gallium drops formed on preliminary heating of the substrate as a result of incongruent evaporation of gallium arsenide at relatively high temperatures. These drops later disappeared on cooling, in accordance with the phase diagram of the Ga-As system. A quasiliquid phase in dynamic equilibrium with the gas phase could disappear on cooling in a similar manner. The hypothesized quasiliquid aggregates facilitating the formation of pyramids on epitaxial layers of silicon and germanium could be initiated by uncontrollable impurities evolving from the substrate or the gas phase.

Conclusion

A simultaneous analysis of the kinetic behavior of the VLS growth of whiskers and the growth of epitaxial films by the VLS mechanism makes it possible, first, to find the surface energy of the crystal (silicon in the present case) covered with a dense adsorbed layer; second, to confirm the validity of the hypothesis stating that the growth of epitaxial films involving chemical vapor deposition proceeds via two-dimensional nucleation at a sharply reduced edge energy of steps, supporting the quasiliquid layer model. This model is used to explain some morphological features of epitaxial films.

Literature Cited

1. G. Mandel. Surface-limited vapor solvent growth of crystals. J. Chem. Phys., 40, 683-690 (1964).
2. A. A. Chernov and N. S. Papkov. On the mechanism of CVD crystal growth (Si-H-Cl system). Kristallografiya, 22, 35-43 (in Russian) (1977).
3. E. I. Givargizov. On CVD crystal growth. Dokl. AN SSSR, 149, 360-362 (in Russian) (1963).
4. E. I. Givargizov. Experimental investigation of crystallization by vapor deposition. Summary of Dissertation, Moscow, Institute of Crystallography (in Russian) (1965).
5. G. A. Kurov. On growth of germanium layers by vapor deposition. Fiz. Tverd. Tela, 3, 2080-2088 (in Russian) (1961).
6. E. I. Givargizov and A. A. Chernov. VLS growth rate of whiskers and the role of surface energy. Kristallografiya, 18, 147-153 (in Russian) (1973).
7. J. A. van Vechten. Effect of reconstruction of a semiconductor surface on the crystal growth. Appl. Phys. Lett., 26, 593-596 (1975).
8. E. I. Givargizov. Vapor Growth of Whiskers and Platelet Crystals. Nauka, Moscow, (in Russian) (1977).
9. E. Sirtl, L. P. Hunt and D. H. Sawyer. High-temperature reactions in the silicon-hydrogen-chlorine system. J. Electrochem. Soc., 121, 919-925 (1974).
10. A. H. Stepanova. Effect of $AsCl_3$ and BBr_3 impurities on the mechanism of CVD growth of autoepitaxial germanium layers. Summary of the Dissertation, Moscow, Institute of Crystallography (in Russian) (1974).
11. D. Shaw. Mechanisms in vapor epitaxy of semiconductors. In: Crystal Growth, Theory and Techniques, vol. 1, ed. C.H.L. Goodman. Plenum Press, London-New York, 1-48 (1974).
12. R. Wagner. Vapor-Liquid-Solid Mechanism of crystal growth. In: Whisker Technology, ed. A.P. Levitt. Wiley, New York, 147-219 (1970).
13. V. V. Voronkov. Supercooling on the face arising at the rounded crystallization interface. Kristallografiya, 17, 909-917 (in Russian) (1972).
14. R. J. Jaccodine. Surface energy of germanium and silicon. J. Electrochem. Soc., 110, 524-527 (1963).
15. E. Yoda. Anomalous growth of MoO_3 crystals. J. Phys. Soc. Jap., 15, 821-829 (1960).
16. C. H. Li. Epitaxial growth of silicon and germanium. Phys. Status Solid, 15, 419-450 (1966).
17. Y. D. Chistyakov. Mechanism of oriented growth of crystals (epitaxy). In: Growth of Crystals, Vol. 8, ed. N.N. Sheftal', Consultants Bureau, New York (1969).
18. G. G. Lemmlein, E. D. Dukova, and A. A. Chernov. Crystal growth from vapor in the vicinity of the triple junction. Kristallografiya, 5, 662-665 (in Russian) (1960).
19. J. Nishizawa, J. T. Terasaki, and M. Shimbo. Layer growth in silicon epitaxy. J. Cryst. Growth. 13/14, 297-301 (1972).

20. J. Nishizawa, J. T. Terasaki, and M. Shimco. Silicon epitaxial growth. J. Cryst. Growth, 17, 241-248 (1972).
21. M. Shimco, J. Nishizawa, and J. T. Terasaki. Defect-free nucleation of silicon on (111) silicon surfaces. J. Cryst. Growth, 23, 267-274 (1974).
22. A. A. Chernov. Growth kinetics and capture of impurities during vapor crystallization. J. Cryst. Growth, 42, 55-76 (1977).

KINETICS AND MECHANISM OF GALLIUM ARSENIDE GROWTH IN GAS-TRANSPORT SYSTEMS

L. G. Lavrentyeva, I. V. Ivonin, and L. P. Porokhovnichenko

The Siberian Physico-Technical Institute, Tomsk

General Approach to the Problem

The kinetics and mechanism of surface processes in vapor phase crystal growth are complicated, involve a series of consecutive and parallel stages, and have not been intensively studied [1]. Direct methods of growth rate measurement at individual stages of epitaxial growth are not yet available; new information on the growth mechanism and its rate-limiting stages can be obtained only from investigative techniques which can isolate specific stages in the crystal growth. No theory of crystal growth has been developed for CVD systems. The theoretical model of layer growth, generally accepted at the present time, has been applied to condensation from vapor [2, 3]. To become applicable to gas phase epitaxy, this theory must be extended, at least as far as the limiting role of the chemical reaction and the diffusion in a relatively dense adsorption layer are concerned. An attempt to take these specific features into account is known, although no constructive solution has been obtained [4].

All approaches to systems involving chemical transport of materials are based on the concept of a layered mechanism of crystal growth; consequently, this process is assumed to have much in common with crystal growth by other processes [5]. This concept is based on numerous experimental observations which indicate that the heterogeneous reaction stage is not rate limiting under normal conditions for growing reasonably perfect crystals by gas phase epitaxy. Assuming this to be true, the main kinetic relations characterizing the gas phase epitaxy can be compared with similar relations derived from the theory of the layered growth of crystals, with a view to determining to what extent and under what conditions this theory is valid for more complicated systems, and to distinguish those features of the process which are basic for a theoretical model describing the actual process. The most important aspect is the analysis of growth rate as a function of temperature and supersaturation. In addition to growth rate, the morphology of the growth surface should be known since its parameters are included in the theoretical model. As both the rate of a heterogeneous reaction and its mechanism depend on temperature and concentration of components in the gas phase and in the adsorption layer, it appears expedient to begin the analysis of kinetics and mechanisms of growth in gas phase systems by studying the anisotropy of crystal growth and surface morphology at constant supersaturation and constant temperature, and to continue with a study of the effects of supersaturation, process temperature, and ratio of components in the gas phase on the growth kinetics and surface morphology.

Experimental data on growth rate anisotropy make it possible to establish immediately whether growth proceeds in the kinetic mode, or a stepped layer structure of the growth surface is involved. Then, in accordance with prevalent concepts, one can simulate the structure of a surface deviating from a singular face as an echelon (train) of steps bounded with segments of close-packed facets [6]. This allows the calculation of the growth rate of a crystal as a function of step density (crystallographic orientation) [7, 8]. A comparison of an experimentally measured growth rate as a function of misorientation angle θ with respect to a singular face with a calculated curve makes it possible to evaluate the role of surface diffusion in growth, and to eliminate from the analysis, in a specific crystallographic range, such stages as adsorption, desorption, and attachment at kink sites; indeed, the accepted geometric model of high-index planes assumes that such planes consist of singular terraces whose length will depend on the misorientation angle, and monatomic height steps with end facets whose parameters are independent of θ.

By slightly changing the experimental conditions, one can also analyze the role of other surface processes. Thus, for example, the attachment rate can be affected by varying the structure of the end facet or the density of kinks on this facet, with the rates of the remaining stages left unaltered. The first is achieved by varying the direction of the misorientation, and the second by poisoning a fraction of the kinks with impurity atoms. Finally, a comparison of anistropy in crystal growth rate for a number of gas transport systems leads to a qualitative estimate of the effect produced by a chemical deposition reaction on the kinetics of crystallization. The purpose of studying the supersaturation effect on the kinetics and mechanism of growth as applied to vapor deposition is to find whether the formation of a dense adsorption layer can modify the process mechanisms and change the rate limiting stage. Moreover, temperature dependences give information on the energy distribution between the processes; however, data for the rate limiting stage are required to interpret the values of the activation energy obtained. With these arguments in mind, let us discuss the experimental results.

Anisotropy of Growth Rate and the Structure of Growth Surfaces of Gallium Arsenide Homo-epitaxial Layers

Systematic investigations of growth rate anisotropy and the structure of growth surfaces of gallium arsenide homo-epitaxial layers have been carried out for a number of gas transport systems under various experimental conditions, in the range of crystallographic orientations (111)A, B to (100) [1, 8-10]. The table below lists the growth conditions for four systems (Q is the inflow concentration of the transport agent in hydrogen, and F is the linear flow velocity of the gas mixture), and Fig. 1 shows a plot of the experimentally obtained anisotropy of growth rate in these layers.

TABLE 1. Conditions of Epitaxial Growth of GaAs Layers

Gas transport system	Temperature, °C		Q, mol.%	F, cm min^{-1}
	Source	Substrate		
GaAs–AsCl$_3$–H$_2$*	830	750	0.3	60
Ga–AsCl$_3$–H$_2$	830	725	0.4	40
GaAs–HI–H$_2$**	750	700	0.2	35
GaAs–I$_2$–H$_2$	830	770	1.1	15

*The GaAs source was not doped; GaAs layers were also grown under the same conditions from a GaAs source doped with zinc ($2 \cdot 10^{19}$ cm^{-3}), tellurium ($1 \cdot 10^{18}$ cm^{-3}), as well as with a direct supply of zinc vapor to the crystallization zone ($T_{Zn}=490°C$).
**GaAs source doped with tellurium ($1 \cdot 10^{17}$ cm^{-3}).

A comparative analysis of the kinetic curves $V_g(\theta)$ shows that curves obtained for different gas transport systems and under different conditions are qualitatively similar. Hence, the growth rate anisotropy is mainly determined by the crystallographic characteristics of the growing surface and, to a much smaller extent, by the selected transport agent. The growth rate minima correspond to close-packed planes (111)A, B and (100), and we find $V_g(111)A > V_g(100) > V_g(111)B$.

The observation of minima on the growth rate curve points to energy barriers for the nucleation of a new layer on each of the (111) faces, while the (100) face shows no barrier [6]. The fact that an orientational departure from a close-packed face increases the growth rate can be interpreted by the suggested model as resulting from the effect of an echelon of steps introduced by this departure, ensuring barrier-free growth. Assuming monatomic steps (or steps of identical height), the orientational dependence of growth rate due to these steps can be written in the following form [7, 8]:

$$V_g = C\lambda_s \tanh(y_0/2\lambda_s) \sin\theta$$

where C is a constant, λ_s is the diffusion length of adsorbed molecules, and y_0 is the spacing between steps. As the misorientation angle increases, the growth rate reaches a maximum when the distance between the steps in the echelon becomes equal to twice the diffusion length of adsorbed molecules. One may assume by comparing such curves that λ_s on plane (111)B is smaller than λ_s on (111)A.

In order to find to what degree the model is adequate to describe the real structure at the growth surface, the microstructure of the crystallization interface was investigated. The range of crystallographic orientations considered naturally subdivides into three intervals, each of which starts with a close-packed face: (111)A–(311)A (A range), (111)B–(311)B (B range), (100)–(511)A,B (C range). The main results are as follows [9]:

1. In all of the systems investigated and under all conditions, the best agreement between the predicted and the real structure of the growth surface is achieved for layers from the B range where growth steps are nearly monatomic (h ≈ 10 Å) and their density increases as θ rises. At the same time, kinematic density waves of steps and polyatomic steps appear on the surface; this may be caused by overlapping of their diffusional fields and by unequal arrival rates at the step from each side [3, 11].

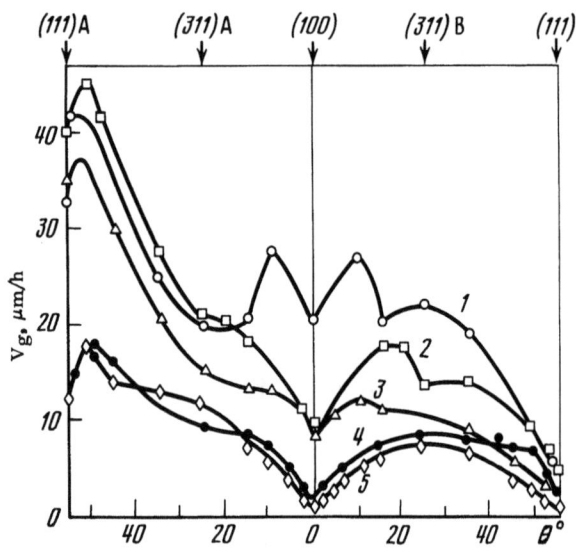

Fig. 1. Anisotropy of gallium arsenide layered growth rate in different systems: (1) GaAs-AsCl$_3$H$_2$; (2, 3) Ga-AsCl$_3$H$_2$ (two series of experiments); (4) GaAs-I$_2$-H$_2$; (5) GaAs-HI-H$_2$.

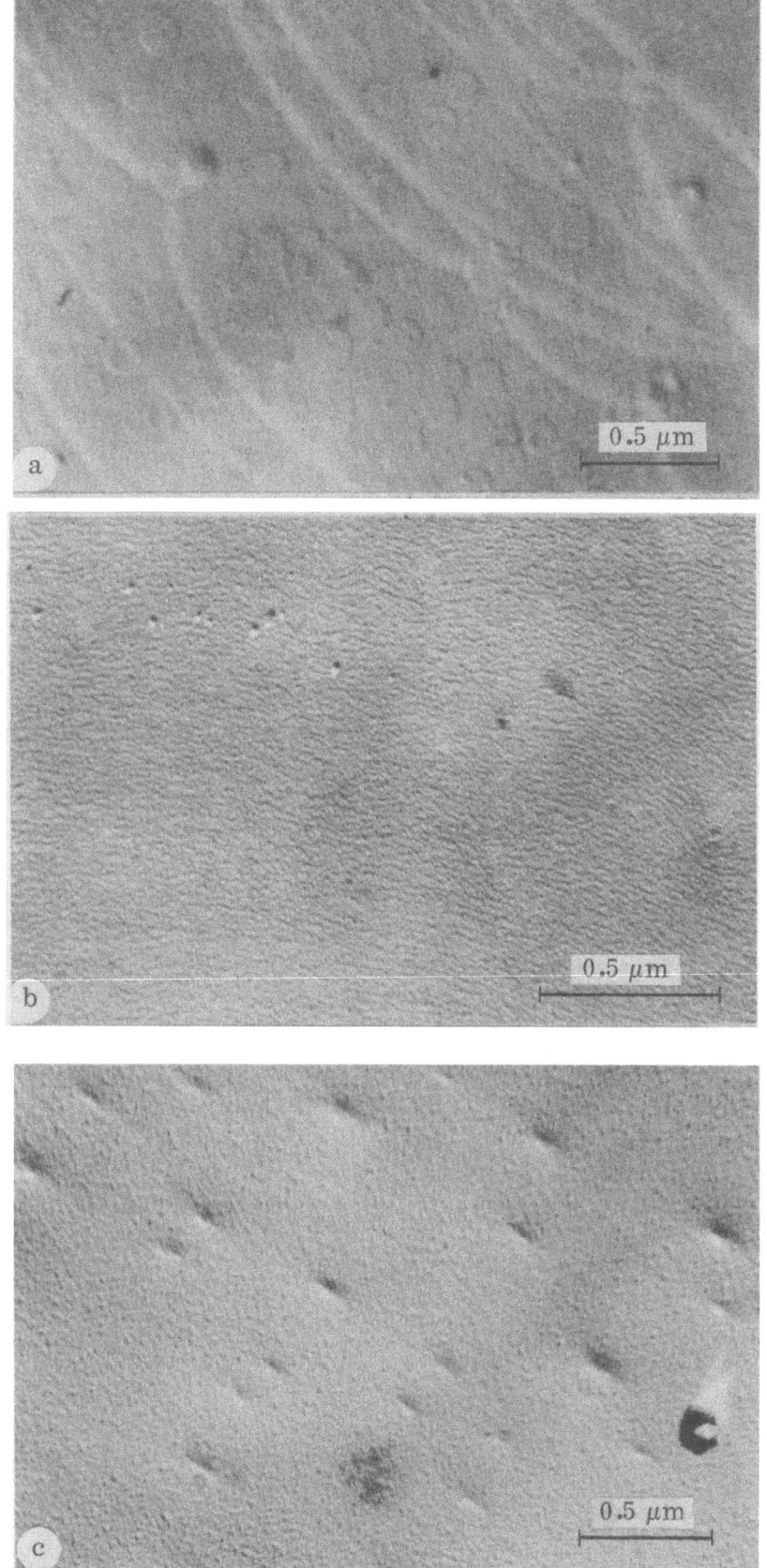

Fig. 2. Morphology of (100) gallium arsenide layers grown in systems (a) $GaAs$-I_2-H_2 and (b, c) $GaAs$-$AsCl_3$-H_2.

2. The structure of surfaces belonging to the C range is more sensitive to growth conditions. The step-layer structure is typical for growth surfaces at low growth rates in the iodide (Fig. 2a) and in the chloride (Fig. 2b) systems. At high growth rates the surfaces become microscopically rough (Fig. 2c) which points to modifications of the growth mechanism, that is to the possibility of "normal" growth.

3. Surface structure in the A range proves to have the maximum sensitivity of growth conditions. Although the layered structure of the crystallization interface occurs for all growth rates, the shape and slope of growth hillocks differ greatly. Vicinal hillocks with gentle slopes (the slope at the base from 2 to 5°) are observed on the surface at low growth rates, but at higher growth rates the hillocks are polygonal and the slopes reach 10-20° (depending on the slope orientation), although there is always a well-pronounced nucleation center of new layers at the top of the hill (Fig. 3). Since the rate of formation of new layers at the hillock top is high, that is, the nucleation barrier is low, deviations from the (111)A plane do not result in an appreciable increment of growth rate, and the hillocks on the growth surface are retained even if the surface is considerably tilted from (111)A orientation (up to 20°).

In all of the typical conditions investigated for the systems discussed, the crystallization interface thus had a predominantly step-layered structure which is characteristic of growth processes limited by surface diffusion.

The difference between the actual structure of the crystallization interface and the model surface consisted in the following. First, under certain conditions the steps on all vicinal surfaces interact to form kinematic waves and polyatomic steps of various structures. Second, the experiments show that the (100) plane, and planes close to (100), grow by the layer mechanism at low growth rates, and a transition to the quasinormal mechanism sets in at relatively high crystallization rates (about 20 μm/h). Third, a model can be postulated based on a high nucleation barrier on (111)A surfaces at low growth rates. At higher growth rates the nucleation barrier may decrease, thus eliminating the growth rate minimum for (111)A and producing a complex surface morphology. Note that grouping of steps, which is ignored by the Burton-Cabrera-Frank model [2], is not a typical feature only for vapor growth systems; it has also been observed in simpler systems [12]. At higher step densities the surface relief may be completely restructured, forming stable (311)- and (511)-type facets.

Fig. 3. Morphology of (111)A+2° GaAs layers grown in the GaAs-AsCl$_3$-H$_2$ system.

The experimental data reviewed above enables us to conclude that the growth of gallium arsenide layers in gas transport systems with halogens is mainly governed by the general considerations relating to the layered growth of crystals. In other words, for the stated conditions of the chemical reaction, although quantitative changes of the characteristic parameters of the surface processes may occur, these do not qualitatively affect the essential features of the process.

With this conclusion, we can estimate in terms of crystal growth theory [2] the following parameters of the surface processes: the activation energy of surface diffusion E_d, surface diffusion coefficients D_s, and tangential displacement velocities of the steps, V_t. The results for the polar faces (111) of gallium arsenide are:

	V_t, Å s^{-1}	λ_s, Å	D_s, cm^{2}s^{-1}	E_d, eV
(111)A	5·10^3	500	1.7·10^{-5}	0.54
(111)B	1·10^3	170	2.0·10^{-6}	0.73

The following formulas were used for calculations:

$$V_g = V \rho h; \qquad E_d = kT \ln(a^2 \nu / 2 D_s); \qquad D_s = 0.5 \lambda_s^2 \nu \exp(-E_a/kT),$$

where ρ, h are experimentally determined density and height of the steps, k is Boltzmann's constant, a is the elementary jump for an adsorbed particle (4×10^{-8} cm), T is the crystallization temperature (1023 K); and E_a is the adsorption energy (made on the basis of available data on bonding energies in diatomic molecules [13, 14]; i.e., it is assumed that the adsorbed particles are singly bonded to the surface). The values of λ_s were found by using data on the surface structure, assuming that the diffusional interaction between steps in the echelon ceases, and that a nucleus of a new layer at the hillock top appears if $y_0 \geqslant 2\lambda_s$.

As the energies of As-As, Ga-Ga, Ga-As bonds in molecules are unequal, with $E_{Ga-As} > E_{Ga-Ga} \approx E_{As-As}$ (50 ± 2, 32 ± 5, 32 ± 1) kcal mol^{-1}, respectively [13, 14], by taking into account the nature of the adsorption centers and adsorbed particles, it can be assumed that the derived diffusion parameters (λ_s, D_s, E_d) refer to arsenic atoms in the case of (111)B planes and to gallium monochloride molecules for (111)A planes.

The effect of changes in the attachment rates at a step on the growth kinetics may be determined experimentally by introducing a dopant or by changing the misorientation direction with respect to a singular plane. In the first case the density of sinks at the end facet of a step decreases because of blocking by impurity atoms (Fig. 4, curves 1, 2), and in the second case the kink structure of the steps is modified (curves 3, 4). It is essential that the rates of other surface processes (adsorption, surface diffusion) remained unaltered in these experiments because the properties of the terraces, of principal importance for these processes, were unchanged.

Effect of Concentration of the Vapor-deposited Species on Growth Rate and the Surface Structure of Close-packed Faces of Gallium Arsenide

The next stage of investigating the kinetics and mechanism of layered growth deals with the effect of concentration changes of the vapor-deposited species. In contrast to simple vapor-solid systems in which the vapor pressure of the deposited component over a crystal can be varied by varying the difference between the evaporation temperature and substrate temperature, in chemical vapor deposition systems this pressure can be controlled in two ways: either by changing the temperature difference between the evaporation and condensation zones, or by changing the concentration of the transport agent. The first of these cases

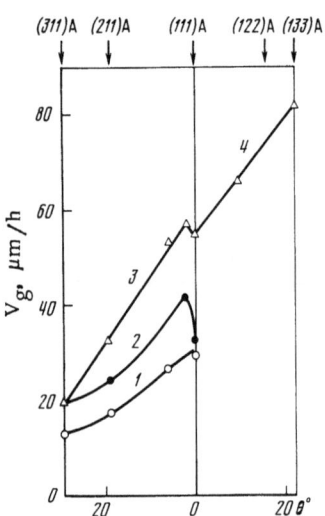 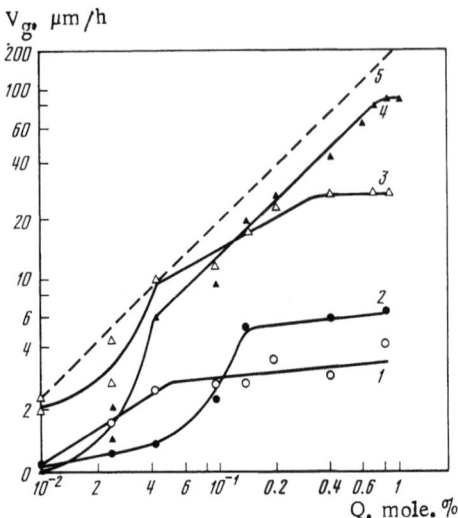

Fig. 4. Growth rate of gallium arsenide layers obtained in the $GaAs$-$AsCl_3$-H_2 system as a function of orientation with respect to (100) planes (1-3) and (011) (4), and the effect of doping (1 doped with zinc; 2-4 undoped).

Fig. 5. Gallium arsenide layer growth rate as a function of input supersaturation for various orientations: (1) (111)B; (2) (110); (3) (100); (4) (111)A; (5) thermodynamic yield of the process.

corresponds to the variation of supersaturation in its classical definition as the ratio of the rates of the direct reaction (deposition) and reverse reaction (evaporation). In the second case the yield of the process varies with the practically constant ratio of the direct and reverse reaction rates. In both cases a change in the pressure of the vapor components results in a corresponding change of the adsorption layer density. The second technique was employed in our experiments, because of its practical advantages. The concentration of the transport agent in a chloride system is fixed by the ratio $Q = P_{AsCl_3}/P_{H_2}$ where P_{AsCl_3} and P_{H_2} are partial pressures of $AsCl_3$ and H_2 at the input of the system. In order to simplify the exposition, we shall refer to Q as the input supersaturation, or simply supersaturation. The basic purpose of the experiment was to determine how the growth kinetics of the rate limiting stage changes as a result of changing the density of molecules in the adsorption layer, if the molar fraction of the deposited atoms in the gas phase (in hydrogen) varies from 10^{-4} to 10^{-2}. Taking into account the results presented in the preceding section, we have used three orientations: (111)A, (111)B, (100) [15]. Experimental curves plotting growth rate and morphology as functions of supersaturation Q are shown in Figs. 5, 6. There are several regions where a distinctive behavior of $V_g(Q)$ and typical structure of the growth surface are revealed in the range of supersaturation Q for each of the singular planes of gallium arsenide, and the width of these ranges and their boundary points are specific to each of the investigated faces. In the case of the two fast-growing (111)A and (100) planes three sections can be identified in the $V_g(Q)$ curves: a nonlinear section (I), a linear section (II), and a section in which the growth rate is only slightly dependent on supersaturation (III). The appearance of section II of the $V_g(Q)$ curves is the easiest to explain. On the basis of the similarity between the experimentally determined and thermodynamically calculated curves for growth rate as a function of supersaturation one can conclude that the slowest stage in this section is the diffusion-driven transport of material to the growing crystal through a boundary layer. Under these conditions the fluctuations of the surface relief are enhanced by the non-uniform concentration field existing in the region adjacent to the crystal. As a result, a complex surface structure appears, with pyramidal protrusions and pits.

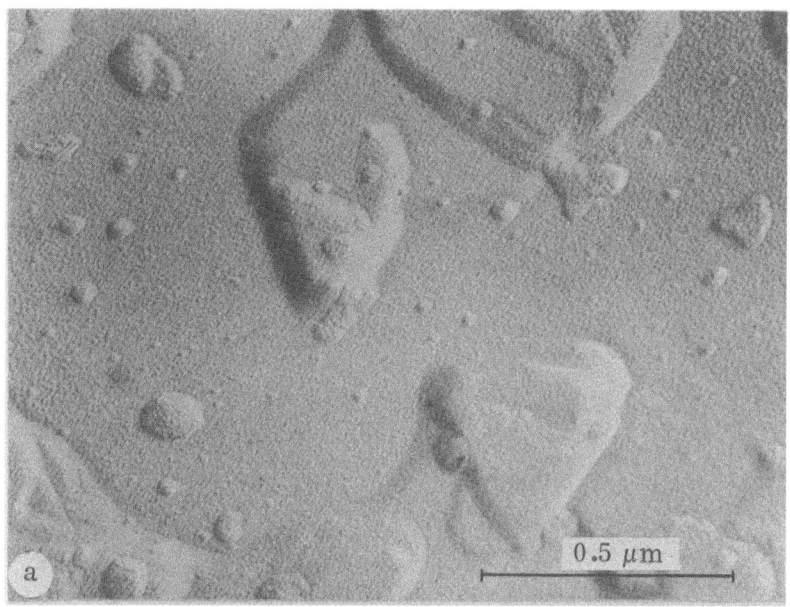

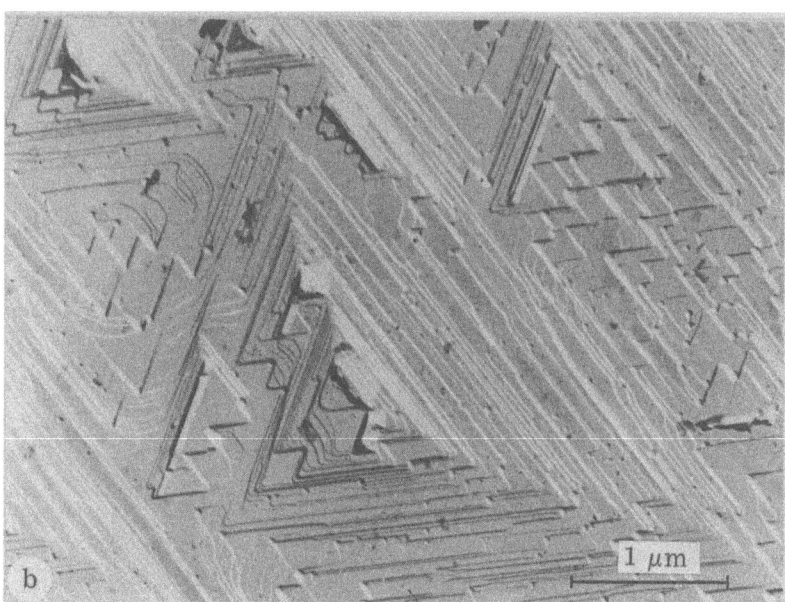

Fig. 6. Micromorphology of (111)A gallium arsenide layers at different input supersaturation Q, mol.%: (a) $1 \cdot 10^{-2}$; (b) $4.6 \cdot 10^{-2}$ (continued next page).

Sections I and III are caused by the limiting influence of surface processes. There are reasons to believe that at low supersaturation the slowest process is that of generation of nuclei (sinks), and at high supersaturation presumably adsorption-desorption is the limitation [15]. Indeed, the study of the morphology of the (111)A growth surface shows that it differs drastically for sections I and III [16]. The growth surface relief for section I has much in common with that for other systems and other conditions. As the supersaturation rises nuclei appear, their density increases, and their size becomes limited; this is evidence of the important role played in the growth morphology by adsorption, surface diffusion, and attachment. By varying the supersaturation one can observe a gradual change in the nature of the active growth sites: spontaneous nucleation and growth of coherent islands dominate at low supersaturations (see Fig. 6a), but at higher supersaturations ($Q \approx 5 \cdot 10^{-4}$) twin islands appear as well; these form large-size growth hillocks with a layered structure of the slopes

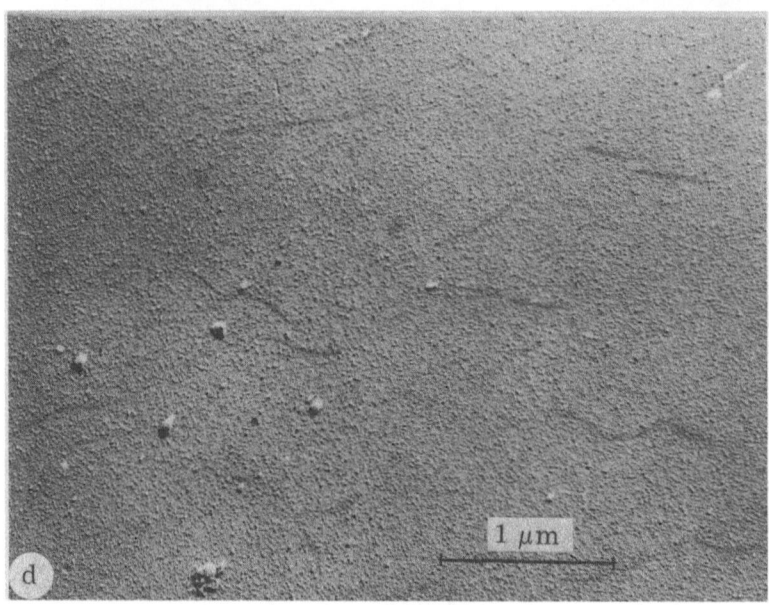

Fig. 6, continued; (c) $1.9 \cdot 10^{-1}$; (d) 1.

consuming most of the growth material and suppressing numerous but ineffective coherent nuclei (see Fig. 6b). A similar situation is observed on (111)B faces. It can be assumed therefore that nearly classical growth kinetics and growth mechanism apply in the system under discussion in the range of low supersaturation when the surface covered with active components (As, GaCl) does not exceed 10% and the remaining part of the adsorption layer is covered with atomic hydrogen (these estimates are based on the Langmuir isotherm and the values of the adsorption energy).

Section III is characterized by a microscopically rough surface, with growth sites either revealed as very flat and large pyramids (see Fig. 6c) or not revealed at all (see Fig. 6d). It is occasionally possible to observe step generation sites in the form of a microtwinning band; as a rule, the steps are very small. Estimates of the adsorbed layer composition (by a method suggested in [17] at the points corresponding to the beginning of the saturation section for gallium arsenide (111)A and (111)B faces) have shown that the density of adsorbed atomic hydrogen considerably diminishes in both cases. Taking into account that three particles take part in the chemical reaction, one can assume that the adsorption layer of section III consists mostly of GaCl and As and that hydrogen arrives at the reaction sites directly from the vapor. This means that the saturation range corresponds to a change in the reaction mechanism (from a Langmuir to a Riedel mechanism), and the growth rate is mostly limited by the desorption of HCl; this is in agreement with the results of [18].

Effect of Deposition Temperature

Earlier investigations of the temperature dependence of gallium arsenide growth [1] yielded the activation energies of the process, and revealed the range of temperature and pressure in which the growth rate is limited by surface kinetics. However, an analysis of growth surface morphology together with growth kinetics, aimed at determining the temperature range in which the layer growth mechanism occurs, has not been seriously attempted. The data on the temperature dependence of growth morphology are available only for the iodide-hydrogen vapor deposition system: typical layer growth on the (111)+2° surface is observed at $T \geq 700°C$ [19]. At lower temperatures (675-650°C) an undulating relief without pronounced steps occurs, and at still lower temperatures the surface is microscopically rough. It is difficult to evaluate the dependence of the main parameters of the layer morphology on temperature at $T > 700°C$, as at $T \approx 725°C$ microscopic drops of eutectic composition are formed on the substrate surface and grow through the epitaxial layer by the VLS mechanism [19]; this disturbs the regularity of the step morphology and generates clustering of the steps whose motion is slowed down by these defects. Only a tendency to step height reduction with increasing temperature of epitaxial growth is observable.

One can assume, therefore, that although chemical vapor deposition of gallium arsenide is possible in the range 600-1000°C [20], layered growth occurs in a much narrower range (from 700 to 800°C). Note that the slope of temperature dependence of gallium arsenide layered growth rate in the kinetic region is different for different orientations; this presumably reflects specific features of the growth mechanisms [1]. This aspect, however, needs further investigation.

Conclusion

The study presented above shows that the growth rate in chemical vapor deposition of gallium arsenide is determined, under a wide range of conditions, by the rate of the surface processes. The quantitatively varying anisotropy of growth rate retains the basic features which can be interpreted qualitatively in the framework of a simple model based on the step-layer structure of the growth surface, the existence of a barrier for layer nucleation and for attachment of the atoms to steps, and surface diffusion which limited the supply of material to the steps. At the same time, several features of the process are revealed, which must be taken into account in the model: reduction of the energy of nucleus formation on (111)A plane under certain conditions; strong interaction between steps leading to their clustering; and the transition from layered to quasinormal growth mechanism.

A quantitative analysis of growth kinetics must take account of the complex composition of a fairly dense adsorption layer in which material transport is accompanied by a chemical reaction leading to deposition of the crystallized components. This complicates calculations relating to the surface processes, as well as the evaluation of the parameters (nucleus formation energy, surface energy, the energy of particle bonding in the adsorbed layer, etc.).

The analysis of growth rate and growth morphology as functions of growth direction shows that the growth rate remains anisotropic in a wide range of variation of the transport agent concentration; hence the limiting effect of surface processes is retained. The study of the growth morphology shows that the nature of the rate-limiting surface stage may change: it is related to crystallization at low concentrations of the transporting agent, and to the adsorption-desorption processes at high concentrations. A similar situation seems to be found when the temperature dependence of growth rate is studied: the nature of the surface stage limiting the growth rate may change, depending on the temperature, supersaturation, and substrate orientation. Some parameters of the surface processes can be estimated from the growth rate and the structure of the growth surface if the growth rate is limited by crystallization, while the correspondence between the parameters and specific components remains a more complicated problem.

Literature Cited

1. D. W. Shaw. Kinetic aspects in the vapor phase epitaxy of III-V compounds. J. Cryst. Growth, 31, 130-141 (1975).
2. W. Burton, N. Cabrera, and F. Frank. The Growth of Crystals and the Equilibrium Structure of Their Surfaces. Phil. Trans., Roy. Soc., A243, 299-358 (1951).
3. A. A. Chernov. The Spiral Growth of Crystals. Soviet Physics Uspekhi, 4, 115-148 (1961).
4. J. Mandel. Surface-limited vapor solvent growth of crystals. J. Chem. Phys., 40, 683-690 (1964).
5. L. G. Lavrentyeva. The mechanism of epitaxial layer growth in the case of chemical transport. In: Growth of Semiconductor Crystals and Films. Nauka, Novosibirsk, 118-136 (in Russian) (1970).
6. R. C. Sangster. Model of studies of crystal growth phenomena in III-V semiconducting compounds. In: Compound Semiconductors, 1, New York-London, 241-253 (1963).
7. L. G. Lavrentyeva, Ju. G. Kataev, V. A. Moskovkin, and M. P. Jakubenya. Effect of substrate orientation on growth rate and doping level of vapor grown GaAs. Interval (111)A—(100)—(111)B. Krist. und. Techn., 6, 607-622 (1971).
8. L. G. Lavrentyeva, and M. P. Yakubenya. Epitaxial growth of gallium arsenide. In: Gallium Arsenide. Tomsk University Publishing House, Tomsk, No. 2, 40-45 (in Russian) (1968).
9. L. G. Lavrentyeva, I. V. Ivonin, L. M. Krasilnikova, and L. P. Porokhovnichenko. Growth rate and surface structure of homo-epitaxial GaAs layers. In: Growth and Doping of Semiconductor Crystals and Films. Nauka, Novosibirsk, Part 2, 84-98 (in Russian) (1977).
10. L. Hollan and C. Schiller. Etude de l'anisotropie de la croissance epitaxiale de GaAs en phase vapour. J. Cryst. Growth, 13/14, 319-324 (1972).
11. R. L. Schwoebel and E. J. Shipsey. Step motion on crystal surfaces. J. Appl. Phys., 37, 3682-3686 (1966).
12. E. D. Dukova. The passage of spiral-layer growth to normal growth on the basal face. 1. Disintegration of elementary layers at the vertices of spiral hillocks. Kristallografiya, 18, 819-825 (in Russian) (1973).
13. L. Pauling. General Chemistry, 3rd edn., Freeman, San Francisco (1970).
14. L. V. Gurvich, G. V. Karachevtsev, V. N. Kondratyev, Yu. A. Lebedev, V. A. Medvedev, V. K. Potapov, and Yu. O. Khozeev. Chemical Bond Energies, Ionization Potentials and Electron Affinities, Nauka, Moscow, (in Russian) (1974).
15. L. G. Lavrentyeva, L. P. Porokhovnichenko, and O. M. Ivleva. Growth of gallium arsenide in the $GaAs-AsCl_3-H_2$ gas transport system at different input supersaturations. 1. Growth kinetics and structure of epitaxial layers. Izv. Vyssh. Uchebn. Zavedenii, Fizika, No. 6, 54-59 (in Russian) (1976).

16. L. G. Lavrentyeva, I. V. Ivonin, and L. P. Porokhovnichenko. Growth of gallium arsenide in the $GaAs-AsCl_3-H_2$ gas transport system at different input supersaturations. 2. Growth mechanism. Izv. Vyssh. Uchebn. Zavedenii, Fizika, No. 12, 24-29 (in Russian) (1977).
17. A. A. Chernov and N. S. Papkov. Adsorption layer and nucleation in crystallization in the Si-H-Cl system. Dokl. AN SSSR, 228, 1083-1086 (in Russian) (1976).
18. R. Cadoret, and M. Cadoret. A theoretical treatment of GaAs growth by vapor phase transport for (001) orientation. J. Cryst. growth. 31, 142-146 (1975).
19. L. G. Lavrentyeva, I. V. Ivonin, L. M. Krasilnikova, Yu. M. Rumyantsev, and M. P. Yakubenya. Effect of crystallization temperature on electrophysical properties and morphology of epitaxial gallium arsenide. 1. Morphology of (1; 1; 1.075)-oriented layers. Izv. Vyssh. Uchebn. Zavedenii, Fizika, No. 6, 68-71 (in Russian) (1973).
20. R. R. Fergusson and T. Gabor. The transport of gallium arsenide in the vapor phase by chemical reaction. J. Electrochem. Soc., 111, 585-592 (1964).

LOCAL EPITAXY UNDER CONDITIONS OF STRONG GROWTH RATE ANISOTROPY

P. B. Pashchenko, G. A. Aleksandrova, and I. M. Skvortsov

Introduction

The possibility of using protective silicon dioxide masks for the localization of growth in the epitaxial deposition of semiconductor materials was established at the beginning of the 60's [1, 2]. Many publications have been devoted since then to the study and development of planar selective epitaxy, a process in which the holes etched through windows in the protective mask are filled in a controlled manner by the epitaxially deposited material. Practical cases of great interest are planar local epitaxial structures in which the surface of the grown layer coincides (is coplanar) with the surface of the original substrate. It has been shown [3-7] that considerable difficulties have to be overcome to achieve coplanarity in the desired structures. The holes may be filled both by upward growth from the bottom (normal growth) and via growth starting from the side walls of the holes (tangential growth); consequently, a specific ratio of the normal and tangential growth rates must be maintained to achieve coplanarity.

The arguments concerning the choice of the optimal value of this ratio and a suggestion for the use of fast-growing faces of the crystal for filling the holes were initially advanced by Shaw [8]. Two variants of the process have been discussed. In the first version the bottom of the hole parallel to the substrate surface is the fast-growing plane, and the hole is mainly filled by the normal growth from the bottom. Growth from the side walls formed by slowly growing faces is suppressed. Certain technological difficulties are encountered if planar structures are grown by this technique. The depth of holes in the substrate and the rate of their filling in the process of selective growth are usually controllable to an accuracy not better than 5-10%. Consequently, the required planarity of the structures is not always achieved by terminating the growth at a predetermined moment of time.

In the second variant the substrate and the bottom must be formed by a slowly growing face, but the side walls of the hole must be fast-growing crystallographic faces. In this case the tangential (lateral) growth is predominant in the process of hole filling. The effective suppression of normal growth enables one to grow highly planar epitaxial structures.

This approach to the problem in question is based on the assumption of a highly pronounced anisotropy of growth rate. In the case of the CVD epitaxy of gallium arsenide with growth rates for typical situations diminishing in the sequence, it has been concluded that nearly planar structures can only be grown on substrates parallel to $\{111\}$B and $\{111\}$A planes, with $\{111\}$B planes being preferred [7-13]. It has also been suggested in the same publications that the probability of growing planar structures in the holes on substrates parallel to $\{100\}$ and $\{110\}$ planes is doubtful.

An adequate planarity of films grown on {111}B substrates was reported in [14]. The upper surface of the selectively grown deposited layers was not more than 1 μm above the mask surface. In the case of structures grown on {100} substrates the protrusions, depressions, and non-parallel segments of deposited layers (with respect to the mask) resulted in a much larger deviation from planarity, ±2 μm [15]. In order to improve the planarity of the hole fillings on {100} substrates, it was suggested that a rectangular-pattern mask for hole etching should be placed at 45° to <110> directions [16], or in the range of directions from <100> to <310> [17].

No data are available in the literature on the growth of planar selective structures on indium arsenide. However, growth patterns for similar structures in $A^{III}B^{V}$ compounds may also prove similar [18, 19]. A general analysis of conditions for planar selective epitaxy is indicative of better prospects for the second method in which lateral growth predominates: in this case it is easier to achieve complete suppression of growth once the holes have been filled. Indeed, the problem becomes resolvable if the growth rate of at least one face of the crystal tends to zero while the growth rates of other crystallographic planes are maintained at an appreciable level. The original substrate surface must then be oriented along the plane with nearly zero growth rate, and filling of the hole must proceed via the lateral growth of side walls advancing at a considerable velocity. Obviously, growth within a hole will automatically stop after the upper surface of the epitaxial deposit coincides with the substrate surface. Experiments on the vapor epitaxy of indium and gallium arsenide point to the technical feasibility of this process.

Experimental

A strong dependence of growth rate on temperature and substrate orientation is observed in the gas transport homo-epitaxy of indium arsenide in the $In-AsCl_3-H_2$ system and gallium arsenide in the $Ga-AsCl_3-H_2$ system. These experimental results are given in Fig. 1 and show that growth rates of {100}, {110} and {111}B faces, at temperatures above 700°C for indium arsenide and 870°C for gallium arsenide, may exceed by several orders of magnitude those of {111}A. This strongly pronounced growth anisotropy can be employed for the selective planar epitaxy.

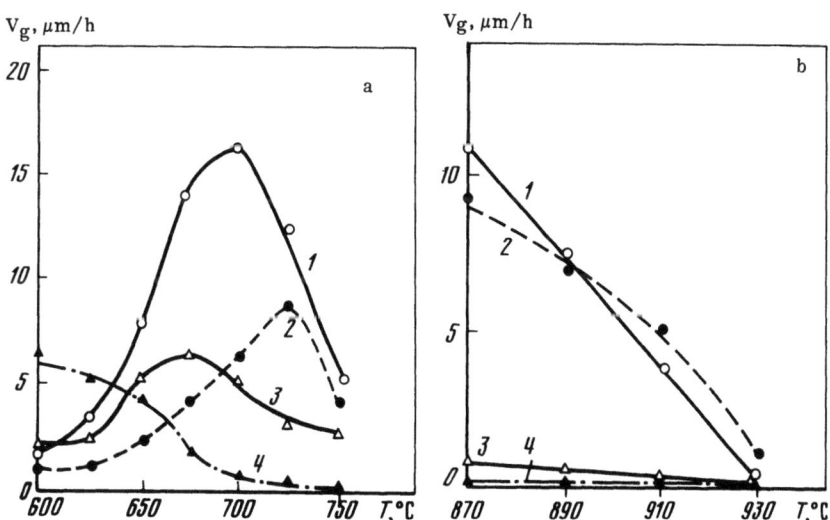

Fig. 1. Epitaxial layer growth rate as a function of temperature and substrate orientation for (a) indium arsenide at a source temperature of 850°C and (b) gallium arsenide at a source temperature of 1020°C; input $AsCl_3$ vapor pressure 0.5 mm Hg. (1) <100>; (2) <111>B: (3) <110>; (4) <111>A.

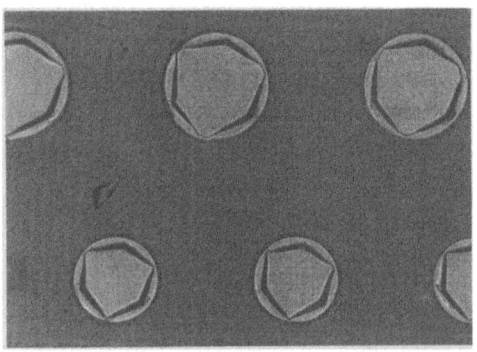

Fig. 2. Formation of an advancing step front for the fast-growing {320} planes in the early stages of hole filling (hole diameters 80 and 100 μm.

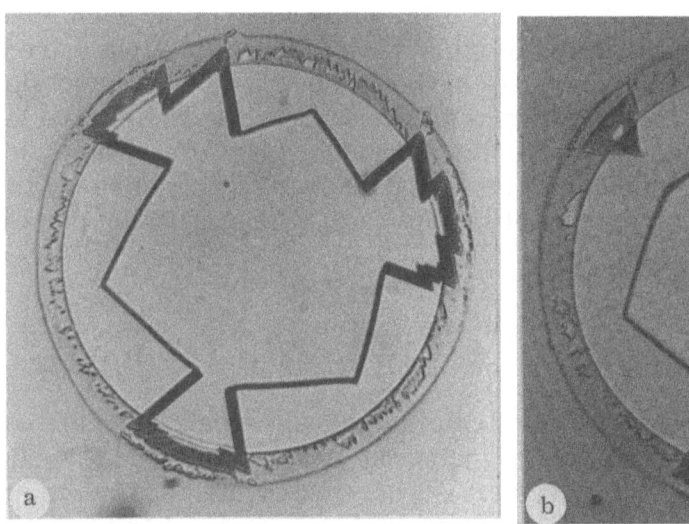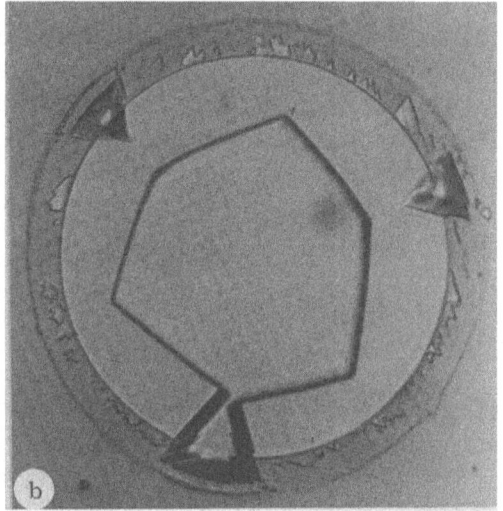

Fig. 3. Formation of an advancing step front of {321} planes (a) initial stage of filling (hole diameter 40 μm); (b) a later stage (hole diameter 30 μm).

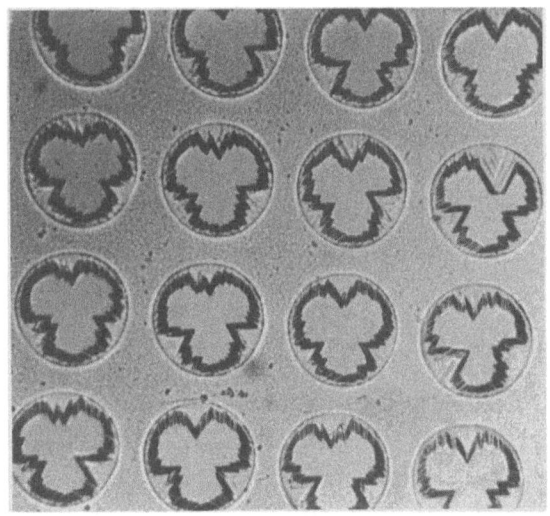

Fig. 4. Elimination of the fast-growing planes from the faceting of the wedge (hole diameter 30 μm).

Holes with diameters from 20 to 200 μm and depth from 1 to 20 μm were etched through an oxide (SiO$_2$) mask in {111} A substrates. Because of the strong growth anisotropy, the holes were filled by a planar epitaxial deposit by the growth from the side walls of the hole.

Some observed deviation from planarity of the grown structures was presumably caused by deviations of the original substrate surface from the {111} A plane, so that a smooth surface of the as-grown layer, being a singular {111} A plane, formed with the substrate an angle nearly equal to the deviation of the substrate surface from the {111} A plane. The measured height Δ of a step at the epitaxial layer edge is in good agreement with the value given by the formula Δ = d tan α where d is the hole diameter, and α is the angle of deviation of the substrate surface from the {111} A plane. For hole diameters below 70 μm and α ≃ 5' the step height did not exceed 0.06 μm.

The experiments on filling the holes of different diameters in a single process demonstrated that holes of smaller diameter were filled earlier, and that the height of a step at the epitaxial layer edge increased with increasing diameter of the hole. We thus conclude that there was practically no normal growth in the filled holes in the chosen conditions of epitaxial deposition.

It was also of interest to observe successive stages of hole filling. A crystal growing from a center tends to show faceting by slowly growing faces, the fast-growing faces vanishing. The growth in the holes proceeds from the periphery to the center. Observations show that in this process the lateral surface of the growing crystal is composed of fast-growing faces, and that the slow-growing faces disappear. In this respect our observations of the lateral faceting of the holes in the process of complete filling agree with the results of refs. [20-22] on filling of holes in synthetic quartz crystals where the growth from the walls towards the center was shown to reveal fast-growing faces. The block interfaces are most often formed by the fast-growing faces {320}, {211} B, or {321}.

As the ambient conditions of epitaxial growth change (temperature, vapor composition, etc.), the composition of the block interfaces is modified, but the tendency for the dominance of fast-growing faces remains and is observed even in the initial stages of hole filling.

Observations of the initial stages of hole filling make it possible to follow the formation of the advancing front of the deposit moving away from the side walls. The lateral surface of a hole produced by isotropic etching can be considered as a set of different

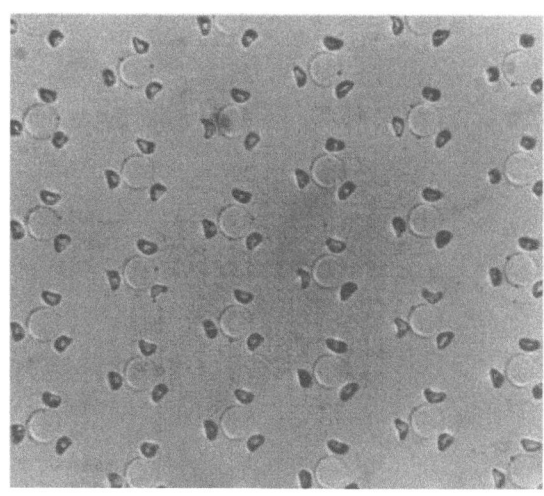

Fig. 5. Unfilled ("bare") areas at the hole periphery (hole diameter 25 μm).

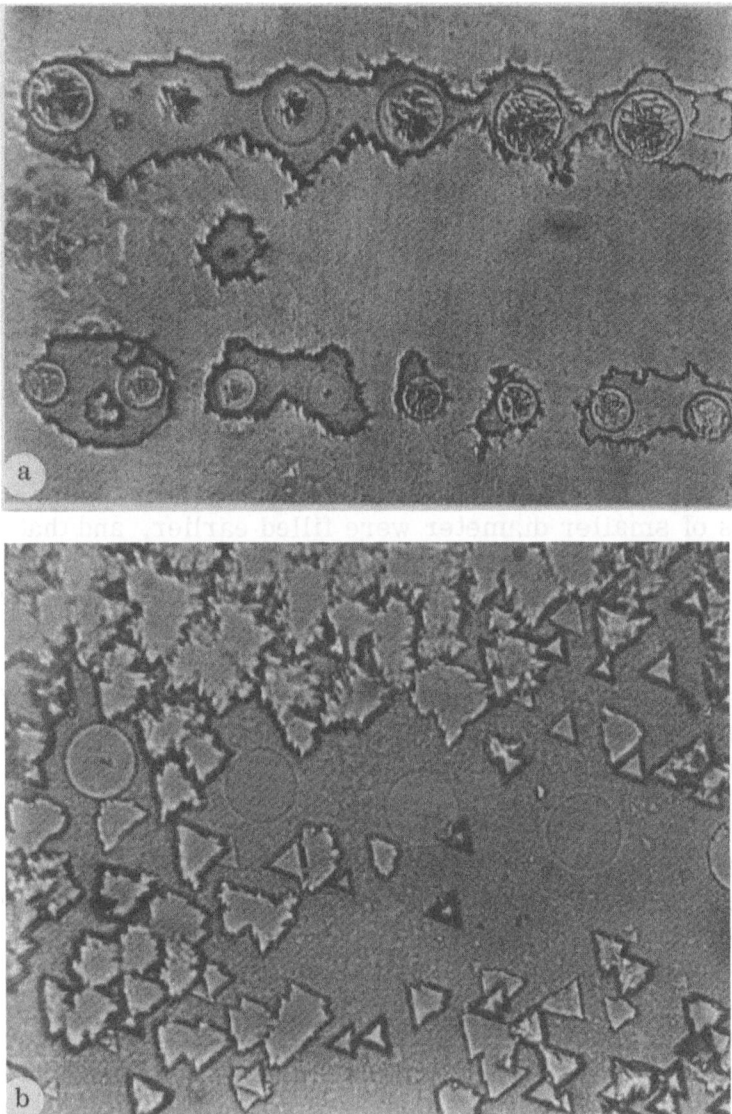

Fig. 6. Deposition of the material in the spaces between holes in the course of epitaxial growth on unmasked substrates (for two values of growth rate).

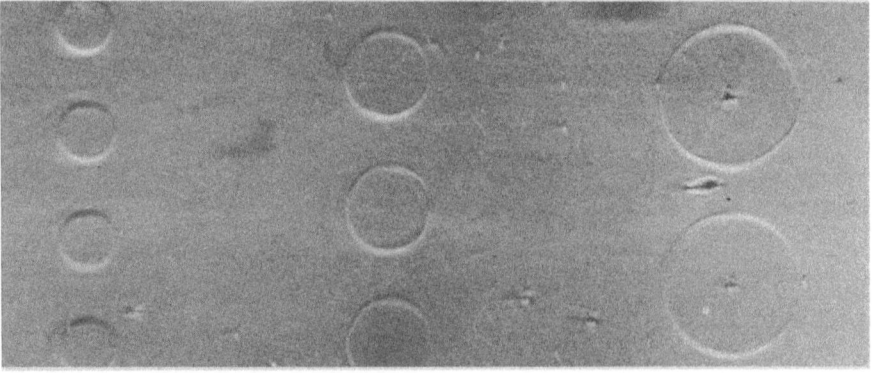

Fig. 7. Selective planar epitaxial layers on unmasked substrates observed by phase contrast microscopy (hole diameters 30, 50, 80 μm, respectively).

crystallographic planes. Having marked on the lateral surface the areas where slowly growing faces close to $\{211\}$A and fast-growing faces close to $\{110\}$, $\{321\}$, and $\{211\}$B appear on the surface, we can follow their evolution. Experiments have demonstrated that a reduction of In/As and Ga/As ratios in the vapor composition can make the planes close to $\{211\}$B grow at a slow rate.

The observations of the initial stages of the hole filling process show that "bare" patches appear at the points where slowly growing planes emerge on the lateral surface; the formation of such patches can be explained by a competition for the supply of material between neighboring segments having different orientations [23]. Removal of the supply from the segments of slowly growing faces at the side wall inhibits the formation of a continuous advancing front. The formation of such a front was observed in the early stages of hole filling at relatively high supersaturation (Fig. 2). As the supersaturation diminishes, competition for the material supply between the neighboring segments of the side walls intensifies. "Bare" patches grow in size, and the epitaxial layer advances from three segments of the side walls and forms truncated faceted wedges. The lateral surfaces of each wedge are slowly growing planes, while the surface of the wedge facing the hole center is formed by fast-growing planes (Fig. 3a). In this case the continuous advancing front is formed in the later stages of hole filling (Fig. 3b). The bare patches left at the periphery are appreciably filled only after the advancing front leaves them far behind.

A further decrease in supersaturation may eliminate the fast growing planes from the wedge faceting (Fig. 4). As a result, a continuous advancing front formed by fast-growing planes will not occur. In this situation the rate of hole filling becomes very slow, and the process of selective epitaxy becomes technologically impracticable.

The formation of bare patches at the hole periphery is especially undesirable in the case of epitaxy on substrates with protective masks because these areas lie under a mask overhang and remain practically bare even in the later stages (Fig. 5).

Defects of this type can be eliminated by selective epitaxy on unmasked substrates. Under these conditions, the crystallizing material is deposited in the space between the holes if the growth rate on the substrate surface plane is sufficiently high. However, the substrate remains free of the deposited material in the immediate vicinity of the holes (Fig. 6). The competition for the material supply to the growth surface is thus apparent not only within the holes, but also in their neighborhood. Once the holes are filled, the material is deposited over the whole surface of the wafer.

The problem of growing selective epitaxial structures on unmasked substrates is hard to solve as it involves the necessity of determining the moment of completion of hole filling. A decrease in supersaturation and a corresponding decrease in growth rate on the substrate surface increases the area free of the deposited material, and ultimately leads to no deposition between the holes (Fig. 7). In this case growth is totally arrested once the holes have been filled.

The selective epitaxial growth of indium arsenide on unmasked substrates proved extremely sensitive to the substrate surface treatment prior to epitaxy. Imperfections on the unmasked surface are often the local sites for selective deposition, with no deposition on the more perfect surfaces. Imperfections at the hole bottoms may also lead to a lack of planarity of the growing deposit.

Conclusion

Gas-transport epitaxy of indium and gallium arsenide, characterized by a strong anisotropy of growth rates, can be effectively used for selective epitaxy and planar filling of substrate holes by the deposited material. Under these conditions selective growth has been effected on {111}A substrates, where holes are filled by growth of the side walls, and the normal growth stage is suppressed.

Planar selective structures can be grown in holes formed by isotropic etching. No preliminary faceting of the lateral surfaces by fast-growing planes is required.

Selective structures with planarity better than 0.1 μm with respect to the initial substrate surface have been obtained by filling holes with diameter ~70 μm.

The possibility is established of effecting selective epitaxial deposition with total suppression of growth after the holes have been filled, as well as selective epitaxy on substrates without protective masks.

Literature Cited

1. B. D. Joyce and I. A. Baldrey. Selective epitaxial deposition of silicon. Nature, 195, 485-486 (1962).
2. M. Takabayashi. Epitaxial vapor growth of single crystal Ge. Jap. J. Appl. Phys., 1, 22-24 (1962).
3. O. S. Bulatov, A. V. Cherepovskaya, and Yu. D. Chistyakov. Mechanism of growth filling of faceted holes in selective epitaxy of gallium arsenide. Elektronnaya Tekhnika. Ser. 3. Mikroelektronika, No. 5(45), 83-92 (in Russian) (1973).
4. E. Mehal. Compound semiconductors for integrated circuitry. Trans. Met. Soc. AIME, 242, 452-462 (1968).
5. Y. Isibashi and M. Yamaguchi. Anisotropy in etching and deposition of selective epitaxial growth of GaAs. Jap. J. Appl. Phys., 9, 1007-1008 (1970).
6. S. Iida and K. Ito. Morphological studies on selective growth of GaAs. J. Cryst. Growth, 13/14, 336-341 (1972).
7. D. W. Shaw, R. W. Conrad, E. W. Mehal, and O. W. Wilson. Gallium arsenide epitaxial technology. In: Proc. Intern. Sympos. on gallium arsenide, Reading (1966) London, 10-15 (1967).
8. Pat. 4325879 (USA). Method of making shaped epitaxial deposits/D. W. Shaw. Feb. 04. 148-175 (1969).
9. L. G. Lavrentyeva, V. D. Dedkov, N. N. Bakin, and V. A. Yermolayev. Growth rate of epitaxial gallium arsenide layers on different crystallographic planes. In: Gallium Arsenide. Tomsk University Publishing House, Tomsk, No. 2, 186-188 (in Russian) (1969).
10. L. G. Lavrentyeva and M. P. Yakubenya. Epitaxial gallium arsenide. 1. Anisotropy of growth rate. In: Gallium Arsenide, Tomsk University Publishing House, No. 2, 40-45 (in Russian) (1969).
11. D. W. Shaw. Influence of substrate temperature on GaAs epitaxial deposition rates. J. Electrochem. Soc., 115, 405-408 (1968).
12. D. W. Shaw. Selective epitaxial deposition of GaAs in holes. J. Electrochem. Soc., 113, 904-908 (1966).
13. D. W. Shaw. Effect of vapor composition on the growth rates of faceted GaAs hole deposits. J. Electrochem. Soc., 115, 777-780 (1968).
14. R. Cox and E. W. Mehal. Planar Gunn oscillator for microwave. Trans. Met. Soc. AIME, 242, 461-464 (1968).

15. O. S. Bulatov, A. V. Cherepovskaya, and Yu. D. Chistyakov. Selective epitaxy of gallium arsenide in substrate holes. Mikroelektronika, 2, 180-183 (in Russian) (1973).
16. N. N. Bakin. Selective epitaxial growth of gallium arsenide. In: USSR Symposium on Generation of Microwave Oscillations Using the Gann Effect, Abstracts. Nauka, Novosibirsk, 75 (in Russian) (1973).
17. Pat. 3752714 (USA). Selective epitaxial deposition of semiconductor intermetallic compounds/K. Ito and S. Iida, 148-175. 4. 08. 69.
18. G. A. Aleksandrova, E. N. Korskaya, G. F. Lymar', A. V. Marchukov, and A. E. Shubin. Growth of epitaxial indium arsenide layers in open chloride systems. In: Growth and Doping of Semiconductor Crystals and Films. Nauka, Novosibirsk, Part 1, 45-50 (in Russian) (1977).
19. O. Mizuno, H. Watanabe, and D. Shinoda. Vapor growth of InAs. Jap. J. Appl. Phys., 14, 184-191 (1975).
20. Xenogenic and artificial liquid inclusions in synthetic quartz. Proc. VNIIP, 2, 81-84 (in Russian) (1958).
21. V. T. Ushakovsky, K. F. Kashkurov, and A. V. Simonov. Growth filling of holes in artificial quartz crystals. Kristallografiya, 13, 559-560 (in Russian) (1968).
22. V. T. Ushakovsky. Selected aspects of quartz crystals growth. Zap. Vsesoyuz. Mineral. Obshchestva, Ser. 2, Part 97, 571-581 (1968).
23. B. D. Joyce. Growth and perfection of chemically deposited epitaxial layers of Si and GaAs. J. Cryst. Growth, 3/4, 43-59 (1968).

ELECTRON MICROSCOPIC OBSERVATION AND COMPUTER SIMULATION OF STEP REDISTRIBUTION IN STEP TRAINS DUE TO CHANGES IN STEP DENSITY

K. W. Keller, D. Katzer and H. Höche

Central Institute for Solid State Physics and Materials Science, Institute for Solid State Physics and Electron Microscopy, Halle, GDR

The theory of crystal growth is based on molecular-kinetic models of surface processes taking place during growth. Direct experimental testing of theoretical models in the general case is, however, a difficult problem. One possible method, suggested by Basset [1], involves decoration of the surface. In this technique steps with height equal to only one interplanar spacing can be observed on the surface of alkali halide and some other crystals. The surface decoration technique yields both qualitative observation of the layered mechanisms of crystal growth [2] and quantitative data on step motion. The conclusions on step motion kinetics are obtained not only by measuring step velocities but also by observing changes in the arrangement within step sequences (trains). Such changes are the subject of the kinematic theory of crystal growth. The theory was first developed to study macroscopic crystal forms involving different arrangements of steps [3, 4]. A modification of this theory treats the arrangement of elementary steps on the surface and their motion [5]. The starting point of the kinematical theory is the dependence of step behavior on the distance to a neighboring step, as given by the Burton-Cabrera-Frank equation [6] (the moving steps are assumed to remain equidistant, with time-independent spacing, within the step train). An equation was derived which allows for nonequidistant steps [6], and a numerical method was developed for the observation of the temporal evolution of step train structures [7]. Electron microscopic study of evaporation structures on NaCl crystals [8] yielded results in good qualitative agreement with the theoretical conclusions on step trains [5, 9], and the results of these observations [10] were compared for the first time with calculations based on the method of Hullet and Young [7]. However, the observations were either unsystematic or conducted on comparatively simple step structures in the initial stages of evaporation. The present communication briefly outlines the progress of these studies: special experiments were conducted in order to compare the experimentally measured changes in step arrangement with the results of numerical simulation of comparatively fast evaporation.

We studied the temperature dependence of the spacing between steps in evaporation structures on a crystal surface [11], using a step-wise temperature jump to generate a sharp change in the step train density and generating, in turn, a redistribution of the steps. Cleaving and evaporation of crystals was conducted in ultrahigh vacuum conditions. Step structures first produced at a high temperature ($t_1 \approx 450°C$) were typically spirals and concentric circles, with steps separated by $\lambda(t_1)$. The temperature was then lowered to $t_2 \approx 350°C$, whereby the separation between steps, $\lambda(t_2)$, increased.

These effects are illustrated in Fig. 1 for a spiral step. Step separation $\lambda(t_1)$ in the peripheral area is small, and in the central part we observe several steps with a larger spacing, $\lambda(t_2)$. These areas are separated by a region in which step redistribution took

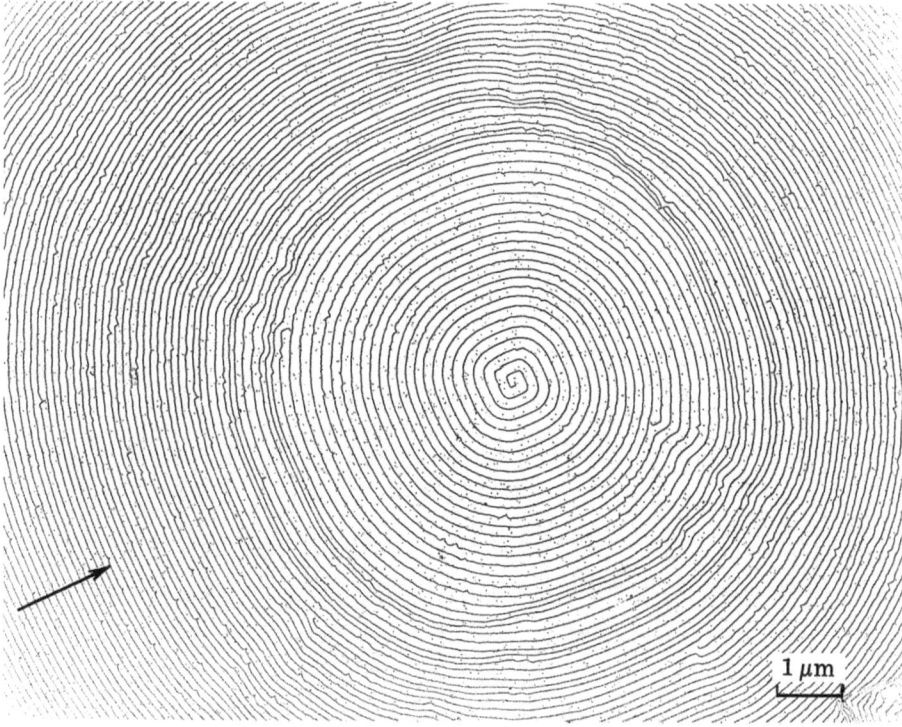

Fig. 1. Redistribution of steps in an evaporation spiral on the surface of NaCl after a jump in temperature.

place and where the steps are not equidistant. Let us compare this pattern with a computer-simulated step arrangement and its temporal changes. The model used (based on calculations by the Hullet and Young method) [7] must describe both the generation and the motion of steps.

The results are shown in Fig. 2 for seven time intervals in the course of annealing. In the lower part of the figure we show a fragment of the spiral indicated by the arrow in Fig. 1. The upper band corresponds to a period when the temperature was stepped. There is a train of equidistant steps spaced by $\lambda(t_1)$; the trailing step of the sequence is marked by a black circle. At the same time, a step spaced by $\lambda(t_2)$ is generated at the step source. The next five bands show the arrangement of the steps after the jump in temperature, as a function of annealing time τ at one hour intervals. The seventh band shows the arrangement of steps after the same time of annealing as for the experimental evaporation spiral (lower band). The experimental data are in good agreement with the results of calculations. The fit depends strongly on the velocity of a single step and the mean diffusion length used in the calculation. This means that experiments of this type and their computer processing provide a good method for the determination of the mean diffusion length at temperatures up to 400°C (this characteristic is 80±10 nm at 360°C).

Numerical simulation can be applied to an analysis of step redistribution as a function of annealing duration. It is essential that numerical simulation allows identification of the steps, and particularly the one generated at the moment of the temperature jump. In the general case the redistribution caused by the change in step density propagates from this point both toward the step source and in the opposite direction. Observations enable us to treat this redistribution as the process in which groups of steps are generated and vanish. The following factors affect the distribution of steps: temperature, which determines the velocity of a single step and the mean diffusion length; time of interaction between the

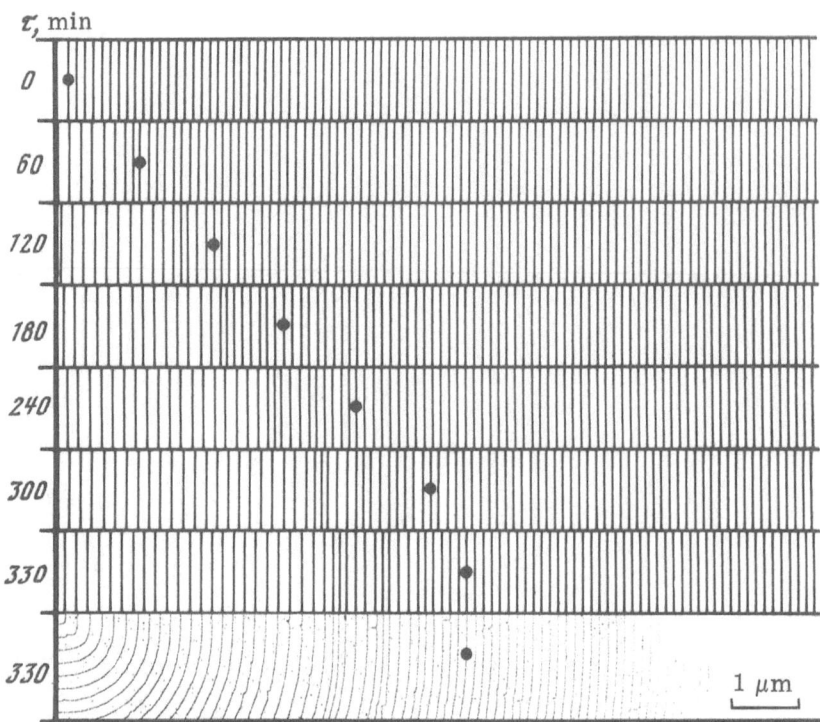

Fig. 2. Computer-simulated temporal evolution of the step structures shown in Fig. 1.

partial trains on both sides of the step density jump; absolute values of spacings between steps, $\lambda(t_1)$ and $\lambda(t_2)$, as well as their ratio.

A good agreement was obtained in a number of cases between the results of numerical simulation and experimental data, and the important parameters were found; this provides a good impetus to an extension of step simulation experiments. In contrast to the laboratory experiments, such computer experiments have an advantage of avoiding inhomogeneities which are usually present on the crystal surface. At the same time, there are additional parameters to be considered which affect the behavior of step trains. Hence, additional information on the motion of step trains can be extracted.

Literature Cited

1. G. A. Basset. Phil. Mag., 3, 1042-1045 (1958).
2. K. W. Keller. In: Crystal Growth and Characterization. North-Holland, Amsterdam, 361-372 (1975).
3. F. C. Frank. In: Growth and Perfection of Crystals. John Wiley, New York, 411-419 (1958).
4. N. Cabrera and D. A. Vermilyea. In: Growth and Perfection of Crystals. John Wiley, New York, 393-410 (1958).
5. W. W. Mullins and J. P. Hirth. J. Phys. Chem. Solids, 24, 1391-1404 (1963).
6. U. K. Burton, N. Cabrera, and F. Frank. The growth of crystals and the equilibrium structure of their surfaces. Phil. Trans. Roy. Soc., A243, 209-358 (1951).
7. L. D. Hullet, Jr., and F. W. Young, Jr. J. Electrochem. Soc., 113, 410-445 (1966).

8. K. W. Keller. Observation of the kinematic interaction of surface steps during evaporation of NaCl. In: Growth of Crystals, Vol. 11, ed. A. A. Chernov. Consultants Bureau, New York, London, 195-201 (1979).
9. T. Surek, G. M. Pound, and J. P. Hirth. Surf. Sci., 41, 77-101 (1974).
10. H. Höche and H. Bethge. J. Cryst. Growth, 42, 110-120 (1977).
11. K. W. Keller. In: Growth of Crystals, vol. 7, ed. N. N. Sheftal'. Consultants Bureau, New York-London (1969).

CRYSTAL GROWTH UNDER THERMODYNAMICALLY METASTABLE CONDITIONS

B. V. Spitsyn

The Institute of Physical Chemistry of the USSR Academy of Sciences, Moscow

Introduction

The crystalline structure of materials occurs because of a minimum in free energy for a rigorously periodic arrangement of the structural element of crystals, namely atoms, ions, or molecules. More than one free energy minimum may exist; that is, two (or more) crystal modifications of identical chemical composition may occur for a given pressure and temperature, with only one of these being stable, i.e., associated with the absolute minimum of free energy. The remaining modifications must be metastable.

The difference between stable and metastable crystal structures can be clarified as follows. The free energy G of a crystal is a function of the arrangement of its structural elements and is determined by a generalized coordinate Q (Fig. 1). The necessary condition of existence of both stable and metastable crystals is the equality to zero of the first derivative of free energy with respect to the coordinate, and the positive value of the second derivative. If a metastable crystal has in some manner been grown, the probability of its transition to the thermodynamically most stable state is determined not by the degree of metastability, $\Delta G = G_2 - G_1$, but by the activation energy G_2 necessary for the transition from phase 2 to phase 1.

Two methods of growing metastable crystals are possible. The first, the simplest in principle, consists in first growing the crystal at a pressure and temperature corresponding to its thermodynamic stability and subsequently transferring (e.g., by quenching) to the range of pressures and temperatures in which the crystal is to be used, but where it is metastable. The second method consists in growing the crystal directly in the range of its thermodynamical metastability. The present paper is devoted to reviewing some phenomena related to the nucleation and growth of crystals under the latter conditions.

The interest in studying the growth of metastable crystals is motivated by a number of reasons; first, the production of materials with identical composition but different crystal structure yields new knowledge on the physics and chemistry of the solid state. Considerable progress has been achieved by now in the synthesis of diamond, a metastable crystal important both in science and technology. As a rule, however, the development of metastable crystal growth techniques involves considerable experimental difficulties.

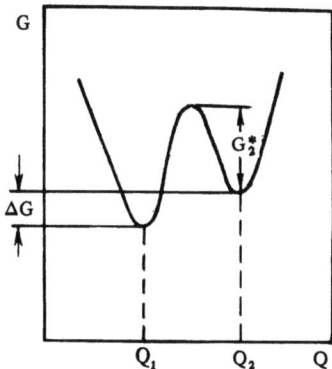

Fig. 1. Free energy of a crystal as a function of generalized coordinate.

The Homogeneous Nucleation of Metastable Crystals

The most important feature which should be taken into account in analyzing the possible nucleation and growth of a metastable crystal is as follows: from thermodynamic considerations the vapor pressure and solubility of a metastable phase possessing an excess of free energy are higher, and its melting point is as a rule lower than those for the stable phase. These differences are determined by the difference in the free energies of the two phases which may lie between several calories to several kilocalories per mole. Consequently, the crystallization ambient (vapor, solution, or melt) in equilibrium with a metastable crystal will be supersaturated with respect to the stable crystal phase.

Consider first the homogeneous nucleation of metastable crystals.

The frequency of nucleation, I, in unit volume per unit time is given by the equation [1]

$$I = 2N_1 \Omega \, (kT\alpha/h)^{1/2} \exp(-\Delta G_a/kT) \exp(-16\pi \Omega^2 \alpha^3/3kT\Delta\mu^2), \qquad (1)$$

where Ω is the volume of one particle of the crystal; N_1 is the density of particles in the nutrient phase; k and h are the Boltzmann and Planck constants; α is the specific free surface energy of the crystal-ambient interface; ΔG_a is the activation energy for the attachment of a new single particle to the nucleus; and $\Delta\mu$ is the difference in chemical potentials of the initial and final phases.

By neglecting the difference in pre-exponential factors, it is easy to derive from equation (1) an expression which is an extension of an equality employed in a number of papers, such as [2, 3], stating the condition of the predominant nucleation of the metastable phase:

$$(\Delta G_{a2} + 16\pi \Omega_2^2 \alpha_2^3/3\Delta\mu_2^2) < (\Delta G_{a1} + 16\pi \Omega_1^2 \alpha_1^3/3\Delta\mu_1^2), \qquad (2)$$

where subscripts 1 and 2 denote the stable and metastable phases, respectively.

If the difference in free energies is small, the main factor of inequality (2) is not the differences in thermodynamic supersaturations $\Delta\mu_1$ and $\Delta\mu_2$, but the differences in free surface energies and free activation energies of the phase transition for the corresponding phases. Usually (though not always) the metastable phase has a reduced surface energy [4], which appears to be one of the reasons for the well-known Ostwald's step rule [5], which states that a phase transition produces, as a rule, not the most stable crystalline modification but one or several successive modifications which approach in a stepwise manner the thermodynamically stable phase.

The additional possibility of spontaneous generation (and subsequent growth) of metastable phases appears in the case of nonstationary homogeneous nucleation; this is caused by a difference in kinetic coefficients of stable and metastable phases [6]. Consequently, the range of applicability of Ostwald's step rule can be considerably broadened compared with the classical treatment [2, 3, 7].

Heterogeneous Nucleation

Because of an inherent potential relief the surface of a solid is a place where the fluctuations of energy and density in the contiguous noncondensed phase are facilitated. As a rule this leads to a higher rate for heterogeneous processes, and in particular heterogeneous nucleation, compared with homogeneous processes. In the most general case of deposition on a foreign substrate, nucleation can be divided into three classes [8], the so-called weak, medium, and strong epitaxy. In weak epitaxy the adsorption energy of atoms of the crystallized material on a given substrate is smaller than the self-adsorption energy; as a result the condensate tends to form three-dimensional nuclei.

In medium epitaxy, which includes homo-epitaxy, the energies of adsorption and self-adsorption are of the same order of magnitude, so that heterogeneous nucleation in a certain range of supersaturation consists in the generation of two-dimensional nuclei which later grow into new crystal planes.

Finally, strong epitaxial interaction between the substrate and the deposit may be so intense that two-dimensional nuclei, and even nuclei of polymolecular thickness, may assume structures different from the bulk structure of the same substance. This is the case of the surface polymorphism much discussed in the literature [9]. Rigorously speaking, the substrate-film system as a whole may be thermodynamically stable and exist for an indefinitely long time. The structure of a thin film may be determined both by the influence of the substrate and by the special thermodynamic properties of small objects, in particular metal films with thickness of several tens of Å [10]. Both of these factors may shift the phase equilibrium curve and lead to nucleation of crystalline phases with a nonequilibrium structure compared with bulk samples of the same composition. However, a film several hundred Å thick tends to gain the structure typical for bulk samples.

Metastable Elemental Crystals

The growth of a metastable crystal is determined by a number of additional factors which are absent (or are unimportant) in the growth of thermodynamically stable crystals.

The following factors are significant, that is, they determine the course and the final result of crystallization under conditions of thermodynamic metastability: relative abundance of the structural type of the metastable crystal; supersaturation of the crystallization medium and its composition; various types of excitation of the crystallization medium components; the means of incorporation of impurities into the crystal and its nonstoichiometry (in the case of compound crystals); and the presence or absence of polymorphic transformation catalysts in the crystallization medium.

Let us consider some specific effects of a number of the above-mentioned factors on the nucleation and growth of metastable crystals with different degrees of metastability and with different types of bonding between the structural elements.

Sulphur. According to the phase diagram of sulphur, an enantiotropic transition of the low-temperature rhombic phase to the high-temperature monoclinic phase takes place at a pressure of 1 atm at 95.6°C. Both modifications are van der Waals crystals built of nonplanar cyclic S_8 molecules. If seeds of both of these modifications are immersed into a sulphur melt supercooled below 95°C, each seed gives rise to a crystal of its own structure, though one of them (monoclinic sulphur) is metastable under these conditions. This is a typical example of homo-epitaxial growth of a metastable molecular crystal.

Phosphorus. The two types of crystalline white phosphorus (α-P and β-P) are composed of symmetric tetrahedral P_4 molecules bound by van der Waals forces. The transition to the stable violet phosphorus having predominant covalent bonding between atoms requires overcoming a high energy barrier, and at 1 atm is completed only in the presence of a catalyst (mercury vapor) at 450°C. White phosphorus provides the most spectacular example of the metastable polymorphic (allotropic) modification obtained by sublimation of stable violet phosphorus [11]. At 25°C the difference in free energies of these two allotropic modifications is 3.2 kcal mol^{-1}. Nevertheless, kinetic barriers, related to polymerization of P_4 molecules into the stable structure of violet phosphorus, make it easy to obtain white phosphorus by sublimation of violet or red phosphorus (the latter is a solid solution of white phosphorus in violet phosphorus), with subsequent condensation of P_4 particles into a van der Waals crystal.

Gallium. Owing to the specific structure of condensed phases including Ga_2 molecules, solid gallium possesses an unusual (for metals) rhombic structure. Since most substances suspended in air as dust particles and acting as potential nucleation centers belong to higher crystal systems, the absence of foreign seed particles isomorphous with solid gallium presumably facilitates the high supercooling of liquid gallium (down to -110°C). It has been reported that metastable γ-Ga with melting point -35.6°C is spontaneously nucleated and grows at this low temperature [12].

Carbon. Because of the different chemical bonding in diamond and graphite crystals where carbon atoms are bonded by hybrid sp^3- and sp^2-orbitals respectively, graphite-diamond and diamond-graphite phase transformations are inhibited. Diamond is metastable at pressures P < 13,800 kg cm^{-2} at any temperature; at atmospheric pressure the difference, ΔG_T^0, between the diamond and graphite free energies increases with increasing temperature from 0.564 kcal mol^{-1} at 0 K to 2.5 kcal mol^{-1} at 2000 K (Fig. 2). Because of the high activation energy of the diamond-graphite transformation (approximately 85 kcal mol^{-1}), graphitization of diamond in high vacuum starts at 1900-2000 K [13]. According to LEED data [14], the surface structure of diamond is the same as its bulk structure at temperatures below 1550-1650 K. Unless special precautions are taken, this factor mainly determines the upper temperature limit on diamond synthesis by homo-epitaxial growth from vapor [15-17].

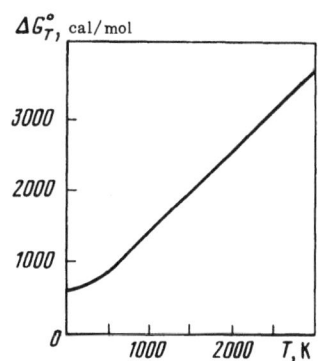

Fig. 2. The difference between isobaric-isothermal potentials of diamond and graphite as a function of absolute temperature at P = 1 atm.

Homo-epitaxial Growth of Diamond

Deposition of carbon in the form of diamond proceeds via a chemical reaction of the type

$$CH_4 = C(diamond) + 2H_2 \qquad (3)$$

The thermodynamically more stable graphite or other non-diamond carbon may be deposited simultaneously:

$$CH_4 \rightleftharpoons C(graphite) + 2H_2$$

Graphite may be removed by a gaseous or liquid selective etchant, after which the (3)-type process can be repeated. In special conditions the thermal decomposition of gaseous carbon-containing compounds may proceed selectively, without the side reaction of graphite deposition [18]. This makes it possible to grow single crystalline diamond films on the surface of seed crystals of natural diamond [19], to grow homo-epitaxial films of semiconductor diamond [20], to achieve nucleation and growth of diamond crystals with controlled habit on foreign substrates [21], and to grow polycrystalline diamond layers [22].

Figure 3 shows the surface morphology of a regrown diamond film. Figure 4 illustrates diamond crystals with cubo-octahedral habit grown on a copper substrate [21].

Excitation of the Vapor Phase

Numerous attempts are known to obtain diamond by the sublimation of graphite. However, the equilibrium carbon vapor at 2500 K is composed of mono, di, and triatomic carbon

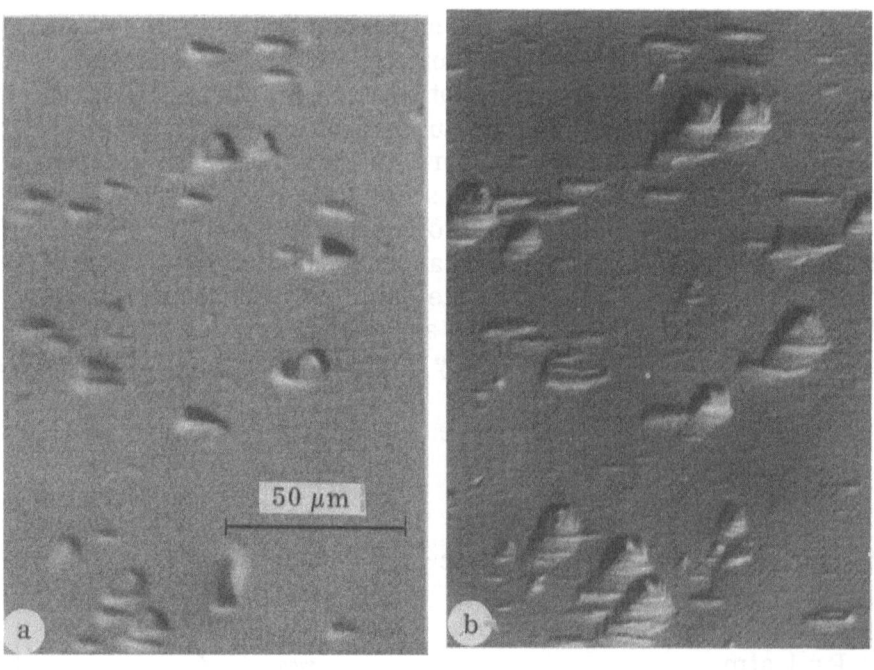

Fig. 3. Successive growth stages of a homo-epitaxial diamond film with (a) 1 μm and (b) 2 μm thickness.

Fig. 4. Crystals with cubo-octahedral habit grown on copper substrate.

molecules, the latter being predominant, while even single carbon atoms do not represent suitable building blocks for diamond growth because in its ground state the carbon atom is divalent. However, excitation of the carbon vapor atoms and molecules by electron bombardment with energy around 50 eV produces excited states which are capable of forming four coordination bonds between deposited carbon atoms [23]. As a result, layers of a so-called diamond-like carbon can be obtained even at sub-zero temperatures.

In principle, a similar method of exciting CH_4 molecules by an electric discharge makes it possible to drastically increase the deposition rate of polycrystalline diamond films [24], compared with the already conventional technique of diamond layer growth via thermal decomposition of methane and its analogs [16, 17].

Gas Transport Chemical Reaction

The main features of the thermodynamic interaction between diamond and graphite and the vapor phase can be generalized to other combinations of stable and metastable phases. Let us consider heterogeneous gas reactions of the same type, involving solid phases C_α and C_β:

$$nC_\alpha + {}^m/_2 X_2 \rightleftarrows C_n X_m;$$
$$nC_\beta + {}^m/_2 X_2 \rightleftarrows C_n X_m.$$

Equilibrium partial pressures of a gaseous compound $C_n X_m$ over a metastable phase α and the stable phase β are given by [25]

$$p^{(\alpha)}_{C_n X_m} = p^{(\beta)}_{C_n X_m} \exp\left[-n(G^0_\alpha - G^0_\beta)/RT\right], \tag{4}$$

where G^0_α, G^0_β are standard isobaric-isothermal potentials of phases α and β, R is the gas constant, and T is the absolute temperature. Equation (4) is valid when the equilibrium partial pressures of all possible compounds are much smaller than the total pressure.

As follows from equation (4), the number of polyatomic molecules over phase β is much smaller than that over phase α, and this effect will increase with the number n of carbon atoms per molecule. Consequently, the role of polyatomic molecules in chemical vapor transport in the direction $\beta \to \alpha$ must be insignificant. This factor is important in the chemical vapor deposition of diamond [17] since molecules of carbon-containing compounds with more than 4-5 carbon atoms are unsuitable for the defect-free growth of diamond.

The development of the CVD growth of diamond under the conditions of strong thermodynamic metastability made it possible to grow diamond layers several hundredths to several tenths of a micron thick, which is quite sufficient for the determination of the basic physical and chemical properties of the overgrowth. The following data indicate that the properties of "artificial" and natural diamond are practically identical:

	Artificial	Natural
Lattice constant a, Å(22°C)	3.5664 ± 0.001	3.5668 ± 0.0001
Carbon content, %	100	100
Refractive index	2.38 ± 0.02	2.40
Microhardness on face (111), kg mm^{-2}	$9,500 \pm 400$	$8,500 - 11,000$
Electric resistance, ohm·cm	$10^{12} - 10^{13}$	$10^{13} - 10^{15}$

Synthesis of Compound Crystals

The growth of metastable polymorphic modifications of chemical compounds may be complicated by nonstoichiometry affecting the position of the phase equilibrium line; this is the case, for example, in sphalerite-wurtzite transition in zinc sulphide [26]. This transition temperature is strongly affected by impurities forming a solid solution in the crystal lattice [26].

Sodium Bromide and Mercury Iodide. Crystallization of sodium bromide dihydrate on the surface of lead sulphide is possible at temperatures above the transition temperature to anhydrous sodium bromide. The growth of metastable mercury iodide is explained in terms of the theory of nucleation of a new phase on a solid surface [7].

Potassium Sodium Tartrate. Perfect crystals of potassium sodium tartrate up to 1 kg were grown by homo-epitaxial techniques at temperatures in the range 41-55°C, that is, in the range of thermodynamic metastability. This compound is unstable above 38°C and decomposes to anhydrous sodium and potassium tartrates [27].

Zinc Sulphide. Hexagonal and cubic zinc sulphide were obtained in identical deposition conditions by vacuum condensation of zinc sulphide on (111) and (110) faces of germanium [28]. In this case the orienting effect of surface forces is evident, with the symmetry of the forces determining the growth of a phase which gives the best fit to the substrate structure.

It should be emphasized in conclusion that the above arguments, though necessarily brief, are sufficiently conclusive in demonstrating the possibility of growing single crystals, and especially single-crystalline and polycrystalline films, in conditions of thermodynamic metastability. The potential of this method is especially great for growing crystals with predominantly covalent bonding by vapor crystallization techniques. The thermodynamic instability is of only nominal significance if a metastable crystal is to operate in conditions precluding its transition to the maximum stability form. Hence, there is no doubt that the development of reliable methods of synthesis of metastable crystals should widen the scope of crystal applications in scientific research and industry.

Literature Cited

1. A. A. Chernov. Crystallization. Ann. Rev. Mater. Sci., 3, 397-454 (1973).
2. L. S. Palatnik and V. S. Zorin. On the theory of transitions of metastable phases. Zh. Fiz. Khim., 33, 1859-1865 (in Russian) (1959).
3. R. Lackmann. Über heterogene Systeme mit Kleinen Abmessungen. Z. Naturfor., 17a, 812-816 (1962).
4. N. N. Sirota. To the theory of crystallization. In: Crystallization Mechanism and Kinetics. Nauka i Tekhnika, Minsk, 6-15 (in Russian) (1969).
5. W. Ostwald. Bildung und Umwaldungfester Köper. Z. Phys. Chem. 22, 307 (1897).
6. J. Gutzow and S. Toschev. Non-steady state nucleation in the formation of isotropic and anisotropic phases. Krist. und Techn., 3, 485-497 (1968).
7. I. N. Stransky and D. Totomanow. Keimbildunggeschwindigkeit und Ostwaldsche Stufenregel. Z. Phys. Chem., 163A, 399-408 (1933).
8. I. Markov and R. Kaischev. Influence of the supersaturation on the mode of thin film growth. Krist. und Techn., 11, 685-697 (1976).
9. L. S. Palatnik and I. I. Papirov. Uniaxial Crystallization. Metallurgy, Moscow (in Russian) (1964).
10. A. I. Bublik and B. Ya. Pines. Phase transitions in metal films, caused by thickness variation. Dokl. AN SSSR, 87, 215-218 (in Russian) (1952).
11. W. E. Addison. The Allotropy of the Elements. Oldbourne Press, London (1964).
12. J. Bosio and A. Defrain. Détermination de la chaleur latente de fusion de la forme γ du gallium. Comptes rendus, 258, 4929-4931 (1964).
13. V. P. Howes. Graphitization of diamond. Proc. Phys. Soc., London, 80, 648-662 (1962).
14. J. J. Lander and J. Morrison. Low-energy electron diffraction study of the (111) diamond surface. Surf. Sci., 4, 241-246 (1966).
15. B. V. Spitsyn and B. V. Derjaguin. Inventor's Certificate 339134 (USSR). A technique of diamond growth on diamond's face. (in Russian) 10.07.56.
16. W. G. Eversole. Pat. 3030187 (USA). Synthesis of diamond (1962).
17. W. G. Eversole. Pat. 3030188 (USA). Synthesis of diamond (1962).
18. B. V. Derjaguin and B. V. Spitsyn. Chemical vapor deposition of diamond. In: Crystal Growth, AN Arm SSR Publishing House, Yerevan, 12, 28-32 (in Russian) (1977).
19. B. V. Derjaguin, B. V. Spitsyn, A. E. Gorodetsky, A. P. Zacharov, L. L. Bouilov, and A. E. Aleksenko. Structure of autoepitaxial diamond films. J. Cryst. Growth, 31, 44-48 (1975).
20. A. E. Aleksenko, V. S. Vavilov, B. V. Derjaguin, M. A. Gukasyan, T. A. Karatygina, E. A. Konorova, V. F. Sergiyenko, B. V. Spitsyn, and S. D. Tkachenko. Charge transfer and the nature of acceptors in semiconductor epitaxial layers of diamond. Dokl. AN SSSR, 233, 334-337 (in Russian) (1977).
21. B. V. Derjaguin, B. V. Spitsyn, L. L. Builov, A. A. Klochkov, A. E. Gorodetsky, and A. V. Smolyaninov. Synthesis of diamond crystals on non-diamond substrates. Dokl. AN SSSR, 231, 333-335 (in Russian) (1976).
22. V. P. Varnin, B. V. Derjaguin, D. V. Fedoseyev, I. G. Teremetskaya, and A. N. Khodan. Peculiarities of growth of polycrystalline diamond films. Kristallografiya, 22, 893-896 (in Russian) (1977).
23. E. G. Spencer, P. J. Achmidt, D. C. Joy, and F. J. Sansalone. Ion-beam-deposited polycrystalline diamond-like films. Appl. Phys. Letters, 29, 118-121 (1976).
24. B. V. Derjaguin, D. V. Fedoseyev, V. P. Varnin, A. E. Gorodetsky, A. P. Zakharov, and I. G. Teremetskaya. Vapor deposition growth of polycrystalline diamond. Zh. Eksp. Teor. Fiz., 69, 1250-1252 (1975).
25. B. V. Spitsyn. On thermodynamics and kinetics of chemical vapor deposition of diamond. In: IVth USSR Conf. on Crystal Growth. Yerevan Univ. Publishing House, Yerevan, Part 1, 97-100 (in Russian) (1972).

26. S. D. Scott and H. L. Barnes. Sphalerite — wurtzite equilibria and stoichiometry. Geochim. et Cosmochim. Acta, 36, 1275-1295 (1972).
27. N. V. Al'avdin, N. N. Sheftal' and Z. I. Frolova. Growth of homogeneous potassium sodium tartrate crystals from highly supersaturated solutions. Kristallografiya, 2 193-195 (in Russian) (1957).
28. Y. Kuniya and G. Shimaoka. Heteroepitaxial growth of CdS films by vapor phase deposition. In: 5th Intern. Conf. on Cryst. Growth: Collected Abstr. Boston, Abstr. 81 (1977).

ROLE OF DEFECTS IN THE NUCLEATION OF WHISKERS GROWING FROM VAPOR

S. A. Ammer and A. F. Tatarenkov

The Voronezh Polytechnical Institute

The role of substrate surface defects in the nucleation of whiskers in the process of vapor deposition is a topic of long standing in the literature. In fact, the beginning was the diffusion-dislocation model of Sears [1] which was an attempt to relate the nucleation and one-dimensional growth of whiskers to the non-disappearing step created by a screw dislocation on a substrate surface. For some time this model was widely accepted because of a number of experimental data which supported the presence of defects in whiskers either directly or indirectly. However, the model had weaknesses which until recently had to be tolerated. The inconsistency of this model was demonstrated by Wagner [2] who substituted the diffusion-dislocation model by the diffusion-droplet model based on the vapor-liquid-solid (VLS) mechanism of whisker growth [3]. Wagner pointed to a good deal of conclusive evidence supporting the VLS mechanism. Many other facts could be mentioned, all pointing to the important role played by impurities in the nucleation of whiskers. Thus, in the reduction of iron halides the iron whiskers grow on magnesium segregations [4]. Selective nucleation of whiskers of iron oxides was observed on the surface of Armco iron doped by bismuth deposited through a mask [5]; the whiskers nucleated and grew in druses only at the impurity sites. The VLS growth of copper whiskers (reduction of halides by hydrogen), with drops of molten salt at the whisker tip, has been observed directly by *in situ* electron microscopy [6].

However, it would be wrong to relate the impurity effect in all cases to VLS growth. Sufficient information is available on the possibility of a directed supply from the vapor to the unidimensionally growing whisker, provided it contains a catalytically active particle at its tip. One example of the catalytic effect of iron particles on the nucleation and growth of carbon whiskers can be found in ref. [7]. The large-scale growth of iron, palladium, and copper whiskers from halide salts and metal oxides was observed after the introduction of carbon particles [8-10]. Porous carbon particles are highly adsorptive and effectively catalyze chemical reactions [11]. Some recent data can be mentioned (e.g., [12-17]) which find no place within the framework of the VLS mechanism.

As the nucleation and growth of whiskers proceed in the kinetic range, they must be strongly affected by the real structure of the substrate, and in particular by centers with low activation barriers for nucleation. It is probable that the points of emergence of dislocations at the surface may be classified as such centers [18].

The present paper describes experimental results which directly confirm the predominant nucleation of whiskers from the vapor at defects on a substrate surface. We have studied the growth of silicon, germanium, and corundum whiskers via chemical transport reactions in open-flow ($SiCl_4$-H_2, Al-H_2-H_2O) and closed (Si-Br, Ge-I) systems. The technological characteristics of the deposition processes and the equipment were conventional [18]. Wafers of high-resistance silicon single crystals, fused quartz, and perfect corundum ribbons were used as substrates. Prior to the growth stage the substrates were degreased, washed, and cleaned by gas etching in a reactor. Defects on the substrates were produced by scratching (by a diamond tip or polishing paste) or by cleaving (for corundum).

Under normal conditions, that is, with no impurities and defects introduced intentionally to the surface, films were deposited and whisker growth was not observed. Scratches on (111) surfaces of single-crystalline silicon substrates generate dislocation half-loops (Fig. 1) consisting of a 60 degree and screw components, with the terminal segments emerging at the surface. The Burgers vector of the dislocations was <110>. Other authors reported these dislocations under similar conditions [19]. Dislocations generated in corundum by mechanical factors were described in ref. [20].

Vapor deposition on substrates thus prepared has shown that at low supersaturation the deposition at scratches is always dominated by the growth of a polycrystalline condensate (Fig. 2). No whiskers nucleated even after many hours of exposure. Consequently, all further experiments involved impurities which were introduced into the systems by a number of techniques: as dispersed particles (Au, Cu, Fe) or films (Au, Cu) deposited on the substrates, or by gas transport reactions, together with the flow of the principal deposited material (Au, Sn). The method of impurity introduction appreciably affected the distribution of whisker nucleation centers on substrate surfaces. In the case of mechanical deposition of particles onto the substrate, the growth of whiskers took place at the points where the particles were sintered onto the substrate (Fig. 3).

A metal film approximately 0.02 μm thick, deposited electrolytically or by evaporation in vacuum, separated after heating in the reactor into small droplets randomly distributed over the surface; some of the droplets were attached to the scratches (Fig. 4), where local growth of whiskers was observed. However, defects produced by mechanical treatment of the surface were not decisive in determining the sites of whisker nucleation.

The results were different when impurities were introduced into the crystallization zone by chemical gas transport from independent sources, together with the flow of the main component in a closed or a through-flow system. Gold and tin were deposited mostly at the substrate defects. The formation of small drops of molten metal and the growth of whiskers took place at the same sites (Fig. 5). The whiskers had a typical shape for VLS growth of prisms with a globule at the tip. Mechanical treatment reduced almost by an order of magnitude the induction period of whisker nucleation on fused quartz substrates. The growth

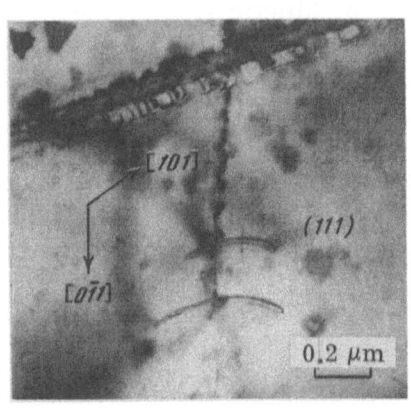

Fig. 1. Dislocation loops at scratches oriented along <110> on the (111) surface of a silicon single crystal wafer.

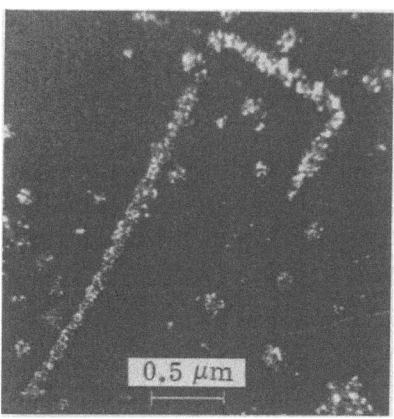

Fig. 2. Vapor deposition of polycrystalline silicon at a scratch on a quartz substrate (no impurities).

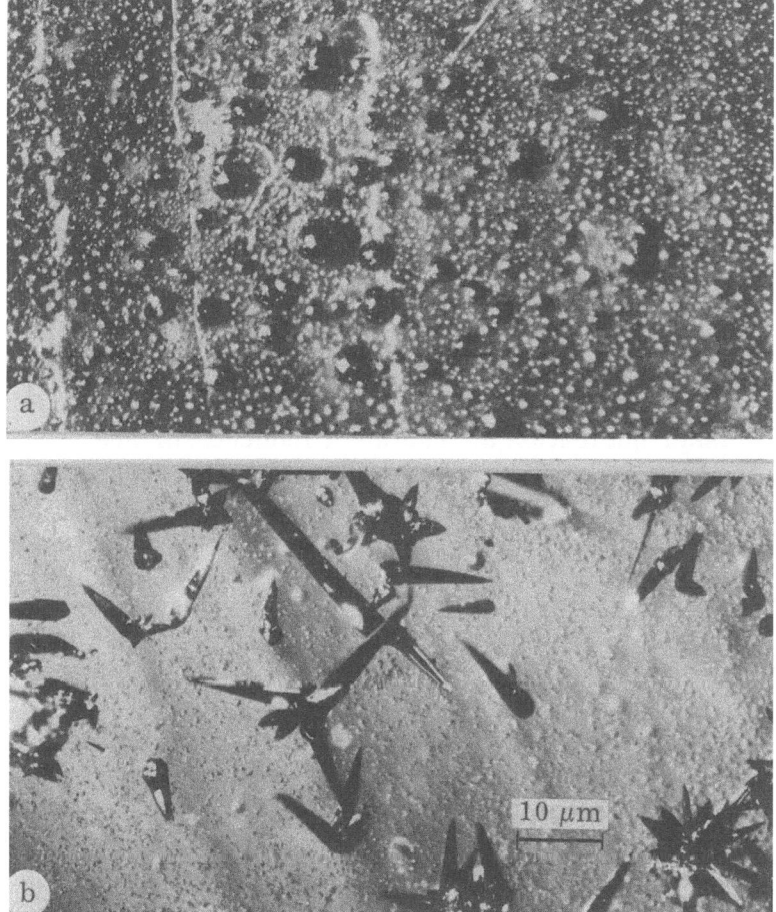

Fig. 3. Distribution of copper drops on a scratched quartz substrate (a), and localized arrangement of silicon whiskers (b).

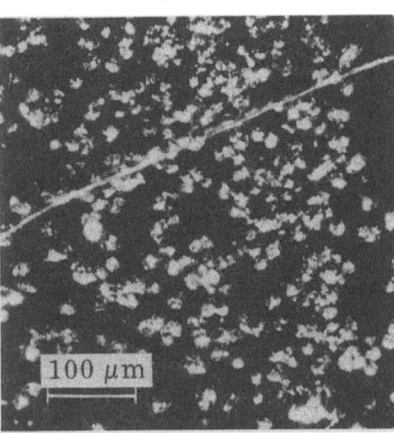

Fig. 4. Distribution of silver drops on a scratched quartz surface after decomposition of a film deposited in vacuum.

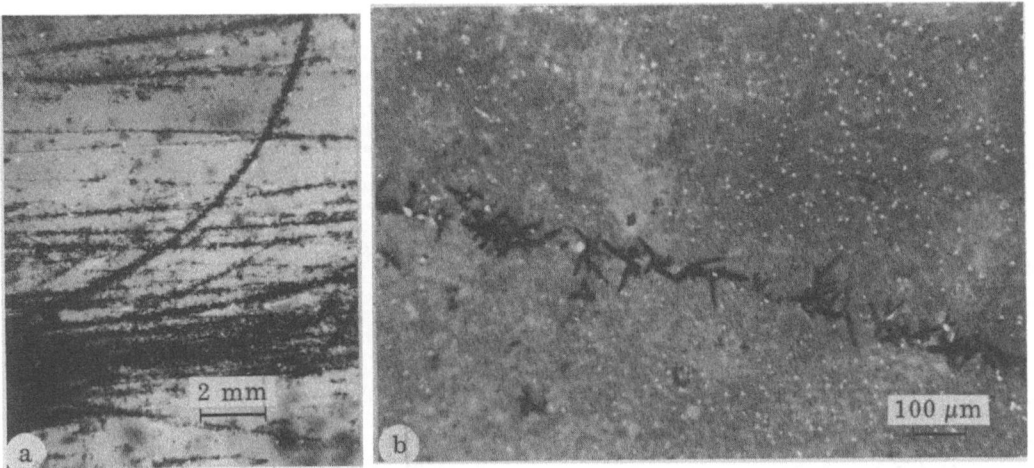

Fig. 5. Silicon whisker nucleation at treatment induced surface defects of a quartz wafer (a), and at a scratch on a single-crystal (111) silicon substrate (impurity: gold) (b).

of whiskers was initiated at lower supersaturation and temperatures than on untreated surfaces. For example, it was observed in the $SiCl_4$-H_2 system at 1050 K, at a mole fraction of tetrachloride as low as 10^{-3}.

When corundum was deposited onto basal substrate ribbons of the same material and containing incipient cleavage defects, whiskers did not grow unless impurities were introduced deliberately, and hexagonal pyramids grew in clusters at the cleavage sites (Fig. 6a). They were also formed at random structural defects in the ribbons. The growth of corundum ribbons occurred only when a metal impurity (Fe) was added (Fig. 6b). In this case very thin whiskers grew out of the pyramids similar to the observations reported in refs. [21, 22].

An analysis of the results obtained makes it possible to state that the presence of dislocations in crystalline substrates is not a decisive factor in the nucleation of whiskers because the same effect was observed after mechanical treatment of fused quartz in which no dislocation structures are formed. Furthermore, no axis dislocations were found by structure analysis in silicon and germanium whiskers. This is not surprising because with the chosen method of mechanical treatment of silicon single crystals the ⟨110⟩ orientation of the Burgers vector of dislocations was not favorable for adoption by the whiskers. The orientation of silicon and germanium whiskers on (111) surfaces was ⟨111⟩.

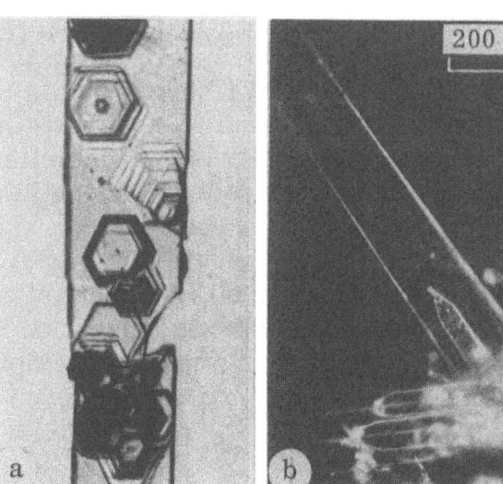

Fig. 6. Growth of hexagonal pyramids on a cleavage surface of a corundum ribbon (0001) with no impurity (a), and nucleation of whiskers with axial defects of pyramids at the cleavage sites with iron introduced as impurity (b).

Vapor deposition onto an energetically inhomogeneous substrate is known to be selective. At low supersaturation the rate of nucleation of a new phase can be described by an equation of a general type [23]:

$$J = AT \exp[-(U + W)/kT],$$

where A is a constant; U is the transition energy of an atom across an interface; k is Boltzmann's constant; T is temperature; and W is the nucleation activation energy. For a spherically symmetric nucleus

$$W \approx \gamma^3/\Delta F^2,$$

where γ is the specific surface energy, and ΔF is the change in free energy due to nucleus formation. The energy of nucleation, and with it the rate of nucleation, are sensitive to changes in the specific surface energy which is determined to a large extent by the structure of the surface on which the heterogeneous nucleation takes place [24]. Structural defects of the surface and impurities may substantially diminish γ and, hence, reduce the energy barrier for nucleation and thus constitute active deposition sites at low supersaturations [25]. A function $f(\theta)$, where θ is the contact angle of the liquid phase nucleus on the substrate [26], must be present as a factor in the formula for the energy change. The effect of surface defects on nucleation will then be partially reflected by a change in θ. The dependence of θ on surface roughness was mentioned earlier in ref. [27].

In the case of CVD the surface roughness makes the real spectrum of nucleation energies even more complicated since an important factor in this situation is the catalytic activity of defects, affecting the microselectivity of heterogeneous reactions [28].

As for some other defects, dislocations are sinks for impurities [29] and thus are capable of accumulating them and facilitating the formation of local areas of impurity sintering to the substrate [30]. In favorable circumstances (points of emergence of dislocations, suitable types of impurities, etc.) they become effective sites of local VLS growth of whiskers.

Summarizing, VLS crystallization theory recognizes the participation of dislocations in growth processes but does not consider the role played by dislocation as being predominant [2, 18]. Whiskers can inherit dislocations only in cases of favorable mutual arrangements of their Burgers vector and the direction of growth of a whisker (this was the case in the experiments on corundum whisker growth, but was not observed for silicon and germanium whiskers).

In addition to dislocations, whiskers may also inherit other substrate defects. For example, together with the usual prismatic whiskers of silicon and germanium on quartz, the growth of <211>-oriented ribbon crystals is observed, with the (111) twinning plane containing the growth axis. Grains of polycrystalline condensate on quartz, on which whiskers are formed, have a high density of twin lamellae [3]. However, whiskers inheriting twinning from the substrates are so far poorly substantiated observations still calling for experimental verification.

Thus, the mechanical treatment of substrate surfaces can be considered as a possible approach to controlling whisker nucleation in vapor deposition, in the case of low supersaturation and with the participation of impurities stimulating the VLS mechanism of growth.

Literature Cited

1. G. W. Sears. Mercury whiskers. Acta Met., 1, 457-459 (1953).
2. E. I. Givargizov. Vapor Growth of Whiskers and Platelets. Nauka, Moscow (in Russian) (1977).
3. R. Wagner. Vapor-Liquid-Solid mechanism of crystal growth. In: Whisker Technology, ed. A. P. Levitt. Wiley, New York, 147-219 (1970).
4. J. V. Laukonis. The influence of impurities on the nucleation and growth of iron whiskers. Met. Sci. Rev. Met., 62, 179-186 (1965).
5. R. F. W. Pease, and R. A. Plok. Growth of FeO whiskers. Trans. Met. Soc. AIME, 233, 1949-1954 (1965).
6. O. Nittono, H. Hasegawa, and S. Nagakura. Growth mechanism of copper whiskers. J. Cryst. Growth, 42, 175-182 (1977).
7. A. Oberlin and M. Endo. Filamentary growth of carbon by benzene decomposition. J. Cryst. Growth, 32, 335-349 (1976).
8. S. Kittaka and K. Kishi. Growth of Cu whiskers from cupric oxide. Jap. J. Appl. Phys., 4, 661-666 (1965).
9. S. Kittaka and T. Koneko. Growth of a large number of iron whiskers by the reduction of halides. Jap. J. Appl. Phys., 8, 860-869 (1969).
10. W. W. Webb. Dislocation mechanisms in the growth of palladium whisker crystals. J. Appl. Phys., 36, 214-221 (1965).
11. D. A. Frank-Kamenetsky. Diffusion and Heat Transfer in Chemical Kinetics. Nauka, Moscow (in Russian) (1967).
12. F. Okuyama. Vapor-growth of tungsten whiskers. J. Appl. Phys., 45, 4239-4241 (1974).
13. E. Schonherr. Photographic observation of the growth of GaP-needles. J. Cryst. Growth, 9, 346-350 (1971).
14. A. A. Nosov, T. A. Poshekhonova, and P. V. Poshekhonov. Mechanism of gold whiskers growth. Radiotekhnika i Elektronika, 18, 1993-1994 (in Russian) (1973).
15. V. A. Shmelev. Diamond whiskers, Khimiya i Zhizn', 5, 15 (in Russian) (1976).
16. S. Simov, V. Gantcheva, and P. Kamadjiev. Study of the morphology of CdTe whiskers by scanning electron microscope. J. Cryst. Growth, 32, 133-136 (1976).
17. S. Motojima, T. Wakamatsu, Y. Takahashi, and K. Sugiyama. Crystal growth and some properties of titanium monophosphide. J. Electrochem. Soc., 123, 290-295 (1976).
18. S. A. Ammer and V. S. Postnikov. Whiskers. Voronezh Polytechnical Inst. Publ. House, Voronezh (1974).
19. V. N. Yerofeyev, V. I. Nikitenko, V. P. Polovinkina, and E. V. Suvorov. Specific features of X-ray diffraction contrast and geometry of slip of dislocation half-loops in silicon. Kristallografiya, 16, 190-196 (1971).

20. V. N. Rozhansky, S. Z. Bokshtein, T. N. Bulygina, M. P. Nazarova, and I. L. Svetlov. Investigation of dislocation microplasticity in sapphire crystals by etching and electron microscopy. In: Whiskers and Nonferromagnetic Films. Voronezh Polytechnical Inst. Publishing House, Voronezh, part 1, 122-129 (1970).
21. S. Mendelson. Growth pips and whiskers in epitaxially grown silicon. J. Appl. Phys. $\underline{36}$, 2525-2534 (1965).
22. N. Holonyak, D. C. Jillson, and S. F. Bevacqua. Silicon, arsenic, whiskers, and tunnel diodes. In: Metallurgy of Elemental and Compound Semiconductors. New York-London 81-92 (1961).
23. J. Hirth and G. Pound. Condensation and Evaporation. Nucleation and Growth Kinetics. Pergamon Press (1963).
24. E. Kaldis. Principles of vapor growth of single crystals. In: Crystal Growth. Theory and Techniques. Vol. 1, ed. C. H. L. Goodman, Plenum Press, 49-191 (1974).
25. A. A. Chernov. Statistical kinetics of crystallization. Vestnik AN SSSR, No. 11, 60-67 (1968).
26. D. Turnbull. Phase changes. Solid State Phys., $\underline{3}$, 225-232 (1956).
27. B. V. Derjaguin. Film Growth. Dokl. AN SSSR, $\underline{51}$, 357-361
28. V. F. Dorfman. Vapor Deposition Micrometallurgy of semiconductors. Metallurgy, Moscow, (1974).
29. N. A. Kolobov and M. M. Samokhvalov. Diffusion and Oxidation of Semiconductors. Metallurgy, Moscow (1975).
30. J. J. Pankove. The effect of impurities on the growth of crystals. J. Appl. Phys., $\underline{28}$, 1054-1059 (1957).

FIELD EMISSION MICROSCOPIC STUDY OF THERMAL FIELD- AND CONDENSATION-INDUCED GROWTH FORMS OF CRYSTAL TIPS

V. N. Shrednik

The Ioffe Physico-Technical Institute of the USSR Academy of Sciences, Leningrad

Introduction

Owing to their small dimensions, conducting crystal tips with curvature radii of 10^{-6}-10^{-4} cm, usually studied by field electron microscopy (FEM) techniques [1], are single-crystalline even if made from polycrystalline wire. As the crystal tip is heated, its surface atoms become mobile, with the mobility increasing with temperature. At temperatures sufficient for an appreciable probability of two-dimensional sublimation of the atoms located at the edges of close-packed planar atomic networks, the tip is blunted (the tip material is transported to its lateral surface). This process, controlled by the Laplace pressure $p_\gamma = 2\gamma/r$, slows down drastically when a certain curvature radius, r, of the tip is reached [2, 3]; this radius is a function of temperature and surface tension γ (for example, in tungsten tips the critical radius is about 10^{-4} cm).

For a crystal to grow at its tip, one has to create conditions in which the flux of atoms towards the growth areas exceeds the flux of atoms from the tip to the lateral surface. This can be achieved by two methods: by condensation from the ambient at a moderate tip temperature, with the tip acting as a substrate, and by applying to the heated tip a sufficiently strong electric field. A field with strength F produces a pressure $p_F = -F^2/8\pi$ counterbalancing the Laplace pressure. At $F > F_0 = (16\pi\gamma/r)^{1/2}$ the diffusional fluxes reverse directions, and transport of material from the lateral surface to the tip vertex becomes energetically advantageous [3, 4] (the detachment of an atom from a step edge and the subsequent transition to the higher atomic layer is facilitated and becomes more probable than the reverse process).

Let us consider the forms and mechanisms of growth in the case of condensation and self-diffusion in strong electric fields. The experiments were conducted by varying the growth conditions (temperature, field strength, and atomic deposition flux) both in field electron and field ion microscopes.

Surface Form Changes Induced by Thermal Fields

General Remarks. The so-called field growth requires higher temperatures than growth by condensation because the material needed to build up new layers must be provided by decomposing old layers. However, an electric field with strength $F > F_0$ inhibits the blunting which would occur by condensation (at F=0) at the same temperature. At present the various stages of shape changes induced by the combined action of heating and electric field (Fig. 1a) are known in reasonable detail. The effects of electric field F and temperature T are fairly reproducible in a given material for the same tip radius and heating time. With a sufficient number of measurements, the results can be plotted on a T vs. F diagram. Figure 1b shows such a diagram for tungsten tips with radii 0.5 μm; the duration of the thermal field treatment was 1 min; F is given as the electric field strength at the tip surface before annealing.

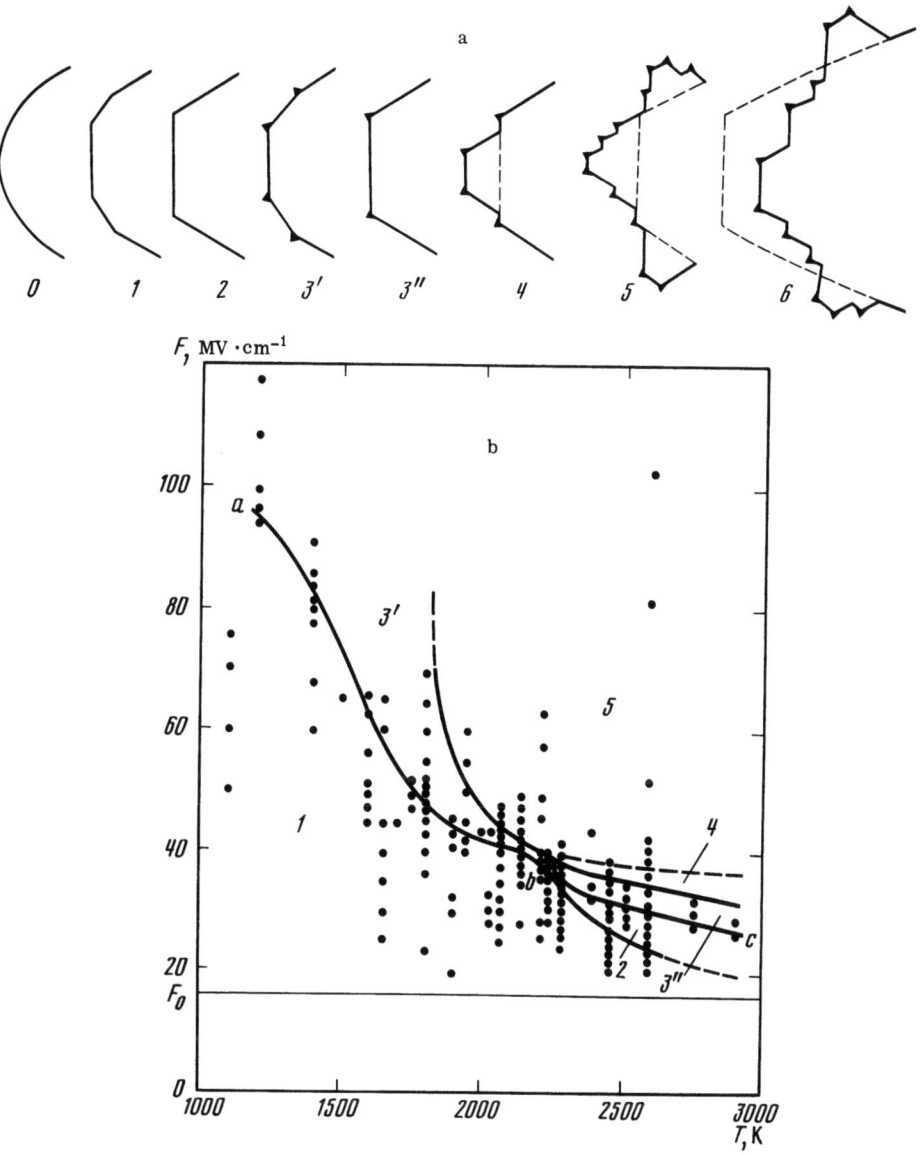

Fig. 1. (a) Schematic representation of tip profiles at different stages of heat treatment in an electric field, and (b) regions of the T-F diagram, corresponding to these profiles. (0) initial rounded annealed shape; (1) usual restructuring; (2) total restructuring; (3' 3") microscopic protrusions on a partially and completely restructured tip, respectively; (4) macroscopic overgrowth on the upper face only (tip elongation); (5, 6) macroscopic overgrowth on a number of close-packed faces without appreciable field-induced vaporization and with appreciable vaporization, respectively; (a, b, c) threshold line.

Field-induced Tip Reconstruction.

Heating in a moderate strength field produces broadening of the close-packed faces. This is the so-called restructuring of the tip surface first mentioned in [5], and interpreted as shape modification [6] (see Fig. 1b, region 1) or as growth via the supply of material to regions with $F > F_0$ from the lateral surface of the tip [3]. At a certain temperature and field strength the {112} W faces vanish and the tip is faceted by broad {110}-type planes; in addition, partly overgrown {100} faces are observed, giving the so-called complete restructuring [7] related to further supply of material to the tip vertex (see Fig. 1b, region 2). A close analogy between restructuring forms and habits produced by vapor deposition confirms the concept of the field being the driving force of the material supply, and makes it possible to estimate the amount of material transported under the conditions of field restructuring using geometric arguments [8].

Thermo-field Overgrowth.

A further increase in temperature and field strength increases the concentration of single atoms and supersaturation on maximum-density close packed faces to a level sufficient for the nucleation of new layers. The faces grow normal to the face plane, beginning with the face at the tip apex. If only this one face grows (and the field is insufficient for the growth of other faces), the tip is elongated and sharpened [4]; formerly this process seemed doubtful to a number of authors [2, 3] (see Fig. 1b, region 4). As the field strength is further increased, the macroscopic overgrowth spreads to other close-packed faces (see Fig. 1b, region 5). The normal growth occurs on the faces which were present on the initial restructured tip. Thus we observed overgrowth on {110}, {100}, and {112} faces. The overgrowth tips are usually faceted as completely restructured tips. The size of these overgrowths (referred to as macroscopic overgrowth [9, 10] in contrast to microscopic protrusions) or of their clusters on a single face is commensurate with the face dimensions.

Microscopic Thermo-field Overgrowth.

In contrast to the vacuum deposition growth, the field growth is characterized by the formation of tiny protrusions on edge and apices of multifaceted overgrowth tips or in rounded areas (while they are still present). Generally the factors causing these small protrusions (microtips) are the same as for macroscopic overgrowth. Hoever, the small dimensions of a microscopic overgrowth (less than 100 Å diameter) enables one to assume that its formation does not require any substantial transport of material and that small local mass displacements are sufficient. Microscopic protrusions appear on restructured tips (see Fig. 1b, regions 3', 3") and grow on the tips of all subsequent forms (regions 4, 5). Nevertheless, the growth of these protrusions is associated with a definite threshold on the (T, F) plane shown by the curve *abc* in Fig. 1b.

Difficulties in Identifying Emission Patterns after Thermo-field Treatment.

Microscopic overgrowth features are the most salient features of the emission patterns observed on the screen. As the substrate on which this overgrowth is located is of a fairly complicated shape (especially after the macroscopic overgrowth is formed), the emission pattern typically resembles a maze of spots. Slight heating, or field-induced evaporation, smooth the microscopic protrusions revealing the structure of the substrate and in particular the macroscopic overgrowth. FEM micrographs obtained after the field evaporation of microscopic protrusions enabled an analysis of complicated, "multi-storeyed" forms of intergrown macroscopic overgrowth on {110} faces (Fig. 2) [9, 10] to be made.

The field emission microscopy (FEM) techniques revealed an ideal defect-free internal structure of the macroscopic overgrowth and its homo-epitaxial nature (this means, in particular, that the field does not generate overly large disturbances, and the crystal grows as in the case of condensation). Once the microscopic protrusions and the edges and

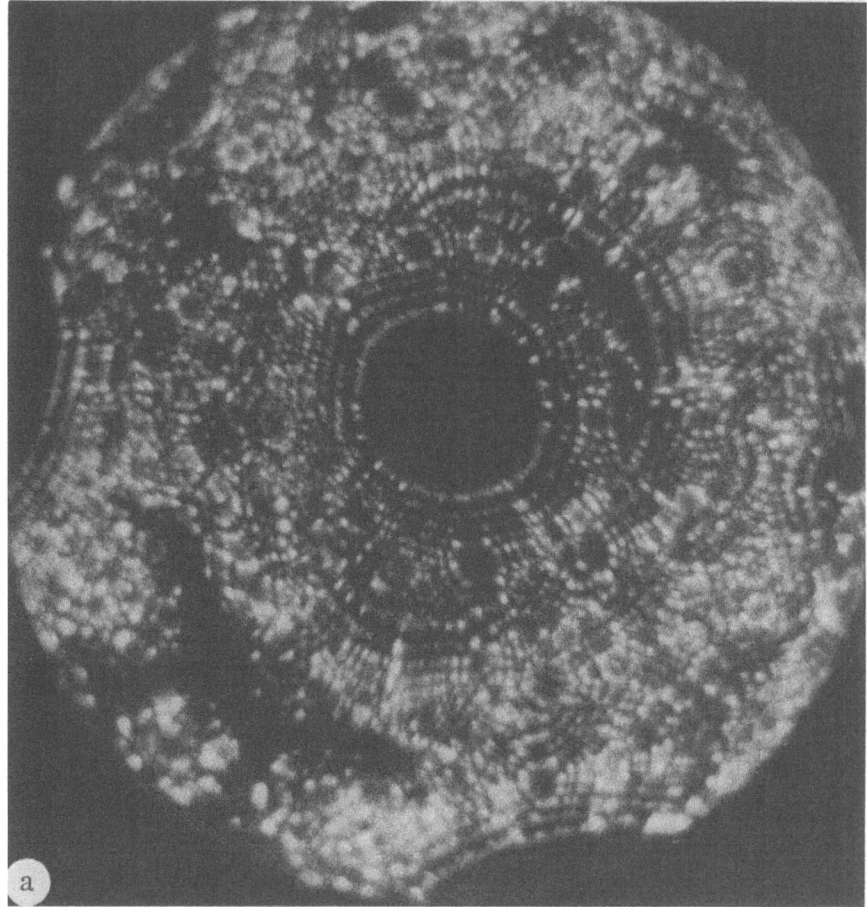

Fig. 2. A complex shape comprising several macroscopic overgrowth features of different heights on the central (110) face of tungsten. (a) field ion image of the tip (initial radius r=0.6 μm) with macroscopic overgrowth revealed after field-induced vaporization (helium ions, solid nitrogen temperature); (b) field ion image of the same surface at increased voltage (details are not resolved, but {110} - type upper faces of macroscopic overgrowth features are clearly defined); (c) a schematic with contour lines clarifying the geometrical features; (d) schematic profile of the tip apex with macroscopic overgrowth. See next page for parts b-d.

apices of macroscopic overgrowth are smoothened by heating, the macroscopic features appear in field electron micrographs as typical annealing forms of appropriate orientation (Fig. 3).

As the macroscopic overgrowth and microscopic protrusions grow, the field at their apices rapidly reaches a level at which the field evaporation at a given temperature becomes equal to the material inflow due to the self-diffusion in the field. The tip form becomes quasi-stationary. The tip in a strong field evaporates through microscopic protrusions and macroscopic overgrowth features. Its length is thereby reduced but the complicated shape is retained because the inflow and outflow of atoms are automatically maintained in the dynamic equilibrium state. This complex process, controlled by self-diffusion and evaporation in the field, and whose main feature is the formation of macroscopic overgrowth, is referred to as field erosion [9]. The latter effect enables the experimenter to blunt the tips

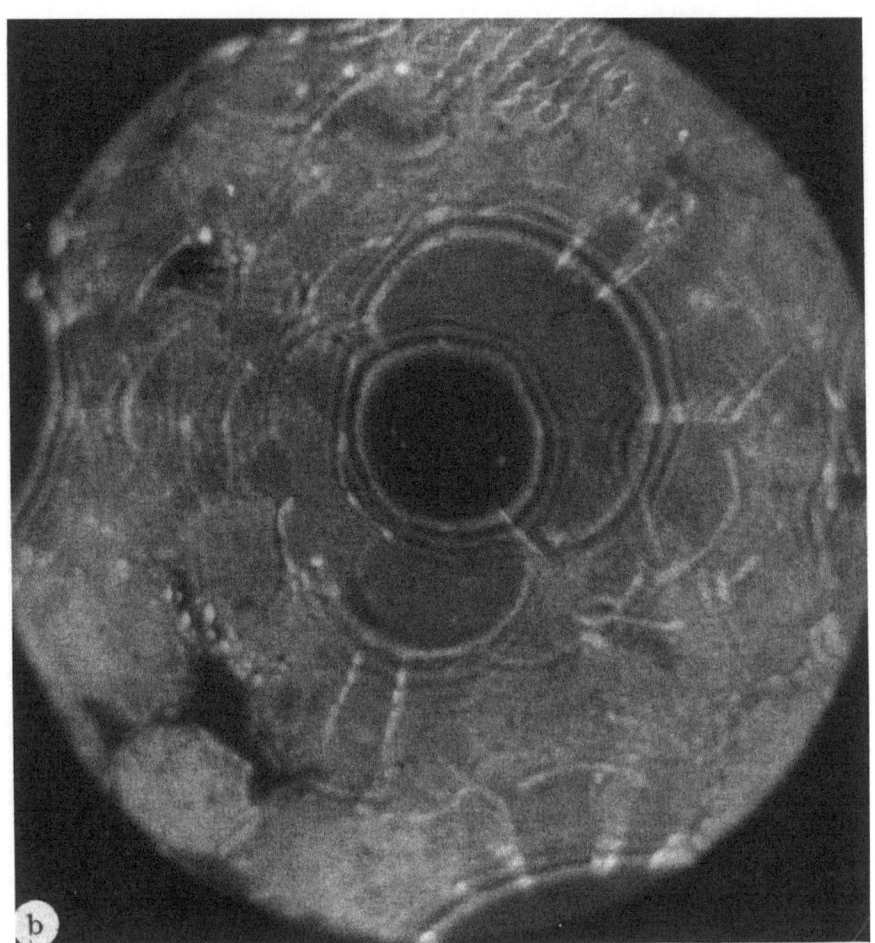
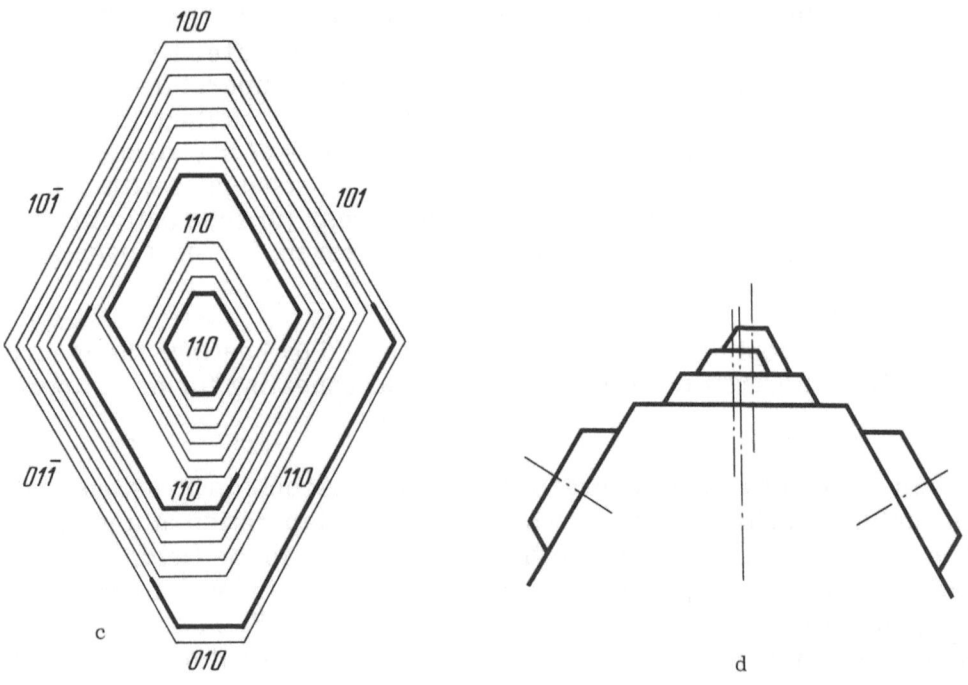

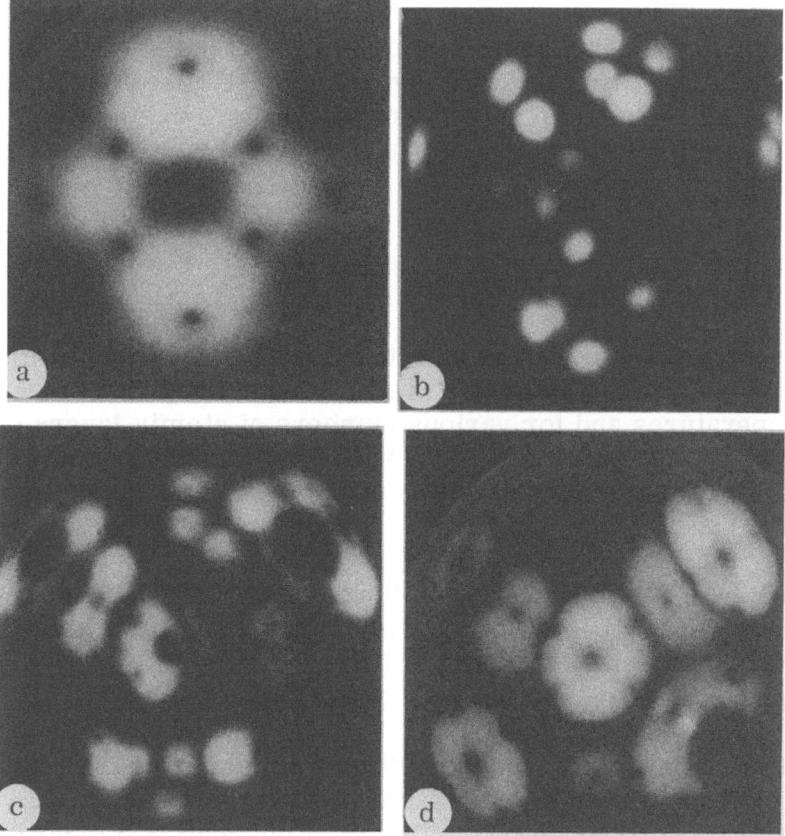

Fig. 3. Thermal smoothing of a tungsten tip of radius 1.2 μm as a result of heat treatment in an electric field (field electron micrographs at room temperature). (a) initial annealed form (voltage u=9kV, current i=0.5 μA); (b) microscopic protrusions on macroscopic overgrowth features after treatment at $F=2.7 \cdot 10^7$ V cm^{-1}, T=2750 K; (c) macroscopic overgrowth revealed by smoothing of microscopic protrusions at T=1400 K for 5 min (u=3.7 kV, i=1.8 μA); (d) further smoothing of macroscopic overgrowth, formed on {110}, {112} and {100} faces, by heating the tip at 1800 K for 30 s (u=6.55 kV, i=7.6 μA).

in a controlled manner, to carry out an effective equalization of curvature in a group of tips using the field as the equalizing factor, and, if a single macroscopic overgrowth feature is to be grown, to sharpen the tip directly in the FEM unit [11-13].

Drechsler studied the growth of steps and microscopic protrusions in an electric field and suggested a mechanism of stabilization of these features based on evaporation in the field [14], similar to that described above. However, the (T, F) ranges in [14] lay, as a rule, to the left of the range of field erosion (see Fig. 1b, regions 4, 5). In some cases, when macroscopic overgrowth was expected to develop, it failed to appear on the micrographs, possibly because of interfering effects of microscopic features.

The General Character of the Discussed Thermo-field Effects.
The main features of the shape changes observed for tips have now been studied for W, Mo, Ta, Nb, Re, and Ir, representing three different metal crystal structures [15]. Similar thermo-field morphology was observed on all these metals; they differed only in the

orientation and symmetry of the tips, as well as in the faceting according to the lattice structure. The cases when the faces grew in the field along the normal are of principal importance for pure surfaces, because for many years these cases have been the subject of much speculation [2, 3, 16]. Impurities, and especially those easily polarized or known to be polar, complicate the self-diffusion pattern, and a poor vacuum causes some other, often uncontrollable, processes to occur.

Condensation on the Substrate from its Own Vapor

The author is of the opinion that an investigation of condensation on a substrate from its own vapor must include an analysis of the growth forms and growth mechanisms for various substrate temperatures and for various numbers of atomic layers of the condensate, n, covering a wide range. A predetermined tip shape for a given amount of condensate means a certain tip curvature radius r, with the process kinetics being a function of the deposition rate.

A large number of experimental data on tungsten-on-tungsten and rhenium-on-rhenium deposition have already been published [8, 18]. Deposition rates were rather low (10^{-2}-1.5 monolayers per second), the tip radii being 1-1.2 μm (in the FEM mode) and 0.25-0.35 μm (in the field electron and field ion emission modes).

An analysis of the tungsten-on-tungsten deposition clarified the conditions for the observation of specific condensation forms. Figure 4a shows schematically the profiles of tips formed under different condensation conditions. Figure 4b is a diagram of log (n/r) vs. T for {110} faces and their terraces. On other faces such changes of shape are found at somewhat different values of n and T [17, 18].

In regions 1-3 the macroscopic form does not change in comparison with the initial one (usually the annealed form), the only difference being in the degree of surface roughness. Region 1 corresponds to a "random arrangement" of atoms during low-temperature deposition. In region 2 the surface is rough but the random structure blocks are enlarged because of limited mobility. In region 3 the roughness is lessened, at least to within the resolution of the field electron emission projector. Regions 4 and 5 correspond to enhanced faceting, with individual overgrowth features appearing in region 4 (this surface includes concave segments) and nearly complete faceting and convex shape appearing in region 5. Region 6 represents the annealed form, or rather the forms indistinguishable from it in emission patterns. Indeed, the closer is the point to the line GH, the closer we are to the transition to the shapes of region 4. Certain precautions are needed in interpreting the schemes of Fig. 4a, which is intentionally drawn with the wrong scale relations. Roughness in shapes 1 and 2 is observed within several (one to four) upper atomic layers, and the condensate may comprise hundreds of layers. The width of flat segments (faces) on a rounded annealed form may be smaller, and the faces may be more numerous than shown in Fig. 4a. Photographs of some condensation forms, obtained in the electron and ion field emission microscopes, are shown in Fig. 5.

Let us analyze the tip shapes represented schematically in Fig. 4a, and the transitions between them. Forms 1-3 follow the surface relief. In these cases the material is not redistributed over the surface, and growth at each point of the surface proceeds only along the surface normal (so-called "normal" growth). In all of the cases discussed we assume the flow of deposited atoms to be homogeneous and isotropic, a state achieved only approximately and locally in the experiments. The bulk structure of the deposit has a maximum defect density in region 1, a smaller defect density in region 2, and is nearly ideal in region 3. With the deposit thickness commensurate with the thickness of the surface

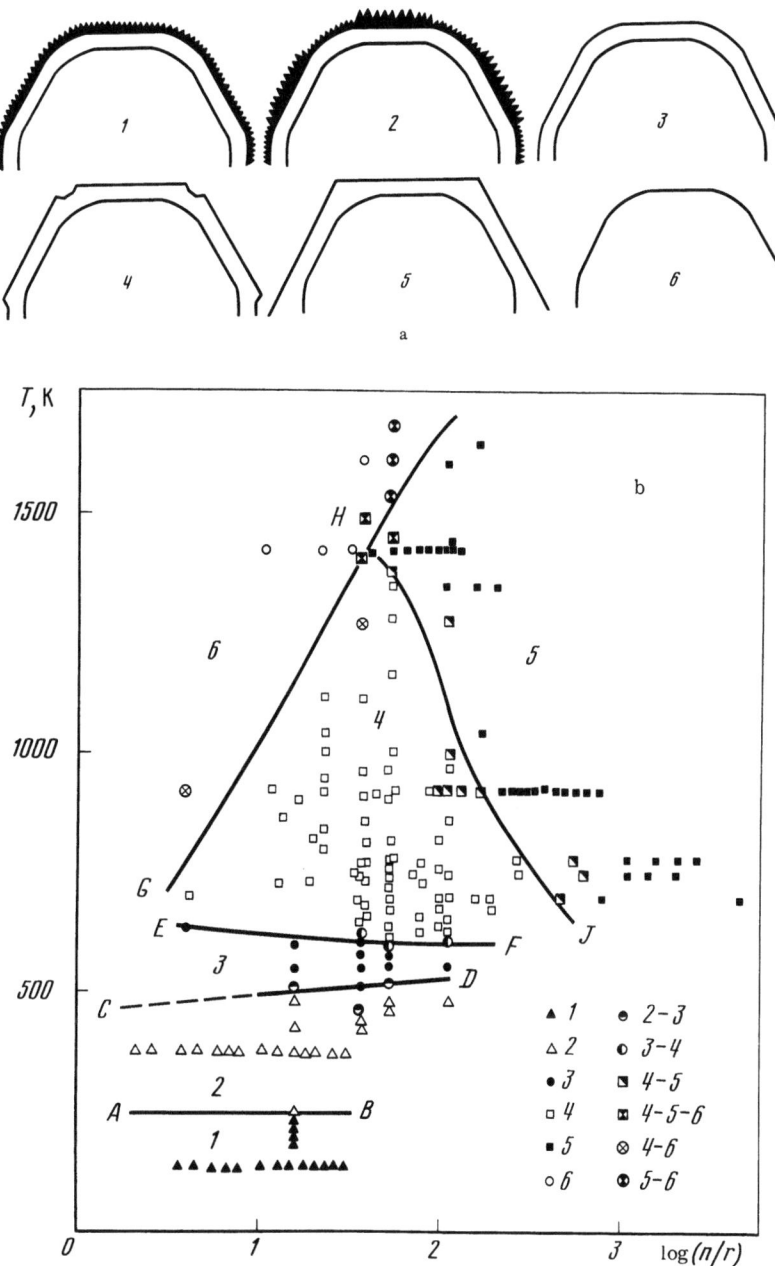

Fig. 4. (a) Schematic tip profiles obtained under different conditions of tungsten-on-tungsten deposition in the vicinity of {110} faces, and (b) regions on log (n/r) vs. T diagram, corresponding to the occurrence of these forms. (1, 2) low-temperature deposit for zero and restricted mobility of atoms, respectively; (3) replication; (4) individual faceted overgrowth units; (5) convex faceted forms; (6) annealed form.

roughness layer, the forms in regions 1-3 are different in principle, with the forms in region 3 being of maximum interest. They are recorded reliably in the electron emission mode in the experiments in which temperature is increased in each subsequent run with a constant number of monolayers, n, in the deposit. Complete identity with the annealed form is observed (in the vicinity of a chosen face) over a wide temperature range after the forms of regions 1 and 2 (which appear to be very different from the annealed form) have been identified. As temperature is further increased, individual overgrowth features 4 are formed.

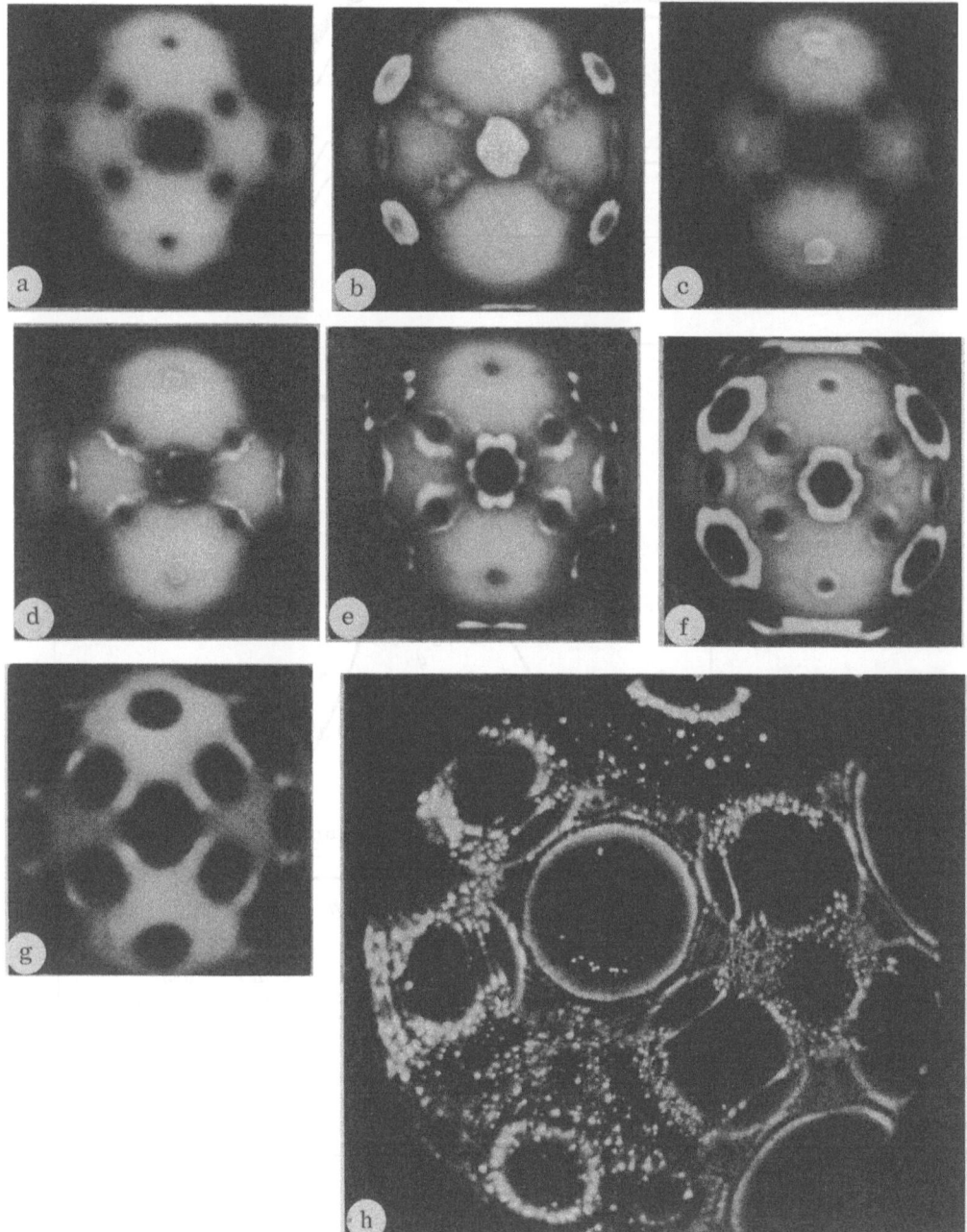

Fig. 5. Various condensation forms for tungsten-on-tungsten deposition. (a-g) field electron images; (a-f) images of a tip with $r=0.8$ μm at $n=30$; (a) initial annealed form; (b) disordered deposit (region 2) on $\{110\}$ and $\{112\}$ faces (T=425 K); (c) replication on $\{110\}$ and $\{112\}$ faces, disordered condensate on $\{110\}$ face (T=550 K); (d) onset of "collar" formation on $\{110\}$ and $\{112\}$ faces, with disordered deposit remaining on $\{100\}$ faces (T=620 K); (e) "collars" on $\{110\}$, $\{112\}$ faces, replication on $\{100\}$ face (T=750 K); (f) "collars" on $\{110\}$, $\{112\}$, and $\{100\}$ faces (T=910 K); (g) depositional restructuring (T=1650 K, image of a tip with $r=0.24$ μm for $n=40$); (h) formation of overgrowth in region 5 at T=1650 K on a tip with $r=0.4$ μm for $n\approx100$; helium field-ion micrograph.

The temperature ranges of region 3 (the replication region) for tungsten {110}, {112}, {100} faces and their vicinities are: 530-600, 490-580, 640-690 K, respectively. These ranges vary only slightly as the number of monolayers in the deposit vary from unity to several tens.

The condition of replication is the absence of material redistribution over the surface, but in region 3 a substantial mobility of adatoms leads to a complete smoothing of the low-temperature roughness. The replication in region 3 can be explained by assuming that adatoms "jump down" into the lower level with a higher probability at the edge of a small cluster (two-dimensional islands) than at the face edge or at the edges of its terraces. This explanation is quite acceptable both from an energetic and a kinetic standpoint (the reflecting barrier at the edge of a small cluster is lower, and the number of collisions of adatoms with this barrier is much larger than at the edge of an extended crystal layer). The edges of the upper layers are sinks for the adatoms from the adjacent layer. The upper face (as well as each subsequent terrace) is extended by incorporating into its edge the atoms adsorbed on the lower terrace. At the same time a new layer is growing on the upper face, with a higher probability of nucleation in its central area. If the "jumps-down" at the face and terrace edges transfer a negligible fraction of the material in comparison with the material deposited from the beam in a given area, the new plane on the upper face should replicate the dimensions of the initial upper face, just as each new terrace should copy the dimensions of the preceding one. A transition upward across the line EF (see Fig. 4b) corresponds to the experimental observation of the precursors of individual overgrowth tips. In the field electron emission pattern they appear as bright, usually symmetric bands around the main close-packed faces. The overgrowth tips formed in region 4 are the best known types of condensation-induced morphology. They were first investigated and reported by Müller [19] and were interpreted as a dynamic form of tungsten crystal frowth (an analysis of forms appearing in region 2 was also given). Müller did not give a detailed discussion of the growth mechanisms. Such an analysis was given by Drechsler [20, 21]. According to Drechsler, the overgrowth features in region 4, referred to as "collars," are formed by the "sweeping" of adsorbed atoms from the upper face, their incorporation into its edge, and the extension of this face and then of the lower terraces [20]. The specific shape of the "collar" is determined by three anisotropic processes: linear diffusion of individual atoms along the face, crossing the reflecting barriers at the face edge, and redistribution of adatoms along the edges of the face and terraces [20]. Segments locally protruding over the original shape of the regions (which are seen on emission patterns because of field enhancement) appear because the faces are extended without growth along the normals and because the lower terraces also undergo extension accompanied with normal growth (but to a thickness such that the terraces reach only the level of the upper face). To distinguish this extension of faces and terraces from the normal growth, we may refer to it as lateral growth. The corresponding process is schematically shown in Fig. 6. Drechsler has postulated only lateral growth for the formation of "collars" [20, 21]. The replication (see Fig. 4, region 3) was first mentioned in [8, 17]. With the replication mechanism taken into account, one must not forget the possibility of normal growth in the formation of individual overgrowth tips (region 4) and convex faceted forms (region 5). As the temperature rises, the kinetics of surface fluxes along faces (and terraces) changes, with the material being transported in appreciable amounts to the lower layer. On the upper face this flux is not balanced by fluxes from upper layers. We thus change from region 3 to region 4, but the lower the temperature (and the stronger the condensate flux from the source), the larger is the contribution of normal growth. The form 4 in Fig. 4a illustrates the appearance of "collars" in the presence of not only the lateral growth necessary for a deviation from the replication growth mode but also of normal growth as shown by two types of experiments. First are the experiments in which the trajectory crosses lines EH and HJ at constant n as T increases (see Fig. 4b). In this case the same amount of condensate is sufficient to build convex faceted forms at high temperatures in region 5 and insufficient at low temperatures in region 4. At a temperature corresponding to crossing line HJ it would be wrong to assume that any appreciable transport

Fig. 6. A schematic of the formation of (a) isolated and (b) convex faceted overgrowth features as a result of purely lateral growth.

of material from the tip apex to its base, or any vaporization of the material, occurs; hence, this experiment confirms normal growth of the tip in region 4. Second are the experiments in which lines GH and HJ are crossed at constant T as n increases. Here the lower the temperature, the higher is the value of n at which a convex faceted form can be obtained (strictly this form cannot always be obtained at low temperatures). For lateral growth only this form requires a relatively small amount of deposit (see Fig. 6b) but in the case of coexisting lateral and normal growth larger amounts of the deposited material are required (see Fig. 4a, form 5).

The faceted convex forms appearing at the appropriate values of n and T go through a stage of extended {110}, {112}, {100} faces (for tungsten) similar to the field restructuring forms; as n increases further, to the forms of the so-called completely restructured tip [7], {112} faces grow over. This process has been termed condensation restructuring [8]. In the course of thermal smoothing the completely restructured forms go through a stage of ordinary restructured ones. A large variety of specific forms is revealed in regions 4 and 5 for various n and T values. As a rule, the edges and apices are the sharper and the faceting is more varied the lower is the temperature. As the temperature is increased, the tendency to smoothing and less diversified faceting is enhanced: the structures unstable at a given temperature do not appear, and the activation contrasts fade out.

Recapitulating, at least two tip forms (3 and 5), previously unknown, have been found in the investigation of tungsten deposition on tungsten under conditions of continuously varied temperature and deposit thickness, and corrections are introduced to the interpretation of the known forms (region 4). Rhenium-on-rhenium deposition reveals the same typical transitions. It seems plausible that such changes in tip forms will be observed in other materials as well: metals, semiconductors (especially germanium and silicon), etc. The ideal layer-by-layer replication (in region 3) which must also occur in the presence of foreign adsorbates (when wetting conditions are satisfied) is an important and attractive approach to field growth.

Literature Cited

1. V. N. Shrednik. Crystal growth and field emission microscopy. In: Problems of Modern Crystallography. Nauka, Moscow, 150-171 (in Russian) (1975).
2. G. Herring. The use of classical macroscopic concepts in surface energy problems. In: Structure and Properties of Solid Surfaces. Chicago: Univ. Chicago Press, 5-81 (1953).
3. W. P. Dyke, F. M. Charbonnier, R. W. Strayer, R. L. Floyd, J. P. Barbour, and J. K. Trolan. Electrical stability and life of the heated field emission cathode. J. Appl. Phys. 31, 790-805 (1960).
4. V. G. Pavlov, A. A. Rabinovich, and V. N. Shrednik. Observation of tip elongation by electric field. Pis'ma v Zh. Eksp. Teor. Fiz., 17, 247-250 (in Russian) (1973).
5. E. W. Müller. Weitere Beobachtungen mit dem Feldelektronenmikroskop. Z. Phys., 108, 668-680 (1938).

6. M. Benjamin and R. O. Jenkins. The distribution of autoelectronic emission from single crystal metal points. Proc. Roy. Soc. London, A176, 262-279 (1940).
7. P. C. Bettler and F. M. Charbonnier. Activation energy for the surface migration of tungsten in the presence of a high electric field. Phys. Rev., 119, 85-93 (1961).
8. O. L. Golubev, B. M. Shaikhin, and V. N. Shrednik. On condensation restructuring of metal tips. Izv. AN SSSR, Ser. Fiz., 40, 1599-1604 (in Russian) (1976).
9. V. N. Shrednik, V. G. Pavlov, A. A. Rabinovich, and B. M. Shaikhin. Effect of strong electric field and heating on metal tips. Izv. AN SSSR, Ser. Fiz., 38, 296-301 (in Russian) (1974).
10. V. N. Shrednik, V. G. Pavlov, A. A. Rabinovich, and B. M. Shaikhin. Growth of tips in the directions normal to close-packed faces by heating in the presence of an electric field. Phys. Status Solidi (a), 23, 373-381 (1974).
11. V. G. Pavlov, A. A. Rabinovich, V. P. Savchenko, and V. N. Shrednik. Inventor's Certificate 493834 (USSR). Method of blunting of tip cathodes. Inventions Bulletin, No. 44, 130, (in Russian) (1975).
12. V. N. Shrednik, V. P. Savchenko, V. G. Pavlov, N. I. Komyak, and A. A. Rabinovich. Inventor's Certificate 425239 (USSR). Method of manufacturing field emission devices. Inventor's Bulletin, No. 15, 166 (1974).
13. V. N. Shrednik, A. A. Rabinovich, and V. G. Pavlov. Inventor's Certificate 464238 (USSR). Method of manufacturing field electron emission cathodes. Inventor's Bulletin, No. 35, 179 (in Russian) (1975).
14. M. Drechsler. Kristallstufen von 1 bis 1000 Å (Herstellung der Stufen in Feldemissionsmikroskop durch elektrische Felder, Messung der Stufenhohen, eine Feldbindungsenergie Theorie der Entstehung der Stufenform und Versetzungen). Z. Elektrochem., 61, 48-55 (1957).
15. U. G. Pavlov, A. A. Rabinovich, and V. N. Shrednik. Field erosion of Mo, Ta, Nb, Ir, and Re. Fiz. Tverd. Tela, 17, 2045-2048 (1975).
16. V. G. Pavlov and N. V. Shrednik. On field crystal growth of refractory metals in high vacuum. In: Proc. of the VII Intern. Symp. on Discharges and Electrical Insulation in Vacuum. Novosibirsk: Acad. Sci. Siberian Branch, 209-212 (1976).
17. V. N. Shrednik. Advances in surface characterization with field electron and field ion microscopy. In: Proc. 7th Intern. Vacuum Congr. and 3rd Intern. Conf. on Solid Surfaces. Vienna: 3, 2455-2466 (1977).
18. O. L. Golubev, V. G. Pavlov, and V. N. Shrednik. On condensation forms of growth of crystal tips. In: 5th USSR Conf. on Crystal growth, Abstracts. The Institute of Cybernetics, AN GSSR, 1, 132-133 (in Russian) (1977).
19. E. W. Müller, Oberflächenwanderung von Wolfram auf dem eigenen Kristallgitter. Z. Phys., 126, 642-665 (1949).
20. M. Drechsler. Vorzugsrichtungen der Oberflächendiffusion auf Einkristallflächen. Z. Elektrochem., 58, 334-339 (1954).
21. M. Drechsler and R. Wanselow. Untersuchung der Temper und Wachstrumformen einiger Metalleinkristalle mit dem Feldelektronenmikroskop. Z. Kristallogr., 107, 161-181 (1956).

Part III
Epitaxy

PECULIARITIES AND MECHANISM OF GRAPHOEPITAXY

N. N. Sheftal' and V. I. Klykov

The Institute of Crystallography of the USSR Academy of Sciences, Moscow

Graphoepitaxy is a new phenomenon which, in contrast to conventional epitaxy, proceeds not on an atomic level but at the level of microscopic crystals. The following arguments led to its discovery. Modelling atoms as cubes [1], that is, as crystals, helped to reveal the mechanism of ordered crystal growth and demonstrated the important role of the repeated translation position. Combination models were designed of parallelohedra to demonstrate the processes of atomic growth of crystals [2]. However, actual growth processes are dominated by microscopic crystals. Will these microcrystals join the substrate in an oriented manner if a microrelief with a parallel arrangement of re-entrant angles, corresponding to their shape, is formed on this substrate (Fig. 1)? By producing such a microrelief on any substrate (amorphous, polycrystalline, or single crystalline but not epitaxial with respect to the crystalline material) it will be possible to attain an analogue of epitaxial growth although not on an atomic but on a microcrystal level. A single-crystalline layer is expected to appear on the substrate as these microcrystals grow and are incorporated into the overgrowth.

The role of microcrystals is well pronounced in a number of processes [3], for example, when polycrystalline silicon films are grown by the gas transport technique [4, 5]. Hydrogen transporting the silicon chloride vapor through a hot quartz tube first deposits silicon as tiny particles when the reaction temperature is reached. A small fraction of these particles is deposited at the entrance to the reaction zone but the main part is carried by the flow along the tube. The suspended crystals continue to grow as they are transported and, with usual growth rates and vapor flow rates, reach at the end of the reaction zone about 0.1 μm in size. Only comparatively large crystals (around 200 μm in dimensions) are deposited at this part of the zone, with no new layers and crystals formed. We conclude that large crystals grow, in particular, by incorporating microcrystals.

Experiments have supported the idea of graphoepitaxy and demonstrated that, with no re-entrant angles on the substrate, orientation can also be achieved if two or three intersecting systems of parallel grooves are scratched on the substrate; the angles between intersecting grooves are made equal to the angles between the edges of the basal face of the microcrystal deposited on the substrate [6, 7].

Figure 2 shows an orienting pattern formed by grooves whose width is equal to the spacing between the grooves. The microrelief is formed by triangular islands with sides $a'=a''=a'''$, left between the grooves.

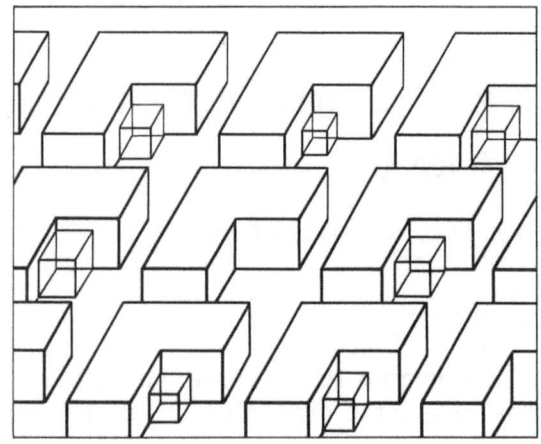

Fig. 1. A schematic of an orienting microrelief formed by re-entrant angles.

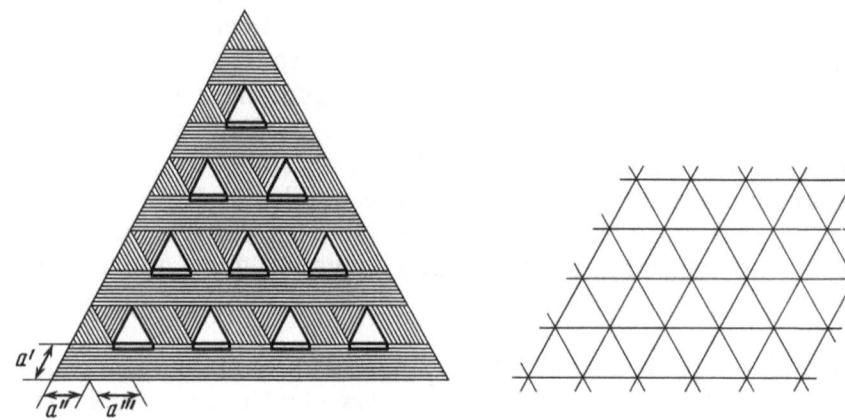

Fig. 2. Orienting relief formed by grooves $a'=a''=a'''$ — triangle sides.

Fig. 3. Hexagonal pattern formed by narrow grooves, leading to twinned growth.

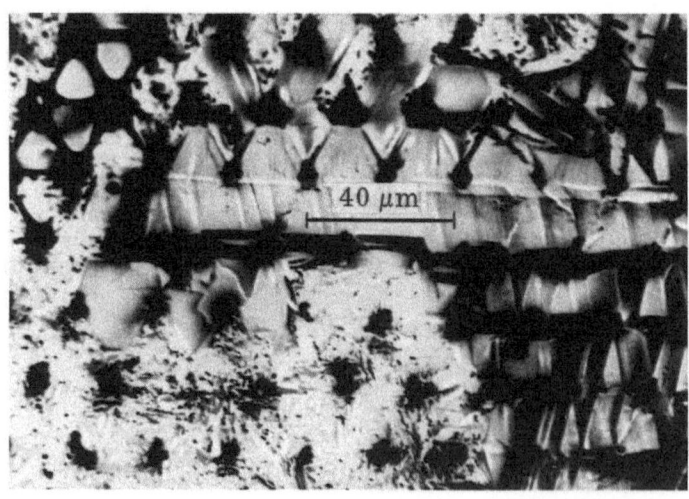

Fig. 4. Germanium monolayer involving the growth of twins.

Experiments on graphoepitaxy were carried out by crystallization from solution (deposition of NH_4 crystals onto diffraction gratings intersecting at 90° and 60° angles), and by vapor-liquid-solid (VLS) crystallization onto a substrate with a microrelief produced either by scratching or by photolithographic techniques. It was invariably found that a liquid is essential to ensure the mobility of microcrystals. The suspended particles move until they are fixed in the most energetically advantageous position. If the amount of liquid is insufficient, it can still play the role of a lubricant facilitating the motion of crystals along the substrate.

The need to wet the substrate was previously established in the case of crystallization from solutions onto diffraction gratings: the addition of soap to the solution improved the orienting effect. This has been confirmed in the VLS experiments. A thin metal layer was first vacuum deposited on the fused quartz plates. The choice of the metal was determined by the composition of the liquid phase which was to be used to wet it. Tungsten was deposited if molten gold was used as the liquid phase, and tantalum or molybdenum in the case of molten silver. Orientation was observed only if the liquid phase wetted the substrate. Geometric features of the microrelief depend on the orientation effect. Thus, a hexagonal pattern of grooved parallel and anti-parallel closed triangles (Fig. 3) forms a monolayer (Fig. 4) in which the neighboring single-crystalline triangular elements are in twinned positions with respect to one another when cubic materials are deposited.

This type of deposition is achieved by preparing three systems of grooves with small widths compared with the intergroove spacings. If, however, the groove width is equal or greater than the spacing, the pattern is formed only by parallel triangular elements, and twinning dictated by the groove pattern is suppressed (see Fig. 2). A hexagonal microrelief produced photolithographically can also consist of "normal" and "inverted" triangles. This relief produces orientation twinning. By alternating these two types of triangles, one can produce a relief with a prescribed twinning periodicity. If only each second triangle is etched, a relief for the crystallization of a single-crystalline layer is formed.

Presumably, it is possible to obtain on a substrate a combination of crystals of two materials, by forming on this substrate two different appropriately combined microreliefs.

The points of intersection of the grooves produce no orienting effect on crystals, so that it is expedient not to extend the grooves to these intersection points. This should facilitate the growth of single crystals from the cells, in which they were nucleated, to the neighboring cells and the formation of a continuous single-crystalline layer.

The experiments with silicon deposition have also demonstrated that the groove pattern need not necessarily have the smallest possible cells, because of the unavoidable distortion of the cells (grooves) owing to technological limitations of the photolithographic techniques. Cells with side grooves about 20 μm in length, which are acceptably perfect at the current level of photolithography, yield good results. Even cells with dimensions up to 100 μm show an orienting effect on the growth of crystals. Cells with larger dimensions produce only a random orienting effect, and at much larger cell dimensions (e.g., of the order of 450 μm) no orientation is observed and often several crystals are deposited within a single cell.

Let us consider the mechanism of orientation using as an example the silicon growth process by the VLS method (vapor-molten gold-fused quartz coated with vacuum-deposited tungsten). First, a drop of silicon solution in liquid gold appears in the cell. It wets the barriers, climbs the three rectilinear edges, and forms in the middle portion of the cell a pit with an approximately horizontal surface. A microcrystal appears and grows at the drop surface since here the temperature is lower than at the substrate and the supersaturation is, consequently, higher.

The orientation mechanism, presumably, operates via the crystallization pressure forces rotating the crystal. This crystal is shown in Fig. 5 in an asymmetrical position within the solution drop (growing layers and crystallization pressure forces, repulsing the crystal from the cell walls, are shown). One edge of each of the three faces of the crystal is somewhat farther from the wall than the other, so that better feeding conditions increase the probability of nucleation of the next layer at this edge. The layer, propagating along the face, meets at the opposite edge a resistance due to the closeness of the cell wall. By producing the crystallization pressure, and being unable to move the wall, the growing crystal is forced to rotate, all the more so since the pressure applied to all three faces produces a torque. An equilibrium is achieved when the crystal is symmetrical within the cell, when the total repulsion at each segment of the face and on the crystal as a whole is balanced. The crystal grows with the basal face parallel to the substrate and later is fixed to it. Until the last moment the liquid film functions as a lubricant.

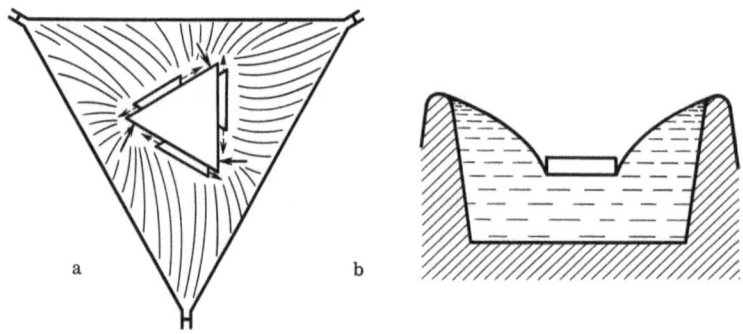

Fig. 5. Schematic representation of the microcrystal orientation mechanism: (a) top view (non-equilibrium position); (b) sectional view.

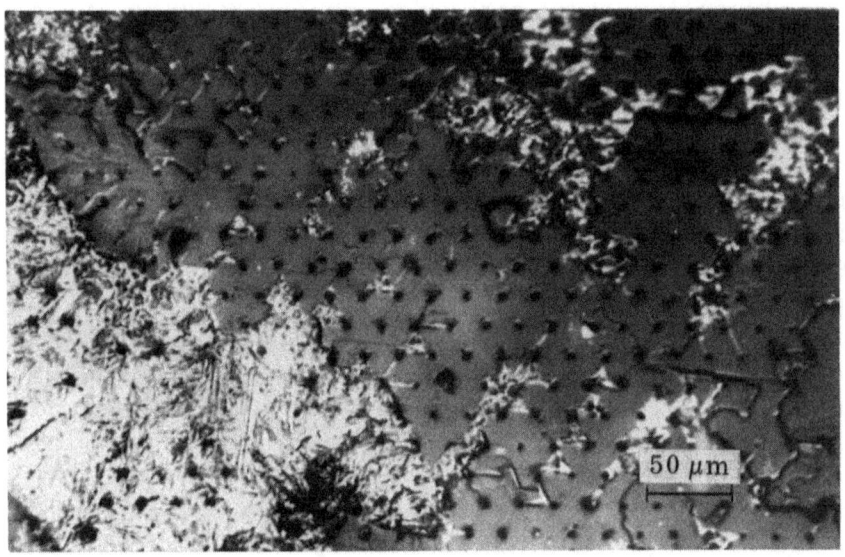

Fig. 6. A segment of single crystal silicon layer on fused quartz.

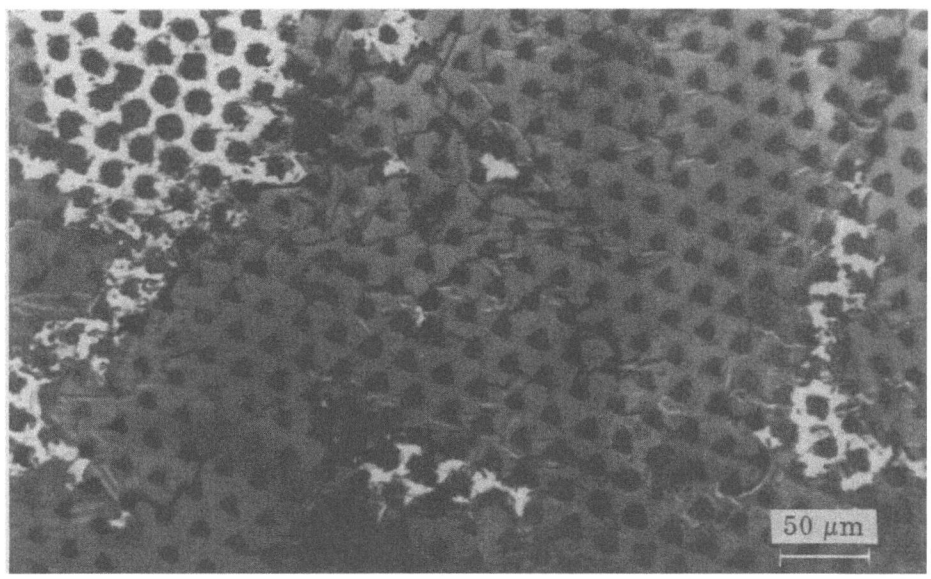

Fig. 7. Initial stage of growth of a single crystal layer.

Capillary forces also may play a role in rotating the crystal.

Single crystalline silicon layers with area exceeding 15 mm² can be grown on fused quartz substrates (Fig. 6). The initial stage of this growth is shown in Fig. 7 (each second triangle is etched out, so that no twins are formed).

The difference between the conventional eiptaxy and graphoepitaxy must be emphasized in conclusion. Ideally, conventional epitaxial growth leads to a successive atomically ordered transition from the structure of a crystal substrate to that of the epitaxial layer. This transition is the less ordered, the poorer is the matching of the substrate and overgrowth structures. Graphoepitaxial growth is totally unrelated to this matching, and bonding forces uniting the substrate and the overgrowth are only statistically ordered.

We believe that graphoepitaxy opens a field of interesting new phenomena [6-9]; the research into this field is expanding.

Literature Cited

1. W. Kossel. Zur Theorie des Kristallwachstum. Nachr. Ges. Wiss. Göttingen math.-phys. Kl., 135-145 (1927).
2. T. A. Smorodina and N. N. Sheftal'. Growth and dislocation models of crystals. In: Growth of Crystals, Vol.8, ed. N. N. Sheftal', Consultants Bureau, New York (1969).
3. N. N. Sheftal'. Real crystal growth processes. In: Crystal Growth, AN SSSR Publishing House, Moscow, 1, 5-31 (in Russian) (1957).
4. N. N. Sheftal', N. P. Kokorish, and A B. Krasilov. Chemical vapor deposition of single-crystal layers of silicon and germanium. Izv. AN SSSR, Ser. fiz., 21, 146-152 (in Russian) (1957).
5. N. P. Kokorish and N. N. Sheftal'. Morphology of polycrystalline silicon films. In: Crystal Growth, Nauka, Moscow, 3, 351-356 (in Russian) (1961).

6. N. N. Sheftal' and A. N. Buzynin. Preferential orientation of crystals on the substrate and effects of grooves. Vestnik MGU, Ser. Geol., No. 3, 102-104 (in Russian) (1972).
7. N. N. Sheftal'. Real crystal growth processes and some principles of single crystal growth. In: Growth of Crystals, Vol. 10, ed. N. N. Sheftal', Consultants Bureau, New York (1976).
8. V. I. Klykov, R. N. Sheftal', and N. N. Sheftal'. Oriented crystallization on amorphous and polycrystalline substrates (graphoepitaxy). In: Real Crystal Growth Processes, Nauka, Moscow, 144-150 (in Russian) (1977).
9. R. N. Sheftal'. Epitaxial growth of single crystal films (review paper). In: Growth of Crystals. Vol. 10, ed. N. N. Sheftal', Consultants Bureau, New York (1976).

APPLICATION OF ELECTRON MICROSCOPY TO A STUDY OF KINETICS AND MECHANISM OF CRYSTALLIZATION

Gl. S. Zhdanov

The State Optical Institute, Leningrad

The maximization of supersaturation in a liquid is a problem of particular interest; it has already attracted considerable attention for several decades. In the 30's and 40's it was erroneously assumed that supersaturation of a liquid metal by several tens of degrees or even several degrees would be sufficient for the observation of homogeneous crystallization. At the same time, experimental evidence was available indicating the possibility of maintaining small droplets of metals in the liquid state even at much larger supercoolings [1]. Turnbull was one of the first to give proper attention to these results. He conducted a systematic study of crystallization in a number of metals at an average relative supercooling of about 18% [2-4]. He worked with micron-size droplets in which the presence of active impurities was assumed to be highly improbable. Turnbull's work was widely accepted as reporting reliable observations of homogeneous crystallization.

However, numerous papers have been published recently [5-9] which seriously question the validity of the interpretations offered. It is now safe to say that in many cases the crystallization of samples in Turnbull's experiments was heterogeneous in nature. Ovsiyenko stated in his review papers [10, 11] that reliable evidence of homogeneous crystallization must be considered as totally lacking. The purpose of the present paper is to attract attention to the potential advantages of electron microscopy in the study of crystallization; this technique is so far very much in the shadow of the classical methods. The high resolution of the electron microscope makes it possible to decrease sample volumes (and with it, the influence of active impurities) by many orders of magnitude, which increases the probability of achieving larger supercooling.

Certain concepts from the theory of crystallization are required for an analysis of the experimental results.

Fundamentals of the Crystal Nucleation Theory

The Gibbs' free energy G plays an important role in the theory of phase transitions. In a system with volume V at temperature T and pressure p it is given by [12]

$$G = U - TS + pV + \sigma A = V G_V + \sigma A, \tag{1}$$

where U is the internal energy, S is entropy, σ and G_V are the specific surface and bulk free energies, and A is the surface area of the interface. Only transitions satisfying the condition

$$\Delta G \leqslant 0 \tag{2}$$

are thermodynamically allowed.

As in all subsequent relations, the symbol Δ stands for an increment of a variable. In normal situations the pressure term can be ignored. The equality $\Delta G_V = 0$ then gives for the melting of a massive object

$$\Delta S = \Delta U / T_{m\infty} = L / T_{m\infty}, \tag{3}$$

where L is the latent heat of melting per unit volume, and $T_m\infty$ is the tabulated melting point. A change in the phase transition temperature yields the following relations:

$$\Delta G_V = \Delta U - T\Delta S = \pm L\Delta T / T_{m\infty} \quad \text{for} \quad p = 0, \tag{4}$$

$$\Delta G_V = \Delta U - T\Delta S + p\Delta V = \pm [L\Delta T / T_{m\infty} + p(1 - \rho_s/\rho_l)] \quad \text{for} \quad p \neq 0, \tag{5}$$

where $\Delta T = T - T_{m\infty}$, ρ_s and ρ_l are the densities of the solid and liquid phases, and L and ΔS are assumed to be independent of T. Signs "+" and "-" refer to crystallization and melting, respectively.

In small systems surface effects must be taken into account. When a spherical nucleus of a solid phase is formed in a supercooled liquid, the increment of free energy has a maximum given by the condition

$$\frac{\partial(\Delta G)}{\partial r} = \frac{\partial}{\partial r}\left(\frac{4}{3}\pi r^3 \Delta G_V + 4\pi r^2 \sigma_{s \cdot}\right) = 0. \tag{6}$$

Equation (6) yields the height of the energy barrier, or the work of formation of a critical nucleus,

$$W = {}^{16\pi}/_3 \sigma_{sl}^3 / (\Delta G_V)^2 \tag{7}$$

and its radius

$$r^* = -2\sigma_{sl}/\Delta G_V = 2\sigma_{sl}/|\Delta G_V|, \tag{8}$$

where σ_{sl} is the specific free energy of the massive solid-liquid interface. It is normally assumed that the free energy of the nucleus-melt interface has the same value. This assumption is one of the weak points in nucleation theory.

The classical theory of nucleation [13] uses the concepts developed to interpret the rate of liquid droplet nucleation in a supersaturated vapor [14-16]. Let n* be the concentration of critical nuclei generated by random fluctuations. The growth of larger aggregates is energetically advantageous but it can be slowed down by random detachment of molecules. The probability of the reverse reaction is taken into account by the infusibility factor Z [16] which is shown by the calculation to be nearly 10^{-2}. By introducing the flux of molecules q* arriving per unit time at the surface of the critical nucleus, we can write the nucleation rate in the form

$$I = Zq^*n^*. \tag{9}$$

Theoretically the rate of transition of molecules across the interface is given by the relation

$$v = (kT/h)\exp(-E_A/kT),$$

where k and h are the Boltzmann and Planck constants, respectively, E_A is the transition activation energy, and kT/h is the frequency of atomic vibrations [13]. The concentration n* of critical nuclei and that of single molecules n are related by the Boltzmann formula

$$n^* = n \exp(-W/kT). \tag{10}$$

Taking into account that $q^* \approx i_S^* v$, where i_S^* is the number of molecules on the surface of the critical nucleus, and assuming $i_S^* Z \approx 1$, we obtain from equation (9)

$$I = nkT \exp(-E_A/kT) \exp(-W/kT)/h.$$

In this equation all factors, with the exception of the last one, can be readily evaluated. Thus, in metals $n \approx 10^{22}$ cm^{-3}, $kT/h \approx 10^{13}$ s^{-1}, exp$(-e_A/kT) \approx 10^{-2}$ (E_A is assumed equal to the activation energy of the viscous flow of the liquid). Consequently,

$$I = K_V \exp(-W/kT), \qquad (11)$$

where the pre-exponential factor $K_V \approx 10^{33}$ cm^{-3} s^{-1}, and W is given by formulas (7), (4), and (5).

In the case of heterogeneous crystallization the pre-exponential factor and the energy of formation of a critical nucleus are reduced. If n_S is the concentration of molecules participating in the formation of nuclei on the catalyst surface, then by analogy with relation (11) we have

$$I_S = K_S \exp(-W_S/kT), \quad K_S = K_V n_S/n \approx 10^{26} \text{ cm}^{-2} \text{ s}^{-1},$$

The value of K_S is calculated on the assumption that the contact area between a unit volume sample and the catalyst is approximately equal to the sample surface area and, therefore, $n_S \approx 10^{15}$ cm^{-2}. Assuming the nucleus to be a spherical cap with contact angle θ, we can write the energy of nucleus formation [13] in the form

$$W_S = Wf(\theta) = {}^1\!/_4 W (1 - \cos\theta)^2 (2 + \cos\theta).$$

In the crystallization of liquids the angle θ, is a formal characteristic; its use is justified because it enables us to evaluate the catalytic activity. The energy W_S can be rewritten in a form similar to relation (7) if we introduce the effective interfacial free energy

$$\sigma' = \sigma_{sl} [f(\theta)]^{1/3},$$

which characterizes the lower limit σ_{sl} but which, similarly to Θ, has no clear physical meaning.

The above exposition shows that the theory of crystal nucleation is based on a number of important simplifications. In particular, formula (10) is not rigorously substantiated [17]. The proposal to take into account the mobility of nuclei in the melt [18], which could increase K_V by a factor of 10^{10}-10^{12}, has led to vigorous discussions, but later one of the authors of ref. 9 considered it to be premature because of the lack of an adequate model of the liquid state. The scatter in experimental estimates of K_V and K_S reaches 20 to 30 percent [3, 7, 9-11].

Crystallization and Melting of Small Particles

A sphere with radius R is subject to an additional pressure

$$p = 2\gamma/R, \qquad (12)$$

where γ is the surface tension. For particles of more complicated shape, with principal curvature radii R_1 and R_2,

$$p = \gamma (1/R_1 + 1/R_2). \qquad (13)$$

Thus, we have $p \approx 10^3$ atm if R = 100 Å and $\gamma \approx 5 \cdot 10^2$ erg cm^{-2}. If, as is usually the case, $\rho_s > \rho_l$, the energy of nucleus formation diminishes as pressure increases (see relations (5), (7)).

The temperature of crystallization of small particles, T_{sR}, is closely related to the change in their melting point, T_{mR}, a fact which is usually overlooked. Several different definitions of T_{mR} can be found in the literature. The approach which seems to be the most reasonable assumes coexistence of the liquid and crystalline phases within a single particle

[19] (an alternative is to consider isolated particles in equilibrium with the gaseous medium [20]).

From the standpoint of the fluctuation theory, the nucleation of the solid phase in the liquid and of the liquid phase in the solid must be described by similar equations. As a rule, it is impossible to achieve any substantial superheating of crystals [21], so it can be assumed that melting starts at the surface. In most metals this conclusion is supported by the fact that solids are effectively wetted by their melts [22], and also by the experimentally measured values of interface energies σ which point to the inequality

$$\sigma_{sv} > \sigma_{lv} + \sigma_{sl},$$

where subscripts s, l, and v denote the solid, liquid, and vapor phases, respectively, and their combinations denote the corresponding interfaces. In particular, σ_{sv}, σ_{lv}, and σ_{sl} in tin are equal to 673, 554, and 59 erg cm^{-2} (at different T) [4, 23].

By setting to zero the change in the free energy of the particle when a liquid coating with thickness t is formed on it, and assuming t ≪ R, we obtain

$$t = [\sigma_{sv} - (\sigma_{lv} + \sigma_{sl})] r^*/2\sigma_{sl} (1 - r^*/R), \tag{14}$$

where r* is given by formula (8). This means that both t and r* increase as supercooling diminishes.

Surface melting leads to the formation of a crystalline nucleus in the melt. Further behavior of the system depends on the ratio of the nucleus radius r=R-t and the critical radius r*. Melting of the crystal is accompanied by a continuous decrease in the free energy, but only for r ≤ r*. Therefore, by using equations (5), (8), and (12), we can derive an expression for the change in the melting point of a spherical particle with radius R:

$$|\Delta T|/T_{m\infty} = (T_{m\infty} - T_{mR})/T_{m\infty} = 2\sigma_{sl}/L(R-t) - 2\sigma_{lv}(\rho_s/\rho_l - 1)/R, \tag{15}$$

where we have assumed $\sigma_{lv} = \gamma$. For t → 0 and $\rho_s \approx \rho_l$ equation (15) is reduced to the Thomson relation. The second term in the right-hand side of the equality (which is generally substantially smaller than the first), follows from the Clausius–Clapeyron equation. The empirically determined value of t is close to 30 Å [24-26].

The change in free energy ΔG accompanying the formation of a crystal nucleus is given by equations (1)-(7) and for each value of supercooling |ΔT|' is plotted by curve edcf (Fig. 1). Segment abc represents the possible change in free energy for the surface melting of a particle with radius r_a at T=T'. A crystalline nucleus with radius r_c melts at T=T". Obviously, the same curve edcf also characterizes the reverse process, i.e., a crystal melting in a drop of the melt. Let a crystal nucleus with radius r=r_c be formed in a particle with radius R=r_a, and assume that its melting leaves the free energy of the system unaltered (ΔG=0). This process requires overcoming the energy barrier and is possible only via fluctuations with sufficiently high amplitudes whose probability at low |ΔT|' is as low as that of

Fig. 1. Variation in the free energy of a particle as a function of the radius of a crystal nucleus in the process of melting or crystallization. (1) at T'; (2) T" > T'.

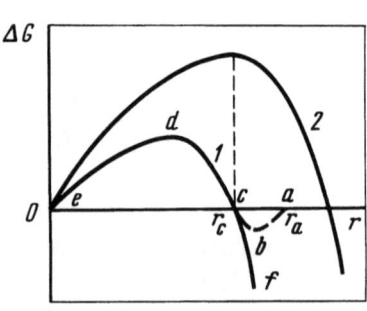

fluctuations leading to homogeneous crystallization in the vicinity of $T_{m\infty}$. By neglecting pressure and using equations (1) and (4), we find supercooling $|\Delta T|'$ corresponding to condition $\Delta G = 0$ for a crystal with radius r:

$$|\Delta T|'/T_{m\infty} = 3\sigma_{sl}/Lr. \qquad (16)$$

In a number of papers [27, 28] the decrease in the melting point of a particle is found from relation (16) or from a similar equation (with σ_{sl} substituted by $\Delta\sigma = \sigma_{sv} - \sigma_{lv}$). In reality these relations characterize not the conditions for the actual occurrence of melting but only the conditions at which melting becomes possible in principle.

A minimum of ΔG may be observed on curve ac (see point b in Fig. 1), showing the change in free energy accompanying surface melting. The minimum corresponds to a steady-state or metastable equilibrium, depending on its position on the abscissa axis. As contact between the solid and liquid phases at $T < T_{m\infty}$ leads only to an unstable equilibrium, point b cannot lie on curve edcf, and the interface must possess singular properties. It would be more correct to refer to the layer coating the particle in this condition as liquid-like. As the temperature increases, this coating grows thicker and its properties approach those of the bulk liquid phase.

It follows from relations (7) and (4) that the energy barrier is lowered when supercooling is increased. The barrier can be overcome only by fluctuations at the temperature of homogeneous crystallization. The reverse transition is equally probable: the melting which started at the particle surface extends to a depth sufficient to reach the critical state, after which the process proceeds with a release of energy. We conclude, therefore, that melting and crystallization of the smallest particles with radius R_{min} must be reversible, and only for these particles will the decrease in the melting point be given approximately by formula (16). It is readily shown that at $r = 1.5 r^*$, ΔG passes through zero, so that $R = 1.5 r^* + t_{min}$, which t_{min} is the minimum thickness of the coating at which it can be treated as a separate "phase." Choosing $r^* \approx 10$ Å, corresponding to the supercooling achievable in metals, and assuming $t_{min} = 10\text{-}15$ Å (several monolayers), we obtain $R_{min} = 25\text{-}30$ Å, in accordance with experimental data [29-31].

Research into Crystallization of Thin Films.

Details of Experimental Techniques

The highly dispersed structure of ultra-thin metal films makes them suitable for studying melting and crystallization of small particles. Extremely high supercooling has been previously achieved in electron diffraction studies of thin tin, bismuth, and lead films [32]. Both in ref. 32 and in a number of subsequent publications [33, 35], the mean film thickness t_{mean}, which for a number of reasons is a rather vague parameter, has been measured instead of the particle size. Later, other techniques, for example the study of island size distribution histograms obtained by processing micrographs [21, 24, 25, 35-37], have been added to the electron diffraction analysis.

An important step forward was the development of an *in situ* electron microscopic technique. The growth, melting, and crystallization of indium islands were observed in ultra-high vacuum [38, 39], and the phase transition was recorded by a microdiffraction technique. Similar studies were carried out with lead [40, 41]. The temperature of crystallization of films containing 100-3,000 Å islands was found from the change in the diffraction pattern in the course of gradually cooling the specimens. Although an accurate measurement of the crystallization rate is hardly feasible because of the scatter in island sizes and the inertial parameters of the temperature control system, equation (11) has been used to evaluate σ_{sl};

the resulting value was twice that obtained by Turnbull. The advantages of dark-field electron microscopy, which make it possible to reduce the volume of processed data by approximately an order of magnitude, have been demonstrated in the study of melting of indium and tin films [26].

The dark-field technique seems to have been applied to the study of crystallization in a supercooled liquid first for mercury [42. 43] and later for other metals [29-31, 44]. This method consists in displacing the aperture stop with respect to the electron beam axis. If the aperture stop singles out the segments of diffraction rings, then each Bragg reflection on the electron diffraction pattern corresponds to a dark-field reflection (bright spot) on the micrograph. The number of reflections is proportional to the content of the crystalline phase in the sample and is a function of temperature. Some of the reflections which appeared when a tin film on carbon was kept for 90 s at 310 K are marked by rings on Fig. 2. These reflections point to the presence of a crystal with a definite orientation with respect to the electron beam.

The specimens were heated by the electron beam which simultaneously produced the electron image. Usually heating by an electron beam is considered to be a serious disadvantage and special measures are taken to reduce it [26]. However, the power released by the electron beam is sufficient for very rapid heating of the object by hundreds of degrees, and in principle there is no reason why electron beam heating should not be used to study nucleation and crystallization phenomena. One advantage of this sort of heating is its short lag time so that a sample can be crystallized or melted in a matter of seconds. With traditional methods this would require about one hour, with the risk of substantial changes in the state of the irradiated object and the ambient medium.

Preliminary studies [42-46] enabled us to choose the irradiation conditions under which the increment in the sample temperature is a linear function of the electron beam current. Calibration was normally established using melting points of large-size islands of low-melting-point metals for which, according to the data of refs. 25-26, deviations from the tabulated values do not exceed a few degrees. The error in the experimentally determined crystallization temperature was below 10 K, which is less than the typical spread in experimental values obtained for thin films. The melting and crystallization kinetics of thin merucry, tin, and indium films deposited on amorphous carbon was studied by *in situ* electron microscopy in vacuum (10^{-5} mm Hg). Tin and indium were vaporized by electron bombardment; mercury penetrated into the microscope column from the vacuum system and was condensed on cooled substrates. The temperature was varied from 110 to 1,000 K.

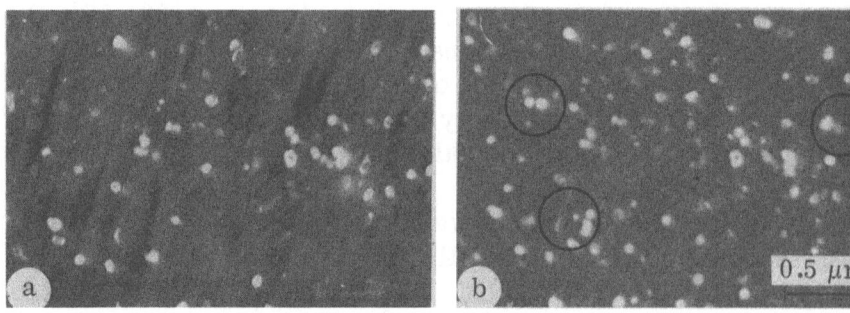

Fig. 2. Dark-field diffraction micrographs of a tin film on carbon at temperatures (a) 315 K and (b) 295 K (temperature was decreased by steps of 5 K).

Results and Discussion

Mercury and tin droplets were crystallized at relative supercoolings up to 36-38 percent [29-31]. The position of dark-field reflections and, hence, the orientation of islands changed in each heating-cooling cycle, which is typical for homogeneous crystallization. If foreign crystallization centers (catalysts) are present one can expect the islands to "memorize" the pre-melting orientation, because of its effect on the nucleus-catalyst interfacial energy. It proved possible to generate such centers by rapidly heating a tin film to 1,000 K, after which many of the reflections appeared at the same places after numerous solid-melt-solid transitions. At the same time, the mean supercooling was reduced by approximately 80 K and the variation of the crystallization temperature distribution increased. Figure 3 plots $\Delta N/\Delta T$ for tin, namely, the number of reflections appearing in the field of view in 1 min for a temperature increment of 10 K, as a function of temperature. Assuming the crystallization to be homogeneous under normal conditions (without intensive preheating), one can estimate the surface energy from relations (5), (7), and (11). The rate of crystallization I was found from the mean expectation time of a reflection, τ. For droplets with diameters around 200 Å, $I=10^{16}-10^{17}$ cm^{-3} s^{-1}. Assuming $K_V = 10^{33}$ cm^{-3} s^{-1}, we found σ_{sl} (Hg)= 24.8±2 erg cm^{-2} and σ_{sl} (Sn)=63.4±3 erg cm^{-2} [29-31]. The Laplace pressure was calculated by formula (12) for the values of γ obtained by linear extrapolation of the data from ref. 47.

If the island orientation is reproducible, crystallization can be definitely considered to be heterogeneous. In this case the dark-field method makes the measurement of I possible for a selected particle at different temperatures. The reflection appearance time τ plotted as a function of the relative beam current $\Delta j / j_0$ (j_0 is the current value at $\tau = 2$ s and T=450 K), which depends on temperature monotonically, shows (Fig. 4) that as the temperature varies by 2 K the value of τ varies by two orders of magnitude (for instantaneous crystallization we assume $\tau = 0.1$ to 1 s). This conclusion is of interest for two reasons. First, the reproducibility of results points to a very small (for electron microscopy) error of measured relative values of temperature (about 1 K). Second, the possibility arises of finding the effective free energy of the interface from the slope of the experimental curve $I_S = f[T(\Delta T)^2]^{-1}$. The value $\sigma' = 37\pm6$ erg cm^{-2} has been obtained for a particle corresponding to the right-hand side of the temperature distribution in Fig. 3; hence, as would be expected, $\sigma' \ll \sigma_{sl}$. For the measured values T=450 K and $I_S \approx 10^{10}$ cm^{-2} s^{-1} we calculate $K_S = 10^{36\pm6}$ cm^{-2} s^{-1}. The error caused by the difficulties of accurately measuring the absolute value of T is too large for unambiguous conclusions to be made. Nevertheless, we feel that the discrepancy with the theory is not spurious. Presumably, in small particles $K_V/K_S = n/n_s \ll 10^7$ since even the total number of atoms can be less than this value. Furthermore, one cannot exclude the possibility of values $K_V \gg 10^{33}$ cm^{-3} s^{-1} [3, 8].

The scattering of crystallization temperatures is to be expected if foreign crystallization centers are present, but in the case of homogeneous crystallization the interpretation is not so obvious. Tin islands, even very similar in diameter, crystallized in the range approaching 30-40 K (this range is typical for thin films [35, 40]). The following explanation appears reasonable. The shape of the islands formed in the process of condensation is often far from equilibrium and may comprise segments of the surface with high curvature leading to increased Laplace pressure (see relation (13)). The pressure-induced shift of the crystallization temperature in mercury and tin reaches 25 K for R=20 Å. Changes in the shape of critical nuclei may introduce an additional contribution.

The temperatures of melting, T_{mR}, and crystallization, T_{sR}, of mercury and tin are plotted in Fig. 5 as functions of island diameter. The difference $T_{mR} - T_{sR}$ is nearly zero at R=25-30 Å. The reversibility of the transition for small values of R is in agreement with the concept of a surface liquid-like layer with minimum thickness t_{min}=10-15 Å [29-31]. Curves 3 in Fig. 5 correspond to equation (15) at t=const. Judging by the discrepancy between the

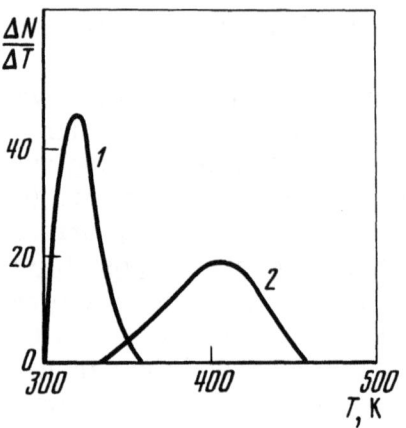

Fig. 3. Relative crystallization rate of tin as a function of temperature. (1) in normal conditions, (2) after briefly heating at 1,000 K.

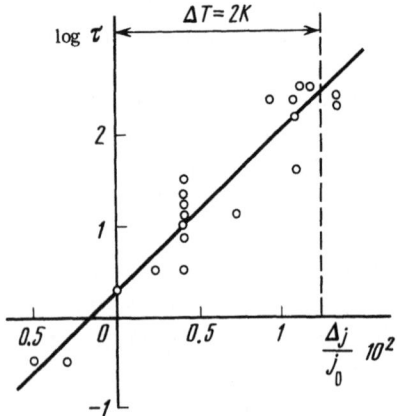

Fig. 4. Crystallization time of a selected tin film island as a function of the relative electron beam current.

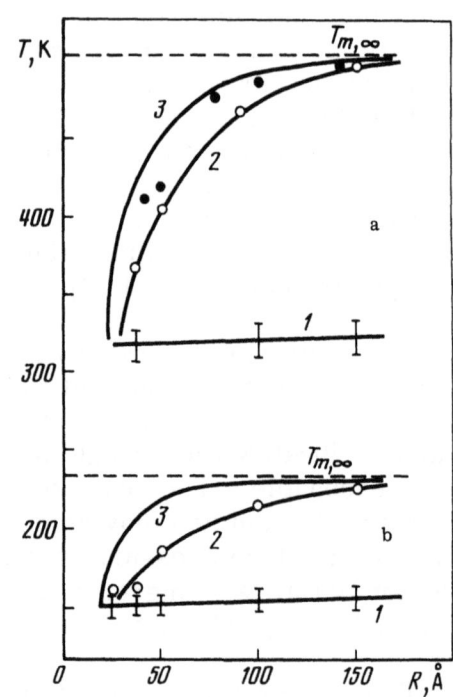

Fig. 5. Crystallization temperatures and melting of (a) tin and (b) mercury as functions of island radius. (1, 2) experimental crystallization and melting temperatures, respectively; (3) results calculated by formula (15) at t=15 Å; black circles represent the data of ref. 25.

theoretical and experimental curves, this assumption is rather crude - a conclusion which also follows from relation (14).

The possibility of surface melting of the islands is confirmed by the instability of the dark-field reflections in the vicinity of the melting point [48]. Positions of these reflections sometimes changed after the crystallization of the islands, especially the smallest ones. Mercury particles up to 100 Å in diameter produced scintillations on the screen, which indicates a rapid change in orientation due to the mobility of the crystalline nucleus. As follows from (14), the thickness of the liquid-like coating diminishes as R increases, so that the mobility of large-size islands must be less pronounced. Indeed, as mercury crystals grew larger, scintillations disappeared and reflection positions were stabilized.

In studying the crystallization of indium, known to be extremely sensitive to traces of oxygen [38, 39], the authors assumed the process to be heterogeneous. This assumption has been borne out by a number of observations [29-31], and in particular by detecting the oxide microcrystals on the dark-field images. In contrast to mercury and tin, indium is characterized by a strong dependence of crystallization temperature T_{sR} on R. This is not surprising because larger crystals undergo a longer exposure to the residual gas which increases the activity of the oxide.

Table 1 given below lists temperatures of melting T_{mR} and crystallization T_{sR} for a number of metals, their maximum deviations $|\Delta T|_{max}$ from $T_{m\infty}$, and also temperature T_g usually defined as the temperature at which one condensation mechanism is substituted by another [49-51]. This definition has the following meaning: the smallest nuclei formed in the process of condensation of a vapor layer are either crystalline ($T<T_g$) or liquid ($T>T_g$), and the threshold temperature $T_g \approx (2/3) T_{m\infty}$ characterizes the triple point in which the two-dimensional phases coexist.

In the case of mercury the transition at T_g does not alter the mechanism of condensation, but changes that of crystallization [42, 43]. Even at $T<T_g$ the smallest mercury particles are not crystallized, but crystallization occurs rapidly as the critical size is reached, while there is no crystallization at $T>T_g$. Consequently, T_g must be interpreted as the limiting temperature for homogeneous crystallization [42, 43, 52]. Later publications sometimes confuse the two definitions [53, 54], although they carry different physical meanings.

Measurements for mercury and tin, as well as the results of refs. 33, 40, 50, 51 for bismuth and lead, show that T_g is practically identical to the upper limit on the distribution of T_{sR}. The crystallization of the main bulk of the condensate at $T<T_g$ proceeds at a higher rate so that in all likelihood the estimates of σ_{sl} in ref. 54 are low. The real supercooling is best found as

$$(T_{sR})_{min} = \lim_{R \to R_{min}} T_{sR} = \lim_{R \to R_{min}} T_{mR},$$

where R_{min}=25-30 Å (a value of 12 Å for R_{min} [21, 37] is hardly acceptable since with such small particles the electron diffraction patterns for the solid and liquid are practically indistinguishable). In mercury and tin the value $(T_{sR})_{min}=(0.62+0.02)T_{m\infty}$ is the upper limit for the existence of the metastable liquid, determined by the high rate of spontaneous nucleation. It seems that tin, gallium, bismuth, silver, and gold films also crystallize homogeneously at appropriate supercoolings. The crystallization of indium and lead, on the other hand, is mostly initiated by an oxide film. This is indicated by an appreciable increase in T_{sR} when lead is heated in an atmosphere containing traces of oxygen [36]. Heterogeneous crystallization is characterized by a large scatter of the results and often by their contradictory behavior.

TABLE 1. Temperature of Melting T_{mR} and Crystallization T_{sR}^{*} of Small Metal Particles

$T_{m\infty}$, K	Melting			Crystallization			$\frac{|\Delta T|_{max}}{T_{m\infty}}$	σ_s, erg·cm^{-2}	Notes	Reference
	T_{mR}, K	R, Å		T_{sR}, K	R, Å	T_g, K				

Mercury

234				155	10^5			31.2	$K_V = 10^{40}$	[3]
				182	10^5–10^6			23.0		[55]
	150–160	25–40		140–150	25–40	160	0.6			[29–31]
				μ 155	150			24.8	$K_V = 10^{33}$	[42, 43]

Gallium

303α	160						0.53		t_{min}	[56]
257β				227	10^5			56.0		[2, 4]
				158	10^5			41.6	β-Ga	[57]
				205	10^5			67.7	α-Ga	[9]

Indium

430				348	10^5			30.8		[55]
	335	50		313–370	75–10^3	293	0.67	110		[35, 39]
	315	30						101		[26]
	340	40		330–400	40–10^3					[29–31]
	290	30		<220	30–50					[36]

Tin

505				383–400	10^4–10^6			58 ± 1		[2, 55, 58]
	475			313					$t_{av} = 50$ Å	[32]
	413	40–50						62.2		[25]
	345–370	30–40		300–340	30–300		0.62	63.4		[29–31]
						348				[50, 51]

Bismuth

544				450	10^5			54–61		[2, 55]
	521			397					$t_{av} = 50$ Å	[32]
						370				[49]
	453–523			343–363		363	0.63		$t_{av} \leqslant 150$ Å	[33]
	370	12		379	μ 200					[21]

Lead

600				520	10^5			33.3		[2]
						408–413		55		[49, 54]
	559			443					$t_{av} = 50$ Å	[32]
	410	30		510	35–500			31.5		[36]
	543–583	50–1500		363–403	50–1500	463	0.6	69		[40]

Silver

1234					803–823			172		[51, 54]
				1006	10^5			126		[2]
				981	10^5			143		[55]
	823					823	0.66		t_{min}	[34]

Gold

1336					891–913			191		[51, 54]
				1106	10^5			132		[2]
	800	12					0.6			[37]

* Typical time of solidification of a particle at T_{sR} is $10^{1 \pm 1}$ s.

At first glance, a considerable difference between the σ_{sl} values calculated for mercury in refs. 3, 29-31 does not support the hypothesis of homogeneous crystallization. However, the discrepancies practically vanish if the value $K_V = 10^{40}-10^{42}$ cm^{-3} s^{-1} is used in calculating σ_{sl} [3]. This result, as well as some others [9], throws doubt on the validity of K_V estimations by the classical theory, and even more so because the experimental value of K_s for tin best fits the value $K_V \approx 10^{40}$ cm^{-3} s^{-1}.

The above analysis thus shows that tiny droplets of molten metals must crystallize instantaneously at a relative supercooling of about 40 percent (deviations observed for gallium are presumably related to the existence of a number of low-temperature modifications). Evidence points to a homogeneous mechanism of crystallization under these conditions and to a heterogeneous one when the supercooling is appreciably lowered. In view of these facts the currently accepted values of the coefficients K_V and K_s have to be re-evaluated.

Literature Cited

1. C. E. Mendelhall and L. R. Ingersoll. On certain phenomena exhibited by small particles on a Nernst glower. Phil. Mag., 15, 205-214 (1908).
2. D. Turnbull and R. E. Cech. Microscopic observation of the solidification of small metal droplets. J. Appl. Phys., 21, 804-810 (1950).
3. D. Turnbull. Crystallization of the supercooled drops of liquid mercury. J. Chem. Phys., 20, 411-424 (1952).
4. D. N. Hollomon and D. Turnbull. Nucleation. In: Progress in Metal Physics, 4, London, 333-388 (1953).
5. G. A. Colligan, W. T. Loomis, and V. A. Suprenant. On the solidification temperature of supercooled metals. J. Austral. Inst. Metals, 10, 89-93 (1965).
6. G. L. F. Powell. Deep supercooling of bulk metallic samples. J. Austral. Inst. Metals, 10, 223-227 (1965).
7. L. Bosio. Surfusion et polymorphisme du gallium à la pression atmospherique. Metaux (Corros. Ind.), 40, 421-435 (1965).
8. W. A. Miller and G. A. Chadwick. On the magnitude of the solid-liquid interfacial energy of pure metals and its relation to grain boundary melting. Acta met., 15, 607-615 (1967).
9. Y. Miyazowa and G. M. Pound. Homogeneous nucleation of the crystalline gallium in liquid gallium. J. Cryst. Growth, 23, 45-47 (1974).
10. D. E. Ovsiyenko. On Homogeneous nucleation in liquid metals. In: Crystal Growth, Yerevan University Publ. House, Yerevan, 11, 11-25 (1975).
11. D. E. Ovsiyenko. Nucleation in supercooled liquid metals. In: Problems in Modern Crystallography. Nauka, Moscow, 127-149 (in Russian) (1975).
12. K. Meyer. Physikalisch-chemische Kristallographie. VEB Deutscher Verlag, Leipzig (1968).
13. D. Turnbull and H. C. Fisher. Rate of nucleation in condensed systems. J. Chem. Phys., 17, 71-73 (1949).
14. M. Volmer and A. Weber. Keimbildung in übersättigten Gebilden. Z. Phys. Chem. 119, 277-301 (1926).
15. R. Becker and W. Döring. Kinetische Behandlung der Keimbildung in übersättigten Dämpfen. Ann. Phys., 24, 719-752 (1935).
16. Ya. I. Frenkel. Kinetic Theory of Liquids. AN SSSR Publ. House, Moscow (in Russian) (1945).
17. R. F. Strickland-Constable. Kinetics and Mechanism of Crystallization from the Fluid Phase and of the Condensation and Evaporation of Liquids. Academic Press, London-New York (1968).

18. J. Lothe and G. M. Pound. Reconsiderations of nucleation theory. J. Chem. Phys. 36, 2080-2085 (1962).
19. H. Reiss and I. B. Wilson. The effect of surface on melting point. J. Colloid Sci., 3, 551-561 (1948).
20. K. J. Hanszen. Theoretische Untersuchungen über den Schmelzpunkt kleiner Kugelchen: Ein Beitrag zur Thermodynamik der Grenzflächen. Z. Phys., 157, 523-553 (1960).
21. S. J. Peppiat. The melting of small particles. II. Bismuth. Proc. Roy. Soc. London A, 345 401-412 (1975).
22. Ya. E. Geguzin, and N. N. Ovcharenko. Surface energy and processes on surfaces of solids. Uspekhi Fiz. Nauk, 76, 283-328 (in Russian) (1962).
23. Kh. B. Khokonov, I. G. Shebzukova, and Kh. N. Kokov. Measurement of surface tension in solid tin, indium, and lead. In: Wetting and surface properties of melts and solids. Naukova Dumka, Kiev, 156-159 (in Russian) (1972).
24. S. J. Peppiat and J. R. Sambles. The melting of small particles. I. Lead. Proc. Roy. Soc. London A, 345, 387-399 (1975).
25. C. R. M. Wronski. The size dependence of the melting point of small particles of tin. Brit. J. Appl. Phys., 18, 1731-1773 (1967).
26. R. P. Berman and A. E. Curzon. The size dependence of the melting point of small particles of indium. Canad. J. Phys., 52, 923-929, (1974).
27. N. T. Gladkikh, V. I. Larin, and V. I. Khotkevich. Determination of surface energy of solids from the melting point of fine-dispersion particles and thin films. Fiz. Met. Metalloved., 31, 786-789 (in Russian) (1971).
28. N. T. Gladkikh, V. I. Larin, and M. N. Naboka. Liquid-solid phase transitions in thin films. In: Whiskers and Thin Films. Voronezh Politechn. Inst. Publ. House, Voronezh, Part 2, 75-80 (in Russian) (1975).
29. Gl. S. Zhdanov. Kinetics of phase transitions in mercury and tin thin films. Fiz. Tverd. Tela, 18, 1415-1418 (in Russian) (1976).
30. Gl. S. Zhdanov. Thermal hysteresis of phase transition and crystallization mechanism in thin metal films Fiz. Tverd. Tela, 19, 229-301 (in Russian) (1977).
31. Gl. S. Zhdanov. Kinetics of melting and crystallization of metal film islands. Izv. AN SSSR, Ser. Fiz., 41, 1004-1008 (1977).
32. M. Takagi. Electron-diffraction study of liquid-solid transition of thin metal films. J. Phys. Soc. Jap. 9, 359-363 (1954).
33. K. J. Hanszen. On the formation of liquid in vacuum-deposited bismuth layers. In: Proc. 6th Intern. Congr. on Electron Microscopy. Tokyo: Maruzen, 527-528 (1966).
34. N. T. Gladkich, R. Niedermayer, and K. Spiegel. Nachweiss grosser Schmeizpunkterniedrigungen bei dünnen Metalschichten. Phys. status solidi, 15, 181-192 (1966).
35. B. T. Boiko, A. T. Pugachev, and V. M. Bratsykhin. On melting of condensed indium films of subcritical thickness. Fiz. Tverd. Tela, 10, 3567-3570 (in Russian) (1968).
36. C. J. Coombes. The melting of small particles of lead and indium. J. Phys. F: Metal Phys. 2, 441-449 (1972).
37. P. A. Buffat. Lowering of the melting temperature of small gold crystals between 150 Å and 25 Å diameter. Thin Solid Films, 32, 283-287 (1976).
38. J. F. Pocza. Forming processes of directly observed vacuum-deposited thin films. In: Proc. 2nd Colloq. on Thin Films. Budapest: Akad. Kiado, 93-108 (1968).
39. J. F. Pocza and P. B. Barna. Formation processes of vacuum-deposited indium films and thermodynamical properties of submicroscopical particles observed by in situ electron microscopy. J. Vac. Sci. and Technol. 6, 472-475 (1969).
40. M. J. Stowell. The solid-liquid interfacial free energy of lead from supercooling data. Phil. Mag. 22, 1-6 (1970).
41. M. J. Stowell, T. J. Law, and J. Smart. Growth, melting and solidification of thin lead films on carbon and molybdenite. Proc. Roy Soc. London A, 318, 231-241 (1970).

42. Gl. S. Zhdanov and V. N. Vertsner. Direct observation of condensation and crystallization of mercury. Fiz. Tverd. Tela, 8, 1021-1027 (in Russian) (1966).
43. Gl. S. Zhdanov and V. N. Vertsner. On heating of samples by electron beam. Radiotekhnika i Elektronika, 11, 1901-1904 (in Russian) (1966).
44. Gl. S. Zhdanov and V. N. Vertsner. Determination of sample temperature in electron microscope. Izv. AN SSSR, Ser. Fiz., 32, 1087-1090 (in Russian) (1968).
45. Gl. S. Zhdanov. Experimental investigation of temperature distribution in thin films heated by electron beam. Inzh.-fiz. Zhurnal, 21, 1084-1090 (in Russian) (1971).
46. Gl. S. Zhdanov. Sensors of sample temperature in electron microscope. Optiko-mekhanicheskaya Promyshlennost', No. 12, 14-18 (in Russian) (1971).
47. V. K. Semenchenko. Surface Phenomena in Metals and Alloys. GITTL, Moscow (in Russian) (1957).
48. Gl. S. Zhdanov. Surface melting of small metal crystals. Kristallografiya, 21, 1220-1221 (in Russian) (1976).
49. L. S. Palatnik and Yu. F. Komnik. On the mechanism of vacuum deposition of metals. Dokl. AN SSSR, 124, 808-811 (in Russian) (1959).
50. Yu. F. Komnik. Electron diffraction study of the formation of metal films. Fiz. Met. Metalloved., 16, 867-871 (in Russian) (1963).
51. Yu. F. Komnik. Characteristic temperatures of thin film condensation. Fiz. Tverd. Tela, 6, 2897-2904 (in Russian) (1964).
52. Gl. S. Zhdanov. In situ electron microscopy investigation of certain features of nucleation, growth, and recrystallization of thin films. Thesis, Leningrad, The State Optical Institute (in Russian) (1969).
53. L. S. Palatnik, M. Ya. Fuks, and V. M. Kosevich. Mechanism of Formation and Substructure of Condensed Films. Nauka, Moscow (in Russian) (1972).
54. N. T. Gladkikh, V. I. Larin, V. M. Severin, and V. I. Khotkevich. Mechanism of vacuum deposition of metals and determination of surface energy at the solid-melt interface. In: Kinetics and Mechanism of Crystallization. Nauka i Tekhnika, Minsk, 126-130 (in Russian) (1973).
55. V. P. Skripov, V. P. Koverda, and G. T. Butorin. Kinetics of crystal nucleation in small volumes. In: The IVth USSR Conference on Crystal Growth, Yerevan, Arm. AN SSSR Publ. House Yerevan, 1, Part 1, 74-77 (in Russian) (1972).
56. B. I. Belevtsev and Yu. F. Komnik. Phase transitions in thin gallium films grown by low-temperature condensation. Fiz. Tverd. Tela, 14, 3240-3244 (in Russian) (1972).
57. V. P. Skripov, G. T. Butorin, and V. P. Koverda. Homogeneous nucleation in crystallization of supercooled gallium. Fiz. Met Metalloved., 31, 790-794 (in Russian) (1971).

THE EFFECTS OF SUBSTRATE-MEDIATED INTERACTION BETWEEN ADSORBED ATOMS ON THE STRUCTURE OF TWO-DIMENSIONAL CRYSTALS FORMED FROM THESE ATOMS

V. K. Medvedev, A. G. Naumovets, and A. G. Fedorus

The Institute of Metal Physics of the UkSSR Academy of Sciences, Kiev

Introduction

In the present paper experimental data on the factors affecting the atomic structure of submonolayer adsorbed films are discussed. Atoms in a submonolayer film may be arranged in one plane, thus forming a two-dimensional system and, in particular, a two-dimensional crystal (at sufficiently low temperature). There are several reasons for the interest in such objects: first, their very small dimensions, which results in phase transitions of unusual types; second, the material on the surface possesses specific chemical properties resulting in unusual interactions within the film; finally, a submonolayer film is the first stage of crystal growth on a foreign surface, and to some extent this stage affects the subsequent stages of the process.

The interaction of adsorbed atoms (adatoms) on the surface can be separated into two components: interaction with the substrate and interaction among themselves. Although this classification is only approximate, it serves very well for illustrative purposes.

The adatom-substrate interaction can be described in terms of a certain potential distribution on the surface. The shape of this distribution depends both on the atomic structure and the chemical nature of the surface, and on the nature of the particles adsorbed. In other words, the distribution characterizes not substrate alone but the adatom-substrate system. Each adatom modifies the potential distribution for another adatom within a certain distance from itself, and this modification is regarded as the interaction between adatoms.

The two interactions referred to above may be caused by different physical mechanisms, and the ratio of their characteristic energies may vary over a wide range. Consequently, the structures of two-dimensional crystals show a considerable diversity. We shall consider some examples which illustrate the effects of a number of factors on the symmetry of two-dimensional adatom lattices.

Systems Investigated

The investigated adsorbed films were those of electropositive elements (lithium, sodium, potassium, cesium, strontium, barium, lanthanum) on various faces of tungsten and molybdenum crystals. An important feature of these systems is the strongly polar adsorption bonding: the adatoms pass to the substrate a considerable fraction of their electronic charge from the outer electron shell and turn into positive ions with a moderate degree of ionization (0.1 to 1). At sufficiently large (on an atomic scale) distances between the adatoms, the tangential interaction between them can be considered as the interaction between identically oriented dipoles formed of charged adatoms, and the electrons which screen them in the substrate. The repulsive dipole-dipole interaction, whose important role has been demonstrated already by Langmuir, affects all the properties of the films under discussion: the dependence of the work function and adsorption heat on the surface concentration of adatoms, surface diffusion, and the structure of the two-dimensional adatom lattices. These structures prove to be especially varied when the repulsive interaction between the adatoms is predominant. As a result of the combination of the attractive and repulsive forces the film undergoes a series of structural transitions when the adatom concentration varies within one monolayer. The systems we have studied demonstrate precisely this type of behavior.

Although the dipole-dipole interaction between adatoms plays an undoubtedly important role, it is not the only one. Evidence accumulated in recent years points to an "indirect" interaction between adatoms (via the electron gas of the substrate), in contrast to the "direct" interaction across the vacuum semispace [1]. Data confirming this interaction and which demonstrate its effect on the structure of two-dimensional crystals are presented below.

Experimental

The structure of the adsorbed films was analyzed by the low-energy electron diffraction method (LEED). The tungsten and molybdenum crystal samples were cut from large single crystals purified by zone melting. Sample surfaces deviated from specific crystallographic planes by not more than 20'. The surface layers structurally distorted by the mechanical treatment of the samples was removed electrolytically. Prior to the experiment the samples were carefully degassed, and carbon-containing impurities were removed by prolonged heating in oxygen. The purity of atomic beams of adsorbates was checked mass-spectrometrically, and in some cases by Auger spectroscopy. The pressure of the active components of residual gases in the LEED unit was below 10^{-11} mm Hg. The atomic beams of adsorbates were calibrated on the basis of structural data, and also by the concentration dependence of the work function measured in some experiments [2]. The accuracy of determination of adatom concentration crystal surfaces was 3-5 percent. In order to single out the factors affecting the structure of submonolayer films, films on specially selected substrates were studied:
— on similar faces of tungsten and molybdenum crystals which have the same body-centered cubic lattice (bcc) and similar lattice constants ($\Delta a < 0.02$ Å);
— on different faces of the same crystal differing in the depth of atomic relief and the degree of anisotropy ((110), (112), and (111) faces of tungsten crystals).

Experimental Results and Discussion

Films on (110) Faces of Tungsten and Molybdenum. The closest-packed faces in bcc crystals are the (110)-type faces. Both the electron adsorption characteristics [3] and the structure of submonolayer films on these faces give evidence of the predominant dipole-dipole interaction between adatoms. As an example, consider the models of two-dimensional strontium lattices on the molybdenum (110) face (see Fig. 1), compatible with the obtained electron diffraction patterns ($\theta = n/n_s$, where n and n_s are the densities of the adatoms and substrate atoms respectively). In these lattices the interatomic spacings are nearly the same in many directions. We conclude, therefore, that the interaction between adatoms on such substrates is essentially isotropic and is a slowly diminishing function of distance; this is typical of dipole-dipole forces. This interaction is responsible for the formation of large-period lattices formed in the early stages of the adsorbate

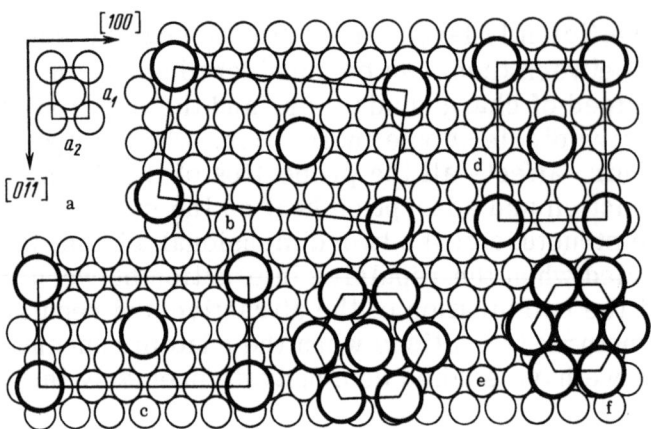

Fig. 1. Two-dimensional lattice models of sub-monolayer films on a Mo (110) face. (a) unit cell of the substrate surface structure $c(1 \times 1)$ with periods a_1 and a_2; (b) one of the possible cells whose statistical mixture represents the structure of the film at $\theta = 0.06$; (c, d) $c(2 \times 6)_2$ and $c(3 \times 3)_2$ lattices, respectively; (e, f) hexagonal structures of layers with different degrees of packing ((f) close-packed monolayer).

deposition (in the range of adatom density of $n \approx 1 \cdot 10^{14}$ cm^{-2}). These lattices are matched to the substrate lattice in the sense that all adatoms occupy identical positions with respect to the substrate atoms, and the film lattice constants are multiples of the substrate lattice constants. If the multiplicity of periods in two directions is given by integers M and N, then the film lattice is denoted by $(M \times N)$, with letter "c" added in front of a centered cell or "p" for a primitive cell [4]. With this notation, Fig. 2 shows a sequence of two-dimensional lattices of submonolayer systems formed by various adsorbates on tungsten and molybdenum (110) faces. The subscript after the parentheses in the structure notations indicates the number of atoms in a non-primitive unit cell. The structure sequence was compiled from electron diffraction patterns for the systems Na-W (110) [3], Sr-W (110) [5], Sr-Mo (110) [10], Ba-Mo (110) [6], and Ba-W (110) [7]. Horizontal arrows indicate the values of θ for which these structures are observed. The symbol Σ denotes the structures which are formed by a statistical mixture of cells, vertical solid lines mark the ranges of phase transitions of the first kind (FTI), and vertical arrows denote a gradual contraction of the cell.

It is clear from the data shown in Fig. 2 that for each adsorbate usually several simple lattices of the types $c(3 \times 7)_2$, $c(2 \times 6)_2$, $(2 \times 10)_4$, $c(3 \times 3)_2$, etc. are observed at low coverage. These lattices correspond to integral values of n_s/n so that θ is equal to, for example, 1/21, 1/12, 1/10, 1/9, etc. For intermediate values of θ, the film remains macroscopically homogeneous (single-phase), and consists of a statistical mixture of unit cells with similar sizes and orientations. This conclusion can be derived by analyzing the typical changes in diffraction patterns caused by deviations from integral n_s/n ratios [8]. In Fig. 2 the transient structures of this type are denoted by Σ. Figure 1b shows one of the possible versions of such cells for $\Sigma_{0.06}$ (for $\theta = 0.06$). This means that an increase in film density does not destroy its structural homogeneity for a wide range of coverage, which agrees with the repulsive nature of the dipole-dipole interaction.

The enhanced interaction between adatoms at a certain density of the adlayer (against the background of the potential distribution of the substrate) destroys the structural fit of the

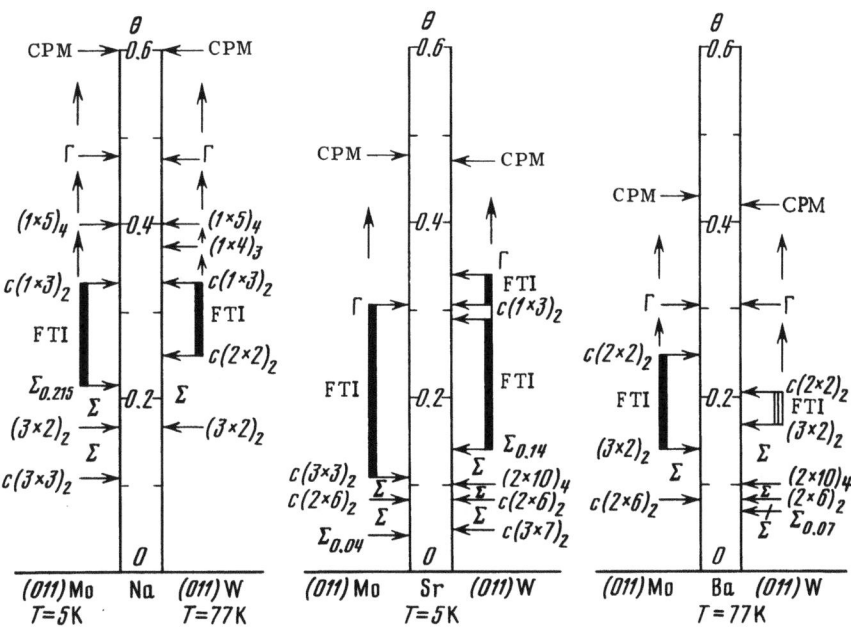

Fig. 2. Structural transition in submonolayer films of sodium, strontium, and barium on (110) faces of Mo and W (CPM — close-packed monolayer; FTI — phase transition of the first kind).

film and the substrate. As a rule, this occurs initially in one direction, and later in other directions as well. Consequently, the adlayer lattice becomes hexagonal, with a period gradually decreasing as the adatom coverage increases. The final stage of this process is the formation of a close-packed monolayer (CPM).

One interesting feature of submonolayer films of electropositive species is the occurrence of first-order phase transitions (FTI) between individual homogeneous structures. This indicates that at a certain critical concentration an effective attraction appears between adatoms; this can also be explained in terms of the elementary dipole model (strong mutual depolarization of adatoms [9]).

The described general character of structural transitions within the first monolayer is standard for all investigated adsorbates on tungsten and molybdenum (110) faces. However, the specific sequence of structures is different in all of the systems under discussion. For example, sodium adatoms on both substrates form $(3 \times 2)_2$, $c(1 \times 3)_2$, $(1 \times 5)_4$ and H (hexagonal) lattices, but for tungsten only $c(2 \times 2)_2$ and $(1 \times 4)_3$ lattices are observed, and FTI on tungsten and molybdenum sets in at different values of θ. These differences are even more pronounced for strontium adsorption: the $c(3 \times 3)_2$ structure is typical for Mo(110), while the $c(3 \times 7)_2$, (2×10), and $c(1 \times 3)_2$ structures are typical for W(110); two regions of FTI are found on W(110), and only one on Mo(110). Barium adlayers on Mo(110) never have the $(2 \times 10)_4$ structure which is formed on W(110).

These differences cannot be explained if adatoms are assumed to participate only in the dipole-dipole interaction modulated by the potential distribution in the substrate. Thus, the ratio of energies corresponding to the possible adlayer lattices for a given θ must be identical both on the tungsten and molybdenum (110) faces, having identical symmetry and almost identical periodicities, provided that only the dipole-dipole interaction is taken into account [10]. A minimum-energy configuration, the same on tungsten and molybdenum, should then be observed experimentally. However, since the experiments show different structure sequences on these substrates, we conclude that other forces, in addition to the dipole-dipole ones, must be introduced into the model. It seems most likely that the mechanism responsible for these differences is the indirect substrate-mediated interaction between adatoms [1]. Here the forces are long-range, sign-variable, and anisotropic, and the asymptotic character of the interaction energy is given by the expression

$$u_\text{к} \sim \cos(2k_F r + \alpha)/r^m, \tag{1}$$

where r is the distance, m=1-3 (depending on the shape of the Fermi surface [11]), k_F is the Fermi momentum in the direction considered, and α is a phase shift.

Adlayers on Tungsten and Molybdenum (211) faces. In contrast to {110} faces, the {211}-type faces of bcc crystals have a highly anisotropic potential distribution: the surface is formed by parallel close-packed atomic strings separated by grooves (Fig. 3). Two-dimensional lattices observed on these surfaces show that the interaction between adatoms here is also strongly anisotropic. Indeed, a number of adsorbates form structures of the type p(1 × N), where N=2-9 (Fig. 4). The sequence of structures was obtained on the basis of electron diffraction patterns given in refs. 12-14. The data for the Li-Mo(211) system were obtained in collaboration with M. S. Gupalo, B. M. Palyukh, and T. P. Smereka. Vertical arrows mark the regions of unidimensional compression of adsorbed films. The strontium and lanthanum films have the centered $c(2 \times 1/\theta)$ structure while the lithium films have a unidimensionally non-coherent structure (UNCS). For the remaining symbols see Fig. 2.

The shortest interatomic spacing along the grooves in such two-dimensional lattices is 6-20 Å, while in the perpendicular direction it is 4.47 Å, determined by the spacing between the nearest grooves. We can conclude that attractive forces appear between the adatoms in the neighboring grooves. It is natural to associate their origin with the exchange interaction of the adatoms via the protruding string of tungsten or molybdenum atoms.

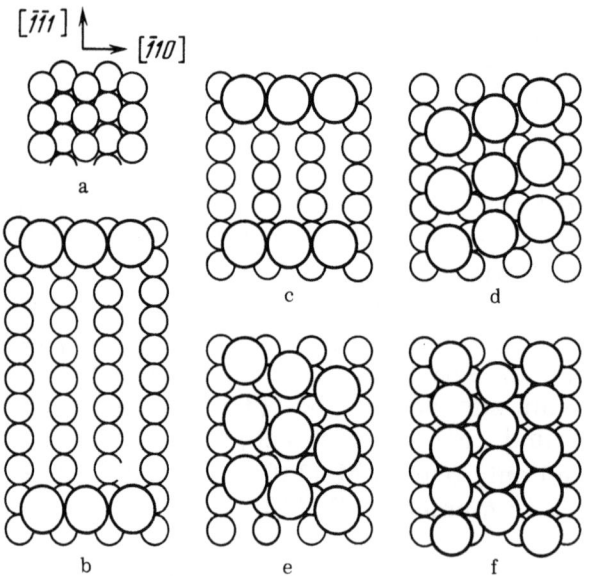

Fig. 3. Two-dimensional lattice models of Sr submonolayer films on Mo(112) face. (a) Mo(112); (b) p(1 × 9); (c) p(1 × 5); (d, e) two orientations of structure $(4 \times 2)_4$; (f) structure of the Sr monolayer film.

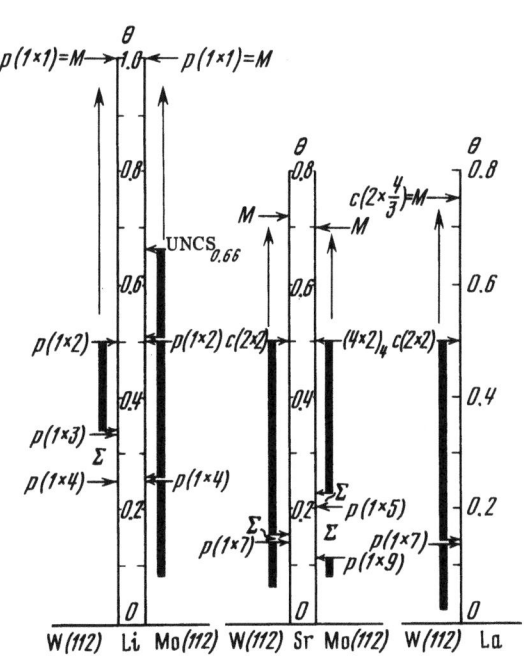

Fig. 4. Structural transitions in submonolayer films of Li, Sr, and La on (112) faces of W and Mo (UNCS—unidimensional non-coherent structure).

Presumably, the indirect interaction of adatoms at small distances (short-range), must strongly depend on the distribution of the electron shells of substrate atoms, that is, on the shape of their unsaturated orbitals. The replication of the structure of the underlying layer by the upper layer atoms on various faces of most refractory metals, as well as the data from field ion microscopy [15], enable us to deduce that the direction of unsaturated orbitals of surface atoms is to some extent similar to the direction of the corresponding orbitals in the bulk material.

The orientations of unsaturated orbitals of surface atoms on the (211) face of a bcc crystal obtained in this approximation is schematically shown in Fig. 5a. We see that two adatoms located in neighboring adsorption sites in the adjacent grooves may experience the exchange interaction via unsaturated orbitals t_{2g} (solid lines) of one substrate atom and orbitals e_g (dashed lines) of the other substrate atom. A chain thus formed is a specific linear surface molecule composed of adatoms and substrate atoms. Exchange-interaction forces compete with the dipole-dipole repulsion. A zigzag shaped chain in which the interatomic spacing is somewhat increased may prove more effective than a linear chain if the dipole moment of the adsorption bond is high. This effect is observed, for example, in the case of barium adsorption on W(211) [16]. The adatoms may be interacting through the unsaturated t_{2g} and e_g orbitals of the same substrate atom. Proceeding from alkali atoms to rare-earth atoms, the role played by the substrate-mediated interaction between adatoms increases. As a result, despite the same dipole moment (3.7 debye), linear chains on W(211) at low θ are formed by strontium adatoms but not by potassium adatoms [13, 17].

At $\theta = 1/2$ many adsorbates on the W(211) face form the $c(2 \times 2)$ lattice characterized by a zigzag arrangement of adatoms transverse to the grooves. The experiments show that the adsorption heat for this lattice is substantially higher than that for a close-packed (110) face at the same concentration of adatoms [13]. This is another argument in favor of the hypothesis of the importance of attractive interactions of adatoms on the (211) face via the protruding substrate atoms.

Now let us discuss the factors which may be decisive for arranging strings of adatoms along the grooves. Since at low θ these strings are at large distances from one another (see Fig. 3), we can conclude that the relative effect of repulsion is much stronger along the

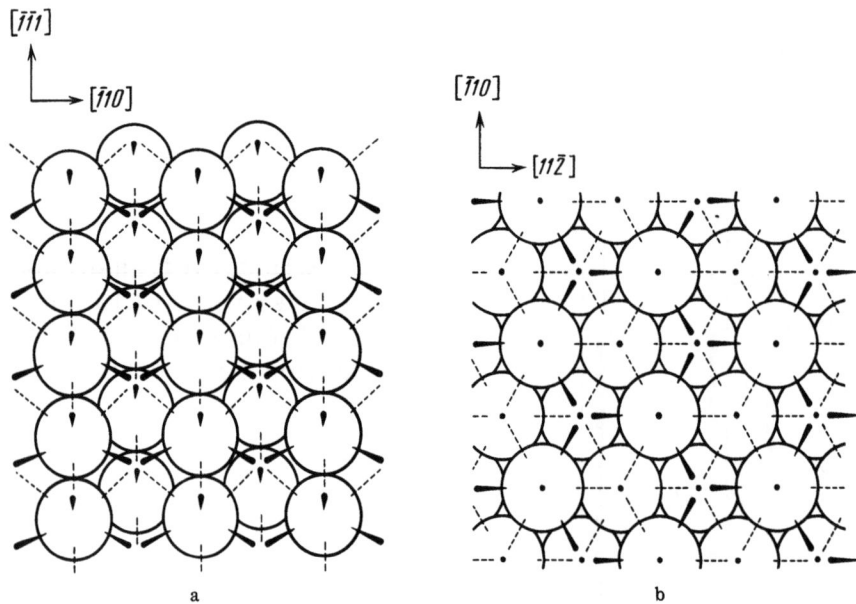

Fig. 5. Schematic arrangement of unsaturated orbitals of surface atoms on faces of bcc crystal: (a) (112), and (b) (111).

grooves than perpendicular to them. One important fact was established by studying p(1 × 7) lattices of strontium and lanthanum adatoms on the (211) face of tungsten: it was shown that these structures grow via an island mechanism [13, 14]. The corresponding superstructure reflections are observed in electron diffraction patterns at coverages smaller than integral fractions by a factor of two to three. The observation of islands with a p(1 × 7) structure proves that at a distance of about 20 Å from one string there is a potential energy well for another string, and obviously, this well can appear only as a result of superposition of long-range forces. It can be assumed that the repulsion is caused by the dipole–dipole interaction while the attraction is a result of the substrate-mediated indirect interaction between adatoms, with the asymptotic behavior given by (1). Interesting data were obtained for the system Sr–Mo(211) in which island growth of the p(1 × 9) structure is observed at low coverage, and the subsequent structure is p(1 × 5). The fact that the intermediate structures at $\theta = 1/8$, 1/7, and 1/6 are not formed is evidence of minima in the interaction potential of adatom strings at distances of 24.6 and 13.7 Å, that is, of the oscillatory character of the interaction. We can conclude, as in the case of adsorption on (110) faces, that the lattices formed by the same adatoms on (211) faces of tungsten and molybdenum are substantially different (see Fig. 4). This is explainable in terms of the difference in the substrate-mediated interaction on the surfaces discussed. Films adsorbed on (211) faces also undergo complex phase transitions. The relevant data are given in Fig. 4. The first monolayer formed by close packing in the grooves (marked by M in Fig. 4) cannot yet be considered as a dense physical monolayer because large areas of the substrate are still "visible" through it.

Adlayers of the (111) Face of Tungsten. Note that this face is atomically very rough (Fig. 5b). The arrangement of orbitals on this face clearly shows that the effective exchange interaction between adatoms may be initiated via these orbitals when the (1 × 1) structure, replicating the substrate structure, is formed. Thus, the structure of films of alkali elements (Li, Na, K, and Cs) on the (111) face of tungsten single crystal has been studied [18]. It was found that all of these adatoms form the (1 × 1) lattice, with high thermal stability. For example, the adsorption heat of cesium adatoms in the (1 × 1) lattice at the density $n = 5.8 \cdot 10^{14}$ cm^{-2} is $q \approx 1.3$ eV (Fig. 6) [16], which exceeds by 0.5 eV the heat of cesium sublimation. This fact is all the more striking because the shortest interatomic

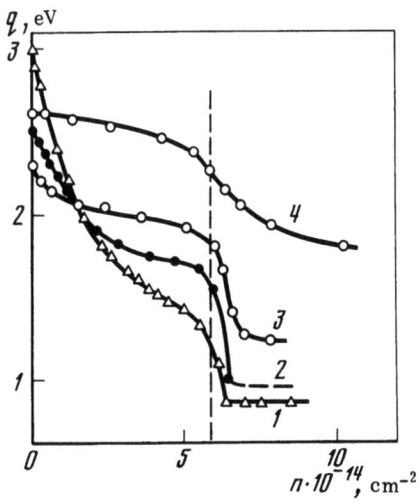

Fig. 6. Adsorption heat as a function of concentration of adatom of alkali metals on W(111) face. (1) Cs; (2) K; (3) Na; (4) Li.

distance in the (1 × 1) lattice, 4.47 Å, is much smaller than in bulk cesium (5.32 Å), that is, the film is substantially compressed. The adsorption heat sharply decreases as the adatom concentration increases above the adatom coverage characterizing the (1 × 1) lattice. Presumably this occurs because the conditions for the substrate-mediated interaction between adatoms drastically deteriorate.

Conclusion

The results discussed demonstrate that on different faces of a substrate crystal the same adsorbate forms two-dimensional crystals with very different symmetry. This is not really surprising since the potential distribution on these surfaces is also very different. However, even on surfaces with practically identical atomic structure (similar faces of crystals having identical symmetry but built of different chemical species) the growth of two-dimensional crystals also differs. Summarizing the experimental facts, one can conclude that the substrate cannot be treated merely as a passive matrix of adsorption sites regularly arranged on the surface. The substrate actively influences the interaction between adatoms, and thus constitutes an additional important factor leading to a wide variety in the structures of two-dimensional crystals formed by adatoms.

Literature Cited

1. T. B. Grimley. Chemisorption theory. Progr. Surface and Membrane Sci., 9, 71-161 (1975).
2. A. G. Naumovets and A. G. Fedorus. Disordering in submonolayer films of electropositive elements adsorbed on metals. Zh. Eksp. Teor. Fiz., 73, 1085-1092 (in Russian) (1977).
3. V. K. Medvedev, A. G. Naumovets, and A. G. Fedorus. Structure and electron-adsorption properties of sodium films on (100) face of tungsten. Fiz. Tverd. Tela, 12, 375-385 (in Russian) (1970).
4. E. A. Wood. Vocabulary of surface crystallography. J. Appl. Phys., 35, 1306-1312 (1964).
5. O. V. Kanash, A. G. Naumovets, and A. G. Fedorus. Phase transitions in strontium submonolayer films adsorbed on (100) face of tungsten. Zh. Eksp. Teor. Fiz., 67, 1818-1826 (in Russian) (1974).
6. A. G. Fedorus, A. G. Naumovets, and Yu. S. Vedula. Adsorbed barium films on tungsten and molybdenum (100) face. Phys. Status Solidi (a), 13, 445-456 (1972).

7. D. A. Gorodetsky and Yu. P. Melnik. Barium on (110) tungsten. Surface Sci., 62, 647-661 (1977).
8. J. E. Houston and R. L. Park. Low energy electron diffraction from imperfect structures. Surface Sci., 21, 209-223 (1970).
9. L. A. Bol'shov. Thermodynamics of adsorbed monatomic films. Fiz. Tverd. Tela, Tela, 13, 1679-1689 (in Russian) (1971).
10. Yu. S. Vedula, V. V. Gonchar, A. G. Naumovets and A. G. Ferdorus. Effect of the electron structure of the substrate on the properties of metal-film substrates. Fiz. Tverd. Tela, 19, 1569-1576 (in Russian) (1977).
11. A. M. Gabovich, and E. A. Pashitsky. Indirect interaction between adsorbed atoms on the metal surface via the substrate electron gas. Fiz. Tverd. Tela, 18, 377-382 (in Russian) (1976).
12. V. K. Medvedev, A. G. Naumovets, and T. P. Smereka. Lithium adsorption on the (211) face of tungsten. Surface Sci., 34, 366-384 (1973).
13. V. K. Medvedev and A. I. Yakivchuk. Structure and electron-adsorption properties of strontium films on (211) face of tungsten single crystal. Ukr. Fiz. Zhurnal, 20, 1900-1908 (in Russian) (1975).
14. Yu. S. Vedula, V. K. Medvedev, A. G. Naumovets, and V. N. Pogorely. Lanthanum adsorption on (211) face of tungsten single crystal. Ukr. Fiz. Zhurnal, 22, 1826-1834 (in Russian) (1977).
15. Z. Knor and E. W. Müller. A refined model of the metal surface and its interaction with gases in the field ion microscope. Surface Sci., 10, 21-31, (1968).
16. V. K. Medvedev and T. P. Smereka. Adsorption of barium on (211) face of tungsten. Fiz. Tverd. Tela, 15, 724-732 (in Russian) (1973).
17. V. K. Medvedev and A. I. Yakivchuk. Potassium adsorption on (211) face of tungsten. Fiz. Tverd. Tela 16, 981-988 (in Russian) (1974).
18. V. K. Medvedev and A. I. Yakivchuk. Structure and electro-adsorption properties of alkali metal films on (111) face of tungsten single crystal. Preprint, Institute of Physics of the Ukr.SSR Academy of Sciences, Kiev (in Russian) (1975).

CRYSTAL GROWTH AND POLYTYPISM IN SILICON CARBIDE

Yu. M. Tairov and V. F. Tsvetkov

The Leningrad Electrical Engineering Institute

Substantial experimental and theoretical knowledge has been accumulated on polytypism in crystals [1-5], but reliable experimental data are still lacking which would reveal the nature of polytypism (thermodynamical or kinetic) and the mechanisms of formation of various polytype structures. In order to solve these problems, it is necessary to develop more effective and more integrated techniques of studying polytype crystals, combining problem-oriented studies of growth with wide-scope structural, morphological, and other investigations. The present work was aimed at an integrated analysis of crystal growth processes and the variety of polytypism, using as an example silicon carbide which is one of the most representative of polytype compounds.

Experimental

The following series of experiments has been performed:

1. To determine the differences in atomic bonding energies in different polytype structures of silicon carbide, a precision method of thermal oxidation of silicon carbide single crystals was developed and oxidation kinetics for various polytypes was investigated [6]. The oxidation mode was such that the oxide film growth rate was limited by the chemical reaction at the oxide-silicon carbide interface. Under these conditions the growth rate of oxide layers depended on the structure of the silicon carbide crystals. By processing the kinetic data for oxide layers grown on (0001) Si faces it was possible to find the reaction constants K_c for different polytypes of silicon carbides. Figure 1 plots ln K in n-type silicon carbide (the difference between the donor and acceptor concentrations $N_d - N_a \approx 2 \cdot 10^{18}$ cm^{-3}) as a function of the degree of hexagonality η which characterizes the sequence and degree of population of coordination spheres in the silicon carbide crystal structures and thus determines the bonding energy. An analysis of experimental data by the least squares method has shown that the curves obtained fit the formula

$$K_c = A \exp\left[\left(-Q_0 - \frac{\partial Q}{\partial \eta} \eta\right) / RT\right], \qquad (1)$$

where $A = 5.6 \cdot 10^3$ Å min^{-1}, R is the gas constant, T is the oxidation temperature, $Q_0 = 21.3$ kcal mol^{-1} is the activation energy of oxidation of 3C-SiC polytype, and $\partial Q/\partial \eta = 17.6$ kcal (mole %).

2. In order to establish the relationship between the conditions of crystallization and the polytype structure of silicon carbide we have conducted a wide study of growth processes

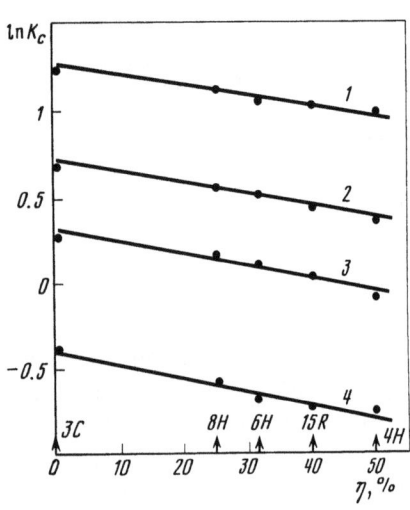

Fig. 1. Oxidation reaction constant K_c (Å min^{-1}) as a function of hexagonality of silicon carbide polytypes at different temperatures (°C). (1) 1180, (2) 1080, (3) 1010, (4) 910.

and crystallization in single crystals and epitaxial layers, under widely varied conditions [7-10]. It has been found that the growth of single crystals and epitaxial layers on (0001) faces of silicon carbide substrates starts with the formation of nuclei and then of thin layers of 3C-SiC polytype structure. During the subsequent stages of growth on (111) faces of layers having 3C-SiC structure, the crystallized layers have the α-SiC structure. A statistical analysis gives the following probability distribution for the formation of epitaxial layers with different α-SiC polytype structures: 6H-SiC — 85%, 15R-SiC — 12%, 4H-SiC — 1-2%. In the remaining cases the growing layers have longer-period structures based on the three given above. Silicon carbide layers growing on 3C-SiC polytype faces other than {111} have a disordered structure.

When silicon carbide crystals are grown by sublimation at temperatures near 2,600°C, crystals grow on "legs" on which nuclei with 3C-SiC structure are formed in the initial stages of the process if the supersaturation is sharply increased [7]. Further growth of the crystal on a 3C-SiC nucleus in directions perpendicular to (0001) proceeds via the formation of a disordered layer (D-layer) [11] (Fig. 2). In directions parallel to (0001) (on both sides of the D-layer) reasonably perfect structures of α-SiC modifications are formed. The probabilities of formation of the different polytypes on nuclei correspond to the probabilities of polytype formation in epitaxial layers.

Sharp changes in the crystallization conditions (mainly in supersaturation) in the course of growth lead to the formation of 3C-SiC nuclei on the growth surfaces, and to the formation of new D-layers with the growth of various α-SiC polytype modifications over them.

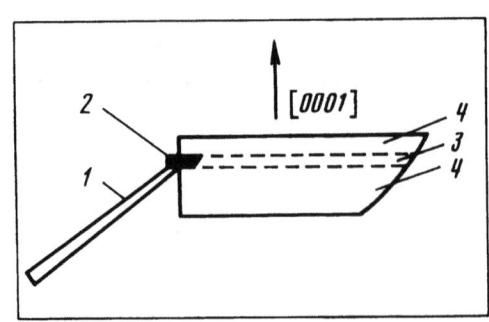

Fig. 2. Schematic cross section along (0001) of a silicon carbide platelet grown by sublimation at a temperature of about 2600°C. (1) "leg" of the crystal, (2) nucleus of 3C-SiC, (3) D-layer, (4) regions of α-SiC polytype structure.

Periodic sharp changes of supersaturation during growth produce single crystals composed of layers of various polytype modifications whose frequency of occurrence is subject to the same statistics. Similar relationships are observed when single crystals of silicon carbide are grown from vapor in vacuum (10^{-3} mm Hg) at temperatures around 1,800°C [12]. Stabilization of the crystallization conditions makes possible the growth of single crystals of one polytype modification (6H-SiC) at growth rates up to 2 mm h^{-1}.

3. The investigation established that impurities have a substantial effect on polytypism in silicon carbide [1, 13-19].

Increased nitrogen pressure in the crystal growth zone increased the relative yield of rhombohedral SiC polytypes and 3C-SiC polytype. At high nitrogen pressures 3C-SiC polytype crystals were found to predominate [15]. An analysis of oxidation and measurement of silicon carbide lattice parameters of the 6H polytype at different levels of nitrogen doping revealed a sharp increase in the oxidation rate (Fig. 3) and an abrupt decrease in lattice parameters (Fig. 4) with nitrogen concentrations in silicon carbide above 10^{19} cm^{-3}. For nitrogen concentrations of 10^{20} cm^{-3} (above which the 3C-SiC structure is stabilized), the rate of oxidation of 6H-SiC polytype crystals exceeds the rate of oxidation of 3C-SiC polytype crystals, which points to a higher stability of the 3C-SiC structure when highly doped with nitrogen.

Doping of silicon carbide crystals by **aluminum** or scandium in the course of growth considerably increases the yield of 4H-SiC polytype crystals [13, 14, 16]. A statistical analysis of batches of vapor-grown silicon carbide crystals, and also some published data [18, 19], have shown that the probability of the growth of crystals with long-period structures increases as the concentration of uncontrolled impurity background in the growing crystals rises from 10^{16} cm^{-3} to 10^{18} cm^{-3}.

4. In order to investigate the phase stability of silicon carbide polytypes, relatively pure ($N_d - N_a \approx 10^{17}$ cm^{-3}) SiC crystals were annealed in the temperature range 1,900-2,450°C. The time of annealing was varied from one to eight hours [20].

The extent of phase transformation was studied by X-ray analysis. Solid-state structural transformations were observed only in 3C-SiC crystals. It was found that the transformation proceeds by the platelet growth of the new phase — disordered 6H-SiC structure. An analysis of the transformation kinetics on the basis of Kolmogorov's theory yielded a value of the transformation activation energy of 53±5 kcal mol^{-1}. It was also found that disordered D-layers in α-SiC single crystals give the same X-ray reflections on Laue diffraction patterns and in diffractometer studies as 3C-SiC crystals after the phase transition. This is further confirmation of the mechanism of D-layer formation from the 3C-SiC structure undergoing a phase transformation.

Discussion of Experimental Results

A physical chemical analysis of oxidation of silicon carbide polytypes has shown that K_c is a function of hexagonality due to the difference in the bonding energy of atoms in the crystal structures of different polytypes [6]. Since the hexagonality of all but one of the known polytypes of silicon carbide does not exceed 50% (the exception is 2H-SiC), one can assume on the basis of the value of $\partial Q/\partial \eta$ obtained that the energies in different silicon carbide polytypes differ by not more than 0.9 kcal mol^{-1}, the bonding energy being maximum in 4H-SiC and minimum in 3C-SiC. The probabilities of formation of crystals and epitaxial layers of different polytypes obtained in near equilibrium conditions do not correspond to the formation probabilities of these structures calculated on the basis of differences in lattice energy.

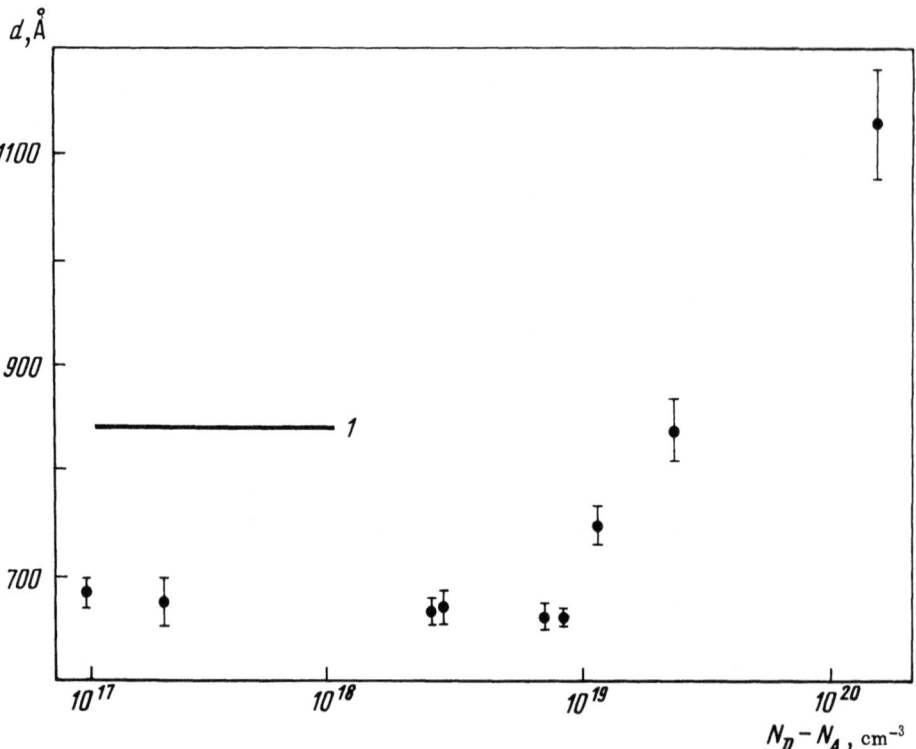

Fig. 3. Thickness of a SiO_2 oxide layer grown on 6H polytype silicon carbide crystal as a function of nitrogen concentration in the crystal (T=1170°C, t=4 hours, dry oxygen). (1) thickness of oxide layers obtained under identical conditions on 3C-SiC crystals.

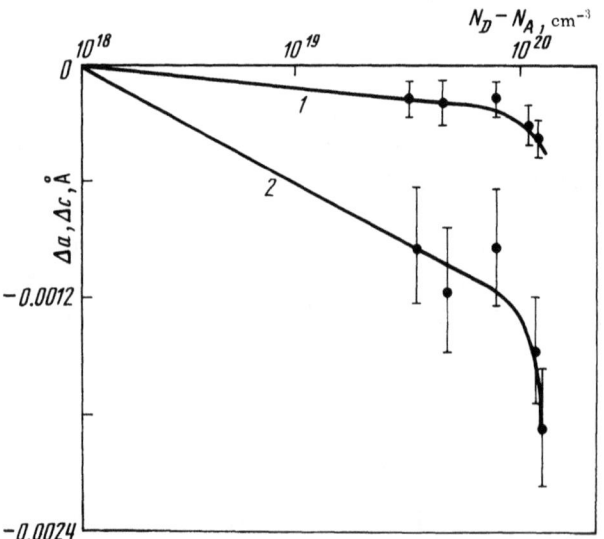

Fig. 4. Lattice parameters of 6H polytype silicon carbide crystals as functions of nitrogen concentration (with respect to lattice parameters of the crystal with nitrogen concentration of 10^{18} cm^{-3}). (1) Δa, (2) Δc.

Therefore it can be stated that in silicon carbide, and presumably in other crystals, polytypism is of kinetic and not thermodynamic origin. This conclusion is supported by the fact that the formation of a metastable modification 3C-SiC is caused primarily by the kinetics of synthesis, and in particular, by supersaturation steps. As the 3C-SiC structure is a high-pressure phase [21], its formation in epitaxial layers and in single crystals grown by chemical vapor deposition or from solutions occurs as follows. It has been shown both theoretically and experimentally [22-24] that in thin layers or in a fine-dispersion state (at the nucleation stage) the material can form phases which are metastable in bulk samples; this is due to the relatively large contribution of the surface free energy in small-size particles. As the nuclei grow and coalesce into a layer, a phase transition to a stable modification must occur. However, in real processes this transformation may be delayed because the transformation rate is determined by the probability of nucleation of the equilibrium phase and by the diffusion activity of the atoms. The longest delays are expected for crystals with covalent chemical bonding, because of the high activation energy barriers inhibiting the transition of a metastable crystal to the stable modification. Owing to this delay of the transformation, epitaxial growth of the initial phase continues in the later stages of condensation. Hence, the initial modification can be found not only in the thinnest layers but in relatively thick films as well, with the probability increasing with decreasing deposition temperature.

This mechanism of the formation of metastable crystals is clearly observed in the crystallization of epitaxial diamond layers [25, 26] (diamond being the closest analog of silicon carbide) and was also observed in the growth of 3C-SiC in relation to the kinetics of synthesis [7, 10, 21]. The nuclei formed after a step change in supersaturation had the 3C-SiC structure, and the probability of maintaining the nucleus structure in the growing epitaxial layer increased as the deposition temperature decreased. With increasing deposition temperatures, the expitaxial layers tended to show polytypic growth; this seems to be caused by phase transitions in the growing nuclei of 3C-SiC at elevated temperatures. Indeed, electron microscopy studies of epitaxial 3C-SiC layers after high-temperature annealing [27] show that the phase transformation in cubic silicon carbide is modified by the formation and motion of stacking faults. This is accompanied by the formation of microscopic regions with stacking sequences 33; 23; 22, etc. (Zhdanov's notation [28]), that is by the formation of nuclei of 6H, 15R, 4H, etc. polytypes.

The phase transformation in 3C-SiC epitaxial layers at the nucleation stage inevitably causes changes in the morphology of the nuclei, followed by the formation of growth centers (e.g., screw dislocations) of the 6H, 15R, 4H, etc. polytypes on their surface. This hypothesis makes it possible to interpret the processes of formation and growth of various α-SiC polytype structures on 3C-SiC layers, nucleated as a result of sharp changes in supersaturation and subsequently undergoing a phase transformation (D-layers), and to conclude that 3C-SiC is the basic structure of silicon carbide from which all other polytypes are formed via phase transformations. This is supported by theoretical studies [29] which show that all important α-SiC polytypes can be transformed into the cubic structure by single dislocations; hence, any polytype structure can be derived from the cubic one. A growth center, such as a screw dislocation, formed on a nucleated layer of 3C-SiC, will subsequently control the structure of the growing layer. This occurs because of the strong orientational nature of the covalent bonding, and is confirmed by experiments on the transfer of structural information from a metastable-structure substrate to the epitaxial layer in directions other than [0001] [9]. The mechanism of the formation and the ordered growth of metastable polytype structures can therefore be explained only in terms of kinetics (taking into account the role of screw dislocations or similar centers). Otherwise, we would always observe a set of regions with different polytype structures with maximum probability of formation close to 4H-SiC, because of the closeness in polytype lattice energies and the statistical nature of the interactions in the growth of an individual crystal.

The formation of series of long-period polytypes based on the 6H-, 15R-, and 4H-SiC structures [1] may be caused by the additional formation of stacking faults and appropriate growth centers in these three structures, nucleated in the process of the phase transition to the 3C structure.

The observed effect of impurities on polytypism is presumably related to distortions of nonspherical symmetry produced in the lattice by imbedded foreign atoms. Even at low levels of doping we can expect the formation of stacking faults, which determine the polytype structure of the growing crystal, depending on the nature of the impurity (taking into account their sectorial incorporation into the crystal [30, 31]). Thus, the 4H-SiC structure is observed when silicon carbide is doped with scandium even at levels as low as 10^{17} cm^{-3}.

The stabilization of the 3C-Si structure at high dopant concentrations (e.g., using nitrogen) can be explained as follows. A decrease in lattice parameters caused by the increased concentration of nitrogen is equivalent to the effect of pressure on the 6H-SiC lattice, resulting in an increased free energy of the lattice. Estimates show that at high levels of nitrogen doping, when stabilization of the 3C-SiC structure is achieved, the compressive stress corresponds, in order of magnitude, to the values of pressure at which the 6H→3C phase transition was observed in ref. 21. At these concentrations an increase in the 6H-SiC lattice energy corresponds to the difference in the 6H and 3C lattice energies; in other words, at these doping levels the 3C-SiC structure is thermodynamically more stable.

Conclusions

1. At normal pressures the stability of polytype structures of silicon carbide is enhanced by an increase in hexagonality of the structure. The least stable structure is 3C-SiC, and the most stable one is 4H-SiC. Annealing of 3C-SiC in the temperature range 1,900-2,450°C leads to a solid-state phase transition to a disordered 6H-SiC structure.

2. When 6H-SiC crystals are doped by nitrogen to concentrations 10^{19}-10^{20} cm^{-3}, the free energy increases to the level of the free energy of 3C-SiC crystals so that the 3C-SiC structure, which is thermodynamically in equilibrium under these conditions, is stabilized. These results agree with the observation of the 6H→3C phase transition under pressure, since nitrogen doping appreciably reduced the lattice parameters, an effect similar to the effect of pressure.

3. In pure silicon carbide polytypism arises basically from kinetic factors, since the probabilities of formation of various silicon carbide polytypes in the nearly equilibrium growth processes of crystals and epitaxial layers are in disagreement with the probabilities of formation of these structures calculated on the basis of differences in lattice energies.

4. The formation of a metastable 3C structure in the growth of undoped crystals and epitaxial layers of silicon carbide is determined by the kinetic conditions of the synthesis. Oscillations of supersaturation in front of the crystallization interface during growth play a decisive role in the formation of the 3C structure. The growth of various α-SiC, nucleated under nonequilibrium conditions, points to the presence of microscopic regions with nuclei of 6H, 15R, 4H etc. polytypes in the phase-transformed 3C-SiC structure, and is confirmed by the electron microscopy studies [27].

5. Overall, the results obtained show that the origin of polytypism in silicon carbide and, presumably, in other polytype crystals is of a dual nature: thermodynamic and kinetic. The 3C-SiC structure is the basic silicon carbide structure from which other polytypes are formed.

Literature Cited

1. A. R. Verma and P. Krishna. Polymorphism and Polytypism in Crystals. John Wiley, New York (1966).
2. D. Pandey and P. Krishna. Influence of stacking faults on the growth of polytype structures. Phil. Mag., 31, 1113-1148 (1975).
3. G. Trigunayat and A. Verma. Polytypism and stacking faults in crystals with layer structure. In: Crystallography and Crystal Chemistry of Materials with Layered Structures. Dordrecht. Holland: D. Reidel Publ. Co., 269-340 (1976).
4. P. Prasad. Present state of polytypism in cadmium iodide crystals. Phys. Status Solidi (a), 38, 11-44 (1976).
5. A. Baronnet. Some aspects of polytypism in crystals. Progr. Cryst. Growth Character, 1, 151-211 (1978).
6. Yu. M. Tairov, V. F. Tsvetkov, and Yu. Laukhe. Oxidation of single crystals as a method of studying polytypism. Fiz. Tverd. Tela, 19, 1777-1780 (in Russian) (1977).
7. Yu. M. Tairov and V. F. Tsvetkov. Investigation of silicon carbide crystal growth from vapor phase. In: Silicon Carbide-1973. Univ. of South Corolina Press, Columbia (1973), pp. 146-1670.
8. Yu. M. Tairov, V. F. Tsvetkov, S. K. Lilov, and G. K. Safaraliev. Studies of growth kinetics and polytypism of silicon carbide epitaxial layers grown from the vapor phase. J. Cryst. Growth, 36, 147-151 (1976).
9. Yu. M. Tairov, S. K. Safaraliev, V. F. Tsvetkov, and M. A. Chernov. On the transfer of structural information in silicon carbide homoepitaxy. Pis'ma Zh. Eksp. Teor. Fiz. 2, 699-701 (in Russian) (1976).
10. Yu. M. Tairov and V. F. Tsvetkov. Effects of kinetic factors on polytypism in epitaxial layers. In: Vth USSR Conference on Crystal Growth, Tbilisi. Abstracts. Institute of Cybernetics of the AN GSSR, 1, 96-97 (in Russian) (1977).
11. J. P. Golightly and L. J. Beaudin. Some aspects of disorder in silicon carbide. In: Proc. Int. Conf. on Silicon Carbide. Univ. Park, Materials Res. Bull., 4, 119-128 (1968, 1969).
12. Yu. M. Tairov and V. F. Tsvetkov. Investigation of growth processes of ingots of silicon carbide single crystals. J. Cryst. Growth, 43, 209-212 (1978).
13. H. Wahner and Yu. M. Tairov. On polytypism of solution-grown SiC(Sc). Fiz. Tverd. Tela, 12, 1543-1544 (in Russian) (1970).
14. Yu. M. Tairov, I. I. Khlebnikov, and V. F. Tsvetkov. Investigation of silicon carbide single crystals doped with scandium. Phys. Status Solidi (a), 25, 349-357 (1974).
15. S. K. Lilov, Yu. M. Tairov, V. F. Tsvetkov, and M. A. Chernov. Structural and morphological peculiarities of the epitaxial layers and monocrystals of silicon carbide highly doped by nitrogen. Phys. Status Solidi (a), 37, 143-150 (1976).
16. W. F. Knippenberg and G. Verspui. The influence of impurities on silicon carbide crystals grown by gas-phase reaction. Proc. Int. Conf. on Silicon Carbide, Univ. Park, Materials Res. Bull., 4, 33-44 (1968, 1969).
17. A. R. Kleffer, P. Ettmayer, E. Gugel, and A. Schmidt. Phase stability of silicon carbide in ternary system Si-C-N. Proc. Int. Conf. on Silicon Carbide, Univ. Park, Materials Res. Bull., 4, 153-166, (1968, 1969).
18. V. G. Fomin, L. A. Shchegol'kova, M. B. Reifman, V. I. Innov, and O. A. Kolosov. X-ray structural analysis of α-SiC crystals doped with nitrogen and boron. In: IIIrd USSR Conference on Semiconductor Silicon Carbide, Moscow ONTI, 268-272 (in Russian) (1970).

19. Yu. Ya. Vodakov, E. N. Mokhov, A. D. Royenkov, and M. M. Anikin. Effect of impurities on polytypism of silicon carbide. Pis'ma Zh. Eksp. Teor. Fiz., 5, 367-370 (in Russian) (1979).
20. Yu. M. Tairov, V. F. Tsvetkov, M. A. Chernov, and V. A. Taranets. Investigation of phase transformations and polytype stability of β-SiC. Phys. Status Solidi (a), 43, 363-369 (1977).
21. A. A. Kalinina, M. I. Sokhor, and L. I. Feldgun. Modification transformations in silicon carbide induced by high pressure. In: High-Temperature Carbides, Naukova Dumka, Kiev, 32-34 (1975).
22. B. Ya. Pines. Essays on Physics of Metals. Kharkov University Publishing House, Kharkov (in Russian) (1961).
23. Yu. F. Komnik. On nucleation of nonequilibrium phases in thin films. Fiz. Tverd. Tela, 10, 312-315 (in Russian) (1968).
24. L. S. Palatnik and V. K. Sorokin. Fundamentals of Semiconductor Film Material Science, Energiya, Moscow (in Russian) (1973),
25. B. V. Derjaguin and D. V. Fedoseyev. Vapor Growth of Diamond and Graphite. Nauka, Moscow (in Russian) (1977).
26. B. V. Spitsyn. Crystal Growth in Thermodynamically Metastable Conditions. In: Vth USSR Conference on Crystal Growth, Tbilisi. Abstracts. Institute of Cybernetics of the AN GSSR, 1, 24-25 (in Russian) (1977).
27. S. Shinozaki and I. Sprys. TEM analysis of long-period polytypes in SiC. Proc. Annu. Meet. Electron Microsc. Soc. Amer., 34, 630-631 (1976).
28. G. S. Zhdanov. Numerical complex of a close-packed sphere and its application to the theory of close-packed spheres. Dokl. AN SSSR, 48, 40-43 (in Russian) (1945).
29. R. S. Mitchell. A correlation between theoretical screw dislocations and the known polytypes of silicon carbide. Z. Kristallogr., 109, 1-28 (1957).
30. Yu. M. Tairov, V. F. Tsvetkov, I. I. Khlebnikov, and M. A. Chernov. Effect of nitrogen on formation of structural imperfections in silicon carbide single crystals. Kristallografiya, 21, 425-426 (in Russian) (1976).
31. V. I. Levin, G. I. Pozdnyakova, Yu. M. Tairov, V. F. Tsvetkov, and Yu. M. Shashkov. Inhomogeneous doping of SiC by nitrogen in relation to conditions of growth of individual crystals. Izv. AN SSSR, Neorg. Mater., 13, 254-257 (in Russian) (1977).

Part IV
Mechanisms and Kinetics of Crystal Growth from the Melt and from High-Temperature Solutions

Part E

MOTION OF LOW-ANGLE MACROSTEPS

V. V. Voronkov

The Institute of Rare Metals, Moscow

The properties of a crystal surface essentially depend on its orientation: there is a finite number of singular surfaces [1-4] requiring for their growth a considerable supersaturation. This supersaturation is a function of the mechanism by which elementary growth steps are generated. As a rule, the closest-packed surfaces are singular surfaces. Other (nonsingular) surfaces are characterized by a high density of elementary steps and require for their growth a relatively low supersaturation. In actual cases the surface of a growing crystal is often of a nearly singular orientation, deviating from it by a small (but appreciable) angle θ_0. Three types can be distinguished among the forms of such a surface on a macroscopic scale.

1. A stable planar shape (Fig. 1a). This requires the effective free surface energy σ_{eff} to be positive [1-3]:

$$\sigma_{eff} = \sigma + \frac{d^2\sigma}{d\theta^2}, \qquad (1)$$

where $\sigma(\theta)$ is the specific surface free energy (angle θ characterizing the surface orientation is measured from the singular orientation).

2. If condition (1) is violated, the surface is thermodynamically unstable, that is, the free energy of the surface decreases when it deviates from planarity. The surface breaks up into facets (segments of singular surface) separated by macroscopic steps, or macrosteps (Fig. 1b). The angle ψ between adjacent segments has an equilibrium value [2, 5], that is, the surface has a saw-tooth shape. The slope of a macrostep AB may be either a nonsingular [5, 6] or a singular surface (faceted macrostep) [7].

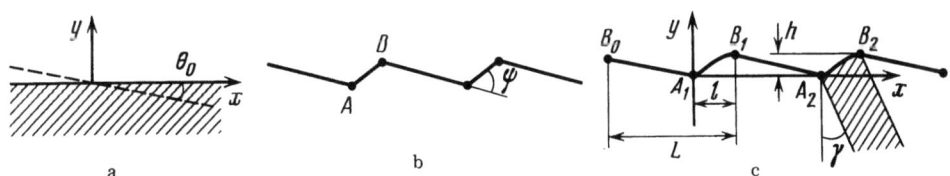

Fig. 1. Surface profile with nearly singular mean orientation of the surface. (a) stable planar from ($\sigma_{eff} > 0$); (b) macrosteps due to thermodynamic instability; (c) macrosteps due to morphological instability ($\sigma_{eff} > 0$).

3. A thermodynamically stable surface ($\sigma_{eff} > 0$) may nevertheless become morphologically unstable because of a temperature or concentration gradient [8, 9]. A small sinusoidal deviation from planarity which increases with time produces surface segments $B_0 A_1$, $B_1 A_2$, etc. (Fig. 1c) with very small slopes θ, that is, with a very low density of elementary steps. The supersaturation (supercooling) necessary for the growth of these nearly flat segments (facets), at a given rate V_n, is considerable and increases with decreasing angle θ. Therefore, if the growth rate V_n (constant along the face) is given, the supersaturation (varying along the face) may be arbitrary: each value of supersaturation corresponds to a specific low density of elementary steps. The source of these steps for the face $B_0 A_1$ is the curved segment $A_1 B_1$ (Fig. 1c) for which slope θ exceeds the initial slope θ_0 (i.e., the density of elementary steps is high). Segments $A_1 B_1$, $A_2 B_2$, etc. can be treated as macrosteps separating the adjacent facets. In contrast to the preceding case (Fig. 1b) a macroscopically stepped surface has a smoothly rounded shape.

The present paper gives an analysis of macrosteps, represented in Fig. 1c, appearing as a result of morphological instability. Since at a given supersaturation the slopes of macrosteps, $A_1 B_1$, $A_2 B_2$, etc., grow faster than the faces, the structure will move to the left at a certain velocity v (and at the same time, upward with the fixed growth rate V). A growing crystal consists of two types of zones, crystallized by the motion of either facets or macrosteps (the zone hatched in Fig. 1c is generated by a moving macrostep $A_2 B_2$). These zones have different impurity concentration levels [10], so that the macrosteps produce in the crystal a number of impurity bands at an angle $\gamma = \arctan(v/V)$ to the growth axis y. The fraction of the crystal volume generated by moving facets equals

$$(V - v \tan \theta_0)(L - l)/VL$$

where l is the macrostep width, and L is the distance between these steps (Fig. 1c). We assume now that $l \ll L$. Two qualitatively different situations can be distinguished:

1. $v \ll V/\theta_0$, so that despite a relatively fast motion of the macrosteps, the crystal is mostly built up by moving faces. Macrosteps leave in the crystal thin impurity striae at considerable angles to the faces.

2. $v \approx V/\theta_0$, so that crystallization mostly takes place at macrosteps. The impurity striae are practically contacting and nearly parallel to the facets.

The tangential velocity v is thus an important parameter influencing the type of inhomogeneity in the crystal. In order to calculate v (for given V, L, θ_0), let us introduce a number of constraints. First, we assume that the macrostep height h is much smaller than its width l, that is, macrosteps are low-angle. The surface geometry is then nearly planar. The surface shape is described by a curve $y = \tilde{y}(x)$, with the deviation $\tilde{y}$ from the initial plane being much smaller than the characteristic size l, and the angle of deviation from the initial plane, $\tilde{\theta}$, being everywhere small compared with unity and approximately equal to $d\tilde{y}/dx$. For a face $\tilde{\theta} = -\theta_0$, and for a macrostep $\tilde{\theta} \gg \theta_0$ (since $\tilde{\theta} \sim h/l$, and $h \approx L\theta_0$ and $L \gg l$). The local normal growth rate $V_n(x)$ for a stepped surface (Fig. 1b) can be written as $V \cdot \cos \tilde{\theta} + v \cdot \sin \tilde{\theta}$, hence an expression linearized in $\tilde{\theta}$:

$$V_n = V + v \frac{d\tilde{y}}{dx}. \tag{2}$$

We calculate velocity v and the steady-state profile of a macrostep, $\tilde{y}(x)$ (in particular, its width l), for nonisothermal binary systems (e.g., a melt with an impurity). However, in order to elucidate the nature of these assumptions, it will be more convenient to begin with the isothermal case.

Growth of a Single-Component Crystal from a Dilute Solution

In a reference frame fixed to macrosteps the concentration of dissolved molecules, $c(x, y)$, is independent of time. Concentration c can be written as the sum of two components: a steady-state distribution for the initial planar surface $c_0(y)$, and a small increment $\tilde{c}(x, y)$ due to the deviation $\tilde{y}(x)$. Component $\tilde{c}$ is mainly associated with an increase in the normal growth rate V_n over an l-long segment, so that the gradient of $\tilde{c}$ is, within an order of magnitude, equal to $\tilde{c}/l$. For the problem under discussion the essential characteristic is the distribution $\tilde{c}(x, y)$ close to a macrostep. We shall neglect the motion of the medium that describes $\tilde{c}(x, y)$ in Laplace's equation. This is justified if D/l (where D is the diffusion coefficient) is much greater than the velocity of motion of the medium. The tangential and normal velocities of the medium are, within an order of magnitude, equal to v and $V\rho_s/\rho$, respectively, where ρ_s and ρ are the crystal and medium densities. If $v \leq V$, the constraint on velocity can be written as

$$Vl/D \ll \rho/\rho_s. \tag{3}$$

The total concentration $c = c_0 + \tilde{c}$ satisfies the standard boundary condition for $y = \tilde{y}(x)$ [2]

$$D\frac{\partial c}{\partial n} = qV_n, \tag{4}$$

where q is the number of molecules per unit volume of the crystal. In the linear approximation in θ the normal derivative dc/dn can be replaced with $\partial c/\partial y$. As the solution is dilute ($c \ll q$), equation (4) neglects the material transfer by the moving medium. Substitution of (2) into equation (4) yields the boundary condition for $\tilde{c}$

$$\frac{\partial \tilde{c}}{\partial y} = \frac{qv}{D}\frac{\partial \tilde{y}}{\partial x}. \tag{5}$$

This condition must hold at the plane $y = 0$ (since we are discussing a linear approximation in $\tilde{y}$ and $\tilde{c}$). The solution of the two-dimensional Laplace equation with boundary condition (5) is known [11]:

$$\tilde{c}(x, y) = \frac{qv}{\pi D}\int \frac{d\tilde{y}}{dx}(x')\ln r\, dx' + \text{const}, \tag{6}$$

where $r = [(x-x')^2 + y^2]^{1/2}$ is the distance from a given point of the medium to a running point x' on the abscissa axis along which the integration is carried out. As $d\tilde{y}/dx$ is much greater for a macrostep than for the faces, integration of (6) must be carried out from 0 to l (see Fig. 1c).

The total concentration c on the slope of a macrostep (segment A_1B_1 in Fig. 1c) deviates from the concentration of saturated solution, c_e, by a quantity which is a function of the local curvature $\tilde{\zeta}$, growth rate V_n, and kinetic coefficient $\beta(\theta)$ [2]. Assume now that the kinetic coefficient is sufficiently large (owing to a high density of elementary steps), so that the contiguous phases are practically in equilibrium. This means that supersaturation $c-c_e$ is related to the curvature $\tilde{\zeta}$ [1-3]:

$$qkT\ln(c/c_e) + \sigma_{\text{eff}}\tilde{\zeta} = 0, \tag{7}$$

where σ_{eff} is given by (1). As the angle of deviation θ is small, the curvature is given by $\tilde{\zeta} = d^2\tilde{y}/dx^2$. We shall assume the relative supersaturation $(c-c_e)/c_e$ to be small, which allows the linearization of equation (7) in $c-c_e$:

$$\frac{d^2\tilde{y}}{dx^2} + \frac{qkT}{\sigma_{\text{eff}}c_e}(c - c_e) = 0. \tag{8}$$

The lower part of the macrostep slope (adjacent to A_1 in Fig. 1c) has a positive curvature; this means, according to equations (7) and (8), that within this segment the solution is undersaturated ($c<c_e$). If the adjacent face B_0A_1 grows with a non-zero rate, the solution close to it is supersaturated. Hence, supersaturation (and macrostep curvature) at the boundary point A_1 becomes zero.

In principle, the growth rate on the facet may be exactly zero, and then the solution at A_1 is undersaturated (consequently, elementary steps do not spread over the facet). We shall show that this situation cannot arise. The difference in concentrations δc between the inflection point on the macrostep (where $c=c_e$) and the boundary point A_1 (where $c \leqslant c_e$) is non-negative. On the other hand, the contribution of $c_0(y)$ to δc is qhV/D (the gradient of c_0 is found from (4)). The contribution to δc due to $\tilde{c}(x, y)$ can be evaluated from relation (6) or directly from equation (5): it is negative and within an order of magnitude is equal to $-qhv/D$. This gives $v \leqslant V$. Consequently, the normal growth rate V_n differs from V only slightly over the whole surface (including the facets); that is, the solution at the faces is supersaturated.

Thus, supersaturation and curvature are zero at the boundary point A_1 (for $x=0$, $y=0$). As a result, supersaturation $c-c_e$ at other points on the surface can be written as $c(x, \tilde{y}) - c(0, 0)$. In the linear approximation $\tilde{c}(x, \tilde{y})$ can be replaced by $\tilde{c}(x, 0)$. We obtain

$$c - c_e = [c_0(\tilde{y}) - c_0(0)] + [\tilde{c}(x, 0) - \tilde{c}(0, 0)]. \tag{9}$$

According to condition (4), the first difference in equation (9) is equal to $qV\tilde{y}/D$. By substituting (9) into equation (8) and using relation (6) for $\tilde{c}$, we obtain an integral equation for the unknown variables v, l and $\tilde{y}(x)$. Introducing dimensionless coordinates $\xi = x/l$ and $\eta = \tilde{y}/h$, and dimensionless combinations λ_1 and λ_2:

$$\lambda_1 = kTVq^2l^2/\sigma_{\text{eff}}Dc_e; \quad \lambda_2 = \lambda_1 v/\pi V. \tag{10}$$

With these notations the integral equation (6) transforms to

$$\frac{d^2\eta(\xi)}{d\xi^2} + \lambda_1\eta(\xi) + \lambda_2 \int_0^1 \frac{d\eta}{d\xi}(\xi') \ln\left(\frac{|\xi - \xi'|}{\xi'}\right) d\xi' = 0. \tag{11}$$

In what follows we assume energy σ_{eff} to be independent of angle, that is, λ_1 and λ_2 to be independent of ξ.

We assumed above that the slope θ_0 of a facet is negligibly small compared with that of a macrostep, h/l. This corresponds to the boundary condition $d\eta/d\xi = 0$ for $\xi = 0$ and $\xi = 1$. In addition, $\eta(0) = 0$ and $\eta(1) = 1$. Let us begin with finding an approximate solution, replacing the unknown function $\eta(\xi)$ by the simplest polynomial which satisfies both the given boundary conditions and equation (11) at point $\xi = 0$ (this demands that the dimensionless curvature $\zeta = d^2\eta/d\xi^2$ equals zero at this point):

$$\eta(\xi) = 4\xi^3 - 3\xi^4. \tag{12}$$

We substitute this polynomial into equation (11), integrating and choosing parameters λ_1, λ_2 such that equation (11) holds for $\xi = 1$ and holds on average for the segment from 0 to 1. We find $\lambda_1 = 15$ and $\lambda_2 = 6$. The left-hand side of equation (11) proves not to exceed unity for all values of ξ, while ζ (the first term of the equation) is approximately equal to 4. The supersaturation (the second and third terms) becomes zero practically simultaneously with curvature ζ (for $\xi = 2/3$). Hence, the solution obtained for $\eta(\xi)$, λ_1, and λ_2 is a fairly good approximation. By using this solution as a first approximation, we can obtain the exact solution by iteration on a computer. To facilitate this, we transform equation (11) by expressing η and $d\eta/d\xi$ in terms of the dimensionless curvature ζ:

$$\zeta(\xi) + \int_0^1 [\lambda_1 K_1(\xi, \xi') + \lambda_2 K_2(\xi, \xi')] \zeta(\xi') d\xi' = 0,$$

where $2K_1 = (\xi - \xi') + |\xi - \xi'|$ and $K_2 = \xi' \ln \xi' + (\xi - \xi') \ln |\xi - \xi'|$. Equation (13) is a generalization of the Fredholm equation of the second kind [12]. Function $\zeta(\xi)$ satisfies the normalization conditions:

$$\int_0^1 \zeta(\xi')\,d\xi' = 0; \quad \int_0^1 (1-\xi')\zeta(\xi')\,d\xi' = 1, \tag{14}$$

the first of which corresponds to zero slope $d\eta/d\xi$ in the end points of the macrostep, and the second to a unit dimensionless height $\eta(1)$.

We divide the range of integration into N segments of equal length, and apply equation (13) to N points (right-hand bounds of these segments). Together with equations (14) we obtain a system of N+2 equations for N+2 unknowns (λ_1, λ_2, and N values of curvature). By representing each unknown variable as a sun of its approximate value (found above) and a small correction term, we obtain N+2 linear equations for these corrections. After solving this system, we shall use the solution thus obtained for the next iteration. Calculations were carried out for N=20, 40, and 60; Simpson's formula was used for integration. A relative accuracy of 0.001 was achieved after two or three iterations. The values of the parameters obtained were λ_1=16.37 and λ_2=5.73, which are not far different from the first approximation. The calculated dimensionless curvature $\zeta(\xi)$ and the macrostep shape $\eta(\xi)$ are plotted in Fig. 2. Plotted for comparison are the curves corresponding to the first approximation (12). The actual profile function of a macrostep is $\tilde{y}(x) = h\eta(x/l)$. From the relations (10) we obtain for the calculated λ_1, λ_2 the required tangential velocity v and macrostep width l:

$$v/V = 1.1, \quad l = 4.05\,(Dc_e \sigma_{\text{eff}}/kTVq^2)^{1/2}. \tag{15}$$

The width l is of the order of tens of microns. The impurity striae of macrosteps are at an angle of 48° to the growth axis.

The theory as developed above is based on the assumption of phase equilibrium at the macrostep slope (eq. (7)). In the general case the supersaturation $c-c_e$ calculated from (8) must be supplemented with a kinetic correction V_n/β (where β is a kinetic coefficient). We have earlier determined $c-c_e \approx qhV/D$. The criterion of validity of equation (8) is the smallness of the kinetic correction:

$$\beta qh \gg D. \tag{16}$$

A theory is thus constructed for the case of low supersaturation: the supersaturation difference along a macrostep is much smaller than c_e, and the kinetic supersaturation is still lower. The main conclusion is that crystallization takes place mostly on faces.

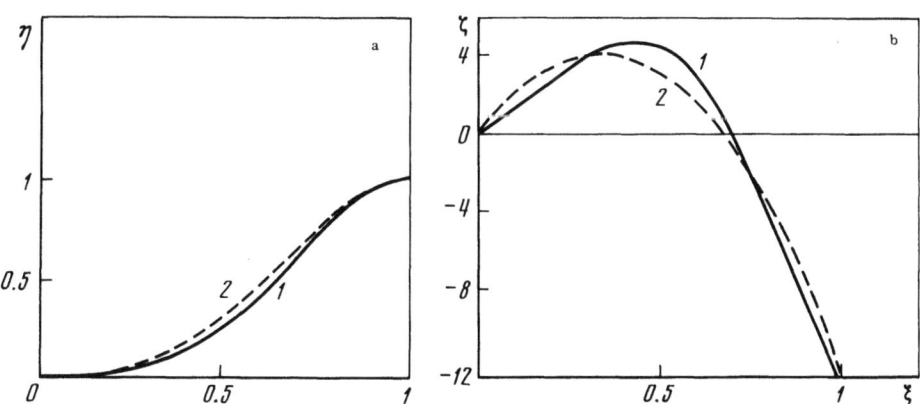

Fig. 2. (a) Dimensionless profile $\eta(\xi)$, and (b) curvature $\zeta(\xi)$ of a macrostep. (1) computer simulation; (2) first approximation.

A situation often discussed in the literature is the opposite one, when supersaturation is reasonably high and constant along the surface; that is, diffusion in the medium is fairly rapid, and an inequality inversed with respect to (16) becomes valid. In this case the behavior of a stepped surface is essentially different [2, 13, 14].

Nonisothermal Binary System

The state of a medium is characterized by its supersaturation $\Delta T = T_e(c) - T$ where $T_e(c)$ is the temperature of phase equilibrium along a planar surface for a given concentration c of one of the components. A planar surface is morphologically unstable if the gradient of equilibrium temperature $\Gamma = \partial T_e(c)/\partial y$ exceeds the temperature gradient $\bar{G}$ averaged over two phases [8, 9]; that is, if

$$\Gamma > \bar{G} = (\varkappa G + \varkappa_s G_s)/(\varkappa + \varkappa_s), \tag{17}$$

where $\varkappa$ and $\varkappa_s$ are the thermal conductivities of the medium and crystal, respectively, and G and G_s are the corresponding temperature gradients $\partial T/\partial y$.

It is convenient to introduce, instead of concentration c, the mass fraction f of the first component (i.e., the ratio of its mass per unit volume to the total density ρ of the medium). Then, neglecting thermal diffusion, we can write the mass flux I_1 of the first component in the form [15]

$$I_1 = f\mathbf{I} - \rho D \nabla f, \tag{18}$$

where D is the interdiffusion coefficient, and I is the total mass flux. As before, we represent $f(x, y)$ as $f_0(y) + \tilde{f}(x, y)$, where $f_0(y)$ is the steady-state distribution for the initial planar surface. Distribution $\tilde{f}(x, y)$ approximately satisfies Laplace's equation if condition (3) is satisfied. From the continuity condition, the normal components of fluxes I and I_1 on the interface are equal to $\rho_s V_n$ and $f_s \rho_s V_n$ (here f_s is the mass fraction of the first component in the crystal). We linearize this boundary condition by substituting V_n from (2) and assuming $\rho, \rho_s,$ and D to be only slightly dependent on the composition of the medium and the crystal. We assume also that the differential distribution coefficient df_s/df is not too large compared with unity (so that the inequality obtained by multiplying the left-hand side of inequality (3) by this coefficient is also satisfied). For $\tilde{f}(x, y)$ we obtain a boundary condition (for y=0)

$$\frac{\partial \tilde{f}}{\partial y} = \frac{\rho_s v}{\rho D} (f_s - f) \frac{\partial \tilde{y}}{\partial x}. \tag{19}$$

This condition differs from the similar condition (5) only in a constant factor. Therefore relation (6) in which q is replaced by $(\rho_s/\rho)(f_s - f)$ will be valid for $\tilde{f}$.

The temperature field $T(x, y)$ in each of the phases can also be written in the form $T_0(y) + \tilde{T}(x, y)$, with $\tilde{T}(x, y)$ satisfying the Laplace equation not only within a step but also over the whole macroscopically stepped surface (since the thermal diffusion coefficient is much greater than D). By virtue of the linearity of the problem we can represent $\tilde{T}$ as the sum $\tilde{T}_1$ plus $\tilde{T}_2$, where the first term is associated only with the deviation of the surface from planarity (for $V_n = V$) and the second only with the additional normal growth rate $V_n - V$ (see eq. (2)). The value of $\tilde{T}_1$ was found in refs. 8, 9; considered simultaneously with $T_0(y)$ it gives the following component of temperature at the interface:

$$T_0(\tilde{y}) + \tilde{T}_1(x, 0) = \bar{G}\tilde{y} + \text{const}, \tag{20}$$

where the mean gradient $\bar{G}$ is given by (17). The component $\tilde{T}_2$ in both phases satisfies the Laplace equation, and at the surface (i.e., for y=0 in the linear approximation) the functions $\tilde{T}_2(x, 0)$ are identical in both phases. If $\tilde{T}_2(x, y)$ is expanded into the Fourier series in x, then each term of the series (its general form being $\exp(i\lambda x - |\lambda y|)$) is an even function of y. Consequently, the distribution of $\tilde{T}_2$ is symmetrical with respect to plane y=0. It is therefore sufficient to analyze $\widetilde{T}_2$ only in the ambient medium ($y \geq 0$). The heat balance equation involves the derivative $\partial \widetilde{T}_2/\partial y$ for the medium and the corresponding derivative for the crystal, which differ only in their sign. Thus, the heat balance equation is written in the form

$$-(\varkappa + \varkappa_s)\frac{\partial \widetilde{T}_2}{\partial y} = Q(V_n - V), \tag{21}$$

where Q is the crystallization heat per unit volume of the crystal. Substitution of V_n from (2) into equation (21) yields a boundary condition for $\widetilde{T}_2$ in the medium, very similar to (5) or (19). Consequently, $\widetilde{T}_2$ is defined by equation (6) in which q/D must be replaced by $-Q/(\varkappa + \varkappa_s)$.

The following condition of phase equilibrium, relating supersaturation ΔT to curvature $\widetilde{\zeta}$ [1-3], is valid on the slope of a macrostep:

$$Q\Delta T/T + \sigma_{eff}\, \widetilde{\zeta} = 0. \tag{22}$$

We have established above that supersaturation at a boundary point A_1 (see Fig. 1c) is zero, so we can write

$$\Delta T = [T_e(f(x,\widetilde{y})) - T_e(f(0,0))] - [T(x,\widetilde{y}) - T(0,0)]. \tag{23}$$

Expanding T_e into a series in increment f, and $f_0(y)$ into a series in $\widetilde{y}$, taking into account that the equilibrium temperature gradient is

$$\Gamma = \frac{dT_e}{df}\frac{\partial f_0}{\partial y}, \tag{24}$$

and using equation (20), we can rewrite the expression for supersaturation in the following form:

$$\Delta T = \Gamma\left\{\widetilde{y} + \left(\frac{\partial f_0}{\partial y}\right)^{-1}[\widetilde{f}(x,0) - \widetilde{f}(0,0)]\right\} - [\overline{G}\widetilde{y} + \widetilde{T}_2(x,0) - \widetilde{T}_2(0,0)]. \tag{25}$$

The composition gradient of the medium will be obtained from the boundary condition on the initial planar surface. By finding the normal component of flux I_1 from equation (18) and setting it equal to $f_s\rho_s V$, we obtain

$$\frac{\partial f_0}{\partial y} = \frac{\rho_s V}{\rho D}(f_s - f). \tag{26}$$

Substituting $\widetilde{\zeta} = d^2\widetilde{y}/dx^2$, and ΔT from (24) into equation (22), and taking into account that $\widetilde{f}$ and $\widetilde{T}_2$ are described by equation (6) when the coefficient q/D is replaced by $(\rho_s/\rho D) \times (f_s - f)$ and $-Q/(\varkappa + \varkappa_s)$, respectively, after the introduction of dimensionless coordinates $\xi = x/l$ and $\eta = \widetilde{y}/h$ the integral equation obtained coincides with equation (11) if we denote

$$\lambda_1 = Q(\Gamma - \overline{G})l^2/\sigma_{eff}, \quad \lambda_2 = (\lambda_1 v/\pi V)[\Gamma + QV/(\varkappa + \varkappa_s)]/(\Gamma - \overline{G}). \tag{27}$$

Consequently, the problem for a nonisothermal binary system is completely reduced to the preceding elementary case. The dimensionless curvature $\zeta(\xi)$ and macrostep profile $\eta(\xi)$ are plotted in Fig. 2, and the parameters are $\lambda_1 = 16.37$ and $\lambda_2 = 5.73$. Relations (26) give the normalized tangential velocity and the macrostep width:

$$v/V = 1.1(\Gamma - \overline{G})/[\Gamma + QV/(\varkappa + \varkappa_s)]; \quad l = 4.05[\sigma_{eff} T/Q(\Gamma - \overline{G})]^{1/2}.$$

If the temperature gradient $\overline{G}$ and the heat released, QV, can be neglected, these formulas transform to (15).

Let us make some numerical estimates for silicon growth from a melt containing an impurity. Low-slope macrosteps are observed at the initial stage of instability of the planar crystallization interface when $\Gamma - \overline{G}$ is not too large (as $\Gamma - \overline{G}$ increases, the interface morphology gets more complicated and transforms into a dendritic structure). For the purpose of evaluation we set $\Gamma-\bar{G}=\bar{G}/2$, $\bar{G}=50$ K cm^{-1}, V=3·10^{-3} cm s^{-1}, σ_{eff}=200 erg cm^{-2} [3], and use known values for constants: Q=4.1·10^{10} erg cm^{-3}, $\varkappa + \varkappa_s$=7.5·10^6 erg cm^{-1} s·K, and T=1683 K. This yields l=23 μm and v/V=0.3. Impurity striae due to macrosteps are at an angle arctan (v/V)≈17° to the growth axis. Figure 3 is a photograph of a silicon crystallization interface with macrosteps, and Fig. 4 is a longitudinal cross-section of a similar ingot after etching. Figure 4 shows conventional impurity striae (corresponding to the averaged shape of the interface), and against this background shows narrow traces of macrosteps at an angle of about 20° to the boule axis. The period of the macrostep structure L≈170 μm, that is, the condition $l \ll L$ is satisfied. The macrostep height is h=Lθ_0 ≈9 μm (mean slope θ_0≈3°). In this case the condition of gentle slope, h≪l, holds if only qualitatively. Condition (3) is satisfied because the left-hand side of this expression for D≈5·10^{-5} cm^2 s^{-1} is of the order of 0.1. Finally, we check the condition of phase equilibrium: the supersaturation difference at a macrostep $(\Gamma-\bar{G})$h must be much greater than the kinetic supersaturation V/β. The kinetic coefficient β for a stepped surface is proportional to the slope angle θ, and approximately equals 20 cm s^{-1} K^{-1} [16]. The equilibrium condition holds very well, since $(\Gamma-\bar{G})\beta h/V \approx$ 150.

We find therefore that in some cases (growth from the melt, liquid phase epitaxy, etc.) morphologically unstable vicinal surfaces develop a macroscopically stepped structure. However, the growth rate of faceted segments of the crystallization interface remains close to the mean growth rate, so that crystallization at the facets is predominant and macrosteps only act as sources of elementary steps for faceted growth.

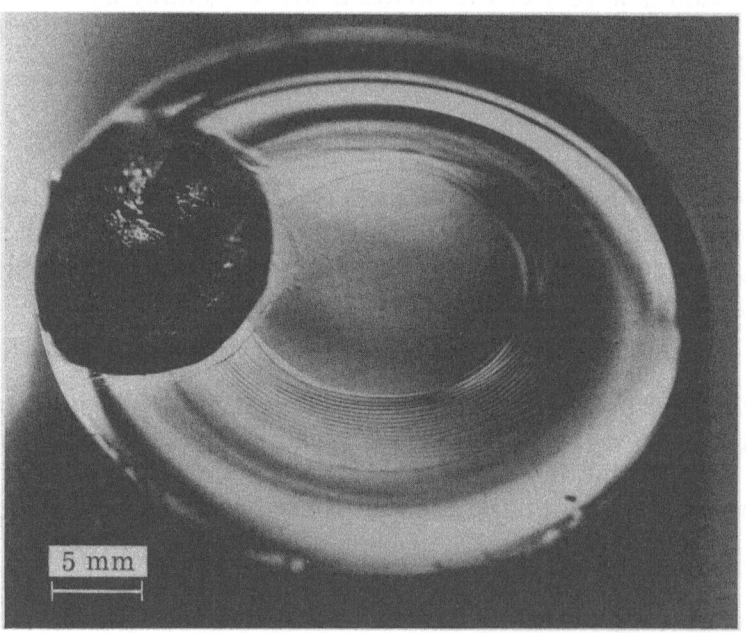

Fig. 3. Silicon crystallization interface with macrosteps (at the center a face 1.7 cm diameter, and to the left a solidified drop of melt).

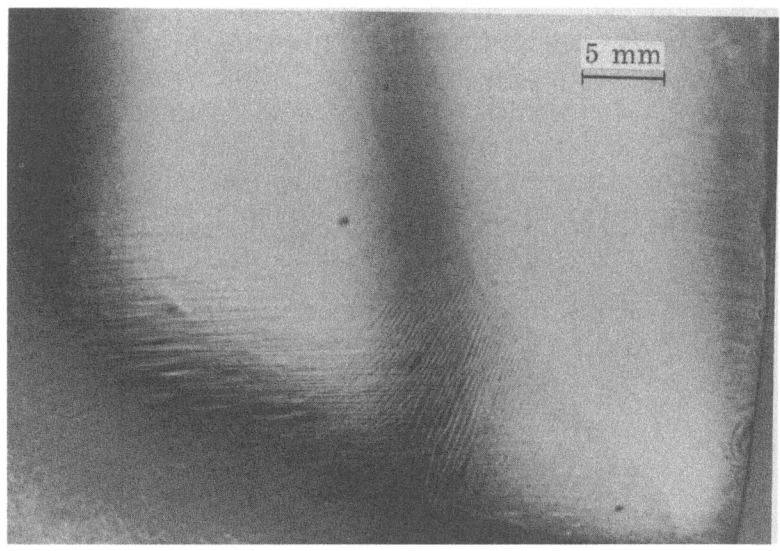

Fig. 4. A longitudinal polished section of a germanium boule with traces of macrosteps.

Literature Cited

1. C. Herring. Surface tension as a motivation for sintering. In: Physics of Powder Metallurgy. New York; Paris: McGraw-Hill Co., 143-178 (1951).
2. A. A. Chernov. The spiral growth of crystals. Soviet Physics Uspekhi, 4, 115-148 (1961).
3. V. V. Voronkov. On angular dependence of free surface energy of crystals. Kristallografiya, 12, 831-839 (in Russian) (1967).
4. V. V. Voronkov. Crystal surface structure in the Kossel model. In: Growth of Crystals, 10, ed. N. N. Sheftal', Consultants Bureau, New York, 1-17 (1976).
5. I. M. Lifshitz and A. A. Chernov. Macroscopic steps on crystal surface. Kristallografiya, 4, 788-791 (in Russian) (1959).
6. A. A. Chernov and E. D. Dukova. Effect of supersaturation on the step morphology of crystal surface and growth rate. Kristallografiya, 5, 655-661 (in Russian)
7. A. A. Chernov and S. I. Budurov. On growth rates of macroscopic steps. Growth of facets at end faces of steps. Kristallografiya, 9, 388-395 (in Russian) (1964).
8. W. W. Mullins and R. F. Sekerka. Stability of a planar interface during solidification of a dilute binary alloy. J. Apply. Phys. 35, 444-451 (1964).
9. V. V. Voronkov. Conditions of formation of cellular structure at the crystallization interface. Fiz. Tvcrd. Tcla, 6, 2984-2988 (in Russian) (1964).
10. V. V. Voronkov. Dopant uptake factor in relation to growth rate and surface inclination. In: Growth of Crystals, 11, ed. A. A. Chernov, Consultants Bureau, New York, 364-373 (1979).
11. H. F. Carslaw and J. C. Yaeger. Conduction of Heat in Solids. Clarendon Press, Oxford, (1959).
12. M. L. Krasnov, A. I. Kiselev and G. I. Makarenko. Integral Equations. Nauka, Moscow (in Russian) (1976).
13. P. Bennema and R. van Rosmalen. Simulation of modes of vibrations in trains of steps. In: Growth of Crystals, 11, ed. A. A. Chernov, Consultants Bureau, New York, 160-165 (1979).

14. A. A. Tikhonova. Formation of macroscopic steps on vinical growth surface of germanium layers. in: Growth of Crystals, 11, ed. A. A. Chernov, Consultants Bureau, New York, 155-159 (1979).
15. L. D. Landau, and E. M. Lifshitz. Mechanics of Continuous Media. GITTL, Moscow (1953).
16. V. V. Voronkov. Supersaturation at a face appearing on a rounded crystallization interface. Kristallografiya, 17, 909-917 (in Russian) (1972).

PECULIARITIES OF MELT GROWTH OF CRYSTALS WITH DIFFERENT ENTROPIES OF MELTING

G. A. Alfintsev and D. E. Ovsiyenko

The Institute of Metal Physics of the UkSSR Academy of Sciences, Kiev

Introduction

In order to grow crystals with controlled structure and properties it is necessary to have a clear understanding of the growth mechanism and growth morphology, because they determine to a large extent the formation of defects, the distribution of impurities and, consequently, the structure-sensitive properties of the crystals. These aspects of growth have attracted much interest in recent years.

At the present time several growth mechanisms have been suggested [1-8], the theory of stability of growth forms has been developed [9-12], and criteria for interface roughness have been proposed [1-15].

According to the current concepts, different mechanisms may be operative in crystal growth: the formation of two-dimensional nuclei and their subsequent growth; dislocation mechanisms; and the random attachment of particles at the interface (the normal mechanism).

The dependence of growth rate on the supercooling at the growth interface is specific for each growth mechanism. The atomic structure of the crystal-melt interface determines which mechanism occurs: the two-dimensional nucleation mechanism is typical for atomically smooth interfaces, the dislocation mechanism for surfaces containing steps formed by emerging screw dislocations, and the normal mechanism for rough surfaces. In addition to the rough interface model, a phenomenological description of the interface, involving the concept of a diffuse interface, is used [8, 13, 14]; "blurring" is characterized by the number of atomic layers within which the crystal gradually transforms to the melt. From theoretical arguments the structure of the interface depends on the entropy of melting. Thus, it has been shown by an analysis of the two-level model of the interface [15] that roughness of the interface between the crystal and its melt is determined by a criterion $\alpha = \xi L/kT_0$, where ξ is the fraction of the total number of nearest neighbor atoms located in the newly formed layer, L is the latent heat of melting, T_0 is the melting point, and k is the Boltzmann constant. If $\alpha > 2$, the interface is smooth, and consequently the formation of steps is required for growth; for $\alpha < 2$ the interface is rough and nucleation is not required for growth. It has been found by analyzing a multi-level model of the interface that the smooth-to-rough-interface transition occurs at $L/kT_0 \approx 3.5$, that is, at high values of the entropy of melting, ΔS [16]. The structure of the interface affects not only the growth kinetics but the growth forms as well. Crystals with atomically smooth surfaces and crystals having stepped surfaces show higher growth rate anisotropy and have faceted habits, while crystals with rough surface grow in rounded shapes.

An analysis of the interface structure shows that the layered growth of atomically smooth surfaces may require higher supercooling than the normal growth of a rough interface because in the latter case the formation of steps is not necessary. However, a different point of view on the conditions under which the various growth mechanisms occur has been presented in a number of papers treating a diffuse interface [8, 13, 14, 17]. For example, according to ref. 13, crystals of all materials grow by a layer mechanism below a certain critical supercooling $\Delta T_{crit} = g\tilde{\sigma} V T_0 / aL$ (where g is the diffusivity parameter of the interface, σ is the surface tension at the solid-melt interface, V is the molar volume of the solid phase, and a is the lattice parameter), and by the normal mechanism above ΔT_{crit}.

Because of uncertainty in the value of the interface diffusivity parameter g, this theory cannot predict the value of ΔT_{crit}. It can be evaluated for specific compounds only on the basis of experimentally measured dependence of growth rate on supercooling $v(\Delta T)$ showing a kink corresponding to ΔT_{crit}. An analysis of both theoretical models demonstrated that a transition from layered to normal growth, on both rough and diffuse interfaces, occurs at a certain ΔT [14, 16, 18]. Testing the existing models is difficult because no direct method of observing the atomic structure of an interface is available, so that the main source of information on the actual mechanisms of crystal growth is the experimental data on kinetics and forms of growth. Many experimental papers are devoted to these problems, but they contain few data suitable for testing the existing theories. Moreover, the theories themselves are unable to predict many features found in actual crystal growth. All of this constitutes a compelling reason for further investigation.

The present paper describes and generalizes the results of experimental work carried out in the Institute of Metal Physics of the Ukr. SSR Academy of Sciences on the study of crystal growth of materials with different values of the entropy of melting.

Experimental Data on Kinetics and Forms of Growth

Selection of Materials and Experimental Techniques.
We shall discuss results concerning the kinetics and forms of growth for crystals of organic compounds and metals as functions of entropy of melting. This choice is determined by their low melting points which make it possible to maintain and accurately measure the temperature, by the possibility of observing the crystallization interface, by chemical purity of the materials, and some other factors.

The study was carried out on thin plane-parallel layers [19-21] which allow efficient heat removal and enable one to observe the growth of individual crystals having a desired orientation. The technique consists in placing a 0.1 - 0.8 mm thick object between two glass or quartz plates heated either by liquids or by a flat electric heater. Thermostats maintained the temperature to an accuracy of ±0.01°C. The temperature at the crystallization interface was measured with a thin (30-50 μm diameter) thermocouple. This method of temperature measurement is not valid for very small and defect-free (perfect) crystals, but for small degrees of supercooling or low growth rates [19], the interface temperature was assumed to be equal to the thermostat temperature.

Stable Growth Forms and Morphology of the Crystallization Interface.
Crystal growth forms are known to depend on the entropy of melting [22, 23]. Crystals of salol, gallium, piperonal (Fig. 1) and benzophenone, having a high entropy of melting ($L/kT_0 > 2$), show clearly pronounced faceting, which is maintained at relatively high supercooling. The faces are normally smooth, although spirals were sometimes observed (Fig. 1a). But in cyclohexanol, succinonitrile, cyclohexane (Fig. 2), camphene, and carbon tetrabromide the crystals are rounded, that is, no faceting occurs. This form is maintained

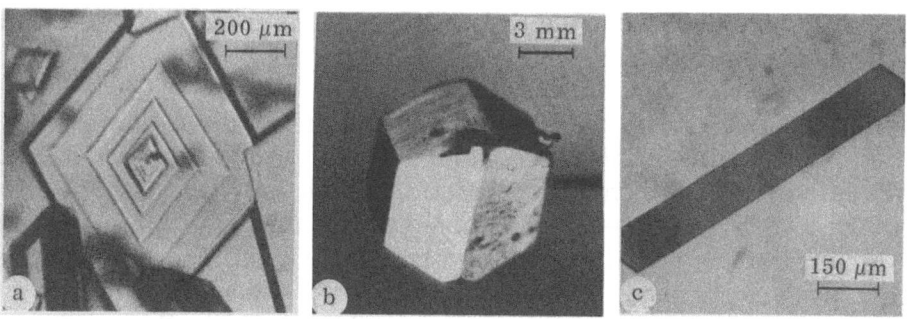

Fig. 1. Faceted growth forms of crystals. (a) salol, (b) gallium, (c) piperonal.

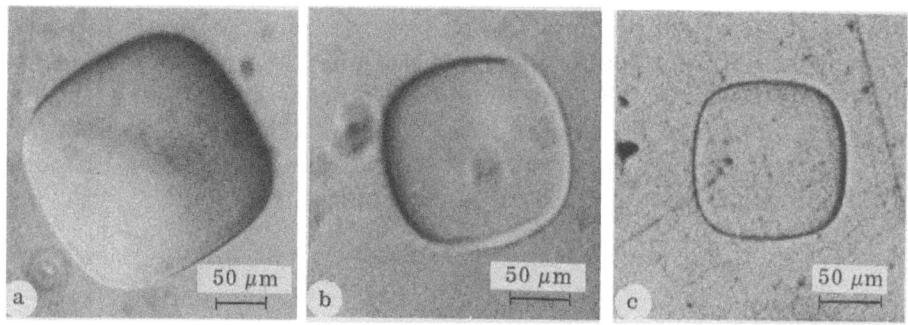

Fig. 2. Rounded growth forms of crystals. (a) cyclohexanol, (b) succinonitrile, (c) cyclohexane.

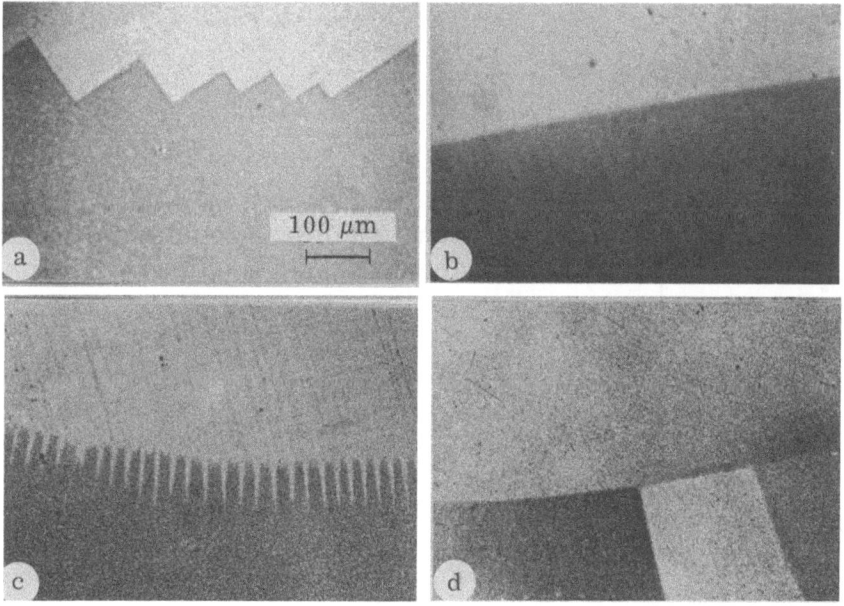

Fig. 3. Forms of crystallization interface. (a) bismuth, $\Delta T = 1°C$; (b) tin at a temperature close to T_0; (c) tin, $\Delta T \approx 0.03°C$; (d) zinc.

only until a certain size of a crystal is reached, when it becomes unstable; this critical size depends on the degree of melt supercooling.

The growth of individual crystals was not observed in metals, but some peculiarities in the morphology of the crystallization interface were noticed. Thus, the crystallization interface in bismuth reveals a rather pronounced faceting which is maintained at high growth rates (Fig. 3a). Contrary to this, no faceting was observed in tin, zinc (Fig. 3b, c and d), and cadmium. When the gradient is positive, the crystallization interface in equilibrium with the melt is smooth. If, however, a negative gradient sets up in the melt (the melt is supercooled), the smooth crystallization interface becomes unstable, and needles protruding into the melt are formed (Fig. 3c); the needle tips are not faceted.

Kinetics of Crystal Growth. The kinetics of crystal growth were studied at relatively low supercooling because under these conditions the role of surface processes becomes dominant, and it is possible to observe the steady-state forms of crystal growth. Combined with the kinetics studies, these observations yield complete information on the growth mechanism. Where possible, experiments were made for similar steady-state shapes of crystals whose size grew from 20 to 500 μm. The data obtained for a large number of individual salol crystals are plotted in Fig. 4 [19, 22]. The graph shows a considerable scatter due to large variations in crystal perfection. It was found that the minimum values of growth rate (in m s^{-1}) are described by an exponential curve (see Fig. 4, curve 1)

$$v = 5 \cdot 10^{-6} \exp(-1100/T\Delta T),$$

which demonstrates that the two-dimensional nucleation mechanism occurs. A straight line, 2, fitted to the maximum values is described by the equation

$$v = 4.6 \cdot 10^{-7} \Delta T^2,$$

which is in agreement with the dislocation growth mechanism [24].

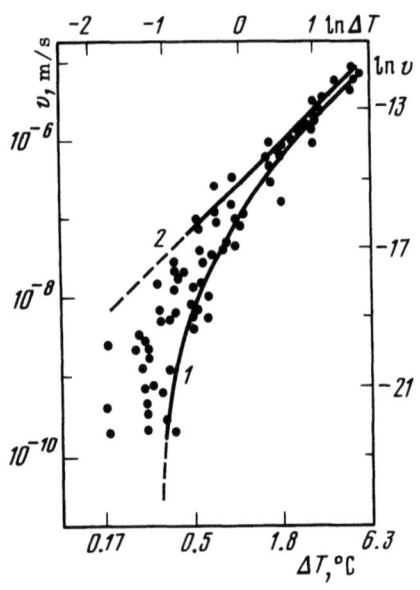

Fig. 4. Growth rate of salol crystals as a function of supercooling. (1) minimum growth rates, (2) maximum growth rates.

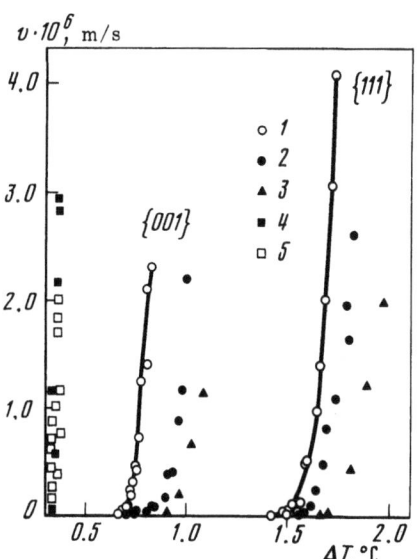

Fig. 5. Growth rate of (001) and (111) faces of pure gallium crystals and gallium alloy crystals as functions of supercooling. (1) pure gallium; (2) Ga + 0.01 wt. % In; (3) Ga + 0.1 wt. % In; (4) Ga + 0.01 wt. % Ag; (5) strained pure gallium crystals.

We find from Fig. 5 that gallium crystal growth is described by exponential kinetics [21, 23, 25, 26]:

$$v_{[001]} = 2.38 \cdot 10^4 \exp(-5500/T\Delta T)$$

and

$$v_{[111]} = 1.05 \cdot 10^5 \exp(-12\,000/T\,\Delta T).$$

From these relations one can calculate the specific surface energies. They were found to be $3.8 \cdot 10^{-7}$ erg cm^{-1} for a step on (001) faces and $5 \cdot 10^{-7}$ erg cm^{-1} for steps on (111). The threshold values of supercooling, below which the growth rates are negligibly small, are 0.58°C on (001) faces and 1.45°C on (111) [26]. It was also found that strain in a crystal due to indentation or bending drastically increases the growth rate and results in a replacement of the two-dimensional nucleation growth mechanism by the dislocation mechanism (see Fig. 5). This is confirmed by the parabolic dependence of growth rate on ΔT for strained crystals:

$$v = 4.2 \cdot 10^{-3} \Delta T^2.$$

A similar curve was also obtained for piperonal and bismuth.

The growth kinetics for imperfect benzophenone crystals is described by a parabolic relation

$$v = 4.68 \eta^{-1} (\Delta T/T)^2,$$

and that of perfect crystals by an exponential relation

$$v = 3.9 \cdot 10^{-3} \eta^{-1} \exp[-6400/T\,(\Delta T T/T_0)],$$

where η is the melt viscosity in poise.

The temperature dependence of growth rate for low-entropy cyclohexanol in the supercooling range 0.005-0.1°C is linear [27-29], with a small scatter of experimental values (Fig. 6). Along the [100] and [110] axes these relations have the form

$$v_{[110]} = 8 \cdot 10^{-7} \Delta T \quad \text{and} \quad v_{[100]} = 8.9 \cdot 10^{-7} \Delta T.$$

It seems likely that these low values of kinetic coefficients can be explained by the effect of soluble impurities.

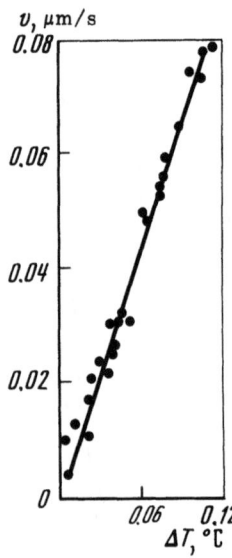

Fig. 6. Growth rate of cyclohexanol crystals along [110] as a function of supercooling.

The velocity at which the crystallization interface in tin advances for ΔT values as small as 0.06°C is quite high (approximately $2 \cdot 10^{-4}$ m s^{-1}). For this reason the type of function $v(\Delta T)$ could not be determined [25, 29, 30].

Effect of Soluble Impurities on Growth Kinetics. These effects were measured in gallium and tin [31-33]. A mixture of 0.01 weight percent of silver to gallium results, similarly to straining the crystals, in a sharp increase in growth rate (see Fig. 5) and in replacement of the nucleation growth mechanism by the dislocation mechanism. In contrast to the growth of strained pure gallium crystals, the addition of silver increasingly slows down the growth the higher is the silver concentration. Unlike silver, indium doping does not change the character of the $v(\Delta T)$ function but only inhibits growth (see Fig. 5). In addition, this dopant somewhat increases the supercooling threshold; however, this effect is eliminated by straining the crystals. This shows that the supercooling threshold is caused by the nucleation mechanism and not by impurities. In the case of alloys the surface energy values were found to be the same as for pure gallium. Hence, the decrease in growth rate is mostly caused by the difficulty of gallium atoms passing from the melt into the crystal through the impurity accumulated at the growth interface. Crystallization of the trapped impurity (silver) in Ga-Ag proceeds simultaneously with crystallization of the principal component (gallium) and is accompanied by segregation of a new phase $AgGa_2$ and the generation of dislocation-type defects. In contrast to this, indium-rich regions in Ga-In alloy remain liquid in the solidified crystal and do not produce stresses required to generate dislocations. Copper mixtures may produce different effects depending on the copper concentration. Small concentrations (of the order of 10^{-3} at.%), not exceeding the maximum solubility of copper in solid gallium, behave as for indium, while the effects similar to those observed with silver doping are produced at high copper concentrations when $CuGa_2$ segregates. A comparison of the growth kinetics in strained alloy crystals shows that, at the same dopant concentration, copper slows the growth rate more effectively than indium. This can be explained by the difference in distribution coefficient k of these impurities in gallium ($k_{Cu} < k_{In}$).

We also studied the effect of 1 at.% indium, tellurium, and bismuth added to tin [33]. However, the method used in this case was not as sophisticated as in the case of gallium, so that only qualitative data were obtained. The method was essentially as follows: after melting most of the thin layer of alloy with the planar heater and achieving a predetermined temperature at which the crystallization interface was stationary, the displacement of the interface was observed in the course of cooling the cell at a constant rate. The results are plotted in Fig. 7. We see that as the crystallization interface in pure tin moves away from the initial equilibrium position, the interface advancement rate increases monotonically, and then rises abruptly at the end of the sample. In the case of a positive temperature gradient the crystallization interface is smooth (see Fig. 3b). In the case of negative gradient (supercooled melt), the solidification of the last fragment of the sample is accompanied by destabilization of the planar interface and the formation of protrusions (see Fig. 3c). This improved the removal of heat and thus increased the growth rate.

A mixture of 1 at. % tellurium does not affect initial stages of growth, but later the growth rate in the alloy becomes higher than that in tin, because of an earlier breakdown of stability of the planar interface. The situation is qualitatively different if the dopants are indium and bismuth. As in the case of pure tin, the growth first increases with time, then diminishes, and after a certain time interval sharply increases again. In addition, in the initial stages of the rise and fall of the growth rate the crystallization interface remains planar; the abrupt increase is accompanied by a loss of stability and the formation of protrusions. The slow-down in growth is explained by accumulation of indium and bismuth impurities at the crystallization interface, and the loss of stability and the associated increase in growth rate are caused by constitutional supercooling depending on the distribution coefficient. The ratio of distribution coefficients ($k_{Bi} < k_{In} < 1$) explains the larger effect of bismuth doping compared with indium doping. No such effects are found in Sn-Te alloys, because $k_{Te} > 1$.

According to Cahn's theory [8, 13], the layer mechanism is expected to be replaced by the normal growth mechanism above a critical supercooling; in some cases we found it possible to evaluate these thresholds on the basis of the data obtained. According to these estimates, $\Delta T_{crit} = 12°C$ for the gallium (111) face. The growth rate of gallium crystals at such supercooling values is very high and, in practice, a check of this estimate is possible.

In the case of low-entropy materials we could not detect faceting, threshold supercooling, or deviations from linear kinetics, that is, any features which would be indicative of a transition from normal to layered growth, induced by diminishing the supercooling.

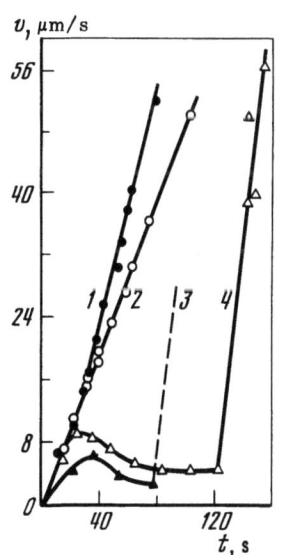

Fig. 7. Effect of soluble impurities on the growth rate of tin crystals
(1) Sn + 1 at. % Te;
(2) Sn;
(3) Sn + 1 at. % Bi;
(4) Sn + 1 at. % In.

The results of the experiments are summarized in Table 1. It shows that the materials investigated can be divided into two basic groups according to their crystal morphology and growth kinetics. The first group comprises salol, benzophenone, gallium, bismuth, and other materials with high values of L/kT_0 (exceeding 2.2). Crystals of these materials show well-pronounced faceting, and their growth is described by exponential or parabolic curves, reflecting a layer growth mechanism.

A typical representative of the second group, with low entropy of melting, is cyclohexanol whose crystals are rounded and whose growth kinetics is linear; i.e., the growth mechanism is normal. The group comprises succinonitrile [34], camphene, carbon tetrabromide, mercury, zinc, tin and other materials with low values of L/kT_0 (below 1.75). No features indicative of layer growth are observed for these materials. The transition from one growth mechanism to the other takes place according to the experimental data obtained, over a narrow range of L/kT_0 values: from 1.72 (tin) to 2.2 (gallium).

TABLE 1. Growth Forms and Growth Kinetics of Crystals of Materials Investigated

Material	L/kT_0	Growth rate, m s^{-1}
Faceted growth form		
Tristearin	63	—
Salol	7	$v_p = 5 \cdot 10^{-6} \exp(-1100/T\Delta T)$
		$v_d = 4.6 \cdot 10^{-7} \Delta T^2$
Benzophenone	6.7	$v_p = 3.9 \cdot 10^{-3} \eta^{-1} \exp(-6400/T\Delta T)(T/T_0)$
		$v_d = (4.68/\eta)(\Delta T/T)^2$
Piperonal	5	$v_d = 1.6 \cdot 10^{-8} \Delta T^{1.75}$
Bismuth	2.42	$v_d = 1.23 \cdot 10^{-1} \Delta T^{1.7}$
Gallium	2.2	$v_{p(001)} = 2.38 \cdot 10^4 \exp(-5500/T\Delta T)$
		$v_{p(111)} = 1.05 \cdot 10^5 \exp(-12000/T\Delta T)$
		$v_{d(111)} = 4.2 \cdot 10^{-3} \Delta T^2$
Rounded crystals		
Tin	1.72	—
Succinonitrile	1.44	$v_{<110>} > 0.01 \Delta T$
Mercury	1.2	—
Zinc	1.16	—
Camphene	1.15	$v_{<100>} > 0.015 \Delta T$
Cyclohexane	1.15	—
Carbon tetrabromide	0.79	—
Cyclohexanol	0.71	$v_{<110>} = 8.0 \cdot 10^{-7} \Delta T$
		$v_{<100>} = 8.9 \cdot 10^{-7} \Delta T$

Note. v_p = growth rate of perfect crystals, v_d = growth rate of deformed crystals.

Trapping of Impurities

It has been noted that materials with different entropies of melting differ not only in the kinetics and morphologies of growth, but also in the characteristics of impurity distribution in the crystals. The experiments with gallium containing indium as dopant (above 0.01 wt. %) have shown that no visible inhomogeneities are found in slowly growing unstrained crystals. However, if growth rates are increased to 10^{-5}-10^{-4} m s^{-1}, regular growth breaks down. This is observed as trapping of the impurity-enriched layer contiguous with the growing face (Fig. 8a). The trapped-impurity layers are parallel to the growth face, and the frequency of their formation increases as the growth rate and impurity concentration rise. In the course of crystal growth the trapped layers remain liquid and do not produce stresses which generate dislocations. Therefore, the mechanism of perfect crystal growth remains unaltered in the presence of indium, and growth is not accelerated, as is the case for strained crystals. Similar behavior is observed when silver and copper are added (Fig. 8b), but in this case the trapped impurity-enriched regions solidify almost simultaneously with the matrix. A new phase, $AgGa_2$ or $CuGa_2$, is segregated and leads to the generation of dislocations which accelerate growth and cause replacement of the nucleation mechanism by the dislocation growth mechanism.

In contrast to gallium crystals, tin crystals possess low value of L/kT_0 and grow by the normal mechanism, with frequently observed cellular growth forms. The type of impurity trapping is quite different also (Fig. 8c). In this case the impurities segregate between the cells, that is, impurity-enriched layers are perpendicular to the crystallization interface.

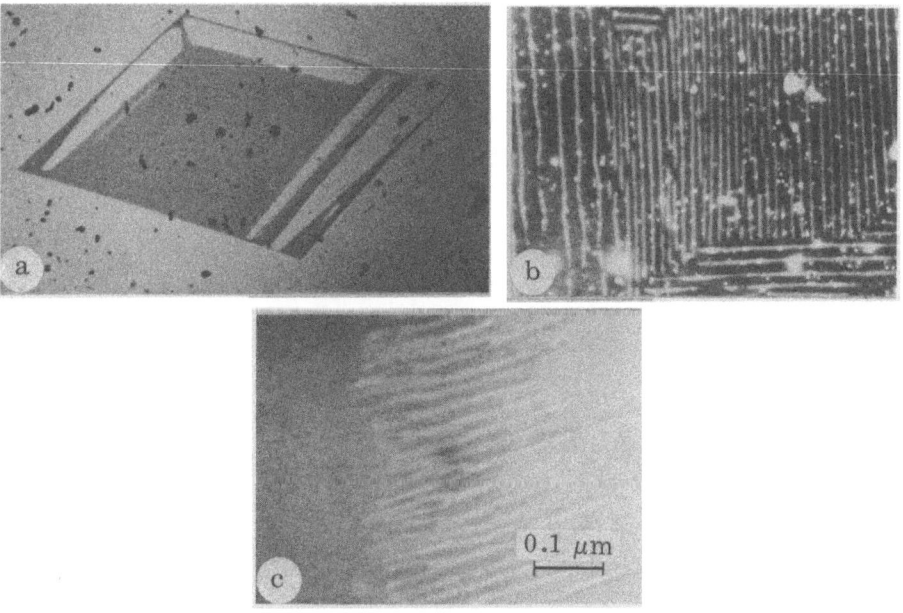

Fig. 8. Micrographs of impurity trapping in a growing crystal.
(a) Ga + 0.01 at.% In; (b) Ga + 0.01 at.% Cu; (c) Sn + 1 at.% Bi.

Stability Breakdown and Unstable Growth Forms

Analysis has indicated that the conditions under which unstable growth forms appear, and their development and morphology, essentially depend on the entropy of melting of the material. Crystals of high-entropy materials retain a normal habit at relatively high supercooling of the melt (exceeding 2°C). For example, piperonal platelet crystals retain the usual habit up to $\Delta T = 6°C$, but then split and form spherulites (Fig. 9). In the case of benzophenone, faceted growth was observed up to $\Delta T > \Delta T_{max}$ corresponding to v_{max}. It must be mentioned that neither cellular nor dendritic growth, typical for materials with low entropy of melting, were observed for high-entropy materials.

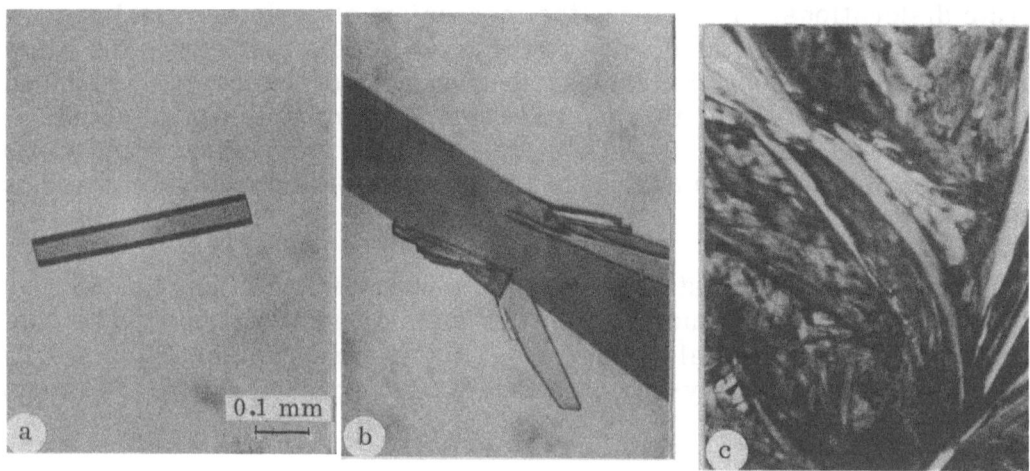

Fig. 9. Growth rates of piperonal crystals at supercoolings of (a) 3°, (b) 7°, and (c) 12°C.

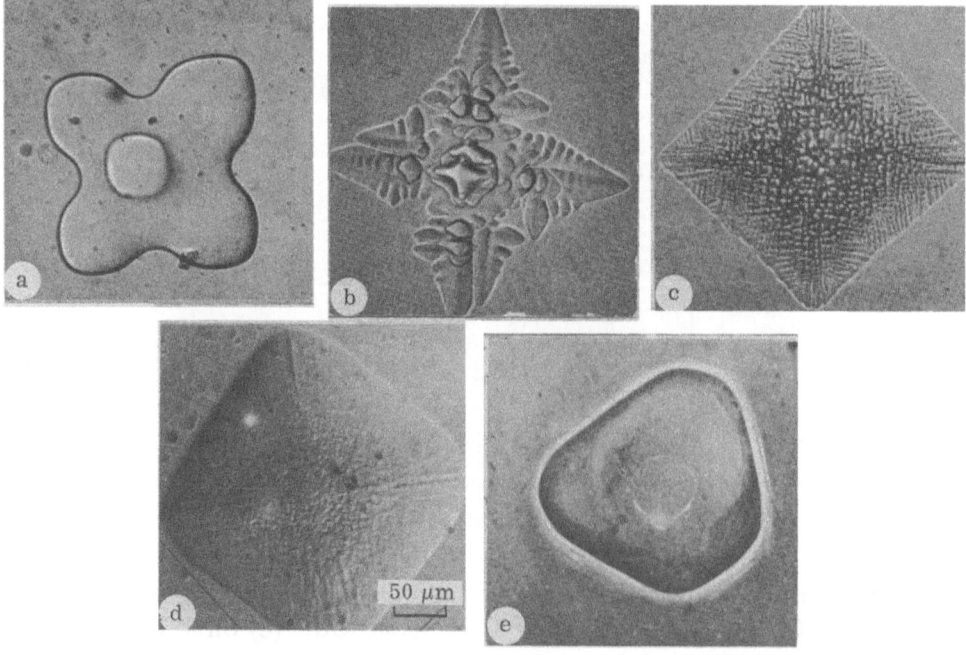

Fig. 10. Modification of growth forms in cyclohexanol crystals at melt supercoolings below (a) 0.3°, (b) 0.45°, (c) 2.35°, (d) 3.5°, and (e) 10.25°C.

One of the important features of crystals with low values of L/kT_0 is the loss of stability at much smaller values of supercooling than for materials with $L/kT_0 > 2$. This aspect was thoroughly analyzed for cyclohexanol crystals. The following growth behavior was found [27, 28]: at low (up to 0.1°C) levels of melt supercooling the crystals are rounded and retain this shape up to sizes ~ 400 μm (see Fig. 2a). At $\Delta T = 0.3$°C, when crystals reach a certain (critical) size, protrusions appear (Fig. 10a) and grow in length continuously. An increase in supercooling to 0.45°C leads to stability breakdown in much smaller crystals and to the formation of protrusions which then transform to dendrites (Fig. 10b). At $\Delta T = 2.35$°C numerous protrusions, later forming small dendrites (Fig. 10c), appear immediately on very small crystals. At $\Delta T = 3.5$°C, the dendritic shape degenerates to a spherulite, or an acicular shape (Fig. 10d), consisting of thin needles without lateral branches. At still larger supercooling (about 10°C) the crystals again assume a rounded, macroscopically smooth stable form (Fig. 10e), similar to that observed at $\Delta T = 0.1$°C. Similar behavior was also observed in other materials: succinonitrile, camphene, and cyclohexane. However, instead of six protrusions typical for cyclohexanol with an fcc lattice (see Fig. 10a), eighteen protrusions are formed in succinonitrile and camphene possessing a bcc lattice, (Fig. 11a). Only six protrusions survive, however, as in the case of cyclohexanol the growth of other protrusions being blocked (Fig. 11b) [30].

Some of the growth forms discussed above were observed for iron ingots ($L/kT_0 = 1.03$) and nickel ingots ($L/kT_0 = 1.26$) and for their alloys [29, 35-38]. Metal samples of about 50 g were supercooled by 300-310°C, and mixtures of silicon, molybdenum, and tin with iron and nickel, up to 5 percent, usually did not lower these values.

Ingots solidified after low melt supercooling always reveal a well-developed dendritic structure (Fig. 12a, b). As the supercooling is increased, this structure becomes finer and is retained up to certain values of supercooling, above which it is transformed to spherulitic, or acicular forms (Fig. 12c, d). In pure iron this transition occurs at $\Delta T = 190$°C, although the observation of a spherulitic (acicular) structure is difficult because of polymorphic transformations and secondary processes (polygonization, recrystallization) destroying the primary structure. These processes are substantially suppressed by mixtures with 3-5 percent silicon, molybdenum, and tin, and it becomes possible to observe the acicular structure. A well-developed dendritic structure is also observed in nickel crystallized at low values of supercooling, and is retained up to $\Delta T = 140$°C. However, recrystallization completely destroys the primary structure in ingots crystallized at high values of melt supercooling, and

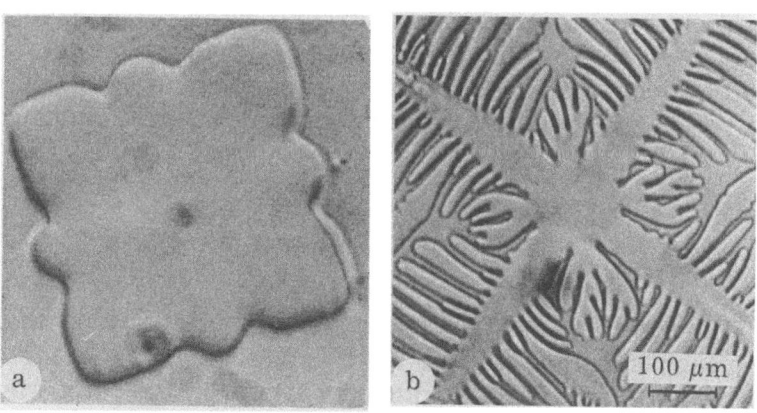

Fig. 11. Unstable growth forms of succinonitrile crystals at supercoolings of (a) 0.35° and (b) 0.8°C.

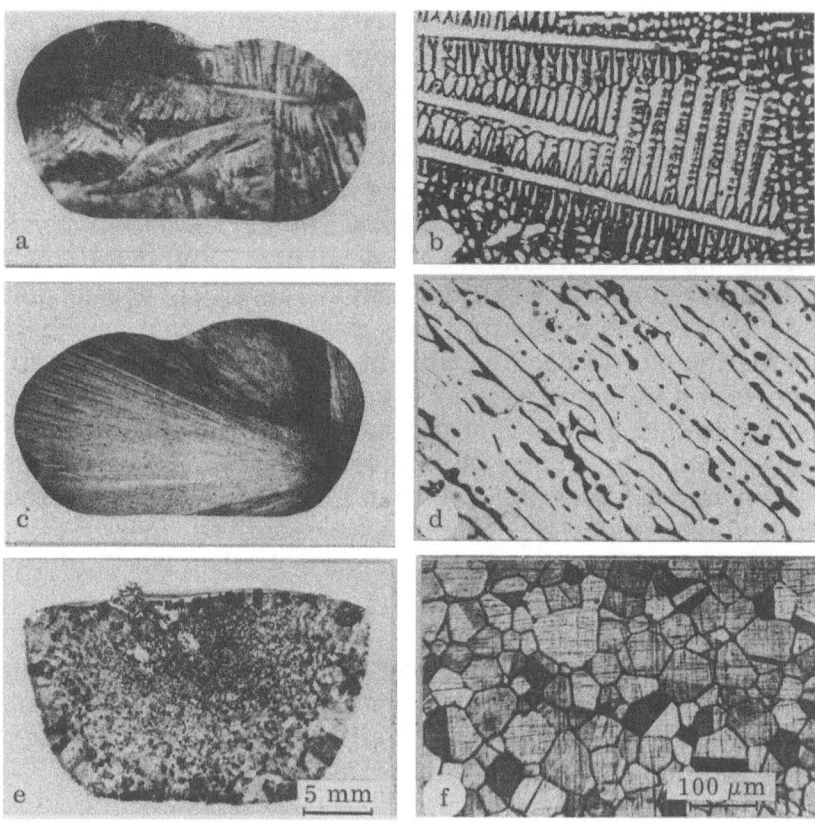

Fig. 12. Structures of boules solidified after different values of supercooling. (a, b) Fe + 5 wt. % Si, $\Delta T=30°C$; (c, d) Fe + 5 wt. % Si, $\Delta T=270°C$; (e, f) Ni, $\Delta T=200°C$.

small-grain structures are formed (Fig. 12e, f). A mixture with 3.7 percent silicon lowers the critical supercooling by 60 to 70°C and drastically reduces the grain dimensions. The acicular structure thus survives and is easily observed.

The above-described dependence of crystal growth forms on supercooling presumably holds for all materials with low entropy of melting and is of the same nature for all of them. The growth of protrusions (cells) and dendritic growth at high values of supercooling are known to result from growth destabilization owing to inhomogeneity of the temperature and concentration fields which are generated at high growth rates. The degeneration of dendritic to acicular form is explained by a higher density of protrusions growing as needles into the melt, at higher levels of supercooling. This produces overlapping thermal fields of the protrusions and, as a result, suppresses the formation of lateral branches and protrusions on their surfaces.

Conclusion

On the whole, the experimental results obtained confirm the predictions of the theories based on the rough interface model. The data show that a transition from layered growth to continuous (normal) growth takes place in a narrow range of melting entropies, ΔS: from 3.44 (tin) to 4.41 (gallium). This range is in good agreement with Jackson's criterion [15]. The present paper offers no confirmation of the phenomenological theories which predict the layered-normal growth transition with increasing supercooling.

It is established that in materials with high values of melting entropy ($\Delta S > 4$) the stable growth forms survive at much higher values of melt supercooling (5-10°C) than in materials with low entropy of melting (a fraction of 1°C).

Specific features of the generation and development, as well as morphology, of crystal growth forms are studied as functions of melting entropy and supercooling of the melt. In high-entropy crystals ($\Delta S > 4$) increased supercooling leads to fracturing of crystals by generated stresses. High values of supercooling lead to multiple successive fracturing and thus to the formation of spherulites. As the supercooling increases the structure of the spherulites becomes finer.

It is shown that stable rounded growth shapes of low-entropy materials, typical at low supercooling, are destabilized at higher supercooling levels: protrusions are formed and later develop to dendrites. At still higher supercooling dendritic growth forms are transformed to spherulites, which differ from spherulites typical of high-entropy materials. A second stability range of rounded growth forms is found at still higher levels of supercooling.

It is found that, depending on the nature and concentration of impurities, different mechanisms are possible for the impurity effects on growth, and that these effects are not related to changes in surface energy. The distribution of impurities is shown to vary. In high-entropy crystals impurity is trapped in a layer fashion, parallel to the growth face, while in low-entropy crystals impurities are localized normally to the crystallization interface between protrusions (cells) as well as between the dendrite axes. Since concentration inhomogeneities can serve as dislocation sources, the differences in impurity distribution to a large extent determine the formation of the dislocation structures; this factor may be important for a number of structure-sensitive properties.

The results presented enable one to achieve a better understanding of the formation of structures in cast metal ingots, the role of soluble impurities, and the mechanism by which they affect the casting process. There is no doubt that any new theory for modifying metals and alloys must take into account the fact that soluble mixtures substantially affect the kinetics and morphology of growth, as well as secondary processes.

Literature Cited

1. M. Volmer and M. Marder. Teorie der linear Kristallisationsgeschwindigkeit unterkühlter Schmelzen und unterkühlter fester Modifikationen. Z. Phys. Chem. A, 154, 97-112, (1931).
2. I. N. Stranski and P. Kaishev. On the theory of crystal Growth and nucleation. Uspekhi Fiz. Nauk. 21, 408-465, (in Russian) (1939).
3. W. B. Hillig and D. Turnbull. Theory of crystal growth in undercooled pure liquids. J. Chem. Phys., 24, 914 (1956).
4. H. A. Wilson. On the velocity of solidification and viscosity of supercooled liquids. Phil. Mag., 50, 238-250 (1900).
5. J. Frenkel. Note on a relation between the speed of crystallization and viscosity. Phys. Z. Sowjetunion, 1, 498-500 (1932).
6. D. Turnbull. The laws of crystallization. In: Thermodynamics in Metallurgy. Cleveland: ASM 282-298 (1950).
7. K. A. Jackson and B. Chalmers. Kinetics of solidification. Canad. J. Phys., 34, 473-490 (1956).
8. J. W. Cahn. Theory of crystal growth and interface motion in crystalline materials. Acta Met., 8, 554-562 (1960).

9. W. W. Mullins and R. F. Sekerka. Morphological Stability of a particle growing by diffusion or heat flow. J. Appl. Phys. 34, 323-329 (1963).
10. S. R. Coriell and R. L. Parker. Interface kinetics and the stability of the shape of a solid sphere growing from the melt. In: Crystal Growth, Proc. Int. Conf. Boston, Pergamon, Oxford, 703-708 (1966).
11. R. L. Parker. Crystal growth mechanisms, energetics, kinetics and transport. In: Solid State Physics, 25, Academic Press, New York-London (1970).
12. A. A. Chernov. Stability of crystal growth forms. In: Growth and Defects in Metal Crystals. Naukova Dumka, Kiev, 79-95 (in Russian) (1972).
13. J. W. Cahn, W. G. Hillig and G. W. Sears. The molecular mechanism of solidification. Acta Met., 12, 1421-1439 (1964).
14. D. E. Temkin. Phenomenological kinetics of interface motion. Kristallografiya, 15, 877-883 (in Russian) (1970).
15. K. A. Jackson. Mechanism of growth. In: Liquid Metals and Solidification, ASM, Cleveland, 174 (1958).
16. D. E. Temkin. On molecular roughness of crystal-melt interface. In: Crystallization Mechanism and Kinetics. Nauka i Tekhnika, Minsk, 86-97 (in Russian) (1964).
17. A. L. Roitburd. On some specific features of crystal growth in condensed media. Kristallografiya, 7, 291-299 (in Russian) (1962).
18. D. E. Temkin. Kinetics of interface motion in phase transitions. Thesis, Institute of Crystallography, Moscow, (1974).
19. D. E. Ovsiyenko and G. A. Alfintzev. Analysis of linear crystallization rate of salol crystals from the melt. In: Problems in Metal Physics. Naukova Dumka, Kiev, No. 19, 170-182 (in Russian) (1964).
20. G. A. Alfintsev. A method of observation of crystal growth from the melt and grain boundary migration in some metals. In: Metal Physics. Naukova Dumka, Kiev, No. 33, 56-59 (in Russian) (1971).
21. G. A. Alfintsev and D. E. Ovsiyenko. Investigation of mechanism of melt growth of gallium crystals. In: Growth of Crystals 5, ed. N. N. Sheftal', Consultants Bureau, New York (1968).
22. D. E. Ovsiyenko and G. A. Alfintsev. On the mechanism of melt growth of salol crystals. Kristallografiya, 8, 796-799 (in Russian) (1963).
23. G. A. Alfintsev and D. E. Ovsiyenko. Investigation of melt growth of gallium crystals. Dokl. AN SSSR, 156, 792-794 (in Russian) (1964).
24. A. A. Chernov. Spiral growth of crystals. Soviet Physics Uspekhi, 4, 115-148 (1961).
25. G. A. Alfintsev and D. E. Ovsiyenko. Investigation of the mechanism of melt growth of some metal crystals. In: Growth and imperfections of metal crystals. Naukova Dumka, Keiv, 40-53 (in Russian) (1966).
26. N. V. Stoichev, G. A. Alfintsev and D. E. Ovsiyenko. Effect of indium doping on gallium crystal growth. Kristallografiya, 20, 823-828 (in Russian) (1975).
27. D. E. Ovsiyenko, G. A. Alfintsev, I. A. Shramchenko and A. V. Mokhort. Investigation of kinetics of melt growth of cyclohexanol crystals. In: Crystallization Mechanism and Kinetics. Nauka i Tekhnika, Minsk, 147-153 (in Russian) (1969).
28. D. E. Ovsiyenko, G. A. Alfintsev and A. V. Mokhort. Gorwth forms of cyclohexanol crystals at different melt temperatures. In: Growth of Crystals, 9, ed. N. N. Sheftal' and E. I. Givargizov, Consultants Bureau, New York (1975).
29. G. A. Alfintsev and D. E. Ovsiyenko. On the mechanism of melt growth of metal crystals. J. Phys. and Chem. Solids, Suppl. 1, 754-762 (1967).
30. D. E. Ovsiyenko, G. A. Alfintsev and V. V. Maslov. Kinetics and shape of crystal growth from the melt for substances with low L/kT values. J. Cryst. Growth, 26, 233-238 (1974).

31. G. A. Alfintsev and D. E. Ovsiyenko. Effect of silver doping on melt growth of gallium crystals. Fiz. Metal. i Metalloved., 20, 401-405 (in Russian) (1965).
32. N. V. Stoichev, D. E. Ovsiyenko, and G. A. Alfintsev. Effect of small copper additives on the growth of gallium crystals from a melt. Krist. und Techn., 11, 905-911 (1976).
33. D. E. Ovsiyenko, V. V. Maslov, and G. A. Alfintsev. On the mechanism of the effect of soluble impurities on tin crystallization. Kristallografiya, 22, 1042-1049 (in Russian) (1977).
34. D. E. Ovsiyenko, G. A. Alfintsev, G. P. Chemerinsky, and S. N. Khevsurishvili. Kinetics and forms of melt-grown succinonitrile crystals. Kristallografiya, 21, 801-806, (in Russian) (1976).
35. D. E. Ovsiyenko, G. A. Alfintsev, and V. V. Maslov. Effect of silicon and manganese on supercooling in iron. In: Metal Physics. Naukova Dumka, Kiev, No. 39, 102-105 (in Russian) (1972).
36. G. A. Alfintsev, D. E. Cvsiyenko, N. V. Stoichev, and V. V. Maslov. Effect of silicon and manganese on supercooling and structure of ingots of iron-silicon alloys. In: Crystallization Mechanism and Kinetics. Nauka i Tekhnika, Minsk, 332-337 (in Russian) (1973).
37. D. E. Ovsiyenko, V. P. Kostyuchenko, V. V. Maslov, and G. A. Alfintsev. Effect of supercooling on structure of nickel ingots. In: Crystallization Mechanism and Kinetics. Nauka i Tekhnika, Minsk, 75-81 (in Russian) (1973).
38. D. E. Ovsiyenko, V. V. Maslov, and G. A. Alfintsev. Effect of silicon doping on supercooling and structure of nickel ingots. Izv. AN SSSR, Metally, No. 4, 92-96 (in Russian) (1974).

KINETIC CONDITIONS AT THE GROWTH INTERFACE OF A MIXED CRYSTAL

D. E. Temkin

The Institute of Metal Science, Moscow

Phenomenological Treatment

The present paper dicusses kinetic conditions which relate the growth rate of a mixed crystal to deviations from equilibrium at the growth surface. In the general case the growth of such a crystal is accompanied by a redistribution of the components and a release of crystallization heat. The kinetic considerations act as boundary conditions in a description of the heat and mass transfer in the crystallizing system. Although in many cases one can treat the conditions as if at equilibrium, situations are possible in which the processes at the interface between the crystal and the initial phase result in appreciable resistance to growth, making it necessary to take into account departures from equilibrium. Kinetic considerations have often been used in analyses of various aspects of crystal growth [1-7]. However, the formulation of these conditions has not been sufficiently general and has been valid only for the growth of a constant-composition crystal [1, 5, 7] or for steady-state growth, when both the crystal composition and the crystal growth rate are time independent [2, 3].

When a single-component crystal grows from its own melt, the temperature continuity condition

$$T_{1s}(t) = T_{2s}(t) \equiv T_s(t) \tag{1}$$

and the heat balance condition

$$Qv(t) = \left(\lambda_1 \frac{\partial T_1}{\partial x} - \lambda_2 \frac{\partial T_2}{\partial x}\right)_s \tag{2}$$

must hold at the interface; the growth rate v is a function of supercooling ΔT_s at the interface:

$$v = f(\Delta T_s). \tag{3}$$

Here x is the spatial coordinate; t is time; T_1, T_2 and λ_1, λ_2 denote temperature and thermal conductivity in the crystal (phase 1) and melt (phase 2), respectively; Q is the crystallization heat per unit volume; subscript s indicates that the corresponding quantity is referred to the interface. Relations (1)-(3), written for a planar interface, are valid for a curved interface as well. In this last case v is the local growth rate along the normal to the interface, x is the coordinate along the normal, and (if the curvature is substantial) ΔT_s is the supercooling corrected for the Gibbs-Thomson effect.

The form of function $f(\Delta T_s)$ under the kinetic condition (3) depends on the structure of the interface and the mechanism of its growth. For an atomically rough interface, growing by the normal mechanism, the growth rate v is a linear function of ΔT_s for a considerable range of supercooling. Obviously, conditions (1) and (2) are also valid for crystallization in multicomponent systems. However, relation (3) must be changed. Furthermore, it will be necessary to add the equations of balance for the components at the interface, by analogy with equation (2).

Let us consider a binary system in which both the crystal and the initial condensed phase are substitutional solutions of components A and B. In binary systems diffusion is characterized by two intrinsic diffusion coefficients D_A and D_B (sometimes called partial diffusivities), which are not equal in the general case, with components thus diffusing at different velocities. Nevertheless, diffusion in such systems is described satisfactorily by the Darken relation [8] which assumes that the total number of particles per unit volume remains constant in the course of diffusion. The diffusion of the two components is described by a single equation

$$\frac{\partial C}{\partial t} = \frac{\partial}{\partial x}\left(D \frac{\partial C}{\partial x}\right), \qquad (4)$$

in which the intrinsic coefficients D_A and D_B are replaced by the mutual diffusion coefficient D equal to

$$D = C D_A + (1 - C) D_B. \qquad (5)$$

Here C is the molar fraction of component B. With the same assumptions as in the Darken relation, the balance of components at the moving interface between phases 1 and 2 is written in the conventional form:

$$(C_{2s} - C_{1s}) v = \left(D_1 \frac{\partial C_1}{\partial x} - D_2 \frac{\partial C_2}{\partial x}\right)_s, \qquad (6)$$

where D_1 and D_2 are the mutual diffusion coefficients in phases 1 and 2 [4]. In deriving equation (6) we assume that partial atomic volumes of the two components are identical and remain constant on transition from one phase to another. Equation (6) is formulated as the balance equation for component B; it is easy to see, however, that it automatically describes the balance of component A whose concentration in the i-th phase is $(1-C_i)$.

The amount of component B removed from phase 2 per unit time when a unit area of the interface moves at a velocity v is $(D_2 \partial C_2/\partial x)_s + v C_{2s}$ and that of phase 1 (i.e., the crystal) is $(D_1 \partial C_1/\partial x)_s + v C_{1s}$. These quantities are equal, and this fact is built into equation (6). On the other hand, it is clear that these quantities must be equal to the flux of the B atoms across the interface from phase 2 to phase 1. Denoting this flux by $I_B \Omega$ (cm·s^{-1}), this yields

$$D_2 \frac{\partial C_2}{\partial x}\bigg|_s + v C_{2s} = I_B \Omega = D_1 \frac{\partial C_1}{\partial x}\bigg|_s + v C_{1s}. \qquad (7)$$

Similarly, we have for component A

$$-D_2 \frac{\partial C_2}{\partial x}\bigg|_s + v(1 - C_{2s}) = I_A \Omega = -D_1 \frac{\partial C_1}{\partial x}\bigg|_s + v(1 - C_{1s}), \qquad (8)$$

where $I_A \Omega$ is the resultant flux of component A from phase 2 to phase 1 across the interface, and Ω is the atomic volume. Relations (7) and (8) are equivalent to three independent relations one of which is equation (6) and the other two can be written in the form

$$v = (I_A + I_B) \Omega, \qquad (9)$$

$$D_1 \frac{\partial C_1}{\partial x}\bigg|_s + v C_{1s} = I_B \Omega. \qquad (10)$$

The kinetic conditions were given in this form in ref. 4. Fluxes I_A and I_B, characterizing atomic transitions from one phase to another across the interface, are functions of the departures from equilibrium at this interface. In the general form

$$I_\alpha = \varphi_\alpha(\Delta C_{1s}, \Delta C_{2s}) \qquad (\alpha = A, B), \tag{11}$$

where φ_α is a function of the indicated parameters, and $\Delta C_{is} = C_i^0 - C_{is}$ is the deviation of concentration in the i-th phase from the equilibrium value C_i^0 at the interface temperature T_s. At equilibrium, $\Delta C_{1s} = 0$ and $\Delta C_{2s} = 0$ and the resultant fluxes I_A and I_B also equal to zero. Relations (9)-(11) fix the kinetic constraints replacing condition (3) for a single-component system.

The relation between fluxes I_α and the conditions at the interface, which is a specific form of relation (11), can be found for each detailed molecular-kinetic description of processes at the interface. When an atomically rough interface grows by the normal mechanism, with particle transitions taking place at all points of the interface practically simultaneously, the above relation can be analyzed in terms of the theory of irreversible processes [2, 3]. This gives a linear dependence of fluxes on the differences between chemical potentials at the interface, for small departures from equilibrium. Neglecting cross effects (i.e., the dependence of the flux of a given component on the difference between chemical potentials of the other component), one can write

$$I_\alpha = B_\alpha(\mu_{\alpha 2} - \mu_{\alpha 1}) = B_\alpha \Delta \mu_\alpha, \tag{12}$$

where $\mu_{\alpha i}$ is the chemical potential of the α component in the i-th phase at the interface, and B_α is a coefficient depending on temperature and equilibrium values of concentrations but independent of $\Delta \mu_\alpha$.

It is necessary to emphaize one important feature which distinguishes multicomponent systems from single-component ones. By fixing the difference in chemical potentials in a single-component system (e.g., $\Delta \mu_A^0$ for a pure A-system), one can immediately determine the direction of interface motion: the crystal will grow if $\Delta \mu_A^0 > 0$, and will melt if $\Delta \mu_A^0 < 0$. This is not always possible in alloys. For example, if we bring into contact phase 1 with composition C_1 and phase 2 with composition C_2 and these compositions are such that $\Delta \mu_A > 0$, $\Delta \mu_B < 0$, then flux $I_A > 0$ (from phase 2 to 1) and $I_B < 0$. As the interface advancement velocity is proportional to the sum of these fluxes (see eq. (9)), it may be positive or negative, depending on the ratio of coefficients B_A and B_B (see eq. (12)). In other words, if B_A and B_B are independent kinetic parameters, it is impossible to find for an alloy such a purely thermodynamic parameter which would act as the driving force for the motion of the interface and would unambiguously determine the direction of interface displacement regardless of the kinetic parameters. The thermodynamic driving forces $\Delta \mu_A$ and $\Delta \mu_B$ only determine the directions of the fluxes I_A and I_B.

If the curvature of the interface is substantial, then as a result of the Gibbs-Thomson effect the difference between chemical potentials contains an additional term compared with $\Delta \mu_\alpha$ for a planar interface. In this case, when the specific free energy of the interface σ is independent of phase compositions, this term is $2\sigma\Omega/R$ [9], where R is the mean radius of curvature of the interface at a given point ($R > 0$ if the interface is concave towards phase 2). Taking this into account, we obtain instead of equation (12)

$$I_\alpha = B_\alpha(\Delta \mu_\alpha + 2\sigma\Omega/R). \tag{13}$$

Each resultant flux I_α into phase 1 can always be written as a difference between the flux of the α species particles from phase 2 to phase 1 and the reverse flux:

$$I_A = q[\overline{w}_{+A}(1 - C_{2s}) - \overline{w}_{-A}(1 - C_{1s})]; \qquad I_B = q(\overline{w}_{+B} C_{2s} - \overline{w}_{-B} C_{1s}). \tag{14}$$

Here q is the surface particle density, assumed to be identical in both phases, $w_{+\alpha}$ is the mean frequency of the transitions of α species atoms from phase 2 to phase 1, and $w_{-\alpha}$ is the frequency of the reverse transitions.

The introduction of transition frequencies shifts the difficulties in determining the fluxes to determining these frequencies (which reflect the structure of the interface and are thus complex functions of phase compositions and temperature). Their calculation must be based on a detailed molecular-kinetics analysis. If one assumes, however, that these frequencies are independent of phase compositions at the interface and are only functions of temperature (via the standard Boltzmann factor), a simple kinetic model is obtained, making it possible to analyze qualitatively the effect of a departure from equilibrium at the interface on the growth kinetics of a mixed crystal. This model is characterized by a cigar-shaped phase diagram. Equilibrium concentrations C_i^0, found from conditions $I_A = 0$ and $I_B = 0$, are given by the relations

$$C_1^0 = C_2^0 \overline{w}_{+B}/\overline{w}_{-B}; \quad C_2^0 = \overline{w}_{-B}(\overline{w}_{+A} - \overline{w}_{-A})/(\overline{w}_{+A}\overline{w}_{-B} - \overline{w}_{-A}\overline{w}_{+B}). \tag{15}$$

The equilibrium distribution coefficient, defined as C_1^0/C_2^0, is equal to w_{+B}/w_{-B}. If T_A and T_B are the equilibrium temperatures of the phases composed of components A and B, respectively, and $T_A > T_B$, then

$$\overline{w}_{+A} < \overline{w}_{-A} \text{ and } \overline{w}_{+B} < \overline{w}_{-B} \text{ for } T > T_A;$$
$$\overline{w}_{+A} > \overline{w}_{-A} \text{ and } \overline{w}_{+B} < \overline{w}_{-B} \text{ for } T_A > T > T_B;$$
$$\overline{w}_{+A} > \overline{w}_{-A} \text{ and } \overline{w}_{+B} > \overline{w}_{-B} \text{ for } T < T_L.$$

It should be remembered that phase 1 is a low-temperature phase. The model outlined was used in refs. 10, 11 to analyze isothermal growth in the single-phase and two-phase regions of the equilibrium diagram and at the boundary between these regions, and in ref. 12 to analyze the stability of a plane interface.

Considering the relations (15), one can rewrite relations (14) in the form

$$I_A = q(\overline{w}_{+A}\Delta C_{2s} - \overline{w}_{-A}\Delta C_{1s}); \quad I_B = q(-\overline{w}_{+B}\Delta C_{2s} + \overline{w}_{-B}\Delta C_{1s}), \tag{16}$$

where $\Delta C_{is} = C_i^0 - C_{is}$, as in equation (11). In this model, therefore, the functions φ_α (see eq. (11)) and the growth rate are linear in ΔC_{is}, and the interface moves by the normal mechanism. The interface advancement velocity is positive if

$$\overline{w}_{+A}(1 - C_{2s}) + \overline{w}_{+B}C_{2s} > \overline{w}_{-A}(1 - C_{1s}) + \overline{w}_{-B}C_{1s}$$

and negative in the opposite case.

Transition frequencies in the model under consideration are independent of phase compositions. On a plane interface between phases consisting only of the α component

$$\overline{w}_{+\alpha}/\overline{w}_{-\alpha} = \exp(\Delta\mu_\alpha^0/kT),$$

where $\Delta\mu_\alpha^0$ is the difference between the chemical potentials of single-component phases. Taking into account the Gibbs-Thomson correction to $\Delta\mu_\alpha^0$ and recalling that $2\sigma\Omega/RkT \ll 1$, one obtains for a curved surface

$$(\overline{w}_{+\alpha}/\overline{w}_{-\alpha})_R = (1 + 2c\Omega/RkT)\overline{w}_{+\alpha}/\overline{w}_{-\alpha}. \tag{17}$$

Relation (17) holds also for alloys if, as in (13), the dependence of Ω and σ on composition is neglected.

Analysis of the Atomic Structure of Growth Interface

Consider the relation between mean transition frequencies and interface structure for a simple model. This aspect is discussed in relation to the growth of mixed crystals in refs. 13, 14, using the elementary cubic lattice and analyzing the structure of the (100) interface between phases 1 and 2. Both phases are substitutional solutions formed by components A and B. The atomic volume is $\Omega = a^3$, and the surface particle density is $q = 1/a^2$, where a is the interatomic spacing.

The system is divided into atomic layers parallel to (100) faces and the state of the system is described by the parameters y_n, C_{1n}, C_{2n}, where y_n is the fraction of sites in the nth layer occupied by phase 1 ($y_{-\infty} = 1$ for phase 1, $y_{\infty} = 0$ for phase 2); C_{1n} and C_{2n} are atomic fractions of component B on sites of the nth layer occupied by phases 1 and 2, respectively. Diffusion in the model discussed proceeds by way of exchanging sites by nearest atoms A and B, with frequency w_i in the ith phase. The intrinsic diffusion coefficients are $D_{Ai} = D_{Bi} = a^2 w_i$, and coincide with the mutual diffusion coefficients D_i (see eq. (5)).

Approximate (neglecting correlations) kinetic equations of interface motion, taking account of atomic transitions across the interface and diffusion of the components, were formulated in refs. 13, 14. It was assumed that type-1 and type-2 sites within each layer are distributed randomly but in such a manner that type-1 sites in the $(n+1)$th layer always lie precisely above the identical sites of the nth layer. Within each of the phases of an arbitrary layer the distribution of atoms of the components is also random. With these assumptions, kinetic equations take the form

$$\frac{d(y_n C_{1n})}{dt} = (y_{n-1} - y_n) C_{2n} w_{+B}(y_n) - (y_n - y_{n+1}) C_{1n} w_{+B}(y_n) + w_1 [y_n(C_{1\,n-1} - C_{1n}) + y_{n+1}(C_{1\,n+1} - C_{1n})]; \quad (18)$$

$$\frac{dy_n}{dt} = (y_{n-1} - y_n)[(1 - C_{2n}) w_{+A}(y_n) + C_{2n} w_{+B}(y_n)] - (y_n - y_{n+1})[(1 - C_{1n}) w_{-A}(y_n) + C_{1n} w_{-B}(y_n)]; \quad (19)$$

$$\frac{d[(1 - y_n) C_{2n}]}{dt} = (y_n - y_{n+1}) C_{1n} w_{-B}(y_n) - (y_{n-1} - y_n) C_{2n} w_{+B}(y_n)$$
$$+ w_2 [(1 - y_{n-1})(C_{2\,n-1} - C_{2n}) + (1 - y_n)(C_{2\,n+1} - C_{2n})]. \quad (20)$$

Here $w_{+\alpha}(y_n)$ is the transition frequency of the α-species atoms (of components A and B) from phase 2 to phase 1 in the nth layer, and $w_{-\alpha}(y_n)$ is the frequency of the reverse transition. It was assumed in refs. 13, 14 that these frequencies depend only on the species of the atoms crossing the interface and on the interface structure at the transition location, and are given by the expressions

$$w_{+\alpha}(y_n) = w^0_{+\alpha} \exp[-\gamma_2(1 - 2y_n)]; \quad w_{-\alpha}(y_n) = w^0_{-\alpha} \exp[\gamma_1(1 - 2y_n)], \quad (21)$$

where $w^0_{\pm\alpha}$ is the transition frequency at a kink on a step; γ_1 and γ_2 are positive quantities such that $\gamma_1 + \gamma_2 = \gamma = 4\varepsilon/kT$; ε is the energy of a kink on a step.

Equilibrium phase concentrations C_i^0 ($i = 1, 2$) satisfy kinetic constraints on the equality of direct and reverse atomic fluxes at a kink:

$$w^0_{+A}(1 - C_2^0) = w^0_{-A}(1 - C_1^0); \quad w^0_{+B} C_2^0 = w^0_{-B} C_1^0$$

and are described by relations (15) in which the mean frequencies $\bar{w}_{\pm\alpha}$ must be replaced by $w^0_{\pm\infty}$.

With the concentration C_{is} fixed at the interface, we introduce mean transition frequencies at a kink:

$$w^0_+ = w^0_{+A}(1 - C_{2s}) + w^0_{+B} C_{2s} \quad \text{and} \quad w^0_- = w^0_{-A}(1 - C_{1s}) + w^0_{-B} C_{1s}. \quad (22)$$

Obviously, the crystal (phase 1) will grow if $w_+^0 > w_-^0$ and will melt if $w_+^0 < w_-^0$. Consequently, we introduce as a unique parameter characterizing the dimensionless driving force for the motion of the interface a quantity β defined as

$$\exp \beta = w_+^0 / w_-^0. \tag{23}$$

In a single-component α-system, $\beta = \Delta\mu_\alpha^0 / kT$ where $\Delta\mu_\alpha^0$ is the difference between chemical potentials of phases 1 and 2 composed of component α. It is clear (and was emphasized above) that in mixed systems the driving force β cannot be expressed purely in thermodynamic terms but depends also on kinetic parameters (in the present case, on the ratio of transition frequencies at a kink). Under equilibrium conditions, that is, when $C_{is} = C_i^0$, the driving force vanishes, i.e., $\beta = 0$. But as concentrations C_{is} deviate from the equilibrium values C_i^0, but such that $w_+^0 = w_-^0$, we again have $\beta = 0$ and, consequently, $v = 0$.

Far from the interface, where y_n is equal to 1 or 0, equations (19) and (20), in which differences are replaced by derivatives with respect to $x = an$, are transformed to the standard diffusion equation (4) with coefficient D_i independent of concentration. In a diffusional quasistationary growth process (when D_i/v is much larger than the effective width of the interface, and concentrations of components at the interface change slightly during a characteristic transition time at a kink) the distribution of concentrations at the interface is practically uniform (independent of n) and $C_{in} = C_{is}$. Taking this into account and adding equations (18)-(20) over n in the vicinity of the interface, we arrive at the constraints at the interface formulated earlier, namely, (6), (9), (10), and (14). This clarifies the physical meaning of the mean transition frequencies $\bar{w}_{\pm\alpha}$ at a planar interface, introduced in relations (14):

$$\bar{w}_{+\alpha} = \sum_n w_{+\alpha}(y_n)(y_{n-1} - y_n); \quad \bar{w}_{-\alpha} = \sum_n w_{-\alpha}(y_n)(y_n - y_{n+1}). \tag{24}$$

Hence $\bar{w}_{\pm\alpha}$ stands for transition frequencies averaged over the transient region at the interface.

A metastable mode, in which kinetic equations (18)-(20) have steady-state solutions related to a fixed interface, was analyzed in ref. 13. In the region of metastability two homogeneous phases with compositions C_1 and C_2 (differing from equilibrium compositions C_1^0 and C_2^0) coexist owing to an "adjustment" of the interface structure; the adjustment is such that the resultant fluxes I_α of each component into phase 1 equal zero. The adjustment is impossible beyond the region of metastability, and the interface moves continuously. Although in reality no such region exists and the interface moves for any, however small, departure from equilibrium, the study of the region of metastability on the basis of approximate equations representing the interface seems to be meaningful, since it provides a qualitative estimate of the transient mode of growth when the two-dimensional nucleation mechanism is replaced by normal growth.

An approximate, continuous solution of the kinetic equations was given in ref. 14; the mean frequencies $\bar{w}_{\pm\alpha}$, the growth rate v, and the effective width l of the interface were calculated as functions of the driving force β. This solution is characterized by the elimination of the region of metastability, and at low values of β yields a growth rate proportional to β. The continuous description is valid for atomically rough interfaces and is obtained by replacing y_n by functions $y(x)$ of a continuous variable $x = an$, substituting derivatives for the differences in relation (18), and integrals for the sums in relations (24). A closed solution was obtained in two particular cases, in the first of which $\gamma_2 = 0$, $\gamma_1 = \gamma$ and the frequencies of attachment to phase 1 are independent of the interface structure (see relations (21)), and in the second case $\gamma_1 = \gamma_2 = \gamma/2$ and attachment and detachment frequencies are symmetric functions of the interface structure. The corresponding expressions are cumbersome, so we give only the final relations for the first case:

$$\bar{w}_{+\alpha} = w^0_{+\alpha}; \quad \bar{w}_{-\alpha} = w^0_{-\alpha}\left[2\gamma/\ln\frac{1+\exp(\beta+\gamma)}{1+\exp(\beta-\gamma)} - 1\right]\exp\beta;$$

$$v/aw^0_+ = 2 - 2\gamma/\ln\frac{1+\exp(\beta+\gamma)}{1+\exp(\beta-\gamma)};$$

$$l = \gamma\left[\gamma + (2 - v/aw^0_+)\ln\frac{1+\exp(\beta-\gamma)}{1+\exp\beta}\right]^{-1}. \tag{25}$$

Here l is the interface width measured in interatomic distances a, and w^0_+ and β, defined by (22) and (23), are functions of the concentrations C_{is} at the interface.

In the case under discussion attachment frequencies are independent of the interface structure, so that $\bar{w}_{+\alpha} = w^0_{+\alpha}$ for any values of β and γ. If $\beta = 0$, we have $\bar{w}_{-\alpha} = w^0_{-\alpha}$, $v = 0$, and $l = \gamma/2\ln\cosh(\gamma/2)$. As the driving force β increases, the interface width grows indefinitely, while the dimensionless velocity v/aw^0_+ and detachment frequency $\bar{w}_{-\alpha}/w^0_{-\alpha}$ grow, tending to limiting values for $\beta \to \infty$ (1 and $\sinh\gamma/\gamma$, respectively). For $\beta \ll 1$

$$v/aw^0_+ = B(\gamma)\beta; \quad \bar{w}_{-\alpha}/w^0_{-\alpha} = 1 + \beta[1 - B(\gamma)],$$

where $B(\gamma) = (2/\gamma)\tanh(\gamma/2)$. The smaller is γ, the larger is the roughness of the interface (the larger is l), the closer is coefficient $B(\gamma)$ to 1, and the weaker is the dependence of mean detachment frequencies on β.

Although the frequencies $w_{+\alpha}(y_n)$ and $w_{-\alpha}(y_n)$ introduced above (see relations (21)) were independent of the phase composition at the transition point, the mean frequencies (see relations (24)) are functions of concentrations at the interface (via parameter β) because the interface structure depends on β. Only in the case of **maximum roughness does interface $\bar{w}_{\pm\alpha}$ become independent of β**, and we pass to that elementary kinetic model which was discussed after relations (14) were introduced. However if $\gamma < \gamma_R$ ($\gamma_R \approx 3.5$ corresponds to the transition point from a smooth to a rough interface [15]), the ratio $\bar{w}_{-\alpha}/w^0_{-\alpha}$ changes, according to (25), as a function of β, within an order of magnitude. Thus, if $\gamma = 3$, this ratio increases from 1 for $\beta = 0$ to 3.3 for $\beta \to \infty$, and equals 1.5 for $\beta = 1$. This, to a certain extent, may serve as a justification for the use of relations (14) (with $\bar{w}_{\pm\alpha}$ independent of concentrations) for a description of the normal growth of mixed crystals.

Adsorption at the Interface

Let us return to the formulation of constraints at a growth face, and take into account the effects of adsorption. In the general case, adsorption takes place in both phases. The simplest model taking adsorption into account is as follows: the crystal and the initial phase contain adsorption layers a_1 and a_2 thick, respectively (a_1 are of the order of the interatomic, or interplanar spacing a). The concentrations of component B within these layers are C^0_{1s} and C^0_{2s}, respectively, and differ from concentrations C_{1s} and C_{2s} in phases 1 and 2 at an interface with an adsorbed layer, so that a jump in concentration, $C^0_{is} - C_{is}$, exists at this interface in the ith phase. When the crystal grows at a rate v, the adsorption layers move at the same rate. Assume that a transition of atom B from phase 2 to layer a_2 at their interface proceeds by way of exchanging places with atom A within this layer, and is characterized by frequency w_{+2}. The reverse transition of atom B from layer a_2 to phase 2 proceeds with frequency w_{-2}. Transitions between layers parallel to the interface, in the case of diffusion within the bulk of phase 2, occur with the same frequency regardless of the direction of the transition, so that $w_{+2} = w_{-2} = w_2 = D_2/a^2$. It is the difference between w_{+2} and w_{-2} at the interface with the adsorption layer which results in the concentration jump referred to above. If $w_{+2} > w_{-2}$, there is a positive adsorption of component B ($C_{2s}^0 > C_{2s}$), and negative adsorption of B ($C_{2s}^0 < C_{2s}$) if $w_{+2} < w_{-2}$.

The resultant flux of component B from phase 2 into the moving adsorbed layer a_2 is

$$I_{B2}\Omega = vC_{2s} + a\,[w_{+2}C_{2s}(1-C_{2s}^0) - w_{-2}C_{2s}^0(1-C_{2s})]. \tag{26}$$

The flux of component A into the adsorption layer a_2 is

$$I_{A2}\Omega = v(1-C_{2s}) + a\,[w_{-2}C_{2s}^0(1-C_{2s}) - w_{+2}C_{2s}(1-C_{2s}^0)],$$

Note that

$$(I_{A2} + I_{B2})\,\Omega = v.$$

Similarly, one can write the resultant flux of component B from adsorbed layer a_1 into phase 1 at their interface:

$$I_{B1}\Omega = vC_{1s}^0 + a\,[w_{-1}C_{1s}^0(1-C_{1s}) - w_{+1}C_{1s}(1-C_{1s}^0)]. \tag{27}$$

Here w_{+1} is the transition frequency of component B from phase 1 into the adsorbed layer a_1, and w_{-1} is the frequency of the reverse transition. Since concentrations in the immediate vicinity of the interface are now C_{1s}^0, and not C_{is}, for calculation of the resultant fluxes I_A and I_B from phase 2 to phase 1 we have to write instead of relations (14) (taking into account the relation $a = q\Omega$),

$$I_A\Omega = a\,[\bar{w}_{+A}(1-C_{2s}^0) - \bar{w}_{-A}(1-C_{1s}^0)]; \tag{28}$$
$$I_B\Omega = a\,[\bar{w}_{+B}C_{2s}^0 - \bar{w}_{-B}C_{1s}^0].$$

If adsorption is allowed for, two additional unknown variables appear, C_{is}^0, so that two additional equations must be formulated to determine them. These equations are the balance conditions in adsorbed layers:

$$\frac{a_1 dC_{1s}^0}{dt} = I_B\Omega - I_{B1}\Omega \quad \text{and} \quad \frac{a_2 dC_{2s}^0}{dt} = I_{B2}\Omega - I_B\Omega.$$

It is easily checked that these constraints automatically provide for the balance of component A. For steady-state growth, when C_{is}^0 are independent of time, or in the quasistationary mode, when quantities $a_i\,dC_{is}^0/dt$ are small compared with each flux in the right-hand side of these equations, they can be written in the form

$$I_{B1} = I_B \quad \text{and} \quad I_{B2} = I_B, \tag{29}$$

where the fluxes are defined by equations (26)-(28).

In the case of adsorption, therefore, the constraints on the growth face are given by the set of equations (6), (9), (10), (26)-(29). We shall now use these equations to analyze the practically significant problem of the dependence of the impurity distribution coefficient K on the crystal growth rate. We shall define K as the ratio of concentrations in the crystal and in the melt at the interface (but outside the adsorbed layer), $K = C_{1s}/C_{2s}$. The effective distribution coefficient K_{eff} is known to be related to K by the formula

$$K_{eff} = C_{1s}/C_{2\infty} = K/[K + (1-K)\exp(-v\delta/D_2)],$$

in which δ is the thickness of the diffusion boundary layer, and C_2 is the concentration in the melt remote from the crystal [16].

In the steady-state growth mode a constant-composition crystal is formed, which means that in equation (10) $D_1 \partial C_1/\partial x = 0$ and $vC_{1s} = I_B \Omega$. Taking account of this and of the relations (26), (27), and (29) we find

$$C_{1s}^0 = K_1 C_{1s}(v + aw_{+1})/(K_1 v + aw_{+1}); \qquad (30)$$
$$C_{2s}^0 = K_2[C_{2s} + (C_{2s} - C_{1s})v/aw_{+2}],$$

where $K_i = w_{+i}/[w_{-i}(1 - C_{is}) + w_{+i}C_{is}]$ ($i = 1, 2$) is the equilibrium adsorption coefficient in phase i, that is, $K_i = C_{is}^0/C_{is}$ for $v = 0$. The required relation between C_{1s} and C_{2s} is found from the condition

$$vC_{1s} = a[\bar{w}_{+B} C_{2s}^0 - \bar{w}_{-B} C_{1s}^0],$$

which follows from (10) and (28). Then

$$K = C_{1s}/C_{2s} = K^0(1 + v/aw_{+2})/[(v/a\bar{w}_{-B}K_1) + (K^0 v/aw_{+2}) + (v + aw_{+1})/(K_1 v + aw_{+1})]. \qquad (31)$$

Here $K^0 = \bar{w}_{+B} K_2 / \bar{w}_{-B} K_1$ is the equilibrium distribution coefficient (for $v = 0$). In the case of melt growth, typically $v \ll aw_{+2}$ since $aw_{+2} \approx D_2/a \approx 10^3$ cm s^{-1}. In this case relation (31) is replaced by

$$K = K^0/[(v/a\bar{w}_{-B}K_1) + (v + aw_{+1})/(K_1 v + aw_{+1})]. \qquad (32)$$

If there is no adsorption ($K_1 = K_2 = 1$), then $K^0 = \bar{w}_{+B}/\bar{w}_{-B}$ and we find from equation (32) $K = K^0/[1 + v/a\bar{w}_{-B}]$, which coincides with the result derived in ref. 17.

In the case of melt growth, when $v \ll aw_{+2} \approx D_2/a$, an equilibrium adsorbed layer builds up in the melt at the interface (according to (30), $C_{2s}^0 \approx K_2 C_{2s}$), and this layer leaves the dependence of K on v practically unaltered. The situation in the crystal is different. Even close to the melting point the diffusion coefficient in the crystal is much smaller than in the melt: $D_1 \approx 10^{-9} - 10^{-12}$ cm^2 s^{-1}. Correspondingly, $aw_{+1} \approx D_1/a \approx 10^{-1} - 10^{-4}$ cm^2 s^{-1}. For low growth rates the equilibrium relation $C_{1s}^0 = K_1 C_{1s}$ will have sufficient time to be restored if $v \ll aw_{+1}$. According to (32), $K \approx K^0$ (here we take into account that $v \ll a\bar{w}_{-B}$). However, as the growth rate increases, the adsorption equilibrium in the crystal breaks down, and for $v \gg aw_{+1}$ relation (30) yields $C_{1s}^0 \approx C_{1s}$. In this situation equation (32) yields $K \approx K^0 K_1$ for $v \ll a\bar{w}_{-B}$. The region of transition from the equilibrium value of K^0 to $K^0 K_1$ corresponds to $v \approx aw_{+1} \approx D_1/a$. A similar transition in layered crystal growth was first analyzed by Chernov [18] and discussed in ref. 19 in relation to the interpretation of experimental data on the dependence of the impurity distribution coefficient on growth rate for semiconductor crystals.

Literature Cited

1. B. Ya. Lyubov. Kinetic Theory of Phase Transitions. Metallurgiya, Moscow (in Russian) (1969).
2. V. T. Borisov. Kinetic diagrams of crystallization in alloys. Dokl. AN SSSR, 142, 69-71 (1962).
3. G. Baralis. Distribution coefficients during the solidification of an ideal binary system in the presence of heat flow. J. Cryst. Growth, 3/4, 627-632, (1968).
4. D. E. Temkin. On diffusion equation in binary interstitial solution. Fiz. Metal. i Metalloved., 24, 207-212 (in Russian) (1967).
5. D. Cahn. On the morphological stability of growing crystals. Proc. Int. Conf. on Crystal Growth Boston (1966), Pergamon Press, pp. 681-690 (1967).

6. R. W. Hopper and D. R. Uhlman. Solute redistribution during crystallization at constant velocity and constant temperature. J. Cryst. Growth, 21, 203-213 (1974).
7. H. Müller-Krumbhaar. Diffusion theory for crystal growth at arbitrary solute concentration. J. Chem. Phys., 63, 5131-5138 (1975).
8. J. R. Manning. Diffusion Kinetics for Atoms in Crystals. D. Van Nostrand Co., Inc., Princeton, Toronto (1968).
9. H. Reiss and M. Shugard. On the composition of nuclei in binary systems. J. Chem. Phys., 65, 5280-5293 (1976).
10. D. E. Temkin. On boundary conditions for the description of diffusional phase transitions in alloys. Kristallografuya, 21, 473-478 (in Russian) (1976).
11. G. M. Kudinov, D. E. Temkin, and B. Ya. Lyubov. Effect of interface on kinetics of isothermal transitions in alloys. Fiz. Metal. i Metalloved., 46, 540-547 (in Russian) (1978).
12. D. E. Temkin. Stability of planar interface in diffusional transformation in binary alloys. Kristallografiya, 22, 924-932 (in Russian) (1977).
13. D. E. Temkin. Metastability region of normal growth in binary alloys. Dokl. AN SSSR, 240, 833-835 (in Russian) (1978).
14. D. E. Temkin. Normal growth on atomically rough interface in binary alloys. Kristallografiya, 23, 1151-1161 (in Russian) (1978).
15. R. H. Swendsen. Monte-Carlo studies of the interface roughening transition. Phys. Rev., B 15, 5421-5431 (1977).
16. J. A. Burton, R. C. Prim, and W. P. Slichter. The distribution of solute in crystals grown from the melt. J. Chem. Phys., 21, 1987-1991 (1953).
17. C. D. Thurmond. Control of composition of semiconductors by freezing method. In: Semiconductors, ed. N. B. Hannay, Reinhold, New York, London, pp. 145-191 (1959).
18. A. A. Chernov. Spiral Growth of Crystals. Soviet Physics Uspekhi, 4, 115-148 (1961).
19. V. V. Voronkov, V. P. Grishin, and Yu. M. Shashkov. Effective distribution coefficient for impurities as a function of growth rate of single crystals. Izv. AN SSSR, Neorg. Mater. 3, 2139-2149 (in Russian) (1967).

GENERAL APPROACH TO MONTE CARLO SIMULATION OF CRYSTAL GROWTH

T. A. Cherepanova

The Latvian State University, Riga

The current theories of crystal growth based on kinetic equations possess a number of important shortcomings. In particular, these equations are capable of describing in single-component systems only individual "extreme" growth mechanisms (normal, nucleation, spiral); the kinetics of crystallization in complex compounds is usually described by analogy with single-component systems, namely, supercooling is replaced in the appropriate equations by supersaturation of the disordered (liquid or vapor) phase. The difficulties encountered in constructing equations describing growth kinetics over a wide range of values of system parameters led to the development of a direct method: Monte Carlo computer simulation of crystal growth. In the present paper, based on the ideas developed in [1-15], we give a statistical-mechanic description of the techniques valid for simulating crystal growth both in single-component and in multi-component crystals.

Consider a lattice model of a two-phase crystal-melt system. For simplicity, we limit the analysis to a binary α - β system. The lattice sites are occupied by the atoms of α or β species, which can belong either to the solid or to the liquid phases. Particle-to-particle nearest-neighbor interaction is characterized by the coupling energies between solid-phase particles ($\varphi_{11}^{\alpha\alpha}$, $\varphi_{11}^{\alpha\beta}$, $\varphi_{11}^{\beta\beta}$), between solid-phase and liquid-phase particles ($\varphi_{10}^{\alpha\alpha}$, $\varphi_{10}^{\alpha\beta}$, $\varphi_{01}^{\alpha\beta}$, $\varphi_{10}^{\beta\beta}$) and between liquid-phase particles ($\varphi_{00}^{\alpha\alpha}$, $\varphi_{00}^{\alpha\beta}$, $\varphi_{00}^{\beta\beta}$) (subscript 0 indicates that a particle belongs to the liquid phase, and subscript 1 — to the solid phase). Each arrangement of atoms will be described by a set of parameters $\vec{g} = \{\xi_j, \eta_j\}$ [7-10]; $\eta_j = 1$ if the j-th lattice site is occupied by a solid-phase particle, and $\eta_j = 0$ if it is occupied by a liquid-phase particle; ξ_j describes the species of particle in this lattice site ($\xi_j = \alpha, \beta$). The thermodynamic potential of the system plays the role of the Hamiltonian $H(\vec{g})$ for the ordering field $\vec{g}$ [7]:

$$H(\vec{g}) = \sum_{\nu, \eta} N_\eta^\nu \tilde{\mu}_\eta^\nu + \sum_\nu N_{10}^{\nu\nu} \omega_{10}^{\nu\nu} + \sum_{\eta, \eta'} N_{\eta\eta'}^{\alpha\beta} \omega_{\eta\eta'}^{\alpha\beta}; \qquad (\nu = \alpha, \beta). \tag{1}$$

Hence $\tilde{\mu}_\eta^\nu$ is the chemical potential of the pure ν-component in phase state η; N_η^ν is the total number of ν-η particles; $N_{\eta\eta'}^{\nu\xi}$, is the number of bonds between nearest-neighbor particles of species ν and ξ, belonging to phase states η and η', respectively; the heat of mixing of the components is $\omega_{\eta\eta'}^{\nu\xi} = \varphi_{\eta\eta'}^{\nu\xi} - (\varphi_{\eta\eta}^{\nu\nu} + \varphi_{\eta'\eta'}^{\xi\xi})/2$. Chemical potentials of the components of the solid and liquid phases are to be found in terms of the mean change in the Hamiltonian for the process of phase transition:

$$\begin{aligned}\mu_1^\nu &= \left\langle \left(\frac{\partial H}{\partial N_1^\nu}\right)_{N_0^\xi, N_1^\xi, N_0^\nu} \right\rangle - T\left(\frac{\partial S}{\partial N_1^\nu}\right)_{N_0^\xi, N_1^\xi, N_0^\nu} \\ \mu_0^\nu &= \left\langle \left(\frac{\partial H}{\partial N_0^\nu}\right)_{N_0^\xi, N_1^\xi, N_1^\nu} \right\rangle - T\left(\frac{\partial S}{\partial N_0^\nu}\right)_{N_0^\xi, N_1^\xi, N_1^\nu}\end{aligned} \qquad (\xi \neq \nu), \tag{2}$$

where S is the configurational part of the system's entropy, and T is the phase transition temperature. The system of equations (2) involves the assumption that $\partial \langle H \rangle / \partial N_\eta^\nu \approx \langle \partial H / \partial N_\eta^\nu \rangle$.

The growth process will be represented by a sequence of elementary events at individual lattice sites, for a fixed configuration of the arrangement of particles which do not take part in the elementary event at the moment it occurs. The characteristic time of the process, τ, is thus chosen to be sufficiently short, so that not more than one event may take place within time τ. An atom of species ξ in the liquid may attach itself to the crystal with a frequency W_{10}^{ξ}, and an atom in a crystal may pass into the melt with a frequency W_{01}^{ξ}. A pair of nearest α-, β-particles in the same phase state η may diffusionally exchange their locations in the lattice with frequency $W_{\beta;\eta}^{\alpha;\eta}$. According to the detailed equilibrium condition [10],

$$W_{01}^{\xi_i}/W_{10}^{\xi_i} = \exp[-\Delta H(\xi_i)/kT] = \exp[-(kT)^{-1} \sum_{\nu,\eta'} l_\eta^{\xi_i\nu'}(\varphi_{0\eta'}^{\xi_i\nu} - \varphi_{1\eta'}^{\xi_i\nu}) + \theta_{\xi_i}],$$

$$W_{\alpha_i,\,\eta_i}^{\beta_j,\,\eta_j}/W_{\beta_i,\,\eta_i}^{\alpha_j,\,\eta_j} = \delta_{\eta\eta_i}\delta_{\eta\eta_j} \exp\{[(l-1-l_\eta^{\alpha_i\alpha}-l_\eta^{\beta_j\beta})(\varphi_{\eta\eta}^{\alpha\alpha}+\varphi_{\eta\eta}^{\beta\beta}-2\varphi_{\eta\eta}^{\alpha\beta}) -$$
$$- l_{1-\eta}^{\alpha_i\alpha}(\varphi_{\eta\,1-\eta}^{\alpha\alpha}+\varphi_{\eta\eta}^{\beta\beta}-\varphi_{\eta\,1-\eta}^{\beta\alpha}-\varphi_{\eta\eta}^{\alpha\beta}) - l_{1-\eta}^{\alpha_i\beta}(\varphi_{\eta\,1-\eta}^{\alpha\beta}+\varphi_{\eta\eta}^{\beta\beta}-\varphi_{\eta\,1-\eta}^{\beta\beta}-\varphi_{\eta\eta}^{\alpha\beta}) -$$
$$- l_{1-\eta}^{\beta_j\beta}(\varphi_{\eta\,1-\eta}^{\beta\beta}+\varphi_{\eta\eta}^{\alpha\alpha}-\varphi_{\eta\,1-\eta}^{\alpha\beta}-\varphi_{\eta\eta}^{\alpha\beta}) - l_{1-\eta}^{\beta_j\alpha}(\varphi_{\eta\,1-\eta}^{\beta\alpha}+\varphi_{\eta\eta}^{\alpha\alpha}-\varphi_{\eta\,1-\eta}^{\alpha\alpha}-\varphi_{\eta\eta}^{\alpha\beta})](kT)^{-1}\}.$$

Here $\delta_{\eta\eta'}$ are Kronecker's symbols; $\Delta H(\xi_t)$ is the change in the Hamiltonian when a ξ_i-particle passes from the liquid phase into the solid; $l_\eta^{\xi_i\nu}$ is the number of ν-species nearest-neighbor particles of the ξ_i particle in phase state η; l is the coordination number; 2). $\theta_\nu = (\tilde{L}^\nu - \Delta\tilde{\mu}^\nu)/kT$; $\tilde{L}^\nu$ is the phase transition heat in the pure ν-component ($\tilde{L}^\nu = l(\varphi_{00}^{\nu\nu} - \varphi_{11}^{\nu\nu})/$ In the case of sufficiently low growth rates the diffusional processes in the solid and liquid phases determine completely the whole spectrum of local equilibrium states of the system in the configuration space, so that $\Delta\mu^\nu = \mu_0^\nu - \mu_1^\nu$ has the meaning of a difference between chemical potentials of the bulk phases. Relations (1) and (2) thus yield

$$\Delta\mu^\nu = \Delta\tilde{\mu}^\nu + \gamma_0^\nu \omega_{00}^{\alpha\beta} - \gamma_1^\nu \omega_{11}^{\alpha\beta} + T\left(\frac{\partial S}{\partial N_1^\nu} - \frac{\partial S}{\partial N_0^\nu}\right).$$

$$\Delta\tilde{\mu}^\nu = \tilde{L}^\nu (T_{eq}^\nu - T)/T_{eq}^\nu.$$

The structure factor of η-phase is $\gamma_\eta^\nu = \langle \partial N_{\eta\eta}^{\alpha\beta}/\partial N_\eta^\nu \rangle$ [5–7]. In the single-particle approximation (the condition of complete mixing in the two phases) we obtain

$$\Delta\mu^\nu = \Delta\tilde{\mu}^\nu + l[(1-C_0^\nu)^2 \omega_{00}^{\alpha\beta} - (1-C_1^\nu)^2 \omega_{11}^{\alpha\beta}] + kT \ln(C_0^\nu/C_1^\nu),$$

where $C_\eta^\nu = N_\eta^\nu/N$ is the concentration of ν-component in the bulk of η-phase; N is the total number of particles in the system; k is the Boltzmann constant.

If the characteristic times of diffusional exchange of atoms in the crystal are much larger than the characteristic times related to growth rate and diffusion in the liquid phase, the composition of the solid phase is determined exclusively by the kinetics of processes on the surface of the growth face of the crystal.

The enthalpy contribution $\Delta\mu^\nu$ is determined by the mean number of solid-solid bonds $\langle l_1^{\nu\xi} \rangle$ (for an atom located at a half-crystal position) broken in the course of growth [7]:

$$\Delta\mu^\nu = \langle l_1^{\nu\alpha} \rangle \Phi_{\nu\alpha} + \langle l_1^{\nu\beta} \rangle \Phi_{\nu\beta} + l\varepsilon_\nu - kT\theta_\nu + T\left(\frac{\partial S}{\partial N_1^\nu} - \frac{\partial S}{\partial N_0^\nu}\right). \qquad (3)$$

In a completely mixed melt

$$\Phi_{\nu\xi} = \varphi_{01}^{\nu\xi} - \varphi_{11}^{\nu\xi} - \varepsilon_\nu. \qquad \varepsilon_\nu = \sum_\xi (\varphi_{00}^{\nu\xi} - \varphi_{10}^{\nu\xi}) C_0^\xi. \qquad \frac{\partial S}{\partial N_0^\nu} = k \ln C_0^\nu.$$

If the statistical weight of the crystalline phase in the configurational space $\vec{g}$ is nearly equal to unity (the symmetry of the distribution of different species of particles in the lattice is practically constant in the course of growth), then the entropy term $\partial S/\partial N_1^\nu$ in equation (3) can be neglected, and

$$\Delta\mu^\nu = \langle l_1^{\nu\alpha} \rangle \Phi_{\nu\alpha} + \langle l_1^{\nu\beta} \rangle \Phi_{\nu\beta} + l\varepsilon_\nu - T\frac{\partial S}{\partial N_0^\nu} - kT\theta_\nu. \tag{4}$$

If the liquid phase can be treated as a mixture of cells each of which consists of a fixed set of particles of species α and β, it must be described as a single-component melt. Multicomponent nature is then taken into account only in the enthalpy terms:

$$\Delta\mu^\nu = \langle l_1^{\nu\alpha} \rangle \Phi_{\nu\alpha} + \langle l_1^{\nu\beta} \rangle \Phi_{\nu\beta} + l\varepsilon_\nu - kT\theta_\nu + T\frac{\partial S}{\partial N_1^\nu}. \tag{5}$$

When such a melt crystallizes into a lattice possessing a specific symmetry of distribution of α- and β-particles [7], we have

$$\Delta\mu^\nu = \langle l_1^{\nu\alpha} \rangle \Phi_{\nu\alpha} + \langle l_1^{\nu\beta} \rangle \Phi_{\nu\beta} + l\varepsilon_\nu - kT\theta_\nu. \tag{6}$$

Relation (3) and all equations derived from it, (4)–(6), establish a relation between the chemical potential difference of the phase and the positions at the half-crystal interface in a binary system. Similar relations can be written for crystallization in multicomponent systems both from the melt and from solution. In the case of growth of a single-component crystal, equation (3) yields $\langle l_1 \rangle = 1/2$.

It is, therefore, necessary to distinguish between two limiting growth modes when describing crystallization in multicomponent systems: a mode limited by diffusion processes in the bulk phases, and a mode limited by the kinetics of surface processes. Let us analyze the latter mode in more detail. For a completely mixed melt [1, 3, 4]

$$W_{01}^{\xi_i}/W_{10}^{\xi_i} = (C_0^{\xi_i})^{-1} \exp\left[-\left(\sum_\nu l_1^{\xi_i \nu} \Phi_{\xi_i \nu} + l\varepsilon_{\xi_i}\right)/kT + \theta_{\xi_i}\right]. \tag{7}$$

In order to find $W_{01}^{\xi_i}$ from relation (7), one has to know $W_{10}^{\xi_i}$. In Ising's kinetic models one normally assumes that $W_{10}^{\xi_i}$ is the following function of the configuration of the nearest-neighbor ξ_i particles [12, 14, 15]:

$$W_{10}^{\xi_i} \approx \sum_\nu \delta_{\xi_i \nu} C_0^\nu \{1 + \exp[-(l_1^{\nu\alpha}\Phi_{\nu\alpha} + l_1^{\nu\beta}\Phi_{\nu\beta} + l\varepsilon_\nu)/kT + \theta_\nu]\}^{-1}. \tag{8}$$

In refs. 1–13 it was assumed that the frequency of attachment of a particle to the crystal is independent of the species of the nearest-neighbor particles, and that

$$W_{10}^\nu = C_0^\nu \omega \exp(-u/kT), \tag{9}$$

where $\omega\exp(-u/kT) = \text{const}$, ω is the frequency factor, and u is the activation energy for the liquid-to-solid transition. We define the characteristic time τ of the process in terms of the maximum possible values of the transition frequencies in the selected model:

$$\tau = [\max(W_{10}^\alpha + W_{10}^\beta + W_{01}^\xi)]^{-1}.$$

The probabilities of transition from the phase state η_i' into η_i in the ith lattice site over time τ are

$$\mathcal{W}_{\eta_i\eta_i'}^{\xi_i} = \tau W_{\eta_i\eta_i'}^{\xi_i}, \qquad \mathcal{W}_{10}^{\xi_i} + \mathcal{W}_{00}^{\xi_i} = \mathcal{W}_{01}^{\xi_i} + \mathcal{W}_{11}^{\xi_i}. \tag{10}$$

Here $\widetilde{W}_{00}^{\xi_i}$, $\widetilde{W}_{11}^{\xi_i}$ are the probabilities of non-events. Once the expressions for the probabilities of elementary events are known it is possible to achieve a Monte Carlo simulation of crystallization [1-4, 6, 9, 13, 16]. The flow chart of the principal simulating procedure is shown in Fig. 1. The equal-probability choice of the coordinates of an event was achieved by means of a generator of pseudorandom numbers, $\varkappa$, distributed with equal probability over the (0,1) interval. For the choice of event type over the chosen event coordinates the procedure employs the next pseudorandom number extracted when the generator is addressed. A comparison of $\varkappa$ with the probabilities of elementary events (10) in the ith surface lattice site defines the type of event:

$$\eta_i = 0;\ 0 \leqslant \varkappa \leqslant W_{10}^{\alpha} - \text{attachment of atom } \alpha;$$

$$\eta_i = 0;\ W_{10}^{\alpha} < \varkappa \leqslant W_{10}^{\alpha} + W_{10}^{\beta} - \text{attachment of atom } \beta;$$

$$\eta_i = 1;\ 0 \leqslant \varkappa \leqslant W_{01}^{\xi_i} - \text{detachment of atom } \xi_i \text{ from the crystal.}$$

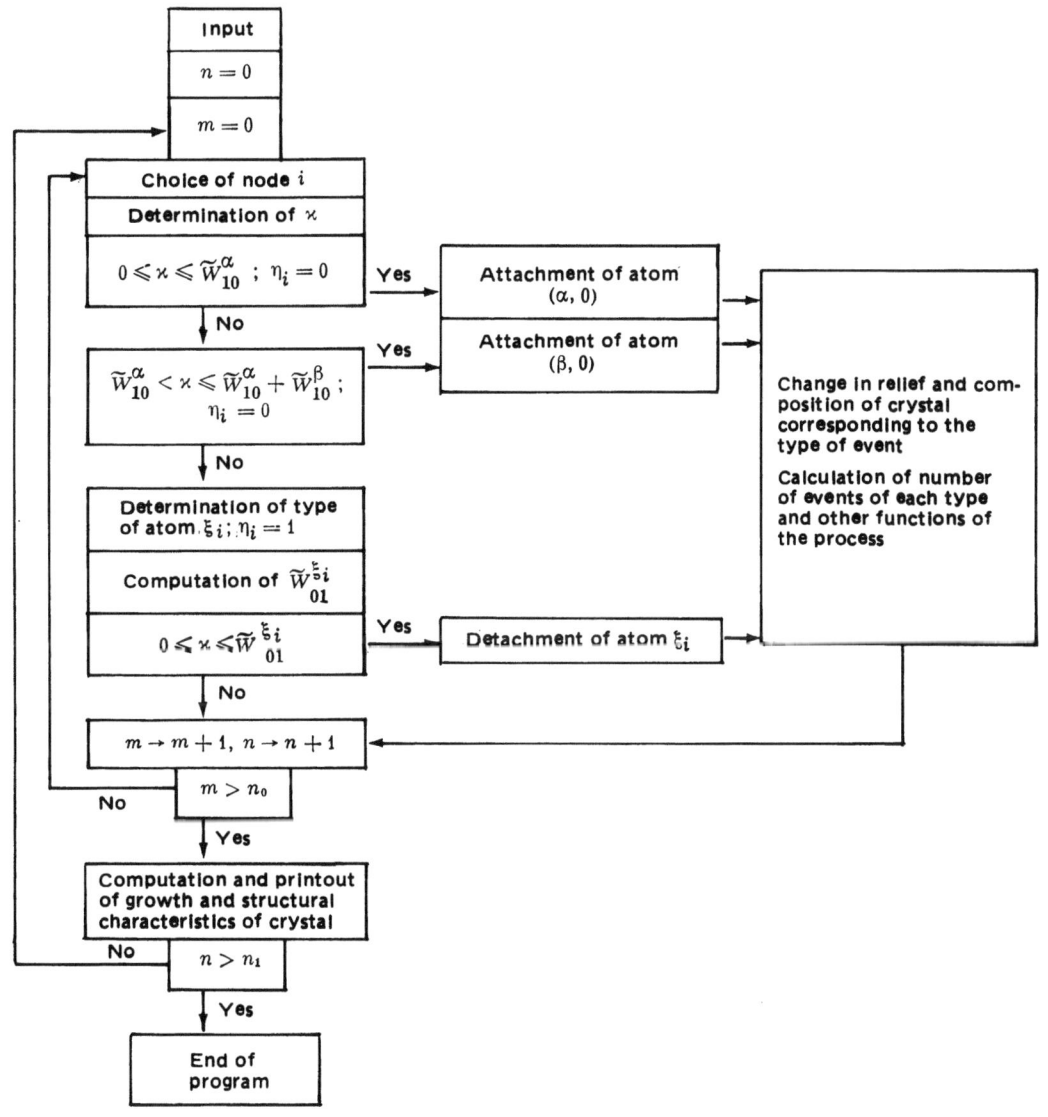

Fig. 1. Flow chart of the principal simulation procedure.

After the event type is determined, the crystal structure is appropriately modified and the process characteristics are calculated. Simulation considers only the states of the system corresponding to interface configurations without overhangs. This means that only one solid-phase atom in each atomic column in the growth direction has a neighbor belonging to the liquid. The physical duration of the process is determined by the number of events n: $t = n\tau$. The number n_0 determines time intervals $n_0 \tau$ separating the moments of printout of the statistical characteristics. All statistical quantities depend on the number of tests, and for a given n are averages over all preceding tests beginning with the onset of the steady-state growth mode. This procedure ensures that a random characteristic tends to its mean value for $n \to \infty$. The number of tests necessary to reach the steady-state mode is adjusted experimentally. The number n_1 determines the total time $n_1 \tau$ of the simulation experiment. The dimensionless growth rate of the crystal is calculated by the formula

$$R = V/a \, (W_{10}^{\alpha} + W_{10}^{\beta}) = \Delta M / n \tau s \, (W_{10}^{\alpha} + W_{10}^{\beta}).$$

Here V is the growth rate, a is the lattice parameter, ΔM is the increment in the number of particles in the crystal over n tests, and s is the size of the simulated face.

A general approach making it possible to describe crystallization of multicomponent systems on the basis of kinetic equations both for a completely mixed liquid phase and for a liquid phase with diffusion processes was developed in refs. 5, 7-10, 14, 15, by using the same physical assumptions as in the case of computer simulation. The system of equations describing crystallization of binary alloys was obtained by the molecular field approximation and by the two-particle approximation of the distribution function.

In conclusion, we shall consider general features of crystallization in metal-type binary systems [growth on atomically rough (001) faces] by using simulation techniques and by solving kinetic equations. For the sake of simplification, we choose as a model those systems with primitive cubic lattice symmetry.

Crystallization of Regular Melts. Growth from the Melt [1, 3, 4].
The difference in the effective energies of bonding, $\Phi_{\nu\xi}$, between particles of the same species and particles of different species is low (the energy parameters of the model system investigated are: $\Phi_{\alpha\alpha} = \Phi_{\alpha\beta} = 300 \, \text{cal mol}^{-1}$, $\Phi_{\beta\beta} = 500 \, \text{cal mol}^{-1}$, $\varepsilon_v = 0; \theta_\alpha = \theta_\beta = 1$). The kinetic phase diagrams shown in Fig. 2 are cigar-shaped; horizontal lines connect liquidus points (on the left) and solidus points (on the right) which correspond to the same melt composition. Unless otherwise stated the lines on all figures denote the results of analytic calculations, and circles and triangles denote the results of simulation. As the growth rate is increased, the phase diagrams remain practically non-deformed and only shift to lower temperatures. The effect of short-range ordering is small. Growth rate remains a linear function of temperature for all values of component concentrations. As the relative supercooling increases, the roughness r and the interface thickness δ diminish. The characteristics R, r, and δ are plotted in Fig. 3 as functions of T for $C_0^\alpha = 0.5$. The results obtained indicate that crystallization kinetics in multicomponent regular alloys are practically the same as in single-component systems. As in those systems, the normal growth mechanism results in a linear dependence of growth rate on supercooling (supersaturation) δ.

Crystallization in Eutectic-type Alloys. Growth from the Melt [7-13]. The effective energies of α-α bonds and β-β bonds are essentially different (the energy parameters of the analyzed model system are: $\Phi_{\alpha\alpha} = 300 \, \text{cal mol}^{-1}$; $\Phi_{\alpha\beta} = 200 \, \text{cal mol}^{-1}$; $\Phi_{\beta\beta} = 1500 \, \text{cal mol}^{-1}$, $\varepsilon_v = 0; \theta_\alpha = 1; \theta_\beta = 3$). Pure α- and β-components are crystallized by the normal mechanism. Figure 4 shows the concentration of the α-component, C_1^α, in the bulk of the crystal plotted as a function of growth rate for different compositions of the liquid phase. We have calculated the structure and composition of the transient zone

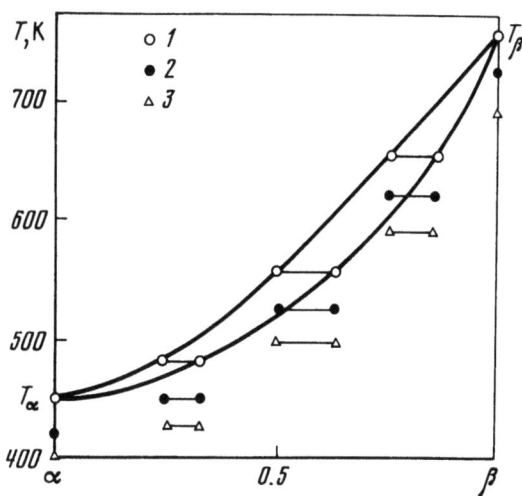

Fig. 2. Equilibrium phase diagram and calculated points of kinetic phase diagrams of a regular alloy for different growth rates. (1) R = 0; (2) 0.0456; (3) 0.0906.

Fig. 3. Roughness r and two-phase zone width δ as functions of temperature, and growth rate R as a function of temperature and supercooling δ.

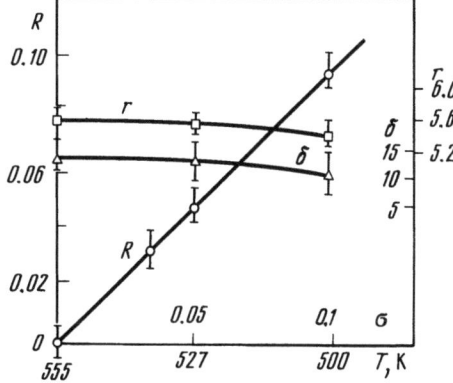

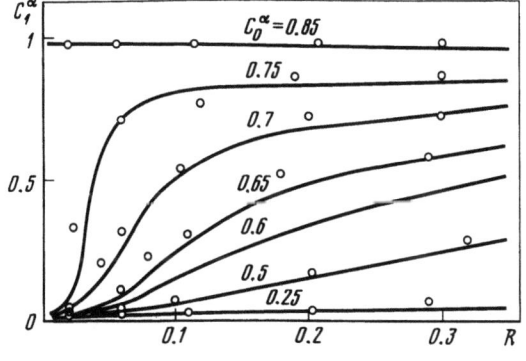

Fig. 4. Concentration of α-component in the bulk of the solid phase as a function of growth rate for different melt compositions.

between the phases for $C_0 = 0.79$ and T = 400 K (Fig. 5a) and for T = 445 K (Fig. 5b) (c_1 is the number of solid-phase particles and c_1^α is the number of solid-phase α-particles in an atomic layer with number z, per surface lattice site of a (001) face; $c_1 = C_1$ and $c_1^\alpha = C_1^\alpha$ for $z \to -\infty$). The results of simulation [12] and of calculations based on the kinetic equations [10] point to an interesting feature in the constitutional structure of the interface. An adsorption layer enriched in the lower melting-point component is formed at the surface of the growth face for low growth rates (see Fig. 5b). We also find that the non-monotonic dependence $c_1^\alpha(z)$ becomes increasingly important, according to the kinetic phase diagram, the larger is the difference between the concentrations of the α-component in the melt and in the crystal. The enrichment of surface layers with α-particles compared with atomic layers

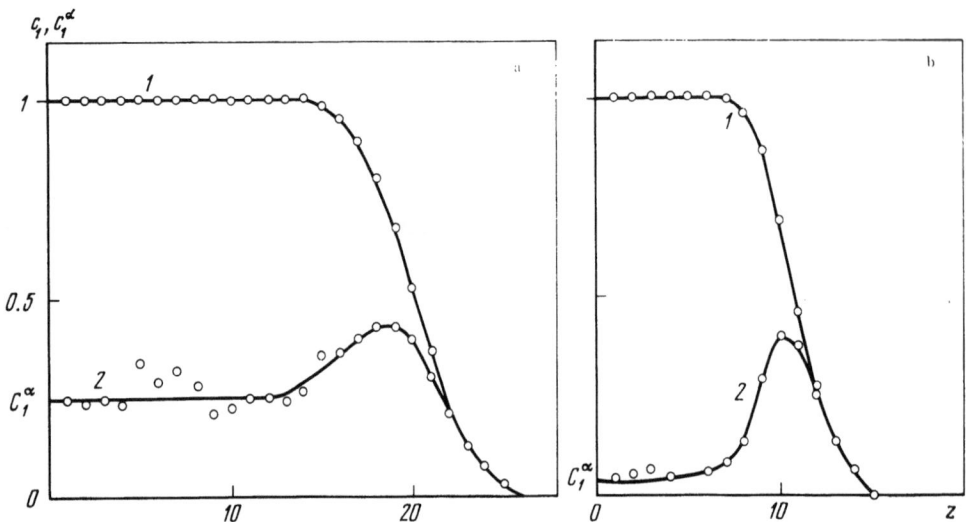

Fig. 5. Structure (1) and composition (2) of the transient zone between phases for $C_0^\alpha = 0.79$ at temperatures (a) 400 K and (b) 445 K.

in the bulk of the crystal diminishes as growth rate increases. In the model under discussion the adsorption layer vanishes for $T < 390$ K, $R > 0.014$. The characteristics shown in Fig. 4 reveal an anomalous dependence of bulk-phase composition C_1^α in the range $0.5 < C_0^\alpha \leq 0.80$ for $R \lesssim 0.1$: as the growth rate decreases, the α-component concentration in the solid phase decreases abruptly. An increase in temperature from 400 K to 445 K reduces C_1^α (see Fig. 5). Anomalous behavior is also clearly seen in Fig. 6, where the phase diagrams are plotted for $R = 0.02, 0.06, 0.2$. The diagram for $R = 0.02$ has a form typical for eutectic systems. As the growth rate increases, the jump in concentrations from the liquid to the solid phase falls, and the kinetic diagrams degenerate to cigar shapes with a minimum. The growth rate and roughness for different concentrations of the melt are plotted as functions of temperature in Fig. 7. Large values of interface roughness and the linear dependence of crystallization rate on temperature in pure α- and β-components are evidence in favor of the normal mechanism of growth in the alloys investigated. Also, the non-linear temperature dependence of growth rate for $C_0^\alpha = 0.065$ and 0.7 points specifically to a normal mechanism operating in multicomponent systems. Consequently, the degree of roughness of growth faces can serve in the general case as a criterion for a specific growth mechanism.

An explicit effect of the attachment frequency W_{10}^ξ on structural and kinetic characteristics of the two-phase zone and the crystal was analyzed in refs. 12, 15. The kinetic diagrams corresponding to W_{10}^ξ found from equations (8) and (9) for $R = 0.1$ are plotted in Fig. 8. These results show that the qualitative behavior of the main characteristics of the system in the course of crystallization is determined by the ratio of transition frequencies W_{01}^ξ / W_{10}^ξ and not by individual values of W_{10}^ξ and W_{01}^ξ.

Up to this point we have discussed a completely mixed melt (see eq. (7)). The diffusion processes at the interface were assumed to be so facile that the component concentration in the melt was independent of the local configuration of the interface. If the characteristic rate of the diffusional transfer of material in the constitutional boundary layer is below the maximum possible growth rate at a given temperature and component concentration in the core of the melt, both the growth rate and the structural characteristics of the interface region are determined by the diffusional mass transfer towards the interface and by the concentration gradient in its vicinity. An analytical description of crystallization in

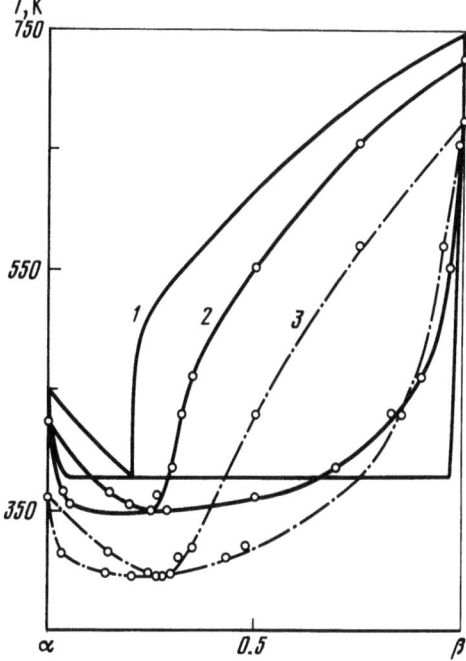

Fig. 6. Kinetic phase diagrams of a eutectic-type alloy for different growth rates. (1) R = 0.02; (2) 0.06; (3) 0.2.

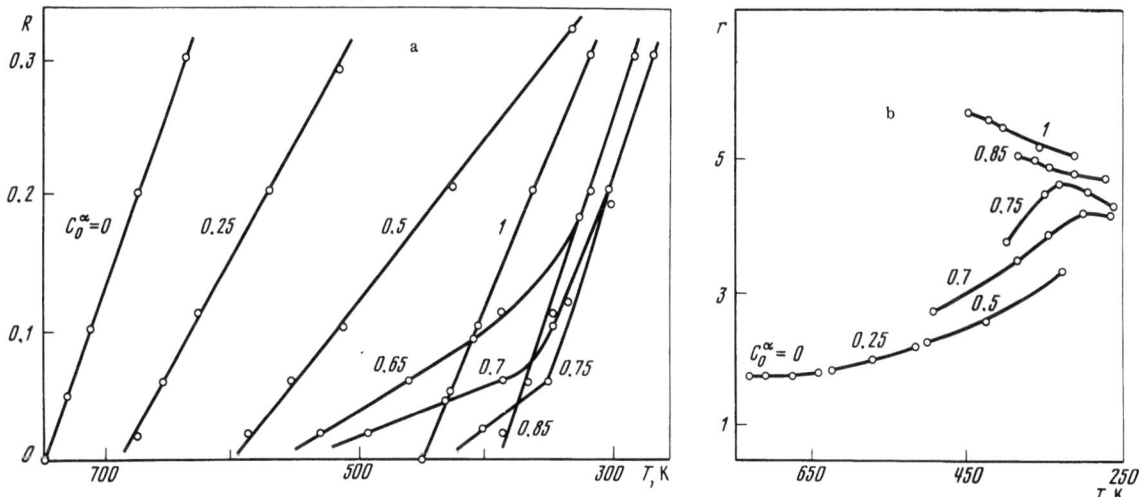

Fig. 7. Growth rate (a) and face roughness (b) as functions of temperature for different compositions of the melt.

binary alloys, taking account of diffusion in the melt for the two-particle approximation of the distribution function, was given in refs. 9, 10, 12. It was shown that the one-particle approximation gives an unsatisfactory description of the process kinetics. Figure 9 gives growth rate R plotted as a function of temperature for two compositions of the melt, $C_0^\alpha = 0.25$ and 0.5, in a completely mixed melt ($W_{\beta 0}^{\alpha 0} \to \infty$) and in a melt with diffusion taken into account ($W_{\beta 0}^{\alpha 0} = W_{10}^{\xi} =$ const.). The results show a considerable effect of diffusion on crystallization kinetics in the two-phase zone. An estimate of concentration gradients limiting the growth can be deduced from Fig. 10 which presents the distribution of atomic concentrations in the solid phase, c_1, in atomic layers of the two-phase zone and of the α-component in the liquid phase c_0^α / c_0 ($c_0^\alpha = C_0^\alpha$ for $z \to +\infty$) and in the solid phase, c_1^α / c_1, in atomic layers along the growth direction. The concentration of the α-component in the melt core is $C_0^\alpha|_{z=h}=0.5$. The thickness of the diffusion boundary layer was measured relative to the atomic layer in which $C_1=0$. In the specific model described above we have chosen $h=4$. For a range of values of system parameters in which disordering in the solid phase is unimportant, both the concentration gradient in the melt and the interface zone width increase with increasing crystallization rate.

The growth kinetics and structure formation mechanism for eutectic alloys with retrograde liquidus kinetic curves were studied in ref. 11. It was demonstrated that as the crystallization rate increases, the kinetic diagrams of such systems first transform to a

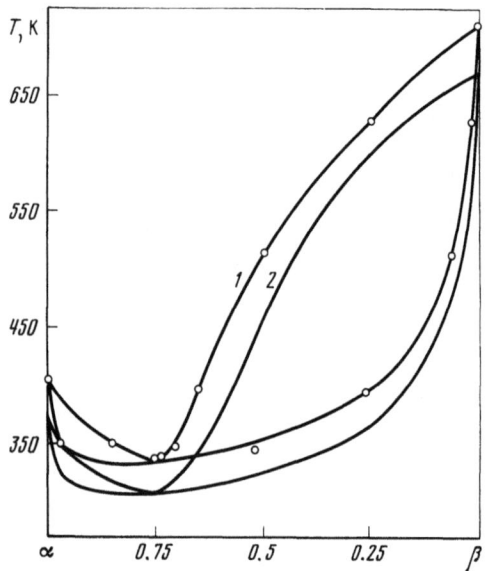

Fig. 8. Kinetic phase diagrams for $R=0.1$. Curves 1 and 2 refer to W_{10}^ξ found from equations (8) and (9), respectively.

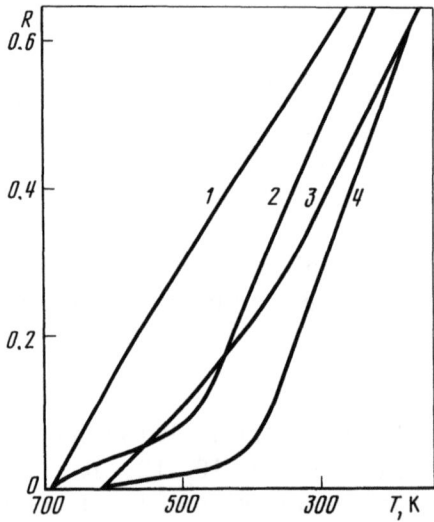

Fig. 9. Temperature dependence of growth rate for different compositions of the melt, in a completely mixed melt and in a melt with diffusion taken into account. (1, 2) $C_0^\alpha=0.25$ and $W_{\beta 0}^{\alpha 0} \to \infty$ (1) and $W_{\beta 0}^{\alpha 0}=W_{10}^\xi$ (2); (3, 4) $C_0^\alpha=0.5$ and $W_{\beta 0}^{\alpha 0} \to \infty$ (3), $W_{\beta 0}^{\alpha 0}=W_{10}^\xi$ (4).

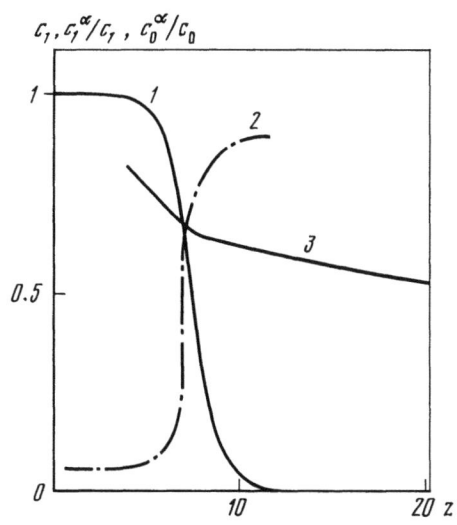

Fig. 10. Structure and composition of the interphase zone for $C_0^\alpha = 0.5$ in the melt core. (1) $C_1(z)$; (2) $C_1^\alpha(z)/C_1(z)$; (3) $C_0^\alpha(z)/C_0(z)$.

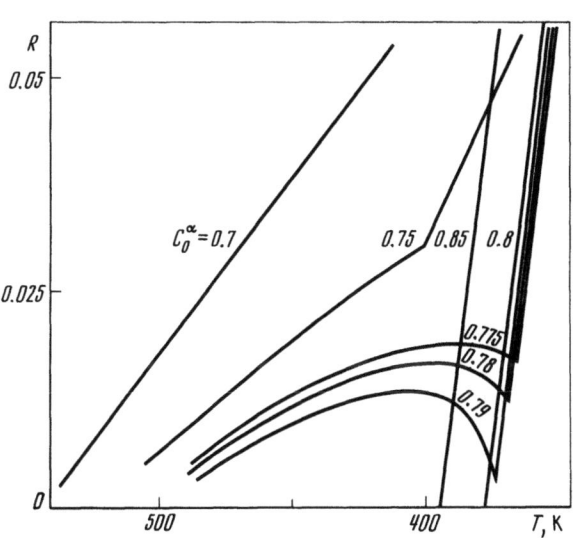

Fig. 11. Temperature dependence of growth rate for different compositions of liquid phase in the range of retrograde liquidus curve.

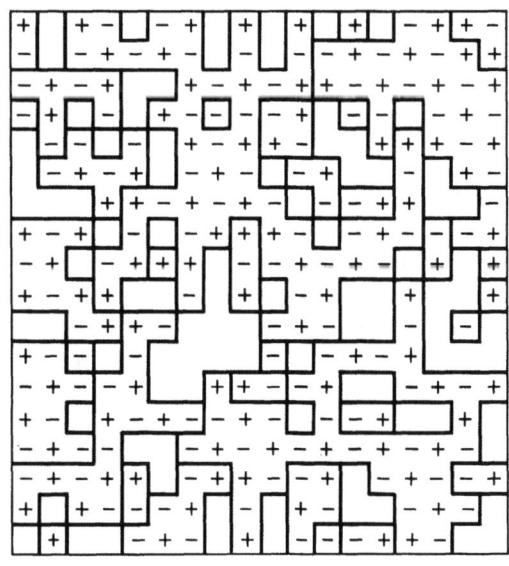

Fig. 12. Cross-section of interphase zone in the plane of a (001) face at high values of supersaturation $\Delta\mu_0^\nu = 3.2$ kT (simulation results).

simple eutectic (of the type shown in Fig. 6), and then at larger values of R degenerate to cigar-shaped diagrams with a minimum. The retrograde behavior of the liquidus curve results in a non-monotonic dependence of growth rate on temperature (Fig. 11).

Growth of Ionic Crystals from Solution [6]. The energy parameters of the model system investigated are: $\Phi_{\alpha\beta} = -\Phi_{\alpha\alpha} = -\Phi_{\beta\beta} = kT$; $\varepsilon_v = 0$; $\theta_v = -3$; the solution is completely mixed ($W_{\beta 0}^{\alpha 0} \to \infty$). The effects of supersaturation in the liquid phase, $\Delta\mu_0^v = \mu_0^v - (\mu_0^v)_{eq}$, on the growth rate and structural characteristics of the interface and crystal were studied. Although the surface of the growth face is atomically rough, crystallization proceeds for NaCl-type structure up to very high values of $\Delta\mu_0^v$. At still higher values of supersaturation the solid phase is disordered by means of domain formation, with a regular structure within the domains. However, the structures at the domain boundaries are not matched (see Fig. 12). As the supersaturation $\Delta\mu_0^v$ increases, the characteristic size of the domains diminishes, thus causing supersaturation in the solid phase. Near equilibrium the value of γ_1^v corresponds to the ordered state: $\gamma_1^v = 3$. At $\Delta\mu_0^v/kT > 2$, γ_1^v deviates from $(\gamma_1^v)_{eq}$ more significantly, reducing γ_1^v to $\gamma_1^v = 1.5$, which corresponds to a totally disordered crystal phase.

Literature Cited

1. T. A. Cherepanova, A. V. Shirin, and V. T. Borisov. Simulation of an equilibrium structure of a crystal-melt interface. Uchen. Zap. Latv. Gos. Univ., 237, 40-59 (in Russian) (1975).
2. T. A. Cherepanova and V. F. Kiselev. On the structure of a crystal-melt interface. Uchen. Zap. Latv. Gos. Univ., 237, 60-67 (in Russian) (1975).
3. T. A. Cherepanova, A. V. Shirin, and V. T. Borisov. Computer simulation of crystal growth from melt. In: Industrial Crystallization. New York: Plenum Press, 113-121 (1976).
4. T. A. Cherepanova, A. V. Shirin, and V. T. Borisov. Computer simulation of binary crystal growth. Kristallografiya, 22, 260-266 (in Russian) (1977).
5. T. A. Cherepanova. General approach to analytical description of multicomponent crystal growth. Acta Cryst., A34, 209 (1978).
6. T. A. Cherepanova, J. P. van der Eerden, and P. Bennema. Fast growth of ordered AB crystals: A Monte Carlo simulation for ionic crystal growth. J. Cryst. Growth, 44, 537-544 (1978).
7. T. A. Cherepanova. Kinetics of binary melt-crystal phase transitions. Zh. Prikl. Mekh. Tekh. Fiz., 5, 122-132 (in Russian) (1978).
8. T. A. Cherepanova. Crystallization kinetics of multicomponent alloys. Dokl. AN SSSR, 238, 162-165 (in Russian) 1978).
9. T. A. Cherepanova. General approach to Monte Carlo simulation of crystal growth. In: Vth USSR Conference on Crystal Growth, Tbilisi, Abstract, Inst. of Cybernetics of AN GSSR Publishing House, 1, 26-27 (in Russian) (1977).
10. T. A. Cherepanova. Analytical description of crystallization in binary alloys. Zh. Prikl. Mekh. Fiz., 6, 96-104 (in Russian) (1978).
11. T. A. Cherepanova and V. F. Kiselev. Kinetics of irregular binary alloy crystallization. Krist. und Techn., 14, 454-552 (1979).
12. T. A. Cherepanova and G. T. Didrihsons. The diffusion influence in melt on the solid solution formation for a system of eutectic type. Krist. und Techn., 12, (1979).
13. T. A. Cherepanova and V. F. Kiselev. Structure of interface in crystal-melt eutectic system. Kristallografiya, 24, 327-333 (in Russian) (1979).

14. V. N. Kuzovkov, B. N. Rolov, and T. A. Cherepanova. Kinetic equations in crystallization theory. Part 1. Single-component system. Izv. AN Latv. SSR, Ser. Fiz.-Tekhn., 5, 43-50 (in Russian) (1977).
15. V. N. Kuzovkov, B. N. Rolov, and T. A. Cherepanova. Kinetic equation in crystallization theory. Part 2. Two-component system. Izv. AN Latv. SSR, Ser. Fiz.-Tekhn., No. 6, 15-21 (in Russian) (1977).
16. J. P. van der Eerden, P. Bennema, and T. A. Cherepanova. Survey of Monte Carlo simulations of crystal surfaces and crystal growth. Progr. Crystal Growth and Character., 3, 219-254 (1978).

PARAMETERS CHARACTERIZING THE KINETICS OF DISSOLUTION OF CRYSTALLINE GERMANIUM IN LIQUID Ge-Au

S. A. Grinberg

The Institute of Crystallography of the USSR Academy of Sciences, Moscow

Formulation of the Problem and Experimental Techniques

This paper describes a quantitative investigation of the kinetics of crystalline germanium dissolution in liquid Ge-Au by observing the motion of droplets in a temperature gradient [1]. This method is a variant of temperature gradient zone melting (TGZM) [2, 3], and consists of the following: if a droplet of liquid alloy (Ge-Au) is placed on the surface of a single crystal (Ge) in a temperature gradient, then diffusional transfer of germanium through the liquid phase towards the lower temperature will take place owing to the difference in equilibrium concentrations of germanium at the high- (C_2) and low- temperature (C_1) ends of the droplet, according to the solubility diagram. As a result, undersaturation produced at the "hot" interface causes germanium dissolution, and supersaturation at the "cold" interface causes germanium crystallization, and the droplet as a whole moves towards the higher temperature.

In order to make quantitative measurements, a reference point was produced on the trace of a droplet which was first moved from a part of a surface coated with a thin gold layer, to a pure germanium area, by means of preliminary anneal (Fig. 1); this was achieved by an abrupt, short-duration lowering of temperature below the eutectic point for the Ge-Au system, after which the temperature was increased to a prescribed level and the crystal was annealed in vacuum or in a flow of purified hydrogen.

A qualitative analysis of droplet migration revealed the following fundamental features of this process [1, 4].

1. Of the two basic stages of TGZM (viz., kinetic atomic processes at the interface and diffusion of the solution components through the bulk of the liquid zone), the rate-limiting stage in droplet migration experiments is usually the interface processes (or a mixed situation occurs). This has been established by analyzing the thermal migration of inclusions [5-7] on the basis of the dependence of droplet migration rate on droplet size (i.e., on supersaturation or undersaturation at the zone boundaries): larger droplets move faster than smaller ones (Fig. 2).

2. The interface processes are more inhibited at the dissolution interface: whenever faceting of a droplet boundary has been observed, it occurred only at the dissolution interface (see Fig. 2).

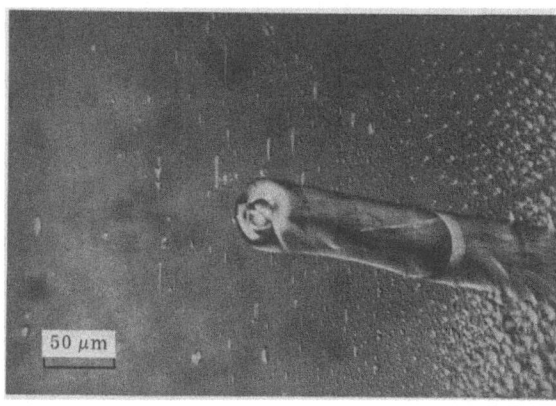

Fig. 1. Reference point on the trace left by a droplet.

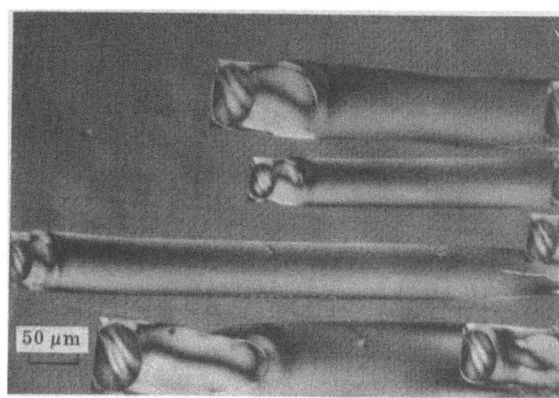

Fig. 2. Increased migration velocity of droplets with larger longitudinal dimensions.

3. When the dissolution interface is faceted (this is observed in droplets moving in a hydrogen ambient on substrates of all orientations, with the exception of (110) faces in the direction [1$\bar{1}$0]), dissolution proceeds by a two-dimensional "nucleation" mechanism.

In the present paper we treat the quantitative characteristics of the dissolution process, namely, the kinetic coefficients of crystallization or dissolution. Until now very little relevant information has been published [8, 9]. The difficulties in measuring kinetic coefficients are mostly related to measuring supersaturation (undersaturation) or supercooling (superheating) directly at the crystallization (dissolution) interface. The TGZM techniques (and among them, the method of droplet motion) make it possible to measure simultaneously the dissolution rate and the driving force for a large number of droplets on the same sample.

Very few experimental measurements of the kinetic coefficients have been made in the Ge-Au system: $\alpha = 2 \cdot 10^{-5}$ cm s^{-1} at temperature 515-625°C [10] and $\alpha = 7.5 \cdot 10^{-6}$ cm s^{-1} on germanium {111} faces at 750°C [8].

The Calculation of Kinetic Parameters of Dissolution

The following equations were derived in the analysis of zone migration velocity v as a function of longitudinal zone dimension l, for three mechanisms of crystallization (or dissolution) with diffusion taken into account (dissolution is assumed to be the rate limiting step) [6]:

(a) for the normal dissolution mechanism

$$v = \beta G/(1 + \beta/l\alpha_1), \qquad (1)$$

where

$$\beta = D\frac{dC}{dT} \Big/ (C_p^s - C_p);$$

G is the temperature gradient, C_p^s is the concentration of the crystal material (Ge) in the solvent, C_p is the concentration of this component at the dissolution interface, dC/dT is the slope of the phase diagram for the (Ge-Au) system, α_1 is a quantity referred to, by convention, as the kinetic coefficient of dissolution for uniform movement of the interface, and D is the diffusion coefficient of the solute (Ge) in the liquid alloy;

(b) dissolution via screw dislocations

$$v = \beta G/(1 + \beta/lv^{1/2}\alpha_2^{1/2}), \qquad (2)$$

where α_2 is the kinetic coefficient for the dislocation mechanism of dissolution;

(c) for two-dimensional nucleation

$$v = \beta G/[1 + \beta A/lv \ln(\alpha_3/v)], \qquad (3)$$

where α_3 and A are kinetic parameters in the case of two-dimensional nucleation.

In the present study we observed two typical forms of $v(l)$ curves:

— curves with a sharp migration threshold (Fig. 3) observed for the migration of droplets in vacuum on (110) and (100) surfaces, and less frequently on (111)Ge, as well as for annealing in hydrogen on substrates of all orientations and in all directions, with the exception of $[1\bar{1}0]$ on (110) faces; that is, for faceted dissolution interfaces (see Fig. 2);

— curves without a sharp migration threshold (Fig. 4), observed when there is no faceting of the dissolution interface (see Fig. 1).

If the motion of droplets is described by a $v(l)$ curve with a migration threshold, we can assume that two-dimensional nucleation is operative. However, it is not easy to determine kinetic coefficients from relation (3) since it implicitly contains the migration rate. Growth (dissolution) for the purely kinetic mode is described by the equation $v = \alpha_3 \exp(-A/\Delta T)$, where $\Delta T = Gl$. In order to evaluate the kinetic parameters α_3 and A, the experimental curves $v(\Delta T)$ are usually plotted in semilog coordinates. However, the range of experimental values of l in droplets with faceted dissolution interfaces is rather narrow (owing to deformation of the droplets), which makes the determination of α_3 and A difficult. An approximate method of evaluating the kinetic parameters was suggested based on a simple analysis of the relation $v(\Delta T) = \alpha_3 \exp(-A/\Delta T)$ [11]. It is easily proved that a tangent to the curve $v(\Delta T)$ at the inflection point intersects the abscissa axis at $\Delta T = A/4$. The tangent to the inflection point of the experimental $v(l)$ curve intersects the abscissa axis only to the right of the threshold characteristic l_0, so that the relation $Gl_0 = A/4$ gives a reliable estimate of the lower limit for parameter A. The shapes of experimental curves characterized by an abrupt rise in growth rate v for droplets with diameters above the threshold value l_0 (see Fig. 3) leads to the conclusion that the derived lower limit for A deviates insignificantly from the actual value of this coefficient. One advantage of the estimate $A \approx 4Gl_0$ (although it gives slightly underestimated A values) is that the threshold value l_0 on $v(l)$ curves is clearly measurable even when there is experimental scatter of data. Note also that in the case $l < l_0$ diffusion is not the limiting stage in the droplet motion, otherwise the migration velocity would be non-zero. In order to evaluate α_3 we take a point l_1 at a distance $l_0/4$ to the right of l_0 on the abscissa axis. Diffusion at this point is also negligible, and the kinetic atomic processes are limiting:

$$v_1 = \alpha_3 \exp(-16/5); \quad \alpha_3 = v_1 \exp(16/5) \approx 25 v_1.$$

When this method of approximate evaluation is used, the parameters α_3 and A are found in a mode explicitly assumed to be kinetic. Satisfactory agreement was found between the estimates obtained by the suggested method and by replotting the data with semilog coordinates.

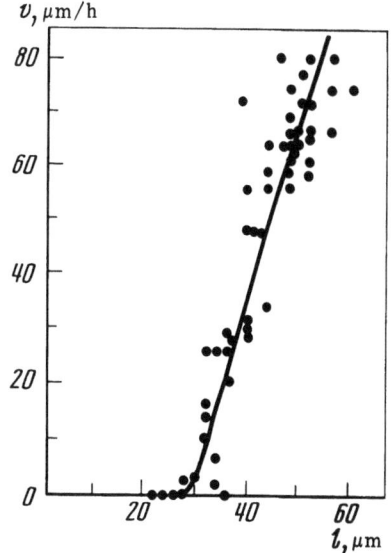

Fig. 3. Kinetic curve with an abrupt migration threshold.

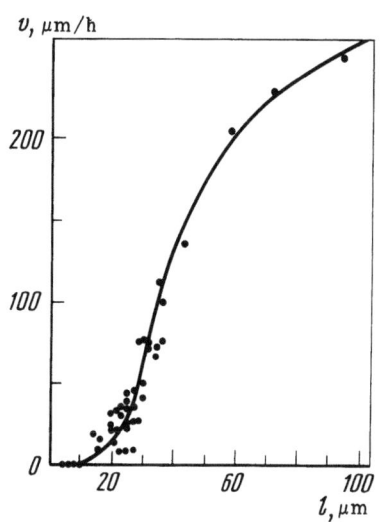

Fig. 4. Kinetic curve without a well-defined migration threshold (direction [110] on a (110) face).

Kinetics of Dissolution with Sharp Migration Threshold

The experimental dependence of the kinetic parameter K on temperature for annealing in hydrogen is shown in Fig. 5. As for annealing in vacuum, the activation energy of migration decreases with increasing temperature. The value of critical "superheating", ΔT_{crit}, changes from 0.04°C at low temperatures to 0.2°C at high temperatures. The critical undersaturation $\Delta C_{crit}/C_e$ is 0.008 and 0.05%, respectively, and $\Delta C_{crit}/C_e$ also rises with decreasing temperature.

A number of parameters characterizing the growth (dissolution) of crystals can be found from the measured values of A. For a description of dissolution kinetics in the liquid alloy on (111) faces of germanium, the experimentally obtained dependence $v = v_0 \exp(-A/\Delta T)$ was compared with the expression for the crystal face growth rate, v, from the melt for polycentric nucleation [12]:

$$v = v_0 \exp(-\pi \varkappa^2/3q_2 Hk\Delta T), \qquad (4)$$

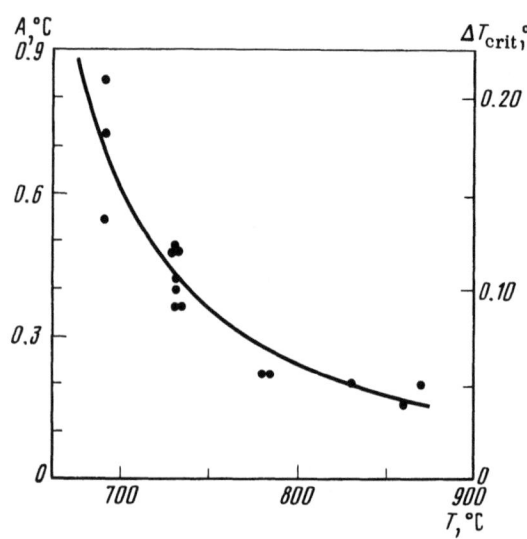

Fig. 5. Parameter A and critical superheating ΔT_{crit} as functions of hydrogen annealing temperature.

where $\varkappa$ is the specific free energy of a step, q_2 is the surface density of atoms in a nucleus, H is the heat of melting per atom, k is the Boltzmann constant, and

$$\Phi = \pi \varkappa^2 / 3 q_2 H k \Delta T \qquad (5)$$

is the work of critical nucleus formation divided by kT. We ignore the dependence of v_0 on ΔT when compared with a steeper exponential dependence. This gives

$$A = \pi \varkappa^2 / 3 q_2 H k,$$

and the specific free energy of a step on a (111) face is

$$\varkappa = (3 A q_2 H k / \pi)^{1/2}.$$

In germanium $q_2 = 1.66 \cdot 10^{15}$ cm^{-2}, $H = 5.5 \cdot 10^{-13}$ erg. This gives $\varkappa = 3.5 \cdot 10^{-7} A^{1/2}$ erg cm^{-1}.

The experimentally derived temperature dependence of $\varkappa$ is plotted in Fig. 6. This temperature effect agrees with the concept of a diffuse step [13]. The values of $\varkappa$ obtained can be compared with the theoretically calculated values for melt-grown germanium [12]. The discrepancy between experimental and theoretical values can be ascribed to roughness of the crystal/liquid-alloy interface, as well as to the possible effect of the three-phase junction line on the kinetic processes [14].

The expression for the specific free energy of a step [12],

$$\varkappa = q_1 \{W/3 - kT \ln [1 + 2\exp(-W/kT)]\},$$

where q_1 is the linear atomic density at the step end face, and W is the excess energy per interface bond, can be used to determine the interfacial free surface energy σ of (111) faces in liquid Ge-Au. For the germanium (111) face, $\sigma = W q_2 / 2$. The experimental temperature dependence of the (111) surface energy for Ge in liquid Ge-Au in the 690-870°C range is plotted in Fig. 7. The fact that the gold content of the melt increased with decreasing temperature, thus lowering the surface energy of the germanium crystal, must be considered. Note that a similar effect of gold doping was observed in the Si-Au system [15]: the surface energy of the silicon/liquid Si-Au interface was found to be 210 erg cm^{-2}, while in pure silicon $\sigma = 300$ erg cm^{-2}.

The results obtained also allow evaluation of the work of formation, F, and radius, R, of a critical nucleus under critical superheating (when droplet migration is possible).

The critical nucleus work of formation is found from (5). In the range 700-900°C we find $F = (5-7) \cdot 10^{-13}$ erg. A well-known relation for this work

$$F = \varkappa L/2$$

yields the perimeter of a critical nucleus, L, and hence, if the nucleus is circular, its radius R.

The radius of a critical nucleus is plotted in Fig. 8 as a function of critical superheating. This curve was obtained for the temperature range 700-900°C under nonisothermal conditions (each black circle corresponds to a temperature at which the given temperature difference is critical for the migration of the droplet).

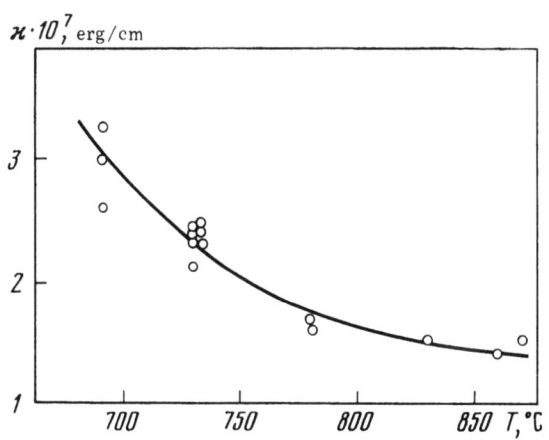

Fig. 6. Temperature dependence of the specific free energy of a step.

Fig. 7. Temperature dependence of the germanium-liquid Ge-Au interface.

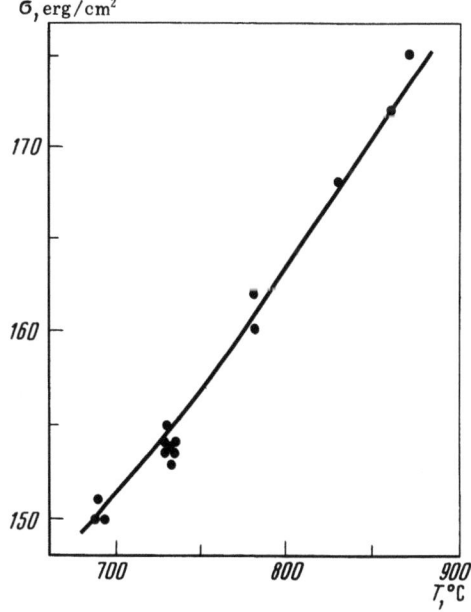

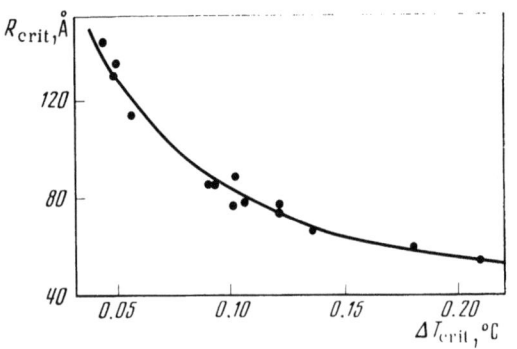

Fig. 8. Radius of the critical nucleus as a function of superheating.

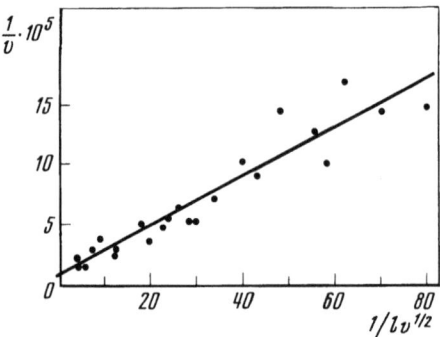

Fig. 9. Kinetic curve without a well-pronounced migration threshold in $(1/v, 1/lv^{1/2})$ coordinates.

Kinetics of Dissolution without a Sharp Migration Threshold

No abrupt migration threshold was observed on $v(l)$ curves for droplets of about ten microns diameter, for droplets moving along $[1\bar{1}0]$ on (110) faces of substrates annealed in hydrogen, on (111) faces in vacuum, as well as for other cases of hydrogen annealing when the dissolution interface was not faceted (see Fig. 4). In these cases the experimental points do not fit the straight line corresponding to the normal mechanism of growth (dissolution). We shall rewrite the quadratic equation of growth (2), taking into account diffusion, in the form

$$1/\beta G + 1/lv^{1/2}\alpha_2^{1/2}G = 1/v.$$

It was found that the $v(l)$ curves were usually linear in $1/v$ vs. $1/lv^{1/2}$ coordinates (Fig. 9). In this case the kinetic coefficient α_2 is found from the slope of the line. In the temperature range 690-910°C, α_2 varies from $5 \cdot 10^{-5}$ to $6 \cdot 10^{-3}$ cm s^{-1} °C^{-2} (the scatter in the experimental points is considerable). The range of α_2 values obtained agrees with the estimate $\alpha_2 = 2 \cdot 10^{-5}$ cm s^{-1} °C^2 for the range 515-625°C [10] (this paper also reports a quadratic dependence of v on l). The diffusion coefficient of Ge in liquid Ge-Au can be evaluated from the intercept on the ordinate axis (see Fig. 9): $D \approx (5-7) \cdot 10^{-5}$ cm^2 s^{-1}.

It is unlikely that the dissolution mechanism operative for the case in question is that usually associated with a quadratic dependence. The surface dislocation density in the initial crystals did not exceed 10^3 cm^{-2}, that is, on an average one dislocation is found in an area of about 300 x 300 μm. Usually the dislocation density was considerably lower. X-ray topography of typical samples in which quadratic $v(l)$ curve was observed showed no dislocations in areas exceeding 1 cm^2.

One can assume that the situation for the experiments described is nearly the same as for the vapor-liquid-solid growth of silicon whiskers [15], where a quadratic growth curve was also observed, with dislocations playing only a minor role. As a quadratic dependence for the initial stage can be approximated by an exponential curve with a low activation energy, it can be assumed that we are concerned with two-dimensional nucleation with a very small activation barrier. For $A \approx 0.1 - 0.3°C$ the exponential curve within the initial segment is very close to a quadratic curve. Obviously, small $\{111\}$ facets are present at the macroscopically rounded dissolution interface, and the mean dissolution barrier in this case is lower than the dissolution barrier for a planar $\{111\}$ interface.

Conclusion

1. A quantitative analysis of Ge dissolution kinetics in liquid Ge-Au carried out by observing the motion of Ge-Au alloy droplets along Ge surfaces revealed two typical dependences of droplet migration velocity v on droplet longitudinal dimension l: with and without an abrupt migration threshold. The $v(l)$ curves correspond to exponential and quadratic dissolution laws. Two-dimensional nucleation is suggested as the mechanism (for quadratic curves this mechanism is assumed to have a significantly lower activation barrier).

2. A number of physical characteristics are determined: critical superheating and relative critical undersaturation, specific free energy of a step, surface energy of the Ge/liquid-alloy interface, work of formation and radius of critical nuclei, and Ge diffusion coefficient in liquid Ge-Au.

Literature Cited

1. S. A. Grinberg and E. I. Givargizov. On the motion of germanium-gold alloy droplets on germanium surfaces in a temperature gradient. Kristallografiya, 18, 380-383 (in Russian) (1973).
2. W. G. Pfann. Zone Melting. J. Wiley and Sons, New York, (1966).
3. V. N. Lozovsky. Temperature Gradient Zone Melting. Metallurgy, Moscow (1972).
4. S. A. Grinberg. Mechanism of germanium crystallization from solution in gold. In: Growth of Crystals, Nauka, Moscow, (1974), 10, ed. N. N. Sheftal'. Consultants Bureau, New York, 176-184 (in Russian) (1976).
5. A. A. Chernov. On the motion of inclusions in solids. Zh. Eksp. Teor. Fiz., 31, 709-710 (in Russian) (1956).
6. W. A. Tiller. Migration of a liquid zone through a solid. J. Appl. Phys., 34, 2757-2962, (1963).
7. Ya. E. Gegusin and M. A. Krivoglaz. Motion of Macroscopic Inclusions in Solids. Metallurgiya, Moscow (in Russian) (1971).
8. E. I. Givargizov. Determination of kinetic crystallization coefficients in experiments with whiskers. In: Growth of Crystals, 11, ed. A. A. Chernov. Consultants Bureau, New York, 136-145 (1979).
9. A. A. Chernov. Theory of stability of faceted growth forms. Kristallografiya, 16, 842-863 (in Russian) (1971).
10. R. G. Seidenstiker. Kinetic effects in temperature gradient zone melting. J. Electrochem. Soc., 113, 152-156, (1966).
11. S. A. Grinberg. Determination of parameters characterizing germanium dissolution kinetics in the liquid Ge-Au. In: Vth USSR Conference on Crystal Growth, Abstracts. Inst. of Cybernetics AN GSSR, Tbilisi, 1, 70-71 (1977).
12. V. V. Voronkov. Supercooling at a face formed on a rounded crystallization interface. Kristallografiya, 17, 909-917 (in Russian) (1972).
13. V. V. Voronkov. Structure of crystal surface in the Kossel model. In: Growth of Crystals, 10. ed. N. N. Sheftal', Consultants Bureau, New York (1976).

14. V. V. Voronkov. Processes at the crystallization front. Kristallografiya, 19, 922-929 (in Russian) (1974).
15. E. I. Givargizov. Vapor Growth of Whiskers and Platelet Crystals. Nauka, Moscow (1977).

Part V
Growth of Crystals from the Melt

STABILITY OF CRYSTALLIZATION IN EDGE-DEFINED FILM-FED GROWTH FROM THE MELT

V. A. Tatarchenko

The Institute of Solid State Physics of the USSR Academy of Sciences, Chernogolovka

Introduction

Capillary effects play a significant role in most current techniques of crystal growth from the melt (Czochralski [1], Stepanov [2], edge-defined film-fed growth (EFG) [3], floating zone growth [4]). The conditions at the crystal-melt interface are common for all of these growth techniques. The differences are found only in the conditions at the lower boundary of the molten column. For Czochralski growth the lower boundary is the free surface of the melt. In the Stepanov method the lower boundary is defined by the melt-wall contact or by the edge-defining die. In EFG the lower boundary is fixed by the die. In the floating zone techniques the lower boundary is represented by the edges of the melting ingot.

It has been stated in some recent publications that the Stepanov technique employs an edge-definer made of a non-wetted material (Fig. 1a, b), while the EFG technique makes use of wetted materials (Fig. 1e) [5]. In fact, a wetted edge-definer was used in Stepanov growth as early as 1958 to grow shaped crystals of alkali halides (see Fig. 1c) [6]. The use of a wetted edge-definer to grow germanium ribbons and tubes was reported in 1967 at the 1st USSR Conference on the Stepanov Growth of Crystals [7]. Another communication at the same conference, dealing with the formulation of boundary conditions for edge-defined film-fed growth, noted that in the case of the edge-definer wetted by the melt an edge anchoring condition is obtained [8]. This clearly shows that the EFG technique represents a particular case of the Stepanov method, having only a few specific design features which are not significant for the description of crystal formation processes, such as the presence of a narrow capillary for melt feeding (see Fig. 1e) [9]. In addition to the terms "Stepanov method" and EFG, the "Toshiba-growth" (classical Stepanov method, Fig. 1d) [10] and "CAST"-technology of edge-defined growth [11] are encountered, but no attempts have been made to specify important design features for these techniques. Therefore, we shall refer to all versions in which edge-definers are used to grow crystals from the melt as Stepanov techniques (see Fig. 1a-d).

Let us consider edge-defined growth in the Stepanov method, comparing it with Czochralski growth. The same treatment can be applied to the floating zone version. However, one additional constraint must be taken into account in the latter case, namely, the change in the volume of the molten zone caused by variations in cross-section and in the rates of crystal solidification and ingot melting.

In all methods of melt crystallization the cross-section of the crystal pulled from the melt is determined by the cross-section of the molten column at the crystallization interface.

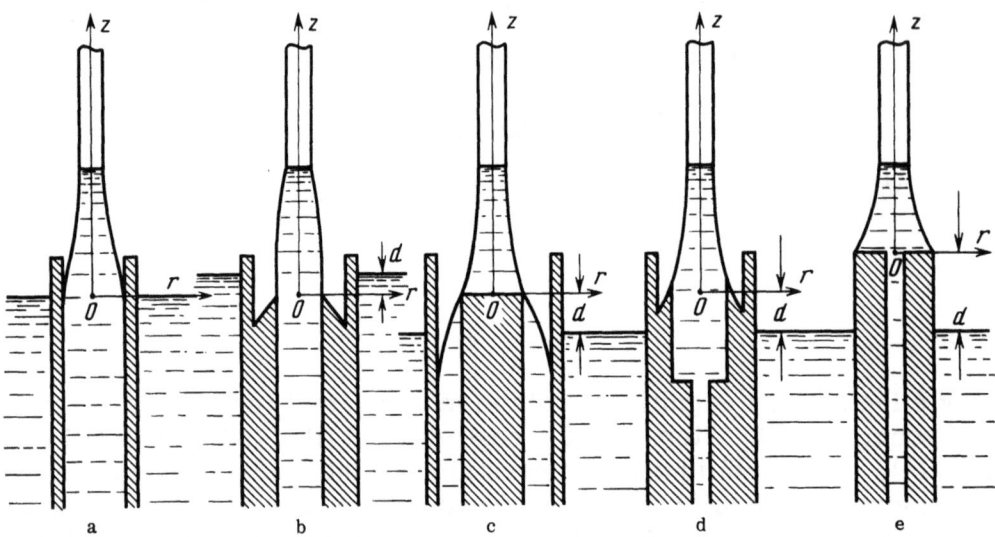

Fig. 1. Variants of edge-defined film-fed growth. (a) wetting boundary condition; (b, d, e) boundary condition of anchoring at the inner (b, d) and outer (e) edges at pressures $d > 0$ (b) and $d < 0$ (d, e); (c) combination of wetting and anchoring boundary conditions.

In Stepanov growth the dimensions and the shape of the pulled sample depend on: the edge-definer geometry; the pressure which feeds the melt into the edge-definer; the location and shape of the growth surface; and the shape of the seed crystal. The latter is important only in the steady-state pulling mode (the seed cross-section must equal that of the desired crystal). Frequently, especially when complicated crystal shapes are needed, the seed may be replaced with a wettable rod introduced to raise a column of melt above the edge-definer level; in this case the crystallization interface must be gradually lowered and the crystal cross-section becomes more and more determined by the cross-section of the edge-definer [12, 13].

The dimensions of the crystal cross-section may be found as functions of the crystallization interface position by taking into account capillarity phenomena in the melt, and heat transfer in the crystal-melt system. As a first approximation, a solution can be found by solving the boundary conditions for the Laplace capillarity equation describing the shape of the meniscus in the melt, and by solving the steady-state thermal flow for the crystal-melt system. This raises a question of the uniqueness and stability of the solution to these simultaneously treated problems.

An analysis of the stability makes it possible to find the mechanism by which the edge-definer affects the process of pulling a constant cross-section crystal, and to distinguish between the Czochralski and Stepanov methods; the latter aspect is not as obvious as might seem at first sight. In the Stepanov method, for example, the pulling of a cylindrical crystal using a circular edge definer usually results in a crystal of smaller diameter than that of the definer. In Czochralski growth, however, the crystal diameter in practice is commensurate with the crucible diameter. The question that follows is: what are the conditions for which the effect of crucible walls on edge-defined growth cannot be neglected, that is, where Czochralski growth "ends" and Stepanov growth "begins"?

Edge-Defined Film-Fed Growth

Equations Describing the Problem. The equilibrium shape of a liquid surface is described by Laplace's capillary equation [14]:

$$\gamma/R_1 + \gamma/R_2 - \rho g z = d, \tag{1}$$

where R_1 and R_2 are the principal curvature radii of the surface; the axis z is directed vertically upward; ρ is the density and γ is the surface tension of the liquid; g is the gravitational constant; and d is a constant whose value depends on the position of the z reference point (the physical meaning of d is the hydrostatic head at $z = 0$). Equation (1) can be transformed to an explicit differential equation of the meniscus surface [15]:

$$\text{div}\{H \text{ grad } z\} - \rho g \gamma^{-1} z + d = 0, \tag{2}$$

where

$$H = \left[1 + \left(\frac{\partial z}{\partial x}\right)^2 + \left(\frac{\partial z}{\partial y}\right)^2\right]^{-1/2}.$$

The transition from equation (1) to equation (2) involves a necessary constraint, namely, the meniscus surface is assumed to possess, for each of its points, unique projections onto the plane Oxy.

Introducing the capillarity constant $a = (2\gamma/\rho g)^{1/2}$, whose dimension is that of length, for units of linear dimension in describing capillarity-based phenomena makes the general description independent of the nature of the melt.

As Laplace's equation in form (2) has no analytical solution, we shall resort to the following simplification. Very complex cross-section crystals can be divided into a finite number of segments, with the contour curvature, r, constant within each one. If the capillarity problem is solved successively within each region, with the axis z placed each time at the curvature center, the surface of the molten column appears as the result of rotating the profile curve z(r) around axis z. The following relation for z(r) is then obtained from equation (2) [8]:

$$z''r + z'[1 + (z')^2] \pm 2(d-z)[1 + (z')^2]^{3/2} r = 0. \tag{3}$$

On flat segments of the crystal, one of the principal curvature radii is infinite and equation (3) transforms to

$$z'' \pm 2(d-z)[1 + (z')^2]^{3/2} = 0. \tag{4}$$

In this case the axis z can be placed at any point on the crystal contour.

On large-curvature segments the weight of the melt column can be neglected compared with the capillary pressure. Equation (3) then gives

$$z''r + z'[1 + (z')^2] \pm 2d[1 + (z')^2]^{3/2} r = 0. \tag{5}$$

The plus sign in equations (3)-(5) corresponds to the case when the crystal contour projection lies within the edge-definer contour, and minus corresponds to the reversed configuration.

In growing a crystal shaped as a thin ribbon, the profile curve for the flat part is given by equation (4), and the edge is described by equation (5). Equation (4) can also be used to find profiles of the small-curvature segments of the crystal.

If the curvature radius of the edge-definer contour is above 5 or below 0.5 (remembering that the unit of length is the capillarity constant), then the use of equations (4) and (5) instead of (3) gives an error in determining the positions of points on the profile curve of 8% or less.

Boundary Conditions Imposed by the Edge Definer. When the boundary problem is formulated for Laplace's capillary equation, it is necessary to fix two boundary conditions because this is a second-order differential equation. One of the boundary conditions is given at the upper boundary, that is, the line of contact between the melt and the crystal pulled, and the second is given at the lower circumference of the molten column, along the line of contact between the melt and the edge-definer. The main role played by the edge-definer for growing shaped crystals consists precisely in the generation of specific boundary conditions.

Two types of boundary conditions must be distinguished: anchoring and wetting [8, 10].

In the case of an anchoring boundary condition the curvature of the meniscus surface is fixed. The edge position is determined by the sharp edge of the definer. In the general case we have for a plane edge-defining contour Γ

$$z(x, y)_\Gamma = -d. \qquad (6)$$

If the plane Oxy coincides with a planar surface of the melt, then d is the pressure under which the melt is fed into the edge-definer.

For a meniscus described by equation (3), condition (6) takes the form

$$z|_{r=r_0} = -d,$$

where r_0 is the radius of the edge-definer contour (Fig. 2a).

In the case of the wetting boundary condition the fixed quantity is the angle between the tangents to the liquid and edge-definer surfaces at the contact point. Here the crystal shape is defined by the surface of the edge-definer, and the melt surface is at an angle θ, that is, the wetting angle, to the definer surface [15, 16].

$$H \frac{\partial z}{\partial n}\bigg|_\Gamma = -\cos \theta, \qquad (7)$$

where H is the same as in (2), and n is the direction of the inward normal to the contour.

In the case of a meniscus described by equation (3), and an edge-definer with vertical walls (Fig. 2b), condition (7) is written in the form

$$z'|_{r=r_0} = -\tan \alpha_1,$$

where $\alpha_1 = \theta - \pi/2$.

If the crystal cross-section definition is realized at the walls placed at an angle β to the horizontal (Fig. 2c), the condition becomes

$$z'|_{r=r_1} = -\tan \alpha_1,$$

where $\alpha_1 = \theta + \beta - \pi$; $r_1 = r_0 - z \cotan \beta$.

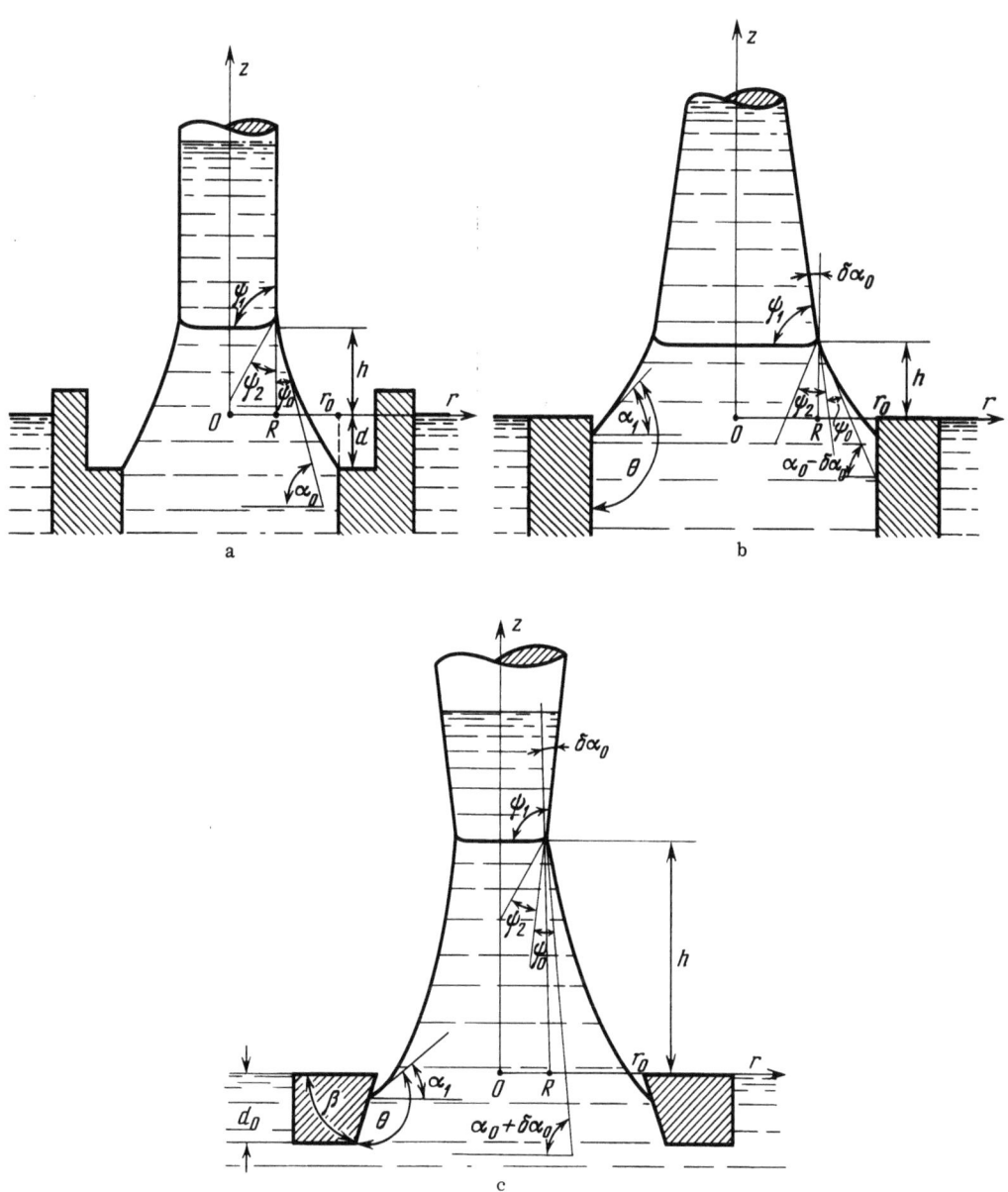

Fig. 2. Schematic of boundary conditions in the Stepanov growth method at the lower ((a) anchoring, (b, c) wetting) and upper (wetting) edges. (a) $\alpha_0 = \pi/2 - \psi_0$ (constant cross section crystal); (b) $\alpha_0 > \pi/2 - \psi_0$ (tapering crystal); (c) $\alpha_0 < \pi/2 - \psi_0$ (inverted-cone crystal).

The anchoring boundary condition is readily attained on an edge-definer wetted by the melt. On a non-wetted material the anchoring boundary condition may be attained by using additional pressure and choosing a seed of sufficient size for the melt column to be supported by the sharp edge of the definer. The pressure may be produced by an appropriate level of the melt. By dipping an edge-definer, shaped as a slab with a cylindrical or conical hole, into the melt it is possible to achieve successively the boundary conditions of anchoring at the lower circumference of the edge-definer, wetting of the walls, or anchoring at the upper circumference [8, 16, 17].

Laplace's equation can be written by choosing the origin of the vertical coordinate at an arbitrary point. This will change the value of parameter d. If the boundary condition at the edge of the definer is that of anchoring, the equation can be written for the origin of the vertical coordinate chosen either at the free surface or at the definer level. In the first case pressure is given as a boundary condition, and in the second case it becomes a parameter in the equation. If, however, the boundary condition is wetting at the lower edge, then the contact line of the edge-definer and melt is not fixed, pressure is not specified, and Laplace's equation must be written for the origin chosen at the free surface level. The pressure is found, therefore, by solving the boundary problem.

The Boundary Condition at the Crystal-Melt Interface. A growing crystal shapes the melt column by its edges and it would seem, therefore, that the anchoring boundary condition is realized at the crystal-melt interface. However, in order to grow a constant cross-section crystal (corresponding to the growth of a crystal with verticle sides), it is necessary for the melt to form a prescribed angle ψ_0 with the surface of the growing crystal; that is, the wetting boundary condition must be fixed at the crystal-melt interface of the crystal with vertical sides:

$$z'|_{r=R} = -\tan \alpha_0, \qquad (8)$$

where

$$\alpha_0 = \pi/2 - \psi_0.$$

Indeed, contact angles of the liquid and solid phases, ψ_0 and ψ_2, at the triple junction must obey a specific relation [18, 19]. From thermodynamic arguments the crystallization interface at the triple junction must be curved (see Fig. 2a-c) [20]. This microscopic curvature must not be confused with the macroscopic curvature of the crystallization interface (which may be absent, as in the case of a planar interface). The micro-curved region covers several microns, and the crystal surface in the curved segment can be described by Laplace's capillary equation in which the capillarity constant is given by the factor $(\gamma/G\Delta S)^{1/2}$ (here γ is the surface tension at the solid-liquid interface, G is the temperature gradient in the solid, and ΔS is the entropy of melting per unit volume of the material.

It can be assumed as a first approximation that the change in angle ψ_1 between the macroscopic growth direction of the crystallization interface and the surface of the growing crystal, caused by the curved segment, must not affect the contact angle ψ_2 between the crystallization interface and the growth surface. In this case the growth direction of the crystal is determined by the contact angle, ψ_0, between the melt and the crystal surface. Condition (8) is then a direct corollary.

If angle ψ_0 differs from the equilibrium value given by (8), by a small quantity $\delta\alpha_0$, then the velocity of the increase or decrease of crystal radius δR is given by

$$\delta \dot{R} = -v\delta\alpha_0, \qquad (9)$$

where v is the pulling rate (see Fig. 2b, c).

The suggested treatment of the boundary mode at the triple junction is in contradiction with a rigid relation between angles ψ_0, ψ_1, and ψ_2 found by analyzing thermal fluxes in the vicinity of the triple junction and then carrying out a limiting transition to the junction itself [21]. This point of view is disproved by direct observation of the constancy of growth angle ψ_0 in the same material in which (a) angle ψ_1 is varied from 70 to 110° (growth of sapphire whiskers with tilted crystallization interface) [22], (b) the thermal conditions and the growth technique are varied [23-25], and (c) the crystal pulling rate is varied (observation of the lateral surface profile) [26].

Experimental Determination of Growth Angle ψ_0. The first results were obtained by direct measurements of the angle between a tangent to the melt and the crystal surface in the Czochralski growth of crystals [23, 24]. Later a method of measuring growth angle on solidified wafers was suggested [25].

Several other methods of measuring angle ψ_0 can be suggested on the basis of the Stepanov method [27]. As the basic error in determining angle ψ_0 on meniscus photographs for crystal growing in a steady-state mode is associated with drawing the tangent, one can increase the accuracy by using the formula for the tabulated derivative of the function involved.

When thin fibers are drawn from the melt and the anchoring boundary condition is used, equation (5) describes the shape of the melt column with reasonable accuracy. The growth angle can be obtained by solving this equation, written in elliptic functions, and by comparing the results with meniscus photographs. This method is similar to measuring the wetting angle from the shape of a drop suspended by a thin fiber [28].

In complete analogy to this, a version with the wetting boundary condition on the vertical or horizontal wall of the edge-definer can be used.

In all of these cases one can employ a non-stationary process; that is, by forming a meniscus with a vertical tangent. The crystal tapering angle will then be equal to the growth angle.

Shape and Height of Profile Curves. With the formulation given above, the boundary problem for equation (3) cannot be obtained. The most frequent approximation in Czochralski growth is the substitution of the actual curvature of the meniscus by a linear function of the vertical coordinate [29]. Attempts to apply this approximation to Stepanov growth [30, 31] cannot be considered successful because both the anchoring and wetting boundary conditions have to be fixed simultaneously at the edge-definer periphery to find the molten column height [32].

Solutions of boundary problems for crystal segments possessing large and small curvature (see eqs. (4) and (5)) can be obtained in the form of a set of well-tabulated Legendre elliptic functions [33-37]. With a set of results for a given edge-definer one can plot a family of profile curves characterized by a growth angle ψ_0. A locus of points which lie at the ends of these profile curves gives the position of the crystallization interface in steady-state growth as a function of crystal dimensions.

Such curves are shown in Fig. 3a for the anchoring boundary condition for pulling a thin fiber. As the pressure is varied, these curves are changed, which allows control of the molten column height by varying the pressure, and prediction of the position of the crystallization interface as the melt level in the crucible drops in the course of crystal pulling.

An analysis of these solutions reveals the conditions in which three types of molten columns are obtained (concave, convex, and convex-concave) and the conditions under which the wetting boundary condition is transformed to the anchoring condition [33-36].

Similar results were obtained by numerically solving equation (3) for regions of crystal with curvature radius of the order of the capillarity constant [8, 33, 38, 39].

The position of the crystallization interface is plotted in Fig. 3b as a function of crystal size for a number of profiles and boundary conditions.

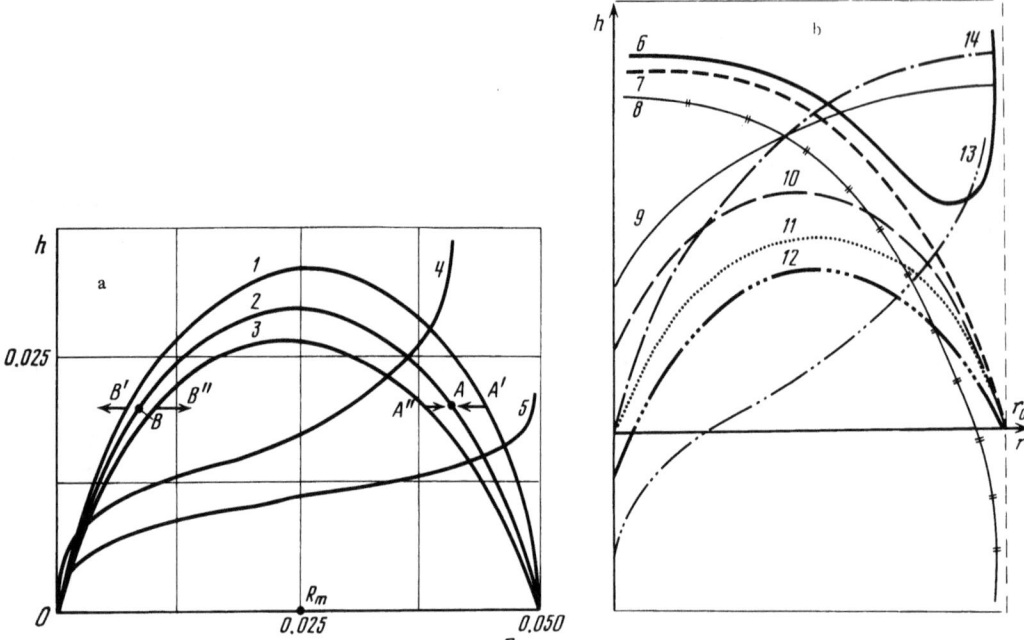

Fig. 3. Vertical coordinate of the crystallization interface as a function of the growing crystal dimensions (a) for a thin fiber, and (b) for platelets and rods. (1-3) solution of capillarity problem, $\alpha_0 = 90$, 80, and 70°, respectively; (4, 5) solution of the thermal problem (4-$v=v_1$; 5-$v=v_2$; $v_1 > v_2$); (A, B) points of capillarity-caused stability and instability, respectively; (6-8) platelet, anchoring boundary condition (7) and wetting boundary condition for $\alpha_0 > \alpha_1$ (6) and $\alpha_0 < \alpha_1$ (8); (9-14) a cylindrical rod, anchoring boundary conditions (9-12) for $d = 0.5/r_0$, $d > 0$, $d = 0$ and $d < 0$, respectively, and wetting boundary conditions for $\alpha_0 > \alpha_1$ (13), and the case of Czochralski growth (14).

Effect of Surface Tension Gradient on Capillary Phenomena. Under real conditions the surface tension of the melt is not constant, and changes may be caused by the gradients of temperature or impurity concentration in the molten column. The effect of the surface tension gradient, λ, which is constant along the meniscus height, on the meniscus form was calculated in ref. 40 for axisymmetrical menisci. In this case equation (3) is transformed to

$$z''r(1 + \xi z) + z'[1 + (z')^2](1 + \xi z - \xi r z') + 2(d - z)[1 + (z')^2]^{3/2}/r = 0,$$

where
$$\xi = \lambda a/\gamma_0.$$

The problem was formulated more correctly in ref. 41 where convection in the meniscus, which is a result of the surface tension gradient and is present even in zero-gravity conditions, was taken into account. An excessive pressure is produced in the meniscus together with convection. The distributions of velocity and pressure and changes in the meniscus shape were found by solving the Navier-Stokes equation for plane flow and Laplace's capillary equation for the meniscus in the case of the melt growth of platelets. The change in the meniscus shape caused by the surface tension gradient is opposite to that due to excessive pressure. The relative contributions of these factors are determined by the ratio of the meniscus characteristic dimension and the capillarity constant.

A film at the meniscus surface also leads to a change in its shape. It was found by analyzing the effect of a dense oxide film on the meniscus shape that the film increases the effective capillarity constant of the melt, by an increment proportional to the meniscus curvature [42].

In conclusion of this section, it should be noted that progress in space technology has generated a renewed interest in the description of capillary phenomena under microgravity conditions. It has been demonstrated in model experiments on equal-density liquids that equation (5) allows an accurate description of the shape and height of liquid menisci to be made under these conditions with zero gradient of surface tension [43]. The results of ref. 42 must be used if this gradient is to be taken into account.

Stability of the Crystal-Melt System

Stability Equation of the Crystal-Melt System. Both the height h of the crystallization interface and the transverse dimension R of the crystal cross-section may vary when a crystal is pulled in the presence of uncontrolled perturbations. Our system thus has two degrees of freedom. In order to analyze the stability of such a system in Lyapunov's sense [44] it is necessary to find linearized expressions for rates of changes $\delta \dot{R}$ and $\delta \dot{h}$ as functions of perturbations. In the general case the system will have the following form:

$$\delta \dot{R} = A_{RR}\delta R + A_{Rh}\delta h;$$
$$\delta \dot{h} = A_{hR}\delta R + A_{hh}\delta h. \quad (10)$$

For the system to be stable, it is necessary and sufficient for the coefficients of the system (10) to satisfy the following conditions:

$$A_{RR} + A_{hh} < 0;$$
$$A_{RR}A_{hh} - A_{Rh}A_{hR} > 0.$$

Capillary Stability. Transformation of equation (9) yields

$$\delta \dot{R} = -v \frac{\partial \alpha_0}{\partial R} \delta R - v \frac{\partial \alpha_0}{\partial h} \delta h,$$

whence

$$A_{RR} = -v \frac{\partial \alpha_0}{\partial R}; \quad A_{Rh} = -v \frac{\partial \alpha_0}{\partial h}.$$

As can be concluded from the solution of the capillarity problem, the signs of coefficients A_{RR} and A_{Rh} depend on the meniscus curvature and boundary conditions [45-47]. Coefficient A_{Rh} is always positive, and A_{RR} is alternating in sign. Both the signs and the values of these coefficients are determined exclusively by the capillarity effects. A system is capillary-stable if $A_{RR} < 0$. Figure 3a clearly shows that capillary stability is attained if $\partial h/\partial R |_{\alpha_0 = \text{const}} < 0$ (point A) because a deviation in crystal dimensions in this situation leads to restoring the crystal to the original state, as shown in Fig. 2b, c. Curves h(R) plotted in Fig. 3b make it possible to find the limits of capillary stability ($A_{RR} < 0$):

Rod	A_{RR}	Platelet	A_{RR}
Anchoring		Anchoring	<0
$R > R_m$	<0	Wetting	
$R < R_m$	>0	$\alpha_0 < \alpha_1$	<0
Wetting	<0	$\alpha_0 > \alpha_1$; R<m	<0
Czochralski growth	>0	$\alpha_0 > \alpha_1$; R>m	>0

An analysis of the above data enables us to conclude that the Stepanov technique is a method of pulling crystals from melts in which capillary stability is provided by the edge-definer. No such stability is inherent in Czochralski growth. When a cylindrical crystal is pulled from an edge-definer producing the anchoring boundary condition, the system is capillary-stable if $R \gtrsim R_m$ (see Fig. 3a); that is, the defining effect of the edge is attained if the crystal-to-crucible size ratio exceeds 1:2, and anchoring of the meniscus at the edge is ensured.

Thermal Stability. The explicit expressions for the coefficients of equation (10) depend on thermal effects and can be derived by solving the non-stationary thermal conduction problem for the crystal-melt system. This approach involves considerable mathematical difficulty. A comparative evaluation of the temperature and crystallization interface relaxation times shows that the latter exceeds the former by approximately two orders of magnitude, and thus quasistatic approximation can be used. From the heat balance equation at the crystallization interface in the non-perturbed and perturbed systems we obtain

$$A_{hh} = \frac{\lambda_2}{B}\frac{\partial G_2}{\partial h} - \frac{\lambda_1}{B}\frac{\partial G_1}{\partial h}; \qquad (11)$$

$$A_{hR} = \frac{\lambda_2}{B}\frac{\partial G_2}{\partial R} - \frac{\lambda_1}{B}\frac{\partial G_1}{\partial R}, \qquad (12)$$

where λ_1 and λ_2 are the heat conductivities of the melt and solid phase, respectively; $B = L\rho$ (L is the heat of melting, and ρ is density); and G_1 and G_2 are temperature gradients in the liquid and solid phases at the crystallization interface, respectively, obtained by solving the steady state thermal problem.

Thermal effects stabilize the process ($A_{hh} < 0$) if the melt is superheated. In this case an increase in the crystallization interface height results in a decreased modulus of temperature gradient in the liquid, and hence, in a diminished heat supply to the interface. The actual crystallization rate increases so that the height of the interface diminishes.

Cross stabilization ($A_{hR} > 0$) in the system is caused by heat exchange between the crystal (molten column) and the ambient. Indeed, the crystallization heat released in the crystal is proportional to its surface area, that is, to R^2, while heat removal from the lateral surface is proportional to its perimeter, that is, to R. As R changes, the thermal balance is violated and the actual crystallization rate is altered (the vertical position of the interface changes).

It is shown by taking thermal effects into account that there are conditions in which a stable mode of Czochralski pulling becomes possible [48]. A detailed analysis of the behavior of coefficients A_{hh} and A_{hR} as functions of growth parameters is given in refs. 47, 49.

The model discussed above can incorporate the crystallization kinetics by assuming crystallization temperature T_0 to be a function of the growth rate, a factor which is most important at high crystallization rates [50]. It seems that B in relations (11) and (12) becomes

$$B = L\rho + \lambda_2 \frac{\partial G_2}{\partial T_0}\frac{\partial T_0}{\partial v} - \lambda_1 \frac{\partial G_1}{\partial T_0}\frac{\partial T_0}{\partial v}.$$

We conclude that it is possible to take into account simultaneously both capillary and thermal effects in crystal growth, not only in stability analysis but also in the analysis of curves h(R) obtained by solving the capillarity and thermal problems (see Fig. 3a). Not more than two such simultaneous solutions are possible, with one of them usually lying in the stability range, the other being unstable [50].

Literature Cited

1. R. A. Laudise. The Growth of Single Crystals. Prentice Hall Inc., New Jersey, (1970).
2. A. V. Stepanov. A new method of growing sheets, tubes and complex-shape rods directly from the melt. Zh. Tekh. Fiz., 29, 381-393 (in Russian) (1959).
3. B. Chalmers, H. E. LaBelle, and A. J. Mlavsky. Edge-defined, film-fed crystal growth. J. Cryst. Growth, 13/14, 84-87 (1972).
4. W. G. Pfann. Zone Melting. J. Wiley and Sons, New York-London, (1966).
5. T. Surek, S. R. Coriell, and B. Chalmers. Crystal shape stability in meniscus-controlled growth processes. See this volume, p. 208.
6. B. M. Goltsman. Pulling of crystalline platelets and tubes from the melt. Optiko-Mekhanicheskaya Promyshlennost', 11, 45-47 (in Russian) (1958).
7. Yu. I. Koptev and A. V. Stepanov. Growth of germanium ribbons and tubes. In: Proceedings of the 1st Conference on Stepanov's Growth of Semiconductor Single Crystals. Ioffe Physico-Technical Institute, Leningrad, 74-78 (in Russian) (1968).
8. V. A. Tatarchenko, A. I. Saet, and A. V. Stepanov. On the shape of molten column in the growth of shaped crystals from the melt. In: Proceedings of the 1st Conference on Stepanov's Growth of Semiconductor Single Crystals. Ioffe Physico-Technical Institute, Leningrad, 83-97 (in Russian) (1968).
9. H. E. LaBelle, Jr. Pat. 3471266 (USA). Growth of inorganic Filaments.
10. Toshiba. Thin and speedy growth of silicon ribbon crystal solar cells achieved. J. Electron. Eng., 5, 46 (1976).
11. G. H. Schwuttke. Low cost silicon for solar energy application. Phys. Status Solidi (a), 43, 43-51 (1977).
12. G. V. Sachkov, V. A. Tatarchenko, and D. I. Levinzon. Control of edge-defined film-fed growth of single crystals from the melt. Izv. AN SSSR. Ser. Fiz., 37, 2288-2291 (in Russian) (1973).
13. V. A. Tatarchenko, S. K. Brantov, and N. V. Abrosimov. Growth of silicon platelets pulled from the melt by the Stepanov techniques. Fiz. i Khim. Obrab. Mater. 1, 83-88 (in Russian) (1978).
14. L. D. Landau and E. M. Liftshitz. Mechanics of Continuous Media. Gostekhizdat, 788 (in Russian) (1953).
15. L. S. Srubshchik and V. I. Yudovich. On asymptotic integration of the equilibrium equation of a liquid with surface tension in gravity field. Zhur. Vychislit. Matemat. i Matemat. Fiz., 6, 1127-1133 (in Russian) (1966).
16. V A. Tatarchenko, A. I. Saet, and A. V. Stepanov. Boundary Conditions in the EDF growth from the melt. Izv. AN SSSR, Ser. Fiz., 33, 1954-1959 (in Russian) (1969).
17. T. Surek. Theory of shape stability in crystal growth from the melt. J. Appl. Phys., 47, 4384-4393 (1976).
18. P. J. Antonov, S. P. Nikanorov, and V. A. Tatarchenko. The growth of controlled profile crystals by Stepanov's method. J. Cryst. Growth, 42, 447-452 (1977).
19. V. V. Voronkov. On thermodynamic equilibrium at the three-phase contact line. Fiz. Tverd. Tela, 5, 571-574 (in Russian) (1963).
20. W. Bardsley, F. C. Frank, G. W. Green, and D. T. Hurle. The meniscus in Czochralski growth. J. Cryst. Growth, 23, 341-344 (1974).
21. G. F. Bolling and W. A. Tiller. Growth from the melt. 1. Influence of surface intersections in pure metals. J. Appl. Phys. 31, 1345-1350 (1960).
22. Yu. F. Shchelkin. Shape of liquid column in Czochralski growth of single crystals from the melt with free surface. Fiz. i Khim. Obrab. Mater., 3, 29-33 (in Russian) (1971).
23. V. A. Tatarchenko and G. A. Satunkin. Analysis of capillarity conditions in crystallization of sapphire rods from the melt. Izv. AN SSSR, Ser. Fiz., 40, 1488-1491 (in Russian) (1976).

24. P. I. Antonov. Analysis of capillarity effects in crystal growth processes. In: Growth of Crystals, Vol. 6, Nauka, Moscow (1965), pp. 158-160.
25. Yu. M. Shashkov and E. V. Mel'nikov. Surface effects in the Czochralski growth of crystals. In: Surface Effects in Melts and Solid Phases Growing in them. Kabardino-Balkar. Publishing House, Nal'chik (1966).
26. T. Surek and B. Chalmers. The direction of growth of a crystal in contact with its melt. J. Cryst. Growth, $\underline{29}$, 1-11 (1975).
27. V. A. Tatarchenko and G. A. Satunkin. Capillary shaping in crystal growth from melts. II. Experimental results for sapphire. J. Cryst. Growth, $\underline{37}$, 285-288 (1977).
28. G. A. Satunkin, V. A. Tatarchenko, and E. M. Tseitlin. Relation between growth angle and the profile of the lateral surface of the growing crystal. Kristallografiya, $\underline{24}$, 134-142 (in Russian) (1979).
29. L. M. Shcherbakov and P. A. Ryazantsev. On one method of measuring contact angles. In: Surface Effects in Melts and Solid Phases Growing in them. Kabardino-Balkar. Publishing House, Nal'chik (1966).
30. G. K. Gaule and J. R. Pastore. The role of surface tension in pulling single crystals of controlled dimensions. In: Metallurgy of Elemental and Compound Semiconductors. New York; London: Interscience, pp. 201-209 (1961).
31. S. V. Tsivinsky. Application of the theory of capillarity to pulling shaped crystals from the melt by Stepanov's techniques. Inzh.-Fiz. Zhur., $\underline{5}$, 59-65 (in Russian) (1962).
32. S. V. Tsivinsky, P. I. Antonov, and A. V. Stepanov. On the shape of molten columns formed under shaped crystals pulled from the melt. Zh. Tekh. Fiz., $\underline{40}$, 372-376 (in Russian) (1970).
33. V. A. Tatarchenko and A. V. Stepanov. On the calculation of molten column height in melt growth of shaped crystals. Izv. AN SSSR, Ser. Fiz., $\underline{33}$, 1960-1962 (in Russian) (1969).
34. V. A. Tatarchenko and G. A. Satunkin. Capillary shaping in crystal growth from melts. I. Theory. J. Cryst. Growth, $\underline{37}$, 272-284 (1977).
35. V. A. Tatarchenko. Edge-defined film-fed growth from the melt. Izv. Vyssh. Uchebn. Zavedenii. Ser. Chernaya Metallurgiya, $\underline{5}$, 145-148 (in Russian) (1976).
36. V. A. Tatarchenko. Edge-defined film-fed growth from the melt. Izv. Vyssh. Uchebn. Zavedenii. Ser. Chernaya Metallurgiya, $\underline{7}$, 136-139 (in Russian) (1976).
37. V. A. Tatarchenko. Edge-defined film-fed growth from the melt. Izv. Vyssh. Uchebn. Zavedenii. Ser. Chernaya Metallurgiya, $\underline{9}$, 141-145 (in Russian) (1976).
38. V. M. Belyakov, R. I. Kravtsova, and M. G. Rapoport. Tables of Elliptic Integrals. AN SSSR Publishing House, Moscow, $\underline{1}$, 2.
39. B. M. Goltsman. Edge definition in sheets, cylindrical rods, and tubes pulled from the melt. In: Growth of Crystals, Vol. 3, Nauka, Moscow (1961), pp. 408-415.
40. J. C. Swartz, T. Surek, and B. Chalmers. The EFG process applied to the growth of silicon ribbons. J. Electron. Mater., $\underline{4}$, 255-279 (1975).
41. V. A. Tatarchenko, B. N. Korchunov, N. A. Gun'ko, and A. V. Stepanov. Effects of surface tension gradient on the form of profile curves. Izv. AN SSSR, Ser. Fiz., $\underline{36}$, 476-480 (in Russian) (1973).
42. E. A. Brener, and V. A. Tatarchenko. Some aspects of the macroscopic theory of uniaxial crystallization in the case of edge-defined film-fed growth from the melt. Izv. AN SSSR, Ser. Fiz., $\underline{43}$, 1926-1934 (in Russian) (1979).
43. V. A. Tatarchenko and B. F. Korchunov. Effect of oxide film on capillarity phenomena. Izv. AN SSSR, Ser. Fiz., $\underline{37}$, 2294-2296 (in Russian) (1973).
44. V. A. Tatarchenko and S. K. Brantov. Model of edge-defined film-fed growth studied in microgravity conditions. Izv. AN SSSR, Ser. Fiz., $\underline{40}$, 1468-1484 (in Russian) (1976).

45. A. A. Andronov, L. D. Vitt, and S. E. Khaikin. The Theory of Vibrations. Fizmatgiz, Moscow, (in Russian) (1959).
46. V. A. Tatarchenko and E. A. Brener. Stability of crystallization from the melt. Izv. Vyssh. Uchebn. Zavedenii. Ser. Chernaya Metallurgiya, 1, 140-144 (in Russian) (1976).
47. V. A. Tatarchenko and E. A. Brener. Stability of crystallization from the melt. Izv. Vyssh. Uchebn. Zavedenii. Ser. Chernaya Metallurgiya, 3, 140-144 (in Russian) (1976).
48. V. A. Tatarchenko and E. A. Brener. Stability of crystallization in edge-defined film-fed growth from the melt. Izv. AN SSSR, Ser. Fiz., 40, 1456-1467 (in Russian) (1976).
49. V. A. Tatarchenko. Effect of capillarity on stability of crystallization in the growth of shaped crystals. Fiz. Khim. Obrab. Mater., 6, 47-50 (in Russian) (1973).
50. V. A. Tatarchenko. Effect of thermal conditions on stability of crystallization from the melt. Inzh.-Fiz. Zhur., 30, 532-539 (in Russian) (1976).

SHAPE AND PROPERTIES OF CRYSTALS GROWN FROM THE MELT BY THE STEPANOV TECHNIQUES

P. I. Antonov

The Physico-Technical Institute of the USSR Academy of Sciences, Leningrad

One of the recent achievements in crystal growth is the progress in direct pulling of shaped crystals (rods, ribbons, tubes, films, etc.) from the melt by various modifications of the Stepanov technique (Fig. 1). All of these modifications are based on creating a molten column outside of the crucible walls by using a novel element, namely, edge-definer (die shaper) which is absent in all other growth methods [1]. Stepanov's techniques are used to grow crystals of semiconductors, dielectrics, and metals. The data for shaped single crystals successfully grown by various authors using these techniques (see Fig. 1) are given in Table 1 [2]. The table shows that a fairly wide range of materials with various thermal and capillary characteristics has been covered.

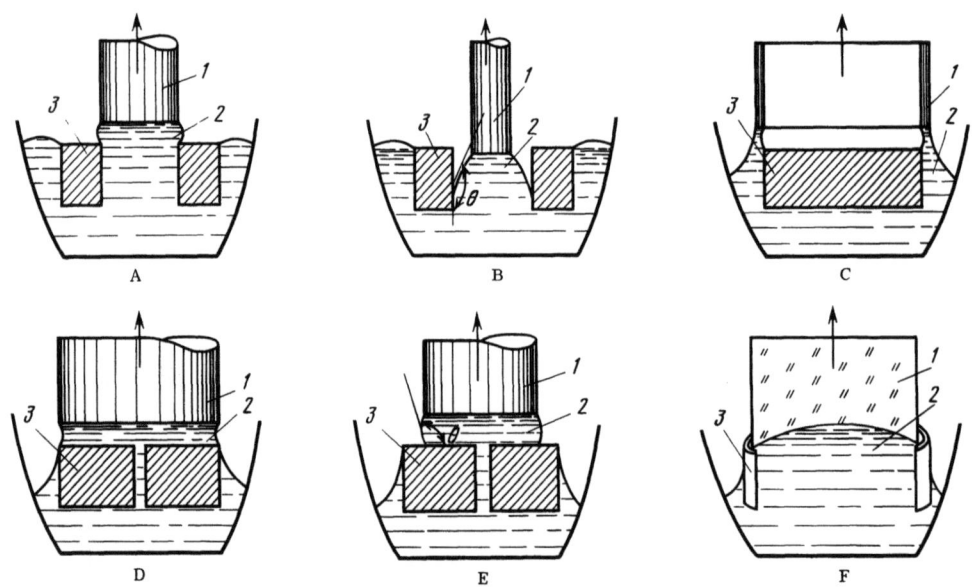

Fig. 1. Modifications of the edge-defined film-fed growth (Stepanov growth) with anchoring of the molten column base line at the edges of the edge-definer (A, C, D, F) and with fixing of the wetting angle Θ (B, E). (1) crystal, (2) molten column, (3) edge-definer.

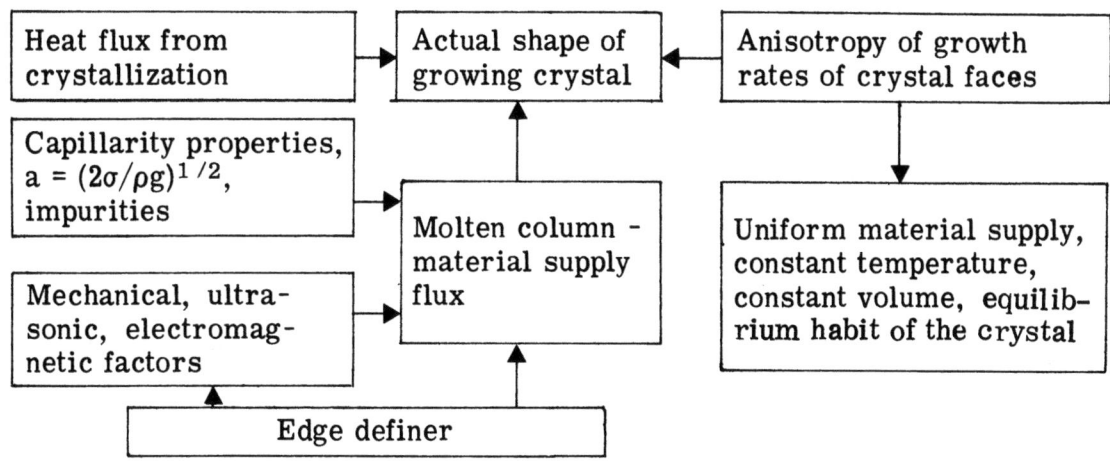

Fig. 2. Basic factors determining the actual shape of a crystal grown from the melt by the Stepanov techniques.

TABLE 1. Shaped Crystals Grown by Stepanov's Techniques

Crystal	Growth Method (see Fig. 1)	Shape	Edge-Definer Material
Semiconductors			
Ge	A, B, C	Rod, ribbon, tube	Graphite,* pyrographite,* vitreous carbon*
	C	Tube	W
	D	Ribbon	W
	F	Ribbon, film	W
Si	A	Ribbon	Quartz,* BN*
	D, E	"	Graphite
GaAs	A	"	Graphite,* BN*
InSb	A	Ribbon, rod	Graphite*
Dielectrics			
LiF, CsI, LiF-NaCl	D, E	Ribbon, rod	Quartz, Ni
AgCl	D, E	Ribbon, tube	Glass, Mo
Al_2O_3	D	Ribbon, rod	Mo, Ir
$LiNbO_3$	D	Ribbon	Pt, Ir
$Ga_3Cd_5O_{12}$	D	"	Ir
$MgAl_2O_4$	D	"	Re
Mn_2SiO_4	D	Rod	Pt
Metals			
Pb, Zn, Sn, Cd, Al, Bi, Hg, Fe	A, B, C	Ribbon, rod	Mica,* quartz,* cast iron,* glass*

*The material is not wetted by the melt.

It would be wrong to consider that the molten column completely determines the shape of the crystal being pulled. Both the shape of the growing single crystal and its properties are results of the interaction between all of the factors involved in the crystallization process. Crystal anisotropy and temperature field also affect the shaping of the single crystal (Fig. 2). Let us attempt an analysis of the interrelated profiles and properties of shaped single crystals and the possibility of controlling these characteristics during crystal growth from the melt.

Liquid Column Shape and Dynamics as Factors Influencing the Growing Crystals

An edge-definer enables the grower to create a desired shape and geometry of the molten column with typical dimensions from a fraction of one millimeter to several millimeters. The possible shapes and stability of molten columns have been investigated [3]. Since the material supply to the growing crystal has to flow through the molten column, all phenomena taking place in the column affect the properties of the growing single crystal.

The complexity of the growth conditions (concentration flows, the presence of inclusions, motion and viscosity of the melt) produce a resultant distribution of impurities in the crystal. When shaped germanium single crystals are grown, the introduction of an edge-definer does not essentially modify the pattern of impurity incorporation, described by the effective distribution coefficient, in the direction of the crystal axis [4]. Consequently, the control of the macroscopic distribution of impurity in shaped crystals is implemented by employing well-known methods which are effective, for example, in Czochralski growth, with the molten column geometry, thickness of the edge-definer walls (capillary length), and crucible rotation taken into account.

Microscopic distribution patterns of impurities (striations) revealed by microprobe or electrochemical methods may be caused by a number of factors. A typical class of periodic microscopic inhomogeneities is purely of equipment-based origin, with the periodicity determined by the pulling and crystal rotation rates. No such inhomogeneities are produced when crystals are grown by the Stepanov techniques. Presumably, this is due to the capillarity-induced stability of the molten column geometry: a negative feedback, damping out external perturbations, occurs in the crystal-melt system.

The use of edge-defined film-fed growth makes it possible to control the flows in the molten column, for example, by passing an electric current through the crystallization interface or by applying ultrasonic vibrations. The introduction of surfactants affects not only the surface tension of the melt but also the structure-sensitive properties, especially in the subsurface layer.

The Role Played by the Seed Crystal in the Growth of Shaped Single Crystals

The use of a seed crystal is essential for the Stepanov techniques. The experience accumulated in growing shaped single crystals shows that the seed geometry is important for shaping the molten column of predetermined cross section only in the initial stages of growth [5]. In later stages a situation is reached in which the crystal cross section is determined only by the design of the edge-definer. The crystal geometry then becomes practically invariant with respect to the seed geometry and dimensions.

In the Stepanov techniques it is usually unnecessary to rotate the crystal. The pulling direction is strictly vertical, that is, oriented along the direction of the gravitational force. Consequently, the lateral surface of the growing crystals can be considered as resulting from the displacement of the crystallization interface contour along the pulling direction. The crystals retain the symmetry planes which include the direction of pulling. Remember that in rotating-seed Czochralski grown crystals the direction of pulling is, strictly speaking, a screw axis of infinite order; this is confirmed by the presence of helical grooves on the surface of the resultant single crystals.

The need for highly perfect shaped single crystals requires the selection of seed crystals with minimum defect content. Necking-in, which is usually employed in Czochralski growth, is equally desirable in Stepanov's techniques when large-sized crystals are grown. Necking is not necessary for the growth of thin crystals. If several single crystals are grown simultaneously, it is possible to use one seed crystal which simultaneously contacts all of the molten columns.

Faceting of Shaped Single Crystals Resulting from Anisotropy of Crystal Properties

Most theoretical treatments of crystal growth from the melt assume the crystallization interface to be planar. This assumption is not valid. Close-packed faces make the crystallization interface non-uniform, due to the anisotropy in growth rate for different faces. In conditions of uniform feed supply, constant temperature, and constant melt volume the single crystal form would be in equilibrium, with typical faceting (see Fig. 2). The Stepanov techniques are versions of the uniaxial crystallization methods, so that the forms of shaped single crystals may be completely or partially faceted.

The real forms of shaped single crystals are characterized by the pulling direction fixed by the seed orientation, by the faceting of the crystal, and by the angular relations between faces on the lateral surface of the crystal. For each specific orientation the lateral surface reveals certain characteristic features, with rough and smooth faces appearing together with rounded segments.

The tendency for faceting in single crystals must be taken into account and utilized when shaped single crystals are grown [6]. Strongly or weakly pronounced faceting may be achieved. The face dimensions are determined by the melt supercooling, temperature gradient in the liquid and solid phases, and by the relative slopes of faces and isotherms. The symmetry of the seed (pulling directions $\langle 100 \rangle$, $\langle 111 \rangle$, $\langle 211 \rangle$) and the crystallization ambient (molten column and temperature field with circular, square, triangular, rectangular, or oval cross section), and the resultant symmetry of the pulled single crystals are listed in Table 2 for germanium. We see that an appropriate choice of single crystal orientation is important for achieving a prescribed shape (for example, the $\langle 100 \rangle$ direction is preferable for growing crystals with circular and square cross sections, and $\langle 111 \rangle$ direction is preferable for a triangular cross section).

Faceting of single crystals affects not only their habits but also their properties, and in particular their microhardness, impurity distribution ("facet effect"), refractive index, etc. Figure 3 shows the variation in the degree of faceting, S, and dislocation density, N_d, along the [211] axis of a germanium rod in a temperature field with tilted isotherms [7]. When the slopes of the isotherms and the (111) face are not identical (Fig. 3a), a weakly faceted single crystal is obtained with the extent of faceting 10%, and a dislocation density about 10^5 cm^{-2}. It was sufficient to turn the crystal seed by 180° (Fig. 3b), with a view to

TABLE 2. Symmetry of the Real Shapes of Crystals Grown by the Stepanov Techniques

Required cross section and its symmetry	Crystallographic symmetry of the seed crystal	Symmetry of the liquid column and temperature field	Actual cross section and its symmetry
Circle or annulus (∞ m)	⟨100⟩ — 4mm	∞ m	4mm
	⟨111⟩ — 3m		3m
	⟨211⟩ — m		m
Square (4 mm)	⟨100⟩ — 4mm	4mm	4mm
	⟨111⟩ — 3m		m
	⟨211⟩ — m		m
Rectangle (2 mm)	⟨100⟩ — 4mm	2mm	2mm
	⟨111⟩ — 3m		m
	⟨211⟩ — m		m
Triangle (3 m)	⟨100⟩ — 4mm	3m	m
	⟨111⟩ — 3m		3m
	⟨211⟩ — m		m
Oval (m)	⟨100⟩ — 4mm	m	m
	⟨111⟩ — 3m		m
	⟨211⟩ — m		m

equalizing the slopes of the isotherms and the (111) face, to increase the degree of faceting of the resultant single crystals to 50%, and to reduce sharply the density of dislocations.

The layered growth of faces generates twinning in shaped single crystals. Twins which are purely of growth origin are frequently observed. When shaped InSb single crystals were grown twinning was usually observed on the curved crystallization interface at the moment of crystal diameter increase, when there was a substantial difference in the sizes of the seed and the orifice in the edge-definer [8]. The growth of single crystal silicon ribbons is accompanied by multiple twinning, so that silicon ribbons are grown only with the

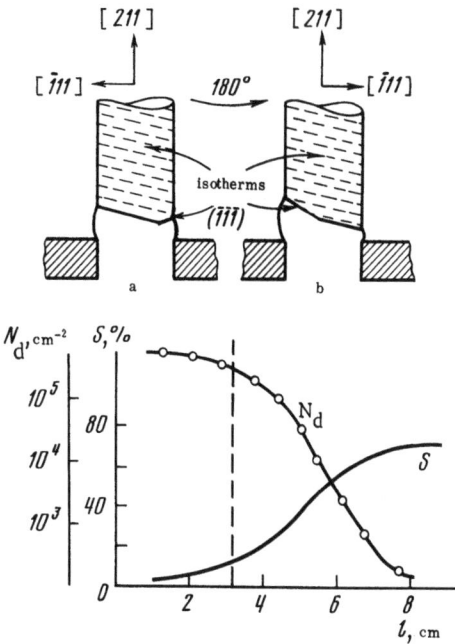

Fig. 3. Variation of dislocation density and faceting along a single crystal grown in a temperature field with tilted isotherms. (a) different slopes of (111) face and isotherms; (b) identical slopes.

orientation [$\bar{1}$12] on the (110) face [9]. Ribbons of sapphire, germanium, and other materials can be grown with arbitrary orientations.

The Stepanov techniques can be used to grow shaped bicrystal structures which are of interest for studying growth and twinning, as well as their physical properties. The twinning plane is reliably fixed by the edge-definer. Such features as artificial twinning planes and symmetric and asymmetric grain boundaries were produced in germanium ribbons [10].

Effects of Thermal Factors on the Shape and Properties of Crystals

The crystal shape and structure essentially depend on the thermal processes occurring. To a certain extent the crystals follow the shape of the heaters and screens. However, the shaping of crystals by the temperature field alone has inherent limitations (only the simplest forms were obtained). Very strict requirements for the temperature field stability must be met, and the capillary and thermal stabilities have essentially different characteristics [2, 3].

The temperature field can be efficiently used to control the crystal shape when used in combination with other factors [6]. An edge-definer determines not only the shape of the molten column but also the heat flux in the region of the crystallization interface. The heat flux can be controlled by installing, for example, additional heaters (coolers) or by partially screening the growing crystal. As a result, the crystallization interface deviates from the horizontal plane and crystallization takes place at different levels of the molten column.

The shape of the final crystal is determined by the shape of the molten column and by the heat flux (see Fig. 2). One possible modification of edge-defined growth can be obtained by producing supercooling of the melt in selected areas within the column. This leads to a combined action of the temperature field (supercooling in the molten column) and anisotropy in growth rates of the faces. This method was applied to growing germanium single crystals with a surface finish of the lateral faces corresponding to a height of irregularities within $R_z = 0.1$ μm.

Thermal factors determine to a large extent the physico-mechanical properties and the structure of single crystals. These effects are especially pronounced in the growth of shaped single crystals. The results of studying radial distribution of impurities in germanium single crystals with circular cross sections, grown by the Stephanov techniques, have demonstrated that the facet effect is not the only factor causing inhomogeneous radial distributions [12]. In some experiments the crystallization interface was nonplanar, and the facet effect was not well pronounced. Figure 4 gives the distribution of specific electric resistivity ρ, concentration n, and mobility μ of charge carriers along the diameter d of a germanium crystal cross-section. In this case the drop in mobility may possibly be explained by additional scattering of charge carriers in residual mechanical stress fields; this is confirmed by direct measurements of these stresses by infra-red polariscopy.

Thermoelastic stresses produced in crystals pulled from the melt generate dislocations (at stresses above a threshold which diminishes as the temperature approaches the melting point). For example, the critical stress for the generation of dislocations in germanium crystals at a temperature in the vicinity of the melting point is only 15 g mm^{-2}. If thermoelastic stresses exceed the threshold value, the crystals are plastically strained and dislocations are generated. Such a treatment of thermoelastic stresses in semiconductor crystals was first discussed in ref. 13 for Czochralski-grown crystals. In shaped single crystals these effects are specifically modified by the crystal shape and the geometry of the temperature fields. It has often been mentioned that in ribbons the density of dislocations is much higher than in rods grown under similar conditions. The cause of this effect remained unknown for a long time, and it was not clear how to achieve controlled changes in the dislocation structure.

Theoretical and experimental studies have demonstrated that the temperature field in the crystals must match the shape of the crystals [14, 15]. Thermoelastic stresses are produced in crystals only by deviations of temperature field from linearity.

It was found that the maximum thermoelastic stress τ_{max} in cylindrical crystals of radius r is proportional to r, and in ribbon crystals to the ratio b^2/t, where b and t are the width and thickness of the ribbon, respectively [16].

The results of calculations were used to plot the distribution of relative temperature T/T_{cr} (Fig. 5a) and profiles of mean squared tangent stresses in germanium rods (Fig. 5b) and ribbons (Fig. 5c) (T_{cr} is the crystallization temperature, z/L is the relative length; r/R is the relative radius of the rod; y/b is the relative width of a ribbon). The dashed curve in Fig. 5a shows the linear temperature distribution. In Fig. 5c the areas in which the mean squared tangent stress exceeds the dislocation generation threshold are shaded. The numbers at the isolines indicate stress levels.

It follows from calculations of thermoelastic stresses that the stress level in thin ribbons exceeds by an order of magnitude the stress level in cylindrical crystals grown under identical thermal conditions. The fact that all attempts to grow low dislocation-density germanium ribbons were until recently unsuccessful can be explained by the failure to take into account the above effect.

As the temperature field in a crystal is mostly determined by the screens which compensate for cooling of the crystals through their lateral surfaces, control of the temperature field requires that thermal radiation from the screens be modified. To achieve this, two types of screens were used in pulling germanium crystals: conventional (unheated) and active (heated with electric currents) screens [17]. The use of heated screens led to a substantial smoothing of the temperature field and to a reduction in the thermoelastic stress in the maximum-danger region in the vicinity of the crystallization interface. It was possible to

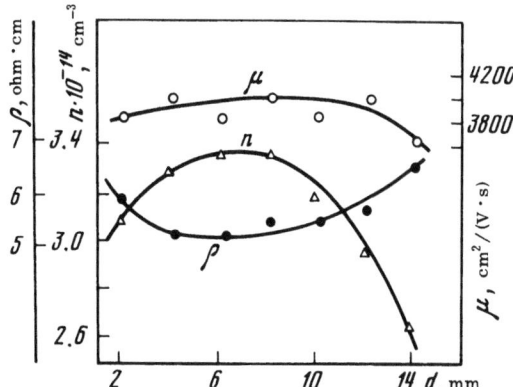

Fig. 4. Distribution of physical properties across the diameter of the germanium single crystal section.

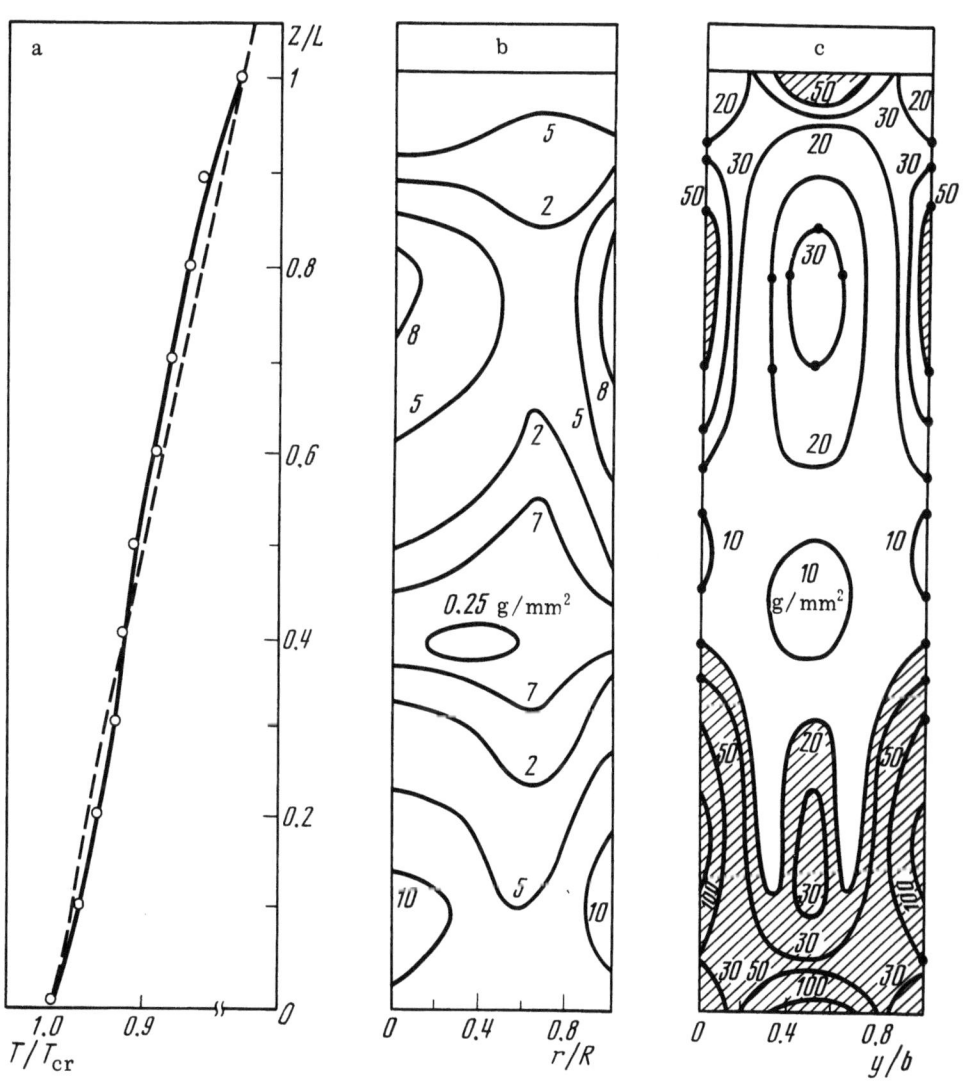

Fig. 5. Distribution of relative temperature (a), and profiles of mean square tangent stress in a germanium rod (b) and ribbon (c) grown under identical thermal conditions.

significantly reduce the dislocation density in ribbons (down to 10^3 cm^{-2}), and to completely eliminate dislocation clusters.

The above examples demonstrate again the interrelation between the shape and properties of crystals. Further investigations should aim at finding a way to control crystallization, taking into account individual features of the crystal to be grown and the effects of the crystallization ambient.

Literature Cited

1. A. V. Stepanov. Growth of shaped crystals. Izv. AN SSSR, Ser. Fiz., 33, 1944-1953 (in Russian) (1969).
2. P. I. Antonov, S. P. Nikanorov, and V. A. Tatarchenko. The growth of controlled profile crystals by Stepanov's method. J. Cryst. Growth, 42, 440-452 (1977).
3. V. A. Tatarchenko. Stability of Crystallization in edge-defined film-fed growth from the melt. See this volume, p.
4. N. V. Bessonova, D. I. Levinzov, V. V. Peller, Yu. M. Smirnov, and L. P. Stolyarchuk. Structure and impurity distribution by Stepanov's techniques. Izv. AN SSSR, Ser. Fiz., 33, 2013-2015 (in Russian) (1969).
5. G. V. Sachkov, V. A. Tatarchenko, and D. I. Levinzon. Control of edge-defined film-fed growth of single crystals. Izv. AN SSSR, Ser. Fiz., 37, 2288-2291 (in Russian) (1973).
6. P. I. Antonov, N. S. Grigoryev, and A. V. Stepanov. Symmetry in germanium single crystals grown from the melt by Stepanov's techniques. Izv. AN SSSR, Ser. Fiz., 35, 447-451 (in Russian) (1971).
7. P. I. Antonov, S. I. Bakholdin, and V. M. Krymov. Effect of faceting on dislocation structure in shaped germanium crystals. Izv. AN SSSR, Ser. Fiz., 37, 2328-2333 (in Russian) (1973).
8. Yu. G. Nosov, P. I. Antonov, and A. V. Stepanov. Twinning and distribution of dislocations in the growth of shaped InSb single crystals. Izv. AN SSSR, Ser. Fiz., 35, 495-498 (in Russian) (1971).
9. T. Surek, C. B. Hari Rao, J. C. Swartz, and L. C. Garone. Surface morphology and shape stability in silicon ribbons grown by the edge-defined, film-fed growth process. J. Cryst. Growth, 124, 112-123 (1977).
10. P. I. Antonov, N. S. Grigoryev, and A. V. Stepanov. Orientational deviation in shaped germanium crystals grown from the melt. Izv. AN SSSR, Ser. Fiz., 36, 486-498 (in Russian) (1972).
11. P. I. Antonov and A. V. Stepanov. Peculiarities of shaped germanium crystals grown from the melt by Stepanov's techniques. In: IVth USSR Conference on Crystal Growth, Yerevan AN ArmSSR Publishing House, Yerevan, 2, 167-170 (in Russian) (1972).
12. N. V. Bessonova, D. I. Levinzon, Yu. M. Smirnov, Yu. I. Sterlikov, A. V. Stepanov and S. V. Tsikhotsky. Analysis of inhomogeneous impurity distribution in shaped germanium crystals. In: Silicon and Germanium. Metallurgiya, Moscow, 2, 84-87 (in Russian) (1970).
13. S. S. Vakhrameyev, V. V. Osvensky, and V. A. Smirnov. Dislocation structure of single crystals in thermal stress field calculated taking into account the radial and axial changes of ingot temperature in the course of growth. In: IVth USSR Conference on Crystal Growth, Yerevan, AN ArmSSR Publishing House, Yerevan, 2, 81-84 (in Russian) (1972).
14. P. I. Antonov, E. V. Galaktionov, V. M. Krymov, and V. S. Yuferev. Analysis of thermoelastic stresses in germanium ribbons grown by Stepanov's techniques. Izv. AN SSSR, Ser. Fiz., 40, 1419-1425 (in Russian) (1976).

15. P. I. Antonov, E. N. Kolesnikova, V. M. Krymov, S. P. Nikanorov, M. P. Proskura, and V. S. Yuferev. Temperature distribution in shaped germanium crystals grown from the melt by Stepanov's techniques. Izv. AN SSSR, Ser. Fiz., $\underline{40}$, 1407-1413 (in Russian) (1976).
16. P. I. Antonov and V. M. Krymov. Effect of geometry of shaped single crystals on thermal stress generated in the course of growth from the melt. In: 1st Symposium on Growth and Synthesis of Semiconductor Crystals and Films, Novosibirsk. Abstracts, Inst. of Inorganic Chemistry, SO AN SSR, Inst. of Physics of Semiconductors, SO AN SSSR, Novosibirsk, 112 (in Russian) (1978).
17. E. V. Galaktionov, V. M. Krymov, E. N. Kolesnikova, S. I. Bakholdin, N. B. Guseva, I. L. Shul'pina and P. I. Antonov. Generation of dislocation structure by thermal stress in germanium ribbon-shaped single crystals. In: Vth USSR Conference on Crystal Growth, Tbilisi. Abstracts, Inst. of Cybernetics of AN GSSR, Tbilisi, $\underline{2}$, 240-241 (in Russian) (1977).

CRYSTAL SHAPE STABILITY IN MENISCUS-CONTROLLED GROWTH PROCESSES

T. Surek,[1] S. R. Coriell,[2] and B. Chalmers[3]

[1]*Mobil Tyco Solar Energy Corporation, Waltham, Mass. 02154, U.S.A.*
[2]*National Bureau of Standards, Washington, D.C. 20234, U.S.A.*
[3]*Division of Engineering and Applied Physics, Harvard University, Cambridge, Mass. 02138, U.S.A.*

Introduction

One of the concerns of crystal growth researchers over the years has been the development of techniques to monitor and control the external shape of melt-grown crystals. Special applications of some single crystal materials have indicated the need to grow these crystals with a pre-determined cross-sectional shape such as ribbons, tubes, or filaments. Considerable progress in this area has been made by Soviet researchers in developing the Stepanov technique [1, 2] which uses a non-wetted shaping device to control the crystal shape. The web-dendritic growth process [3] has been another notable development in the growth of ribbon crystals; various other shaping schemes are also described in the literature [4-6]. A significant advance in shaped crystal growth has occurred with the recognition by LaBelle and co-workers [7] that a die shaper which is wetted by the melt provides a means of effectively isolating the growth from the main melt, and thus a better practical means for controlling the cross-sectional shape of the crystal. Their so-called "Edge-defined Film-fed Growth" or EFG process has evolved, in a relatively short period of time, to become a viable commercial technology for the growth of variously shaped sapphire crystals. There is considerable interest currently in applying the EFG process to the production of ribbon-shaped crystals of silicon for solar cell applications [8, 9].

The fundamental physical processes which determine the external shape of a melt-grown crystal occur at the triple junction of the three interfaces, viz., crystal-vapor, liquid-vapor, and crystal-liquid (see Fig. 1). It has been generally recognized that the shape of the growing crystal is likely to be related to that of the liquid-vapor interface or meniscus at the triple junction; of particular significance is the angle φ_0 in Fig. 1 which denotes the relative orientation of the crystal and liquid free surfaces. Some early speculations (e.g., ref. [4]) regarding this relationship were that φ_0 must be zero, in other words, that the crystal shape must follow the shape of the meniscus during growth. Experimental observations [10, 11] in Czochralski and Stepanov growths of silicon and germanium crystals have shown, on the other hand, that $\varphi_0 > 0°$ during constant diameter growth of these materials. The scatter in the values reported ($\sim 0°$ to $\sim 36°$) is probably the result of inaccuracies in the observation methods.

In this paper, we describe a recent experimental technique which provides a more accurate determination of the relative angle φ_0 than heretofore achieved by direct observations, and examine the implications of our finding that φ_0 is a characteristic material property (Section 2). In Section 3, we extend these concepts to develop a general theory of the time evolution and stability of the crystal shape in meniscus-controlled growth processes. Applications of the theory to Czochralski and floating zone growth, and to techniques which

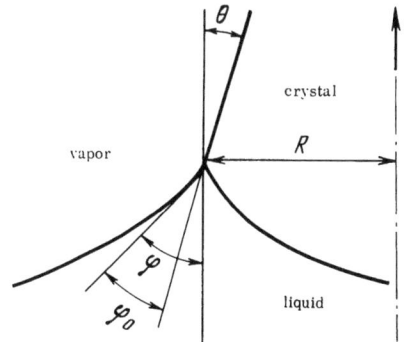

Fig. 1. Schematic of interfacial configuration at crystal-liquid-vapor junction in crystal growth from the melt.

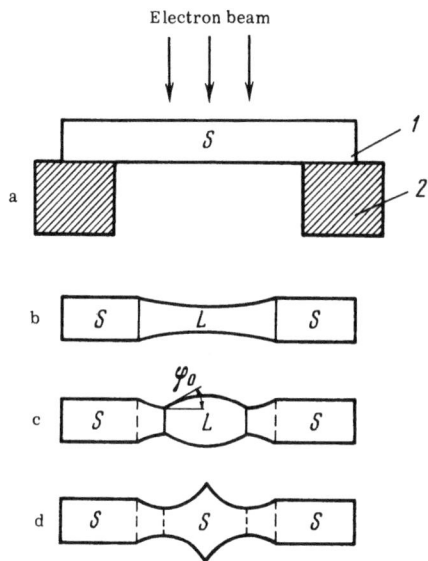

Fig. 2. Schematic cross-section of radially frozen molten zone experiments. Part (a) shows the experimental setup. Parts (b) to (d) depict typical solid (S) and liquid (L) configurations in the wafer during the experiment (1) sample; (2) supporting graphite ring.

utilize a die shaper (such as the Stepanov and EFG methods), are presented in Section 4. Although much of the discussion in this section is applicable to shaped crystal growth of any material, the specific calculations and examples will deal with the growth of silicon crystals.

2. The Meniscus Angle in Crystal Growth from the Melt

A first step toward understanding the growth of shaped crystals from the melt is to obtain a reliable experimental measure of the relative angle φ_0 in Fig. 1. The simple experimental procedure [12] which was developed for this purpose is shown schematically in Fig. 2. A wafer of constant thickness (typically 0.05 cm) of the material under investigation is held in a horizontal plane, and a circular molten region of ~1 cm diameter is formed in the wafer (Fig. 2b). Heating of the zone can be achieved, for example, by focussing an electron beam onto the top surface of the wafer; the experiments reported here were carried out in a vacuum of ~10^{-5} torr. The initial shape (i.e., concave or convex) and size of the molten zone can be varied by adding a known excess amount of the material into the zone. Since the densities of liquid silicon and germanium are greater than those of the respective solids, the initial meniscus shape will be concave for these materials, as illustrated in Fig. 2b, when no excess material is added to the zone.

The molten zone is then frozen at a controlled rate; typical intermediate and final cross-sections are depicted in Figs. 2c and 2d, respectively. The sample is then mounted and sectioned through a diameter of the zone that had been melted to reveal the profile of the resolidified crystal. A comparison of the crystal profile with the calculated instantaneous meniscus shape is made to yield the sought-after relationship between the crystal and liquid free-surface directions (i.e., the angle φ_0 in Fig. 1). The calculation of the meniscus shape involves making a number of simplifying assumptions: (i) that the crystal profile is radially symmetric; (ii) that the solid-liquid interface is vertical (i.e., perpendicular to the crystal growth axis); and (iii) that the effects of gravity on the meniscus shape are negligible.

The results of experiments for {111} orientation silicon and germanium wafers are shown in Fig. 3. The curves show the variation of the relative angle φ_0 with the angle φ between the tangent to the meniscus at the growth interface and the horizontal (or crystal growth) axis (cf. Fig. 1). The data points represent average values of 10 separate experiments each using {111} orientation silicon and germanium wafers, respectively. The results encompass crystal growth rates which vary by nearly two orders of magnitude (from ~0.002 cm s^{-1} to ~0.09 cm s^{-1}), a factor of two variation in water thickness, and various initial meniscus profiles [12, 13].

From the intersection of the curves with the $\varphi = \varphi_0$ line in Fig. 3 (which corresponds to $\theta = 0$, i.e., growth with constant dimension R in Fig. 1), we obtain the characteristic values of φ_0 for {111} orientation surfaces of silicon and germanium, viz., φ_0(Si) = 11°±1° and φ_0(Ge) = 13°±1°. The observed variation of φ_0 with φ in Fig. 3 can be associated [14] with the anisotropy in the crystal-vapor interfacial free energy of silicon (or germanium), in accordance with a theoretical model [15] for the interfacial configuration at the crystal-liquid-vapor junction. This model (as well as those of Voronkov [16] and Pogodin et al. [17]) considers the equilibrium of surface tensions at the triple junction to determine the angular relationship of the interfaces; $\varphi_0 > 0$ is explained as a consequence of the incomplete wetting of the solid by its own melt.

Application to Shaped Crystal Growth. The above findings have important implications on the growth of shaped crystals. Firstly, a crystal cannot grow with a uniform cross-section unless the meniscus angle $\varphi = \varphi_0$ at all times in Fig. 1. In view of the observed constancy of the relative angle φ_0 (aside from the anisotropy effects noted above), the crystal dimension R in Fig. 1 will therefore increase when $\varphi > \varphi_0$ and decrease when $\varphi < \varphi_0$. Secondly, the basic physical processes which determine the shape of the growing crystal are the same in all "meniscus-controlled" growth processes such as Czochralski growth, float zoning, and the Stepanov and EFG techniques. It is the degree of crystal shape control which is different in the various techniques; this, in turn, influences the range of cross-sectional shapes which can be grown, in a stable manner, by each method. As will be shown in the next two sections, the degree of crystal shape control is a function of the detailed shape of the meniscus, and of the details of the heat flow in the crystal growth system. In addition to the $\varphi = \varphi_0$ requirement at the crystal growth boundary for steady state, the manner in which the other end of the meniscus (e.g., the lower end in the schematic in Fig. 1) terminates will be shown to be of prime importance. It is this latter boundary condition which distinguishes the Czochralski method from float zoning and from the EFG and Stepanov techniques. The use of wetted versus non-wetted die-shapers, respectively, in the latter processes has an additional effect on the ability to control the crystal shape [18].

3. General Theory of Crystal Shape Stability

With reference to Fig. 4, which is a schematic of the EFG or Stepanov processes, the following set of equations describes the time evolution of the crystal shape in a meniscus-controlled process:

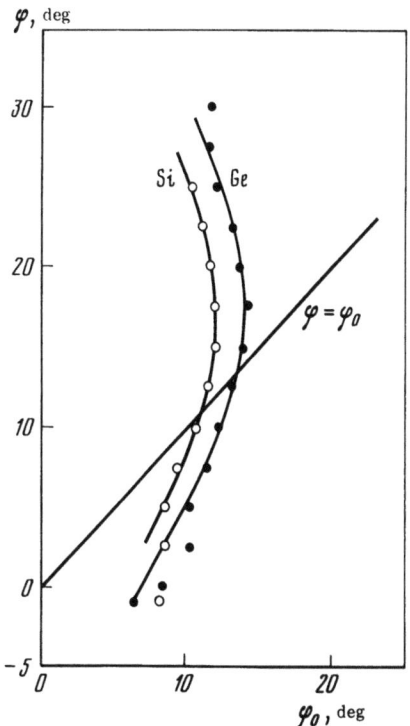

Fig. 3. Variation of the relative angle φ_0 with meniscus angle φ in radially frozen molten zone experiments for {111} orientation silicon and germanium wafers.

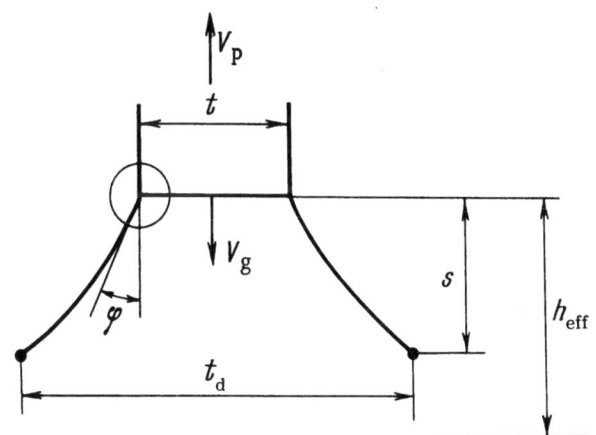

Fig. 4. Schematic cross-section of crystal growth from a die shaper of dimension t_d showing geometrical variables which determine the crystal dimension t. (Fig. 1 is a magnified view of the circled region).

$$\frac{dt}{d\tau} = 2V_g \tan(\varphi - \varphi_0); \quad (1)$$

$$\varphi = \varphi(t, t_d, s, h_{eff}); \quad (2)$$

$$\frac{ds}{d\tau} = V_p - V_g; \quad (3)$$

$$V_g = V_g(t, s, \varphi, V_p, H), \quad (4)$$

Here V_g is the component of the growth velocity along the crystal pulling axis, V_p is the velocity of pulling of the crystal, t is the crystal dimension (say, ribbon thickness or crystal diameter), t_d is the edge-defining die dimension (i.e., the distance between the points of attachment of the meniscus to the die shaper), s is the height of the growth interface above the die top (i.e., the meniscus height), and h_{eff} is a hydrostatic pressure variable which, for the case illustrated in Fig. 4, denotes the effective height of the growth interface above the horizontal liquid level in the crucible. The height h_{eff} corresponds to the pressure difference between the atmosphere and a point inside the meniscus; in the EFG process, it includes the effects of viscous flow of melt in the capillary and in the meniscus film, as well as that of the hydrostatic head [9]. The symbol H in eq. (4) is used to represent heat transfer effects in the system caused by other than the first three variables in the brackets, e.g., external heat input terms. The parameters V_p, h_{eff}, and H may be functions of the time variable τ, in general. The effects of the anisotropy of φ_0 discussed in the preceding section are neglected in this analysis.

Equations (1) to (4) describe the growth of symmetric crystals by the EFG, Stepanov, and Czochralski methods, although the interpretation of some of the terms is different in the latter case (e.g., for the usual Czochralski growth situation, $t_d \to \infty$ and $s \approx h_{eff}$). For the floating zone process, an additional equation describing the time rate of change of the zone volume v is required [19]; in addition, h_{eff} in eq. (2) is replaced by v, V_g in eq. (4) is also a function of v, and some of the terms have different interpretations from the present discussion. For Stepanov growth, the parameter h_{eff} is usually close to or less than zero; in principle (if not in practice), the growth interface is below the liquid level in the crucible for $h_{eff} < 0$, i.e., there exists a positive hydrostatic pressure in the meniscus film. The attachment of the lower end of the meniscus to the die shaper and the effects of die design on meniscus attachment are discussed in the literature [18, 20]; we have also considered the advantages and disadvantages of using wetted and non-wetted dies in the shaping process [18].

Equation (1) is, of course, a direct consequence of the meniscus angle requirement for steady-state growth discussed in the preceding section. The functional form of φ in eq. (2) is obtained from the solution of Laplace's capillary equation for symmetric meniscus shapes. The third equation expresses the time dependence of meniscus height caused by a difference between the pulling velocity and growth velocity; the functional dependence of the growth velocity in eq. (4) is obtained from the solution of the heat transfer in the system and from the condition of heat balance at the solid-liquid interface. Equation (4) is an approximation in the sense that, in principle, the growth velocity will depend on the complete shape of the crystal and the liquid. In other words, the growth velocity $V_g(\tau)$ at time τ may depend not only on the values at time τ of the independent variables in eq. (4) but also on the values of the variables for all previous values of τ. In essence, we are assuming in eq. (4) that the rate of propagation of thermal transients, caused by small deviations of the variables t, φ, and s from their respective steady-state values, is large compared with the rates at which these variables change and with the crystal growth rate. Explicit time-dependent heat flow effects, such as would be caused by fluctuations in external heating sources, are included, of course, in the general parameter H in eq. (4).

Equations (1) to (4) are the basic equations needed to describe the dynamics of a meniscus-controlled crystal growth process. Given the initial conditions, the forms of the functions φ and V_g, and the explicit time dependencies (if any) of V_p, h_{eff}, and H, the equations may be solved for the time evolution of the external shape of the crystal. Our primary interest in this paper is to examine the stability of the crystal shape; we describe below a linear perturbation analysis of the dynamic equations which can be used to determine the conditions for crystal shape stability in meniscus-controlled growth processes.

<u>Linear Perturbation Analysis</u>. In a linear perturbation analysis, one examines whether small deviations from steady state decay with time, i.e., the system is stable and returns to steady-state conditions, or whether the perturbations are amplified and the system is unstable.

We denote the steady-state values of the parameters by the subscript "0" in the following. At steady state, all system variables are independent of time; thus we have $V_{p0} = V_{g0}$, $\dot{s}_0 = 0$, $\dot{t} = 0$, and the steady-state value of the meniscus angle is equal to $\varphi_0(0)$, where the dot denotes a time derivative. We denote small deviations from steady state by the symbol "Δ"; e.g., $\varphi = \varphi_0(0) + \Delta\varphi$, $t = t_0 + \Delta t$, $V_p = V_{p0} + \Delta V_p$, etc. Substituting these expressions into eqs. (1) to (4), expanding the appropriate quantities about the steady-state values, and retaining only linear terms, we obtain a set of three first-order linear inhomogeneous differential equations with constant coefficients for the system variables Δs, Δt, and $\Delta \varphi$. These equations can be solved (for example, by Laplace transform techniques) to determine the time dependence of the fluctuations. Alternatively, we can obtain a second-order linear inhomogeneous differential equation for the system variable of primary interest, i.e., the fluctuation in crystal dimension, Δt:

$$\Delta \ddot{t} + 2\alpha \Delta \dot{t} + (\alpha^2 + \beta^2)\Delta t = g(\tau), \tag{5}$$

where

$$\alpha = \frac{1}{2}\left(\frac{\partial V_g}{\partial s} + \frac{\partial \varphi}{\partial s}\frac{\partial V_g}{\partial \varphi}\right) - V_{g_0}\frac{\partial \varphi}{\partial t}; \tag{6}$$

$$\alpha^2 + \beta^2 = 2V_{g_0}\left(\frac{\partial \varphi}{\partial s}\frac{\partial V_g}{\partial t} - \frac{\partial \varphi}{\partial t}\frac{\partial V_g}{\partial s}\right) \tag{7}$$

and

$$g(\tau) = 2V_{g_0}\left[\frac{\partial \varphi}{\partial s}\left(\Delta V_p - \frac{\partial V_g}{\partial V_p}\Delta V_p - \frac{\partial V_g}{\partial H}\Delta H\right) + \frac{\partial \varphi}{\partial t_d}\Delta \dot{t}_d - \frac{\partial \varphi}{\partial h_{eff}}\Delta \dot{h}_{eff} + \frac{\partial V_g}{\partial s}\left(\frac{\partial \varphi}{\partial t_d}\Delta t + \frac{\partial \varphi}{\partial h_{eff}}\Delta h_{eff}\right)\right], \tag{8}$$

with all partial derivatives evaluated at the steady-state values of the variables.

In the above expressions, the quantities α and β denote the natural frequencies of the system, and, as discussed below, their values determine whether the crystal growth system is stable or unstable. We combined in the inhomogeneous term $g(\tau)$ a number of quantities which may fluctuate (e.g., the crystal pulling speed, ΔV_p, or external power input, ΔH). We regard $g(\tau)$ as given; then, from eq. (5), we can calculate the effect of $g(\tau)$ on the dimensions of the solidifying crystal. We now proceed to solve the homogeneous equation (i.e., $g(\tau) = 0$ in eq. (5)) to derive the general conditions for crystal shape stability, and present a specific example for solving the inhomogeneous equation.

<u>Criteria for Crystal Shape Stability</u>. The solutions of the homogeneous equation are of the form $\exp(\lambda \tau)$, where $\lambda = -\alpha + i\beta$ or $-\alpha - i\beta$. The system will be stable if the real part of λ is negative, since then $\exp(\lambda \tau) \to 0$ as $\tau \to \infty$. Therefore, the criteria for stability are that

$$\alpha > 0 \quad \text{and} \quad \alpha^2 + \beta^2 > 0. \tag{9}$$

From the definitions of α and β in eqs. (6) and (7), we immediately see that the existence of crystal shape stability in meniscus-controlled growth processes is dependent on the explicit functional form of the variation of growth velocity with s, t, and φ, and on the form of the meniscus shape function φ in eq. (2). We will examine crystal shape stability in the various growth processes in the next section.

The most useful form of the solution of the homogeneous equation is given by

$$\Delta t = \exp(-\alpha \tau)[c_1 \cos \beta \tau + c_2 \sin \beta \tau], \tag{10}$$

where the constants c_1 and c_2 are determined from the initial and boundary conditions of the perturbation. A similar expression can be obtained for the time dependence of fluctuations in the other homogeneous variables (i.e., $\Delta \varphi$ and Δs). When β is real, perturbations to a stable system are thus seen to lead to an oscillation about steady state with an exponentially decaying amplitude; the decay time is given by $1/\alpha$, while the period of oscillation is $2\pi/\beta$.

As an example for solving the inhomogeneous equation, we consider a system fluctuation of the form $g(\tau) = A \exp(i\omega\tau)$, where A is the amplitude and ω is the frequency of the fluctuation. Then, from eq. (5), we obtain

$$\Delta t = A \exp[i(\omega\tau - \varepsilon)]/[(\alpha^2 + \beta^2 - \omega^2)^2 + 4\alpha^2\omega^2]^{1/2}, \qquad (11)$$

where the phase factor ε is given by

$$\tan \varepsilon = 2\alpha\omega/(\alpha^2 + \beta^2 - \omega^2). \qquad (12)$$

The complete solution for the time dependence of the fluctuation in the crystal dimension is, of course, given by the sum of eqs. (10) and (11). However, for long times for a stable system, the homogeneous solutions vanish, and eq. (11) represents the effect on Δt of fluctuations in the system as represented by the particular form assumed for $g(\tau)$. It is interesting to note from Eq. (11) that, when the frequency of the imposed fluctuation is much greater than the natural system frequencies α and β, the amplitude of the fluctuation in t is given by $\Delta t = A/\omega^2$.

Finally, we consider a "thermally stable" system, i.e., a system where V_g = constant, independent of s, t, and φ. In this case, the thermal effects on crystal shape are, in essence, neglected, and the stability of the crystal shape is determined only by meniscus shape effects; the stability criteria in eq. (9) reduce to the requirement that

$$\frac{\partial\varphi}{\partial t} < 0. \qquad (13)$$

Qualitatively, the above condition implies that, if a change in crystal dimension t occurs (i.e., whenever $\varphi \neq \varphi_0$), the meniscus angle φ should approach φ_0 again as a result of the shape change in order for the crystal shape to be stable [20, 21].

4. Applications of Theory of Crystal Shape Stability

In this section, we examine briefly the existence of crystal shape stability in the various meniscus-controlled growth processes under both of the conditions where thermal effects are excluded from or included in the analysis (i.e., the stability criteria in eqs. (13) and (9), respectively). Specific calculations will deal with the growth of silicon crystals.

Czochralski Growth. There are a number of published analyses of the meniscus shape (e.g., refs. [22] and [23]) and of the heat flow (e.g., refs. [24] and [25]) in Czochralski crystal growth systems which are useful for the present analysis.

Figure 5 shows the variations of $\partial\varphi/\partial t$ and $\partial\varphi/\partial s$ with diameter t in the growth of silicon crystals; these curves were calculated from the numerical data of Mika and Uelhoff [22] on the meniscus shape in Czochralski growth ($\varphi_0 = 10°$ is assumed for these curves in order to match the data in the reference). Since $\partial\varphi/\partial t > 0$ for all values of t, it can be concluded that the crystal shape in Czochralski growth would be unstable for the case where the thermal effects are excluded from the analysis (i.e., eq. (13)). Perturbations to steady-stage growth will grow exponentially, in this case, with a time constant given by $[V_{g_0}(\partial\varphi/\partial t)]^{-1}$; for the range of crystal diameters in Fig. 5 and for typical growth rates, the time constant varies from $\sim 10^2$ s to $\sim 10^4$ s. It would thus appear that there is ample time in Czochralski growth to impose changes in the growth parameters in order to effectively control the crystal diameter.

The complete analysis of the shape stability problem (i.e., the criteria in eq. (9)) shows that conditions of growth can be attained in the Czochralski process where the stability criteria are satisfied. From eqs. (6) and (7) (with $\partial V_g/\partial\varphi = 0$), the criteria in eq. (9), and

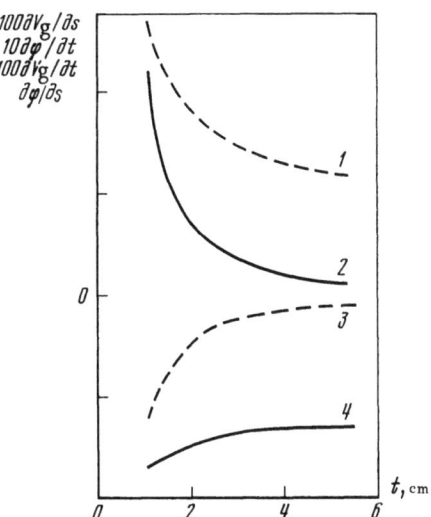

Fig. 5. Variations of the partial derivatives of the growth velocity and of the meniscus angle with crystal diameter in silicon crystal growth by the Czochralski technique.
(1) $\partial V_g/\partial s$; (2) $\partial \varphi/\partial t$;
(3) $\partial V_g/\partial t$; (4) $\partial \varphi/\partial s$

the inequalities $\partial\varphi/\partial t > 0$ and $\partial\varphi/\partial s < 0$ from Fig. 5, it can be argued that the necessary conditions for stability are heat flow situations which lead to $\partial V_g/\partial s > 0$ and $\partial V_g/\partial t < 0$. These conditions can be readily achieved in Czochralski growth systems.

As an example, we consider the following expression for the crystal growth velocity which is obtained from the solution of the steady-state heat transfer in the system and from the condition of heat balance at the crystal growth interface [24]:

$$V_g = L^{-1}[c_s k_s t^{-1/2} + c_r s t^{-1} - c_l k_l s^{-1}], \qquad (14)$$

where L is the latent heat of fusion per unit volume of solid, k_s and k_l are the thermal conductivities of the solid and liquid, respectively, and c_s, c_r, and c_l are constants which depend on material and growth parameters (e.g., emissivities of radiating surfaces, temperature of the melt in the crucible, etc.).

The first term in the brackets in eq. (14) represents the heat flux in the crystal at the crystal-liquid interface; it is obtained from the solution of steady-state heat conduction in the crystal and radiation from the crystal surface to an environment at 0°K [25]. We have taken $c_s = 509$°K cm$^{-1/2}$ and $k_s = 0.287$ watt cm^{-1} °K^{-1} in the calculations. The second term in the brackets denotes the radiation from the meniscus surface [24]; we have assumed $c_r = 90$ watt cm^{-2}. The last term in eq. (14) is a simple approximation for the heat flux entering the meniscus; the complete solution would depend on the details of the conductive and convective heat transfer in the melt. The numerical values of the remaining constants in eq. (14) were assumed to be $c_l = 50$°K, $k_l = 0.67$ watt cm^{-1} °K^{-1} and $L = 4200$ J cm^{-3}. (As noted earlier, values appropriate to silicon crystal growth were used in all calculations).

From Fig. 5, it is seen that $\partial V_g/\partial s > 0$ and $\partial V_g/\partial t < 0$, in other words, that the necessary conditions for stability are satisfied for the particular form of V_g assumed in the analysis. The complete analysis of the crystal shape stability requires the examination of the criteria in eq. (9). For the range of crystal diameters in Fig. 5, we find that $\alpha > 0$ and $\alpha^2 + \beta^2 > 0$, i.e., the crystal shape is stable; values of the decay time (i.e., α^{-1}) are in the range 60 to 500 s. For crystal diameters $\gtrsim 4$ cm, we found that β is imaginary, in which case the appropriate (i.e., longer) decay time, $(\alpha - |\beta|)^{-1}$, was calculated. For very small crystal diameters ($t \lesssim 0.3$ cm), the crystal shape turns out to be unstable ($\alpha < 0$).

In summary, it is seen that the meniscus shape function in Czochralski growth tends to cause instability of the crystal shape, but that heat flow conditions exist for which the crystal shape is stabilized. Specifically, the presence of surface cooling effects (e.g., radiation), which lead to the type of t and s dependences given in eq. (14), can lead to a condition of shape stability.

Growth from Dies. The use of a die shaper which is either wetted [7] or not wetted [1, 2] by the melt has been shown to be an effective method of controlling the cross-sectional shape of melt-grown crystals. In this section, we examine some of the reasons for the stability of the crystal shape in growth from dies. We will limit the discussion to the growth of ribbon-shaped crystals of silicon in view of the growing interest to use such crystals for solar cell applications [8]. Also, we will not attempt to distinguish in this analysis between wetted and non-wetted dies, i.e., we will not consider the details of meniscus attachment at the die top in Fig. 4; these problems were addressed in a recent paper [18].

The problem of the meniscus shape in the growth of ribbon crystals is quite complex; a useful simplification [9] is to solve the separate problems of the meniscus shape at the side of an infinitely wide ribbon, and that against a vertical cylinder of radius t/2, where t is the ribbon thickness. Numerical solutions [9] of Laplace's capillary equation have shown that, at the side of the ribbon, the meniscus radius of curvature is approximately constant and is given by $R_1 = \gamma/\rho g h_{eff}$, where γ is the liquid-vapor interfacial free energy, ρ is the density of liquid silicon, and g is the acceleration of gravity. This approximation holds for the usual case in EFG where $h_{eff} > 1$ cm and $h_{eff} \gg s$; for large negative values of h_{eff} (i.e., $h_{eff} < -1$ cm), the meniscus shape can be similarly approximated by a circular segment of radius R_1.

The left and right insets in Fig. 6 show the meniscus shapes at the ribbon side for $h_{eff} > 0$ and $h_{eff} < 0$, respectively. Attachment of these circular menisci to a die of dimension t_d can be shown to lead to the following relationship (cf. eq. (2)) among the meniscus shape variables in Fig. 4:

$$\varphi = \arctan[(t_d - t)/2s] - \pi/2 + \arccos(z/2R_1), \tag{15}$$

where

$$z = [s^2 + (t_d - t)^2/4]^{1/2}. \tag{16}$$

The meniscus angle $\varphi = \varphi_0 = 11°$ for steady-state growth of silicon crystals. Equations (15) and (16) can be used to obtain the steady-state ribbon thickness as a function of meniscus height for given values of t_d and h_{eff}. Figure 6 shows these relationships for both positive and negative values of h_{eff}; for a given value of t_d, the ribbon thickness increases from left to right in the figure. It can be readily shown that, in the portions of the curves denoted by the solid lines, the partial derivative $\partial\varphi/\partial t < 0$, i.e., the ribbon side is inherently shape stable when thermal effects are neglected (see eq. (13)). The broken lines in the figure denote regions where $\partial\varphi/\partial t > 0$. The maximum meniscus height s_{max} (solid points in Fig. 6) for shape stability is seen to be a function of h_{eff}; the heights correspond to $\theta = 0°$ and $\theta = 180°$ (see Fig. 6) for $h_{eff} > 0$ and $h_{eff} < 0$, respectively. For $h_{eff} < 0$, there may not exist, of course, stable solutions to Laplace's equation beyond some value of meniscus height which may be $< s_{max}$. For typical values of the growth parameters (i.e., $h_{eff} = 2$ cm, $t_d = 0.04$ cm, $t = 0.03$ cm, and $V_{g0} = 0.04$ cm s^{-1}), the decay time of perturbations to steady-state is $[V_{g0}(\partial\varphi/\partial t)]^{-1} = 1$ s.

The value of the derivative $\partial\varphi/\partial s$, which enters the complete shape stability problem (i.e., eqs. (6) and (7)), is also obtained from these calculations. For $h_{eff} > 0$ in Fig. 6, it can be readily shown that $\partial\varphi/\partial s < 0$, including the portions denoted by the broken lines. For $h_{eff} < 0$, on the other hand, $\partial\varphi/\partial s < 0$ only for the portions of the solid lines below the points denoted by the open circles (which corresponds to $\partial t/\partial s = 0$); for the remaining parts of the curves, $\partial\varphi/\partial s > 0$. As shown below, this has an important effect on crystal shape stability for growth with $h_{eff} < 0$.

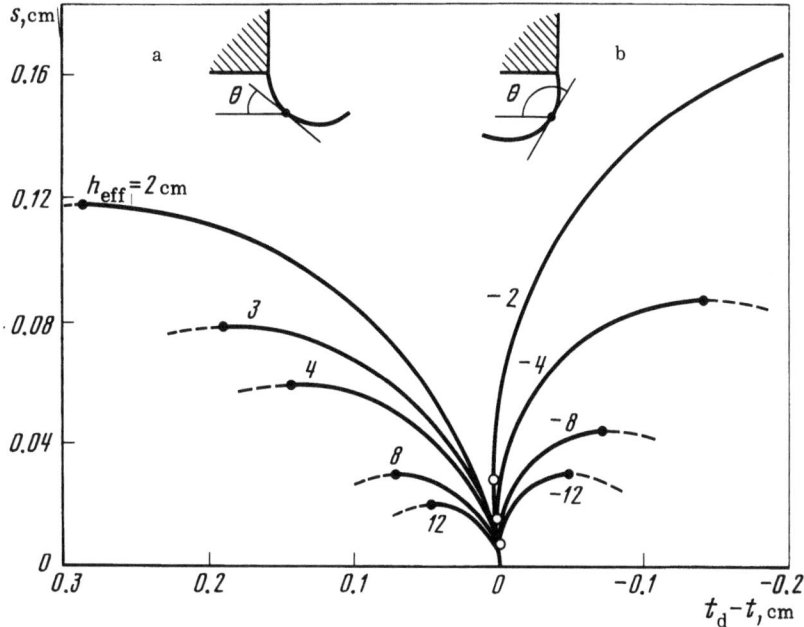

Fig. 6. Meniscus height-ribbon thickness relationships in silicon ribbon growth as a function of h_{eff} (t_d is the die thickness). The insets in the figure show schematically the circular menisci at the side of the ribbon for positive and negative h_{eff}.

Fig. 7. Variation of the steady-state crystal radius with meniscus height and h_{eff} in the growth of rod-shaped silicon crystals from a die of radius $t_d/2$. The curves also represent an approximate solution to the ribbon-edge problem. (1-3) $h_{eff} = -2, 2,$ and 4, respectively.

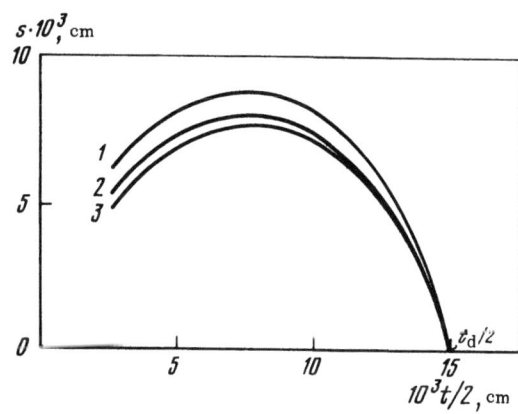

We now consider briefly shape stability at the ribbon edges. An approximation [21] to the problem is to consider a crystal of radius $t/2$ growing from a die shaper of radius $t_d/2$ (cf. Fig. 4); the analysis is thus exact for the growth of rod-shaped crystals. Figure 7 shows the variation of the steady-state crystal radius (i.e., $\varphi = \varphi_0 = 11°$) with meniscus height and h_{eff} for growth from a die with $t_d/2 = 0.015$ cm. In the portions of the curves where $\partial s/\partial(t/2) < 0$, it is found that $\partial \varphi/\partial t < 0$. Shape stability in the growth of rod-shaped crystals (or at the ribbon edges) is thus seen to be conditional; it requires that the edge-defining die radius should not exceed approximately twice the radius of the crystal (or the radius of curvature at the ribbon edge) [20, 21].

To examine the effects of thermal gradients on ribbon shape stability, we consider the following expression for the growth velocity which was derived [9] for the growth of wide, thin ribbons by EFG (cf. eq. (14)):

$$V_g = L^{-1}(c_e k_s t^{-1/2} - \Delta T k_l s^{-1}), \tag{17}$$

where ΔT is the difference between the temperature at the die top and at the growth interface and c_e is a constant which depends on material and growth parameters (e.g., emissivities and temperatures of radiating surfaces). The terms in the brackets represent, respectively, the heat fluxes in the crystal and in the liquid at the crystal-liquid interface; the radiation from the meniscus surface is neglected in this analysis. In the growth of silicon ribbons by EFG, the parameter $\Delta T > 0$; maximum growth rate, for a given ribbon thickness, is achieved under conditions where $\Delta T \to 0$. Growth from a supercooled melt (i.e., $\Delta T < 0$) has been observed to occur in the growth of sapphire filaments by EFG [7].

Since $\partial V_g / \partial t < 0$ and $\partial V_g / \partial s > 0$, it is seen from eqs. (6), (7), and (9) that the stability of the ribbon thickness is further enhanced by the thermal gradients under the usual conditions of growth of EFG silicon ribbons. (The treatment of thermal effects at the ribbon edges is beyond the scope of this paper). For typical values of h_{eff} = 2 cm, t_d = 0.04 cm, c_e = 200°K cm$^{-1/2}$ and ΔT = 4°K, it can be shown that the stability criteria in eq. (9) are satisfied for the entire range of ribbon thicknesses from 0 to 0.04 cm. The decay time (α^{-1} or $(\alpha - |\beta|)^{-1}$, if β is imaginary) is in the range 0.1 s to 1 s.

For h_{eff} = -2 cm (i.e., growth with a positive hydrostatic pressure in the meniscus film) and the above values of the other variables, the ribbon thickness will be stable only for small values of meniscus height. For $s \gtrsim 0.05$ cm, the fact that $\partial \varphi / \partial s > 0$ (as discussed earlier) begins to override the stabilizing effects of the thermal gradients, and $\alpha^2 + \beta^2 < 0$ results. Perturbations to steady-state will grow exponentially in this case; the time constant of the instability is ~10 s. In the limit where the maximum growth velocity is approached, i.e., $\Delta T \to 0$ and $\partial V_g / \partial s \to 0$, the ribbon thickness will be unstable for $s \gtrsim 0.0275$ cm (i.e., where $\partial \varphi / \partial s > 0$ on the h_{eff} = -2 cm curve in Fig. 6). Thus, it is seen that ribbon growth under large negative values of h_{eff} and at increasing growth rates tends to be shape unstable, in spite of the fact that the ribbon thickness is inherently stable (i.e., $\partial \varphi / \partial t < 0$) when thermal effects are excluded from the analysis.

Finally, we consider briefly the intermediate range of h_{eff} values, i.e., $|h_{eff}| < 1$ cm, which is not covered by our approximate analysis of the meniscus shape in eqs. (15) and (16). Accurate calculations of the meniscus shape can be obtained under these conditions (which correspond to the usual conditions of growth by the Stepanov technique) by numerical integration of Laplace's equation. The s-t curves for these cases are expected to be intermediate to the curves in Fig. 6; thus it is expected that the crystal shape will be stable for small values of s, and that it will be unstable for large values of s (where $\partial \varphi / \partial s > 0$), particularly at fast crystal growth rates.

<u>Floating Zone Growth.</u> As pointed out in Section 3, eqs. (1) to (4) are not directly applicable to describe the time evolution and stability of the crystal shape in floating zone growth. The latter problem has been considered in a recent publication [19] for the case where thermal effects are excluded from the analysis. It was found that, for shape stability of the resolidifying crystal, the necessary and sufficient conditions are that

$$\left(\frac{\partial \varphi}{\partial t}\right)_{v,l} < 0 \quad \text{and} \quad \left(\frac{\partial \varphi}{\partial v}\right)_{t,l} > 0 \tag{18}$$

where v is the zone volume, l is the zone length, and t is the dimension of the resolidifying crystal. These conditions are satisfied in the usual float-zone geometries of interest in silicon crystal growth [19]. It is expected that thermal effects, under the usual conditions of growth, would tend to enhance the stability of the crystal shape in float zoning.

5. Summary

A general theory has been presented for the stability of the crystal shape in meniscus-controlled growth processes. Although the basic physical processes which determine the crystal shape occur at the crystal-liquid-vapor triple junction (e.g., the $\varphi = \varphi_0$ boundary condition for steady state), the manner in which the lower end of the meniscus in Fig. 1 terminates has been shown to be of prime importance. The fact that the lower end of the meniscus is constrained by the die shaper in the EFG and Stepanov methods leads to the condition that $\partial\varphi/\partial t < 0$, i.e., inherent shape stability for a wide range of growth conditions. On the other hand, in the usual Czochralski growth situation where the crystal diameter is much smaller than the diameter of the crucible, we found that $\partial\varphi/\partial t > 0$; shape stability in Czochralski growth can only be achieved under appropriate heat flow conditions.

The type of analysis which was presented in this paper is clearly very important to the development of shaped crystal growth techniques, and to methods of automatic diameter control [26]. In addition to identifying the critical growth parameters in the particular crystal growth process, such analyses are useful in the determination of control requirements and in the design of control schemes for the process.

Literature Cited

1. A. V. Stepanov, Izv. Akad. Nauk SSSR, Ser. Fiz. 33, 1946 (1969).
2. P. I. Antonov, J. Cryst. Growth 23, 318 (1974).
3. A. I. Bennett and R. L. Longini, Phys. Rev. 116, 53 (1959).
4. G. K. Gaule and J. R. Pastore, In: Metallurgy of Elemental and Compound Semiconductors, Ed. R. O. Grubel (Interscience, New York), p. 201 (1961).
5. J. C. Boatman and P. C. Goundry, Electrochem. Tech. 5, 98 (1967).
6. C. E. Bleil, J. Cryst. Growth, 5, 99 (1969).
7. B. Chalmers, H. E. LaBelle, Jr., and A. I. Mlavsky, J. Crystal Growth 13/14, 84 (1972); H. E. LaBelle, Jr., U.S. Patent 3,471,266 (October 1969).
8. K. V. Ravi and A. I. Mlavsky, In: Sharing the Sun: Solar Technology in the Seventies, Vol. 6, Ed. K. W. Boer, ISES p. 23 (1976).
9. J. C. Swartz, T. Surek, and B. Chalmers, J. Electron. Mater. 4, 255 (1975).
10. P. I. Antonov, Growth of Crystals, Vol. 6, Consultants Bureau, New York (1968).
11. P. I. Antonov and A. V. Stepanov, Izv. Akad. Nauk SSSR, Ser. Fiz. 33, 1974-1989 (1969).
12. T. Surek and B. Chalmers, J. Cryst. Growth, 29, 1 (1975).
13. T. Surek, Scripta Metall. 10, 425 (1976).
14. J. W. Cahn, private communication (1977).
15. W. Bardsley, F. C. Frank, G. W. Green, and D. T. J. Hurle, J. Cryst. Growth 23, 341 (1974).
16. V. V. Voronkov, Soviet Physics - Solid State 5, 415 (1963).
17. A. I. Pogodin, I. M. Tumin, and A. M. Eidenzon, Izv. Akad. Nauk SSSR, Ser. Fiz. 37, 2292 (1973).
18. T. Surek, B. Chalmers, and A. I. Mlavsky, Proceedings of ICCG-5, J. Cryst. Growth (1978).
19. T. Surek and S. R. Coriell, J. Cryst. Growth 37, 253 (1977).

20. V. A. Tatarchenko, J. Cryst. Growth 37, 272 (1977).
21. T. Surek, J. Appl. Phys. 47, 4384 (1976).
22. K. Mika and W. Uelhoff, J. Cryst. Growth 30, 9 (1975).
23. S. V. Tsivinskii, Inzh. Fiz. Zh. 5, 59 (1962).
24. G. K. Steel and M. J. Hill, J. Cryst. Growth, 30, 45 (1975).
25. T. F. Ciszek, J. Appl. Phys. 47, 440 (1976).
26. D. T. J. Hurle, Proceedings of ICCG-5, J. Cryst. Growth (1978).

NUMERICAL ANALYSIS OF HEAT AND MASS TRANSFER IN THE GROWTH OF LARGE SINGLE CRYSTALS FROM THE MELT

N. A. Avdonin

The Latvian State University, Riga

V. A. Smirnov

The Institute of Rare Metals, Moscow

Semiconductor single crystals, which nowadays are grown in ever increasing quantities to meet industrial demand, must be highly perfect structurally and contain minimum amounts of macroscopic and microscopic inhomogeneities. Analyses of structure and properties of semiconductor materials as functions of growth conditions have been carried out mainly by experimental methods. A new approach based on mathematical simulation is now becoming important. The present review will be devoted to this field.

A complete mathematical model must include a description of temperature and concentration fields, the fields of thermoelastic stress and plastic strain in the crystals, and also the melt hydrodynamics. It is necessary to take into account complex conditions of thermal radiation from the outer surfaces and the latent heat released at the interface. In this formulation such a nonlinear problem is extremely complicated and cumbersome, so that no attempts were made to solve the problem as a whole. However, much has been done in solving particular cases, and considerable success has been achieved.

Whatever the formulation of the problem involving the growth of single crystals, one has first of all to take accurate account of the conditions at the crystallization interface, namely, the release of latent heat and the specific crystallization kinetics. It should be noted that if growth at the crystallization interface proceeds via the normal mechanism and the interface is isothermal, the Stefan problem offers a satisfactory framework for studying thermal diffusion processes in crystal growth:

$$\gamma \rho V_n = \lambda_c \frac{\partial T}{\partial n} - \lambda_L \frac{\partial T}{\partial n};$$
$$T = T_l(c), \tag{1}$$

where γ is the latent heat of melting (crystallization); ρ is the density of the material; V_n is the growth rate perpendicular to the crystallization interface; λ_c, λ_L are the thermal conductivities of the crystal and melt, respectively; T is temperature; T_l is the equilibrium temperature; n is the unit vector of the normal to the interface; and c is the impurity concentration.

In the range of conventional industrial growth rates in the case of normal growth, supercooling at the interface is negligibly small ($\Delta T = T_l - T = 0.01$ to 0.001 °C), and the equilibrium constraint (1) is acceptable [1]. If, however, the crystallization interface includes segments of a plane face, the Stefan condition (1) must be replaced by a kinetic condition describing its layered growth:

$$V_n = f(\Delta T_{max}), \qquad (2)$$

since in this case the maximum supercooling at the face, ΔT_{max}, may be considerable.

Although much has been published on the analytical solution of the Stefan problem, most of the papers were based on such crude simplifications [2-6] that not even a qualitative picture of the processes emerged, let alone any quantitative agreement. An accurate solution of the Stefan problem is possible only in the one-dimensional [7] and quasistationary [1] cases. An analysis of the quasistationary solution of the uniaxial crystallization process [1] yielded a critical value of growth rate V_{crit}. If the crystallization rate V exceeds V_{crit}, supercooling in the bulk of the melt in front of the crystallization interface is produced and the single-crystalline character of growth is lost. The critical growth rate is determined by the thermal characteristics of the material:

$$V_{crit} = 2\lambda k/\gamma\rho,$$

where λ is the thermal conductivity, and k is the temperature gradient on the muffle of the resistance furnace. In the case of crystal growth from a binary melt, the critical value is found from the relation [8]

$$V_{crit} = k[\gamma\rho/2\lambda + (m^{-1}-1)\,a_l D^{-1} c_0]^{-1},$$

where m is the equilibrium distribution coefficient; D is the diffusion coefficient; c_0 is the initial concentration in the melt core; α_1 is the slope of the liquidus curve, with $T_l = T_A - \alpha_l c$ being the liquidus curve equation (T_A is the crystallization temperature of the pure material).

We shall also mention an interesting result reported in ref. 9 in which the solution of the inverted single-phase Stefan problem yields the temperature or flux at the lateral surface of the sample for a given equation of motion of the interface, y(t) (y(t) is assumed to be analytical). The solution is given as a simple series.

We shall illustrate the present state of the art in the numerical analysis of heat and mass transfer processes in the growth of large crystals choosing Czochralski growth, float-zone melting, and uniaxial crystallization as examples.

Figure 1 is a schematic representation of the mathematical model for the Czochralski growth of single crystals (R, R_c denote the housing and crucible radii, respectively, and H is the chamber height), and Fig. 2 gives a similar schematic for the floating-zone melting. The following system of equations must be solved to analyze the distributions of the temperature fields in the crystal and melt, the growth rate, and impurity distribution in the melt:

$$L\rho\left(\frac{\partial T}{\partial t} - V_0 \frac{\partial T}{\partial z}\right) = \frac{\partial}{\partial z}\left(\lambda \frac{\partial T}{\partial z}\right) + \frac{1}{r}\frac{\partial}{\partial r}\left(\lambda r \frac{\partial T}{\partial r}\right), \quad (r,z) \in Q_1;\ t>0; \qquad (3)$$

$$L\rho\left(\frac{\partial T}{\partial t} + u\frac{\partial T}{\partial r} + w\frac{\partial T}{\partial z}\right) = \frac{\partial}{\partial z}\left(\lambda \frac{\partial T}{\partial z}\right) + \frac{1}{r}\frac{\partial}{\partial r}\left(\lambda r \frac{\partial T}{\partial r}\right); \qquad (4)$$

$$\frac{\partial c}{\partial t} + u\frac{\partial c}{\partial r} + w\frac{\partial c}{\partial z} = \frac{\partial}{\partial z}\left(D \frac{\partial c}{\partial z}\right) + \frac{1}{r}\frac{\partial}{\partial r}\left(Dr \frac{\partial c}{\partial r}\right); \qquad (5)$$

$$\frac{\partial u}{\partial t} + u\frac{\partial u}{\partial r} + w\frac{\partial u}{\partial z} - \frac{u^2}{r} = \nu\left(\frac{\partial^2 u}{\partial r^2} + \frac{\partial}{\partial r}\left(\frac{u}{r}\right) + \frac{\partial^2 u}{\partial z^2}\right); \qquad (6)$$

$$\frac{\partial v}{\partial t} + u\left(\frac{\partial v}{\partial r} + \frac{v}{r}\right) + w\frac{\partial v}{\partial z} = \nu\left(\frac{\partial^2 v}{\partial r^2} + \frac{\partial}{\partial r}\left(\frac{v}{r}\right) + \frac{\partial^2 v}{\partial z^2}\right); \qquad (7)$$

$$\frac{\partial w}{\partial t} + u\frac{\partial m}{\partial r} + w\frac{\partial w}{\partial z} = \nu\left(\frac{\partial^2 w}{\partial r^2} + \frac{1}{r}\frac{\partial w}{\partial r} + \frac{\partial^2 w}{\partial z^2}\right) - \beta g\frac{\partial T}{\partial r}; \qquad (8)$$

$$(r,z) \in Q_2;\ t>0;$$

$$\gamma \rho V_n(t) = \lambda_c \frac{\partial T}{\partial n} - \lambda_L \frac{\partial T}{\partial n} \quad \text{for } T = T_l \text{ on } \Gamma_1; \tag{9}$$

$$D \frac{\partial c}{\partial n} + (1-k) c_l V_n(t) = 0 \quad \Gamma_1 \tag{10}$$

with the following boundary conditions at the free surface of the crystal and the melt:

$$-\lambda(T) \frac{\partial T}{\partial n} = \frac{\varepsilon_k}{1-\varepsilon_k} (\sigma T_k^4 - q_k), \tag{11}$$

where L is the heat capacity; V_0 is the growth rate of the crystal; ν is the melt viscosity; β is the volume expansion coefficient; g is the gravitational constant; ε_k and T_k denote the emissivity and temperature of the k-th elementary surface segment; σ is the Boltzmann constant; u, v, and w stand for the radial, tangential, and axial components of the melt motion,

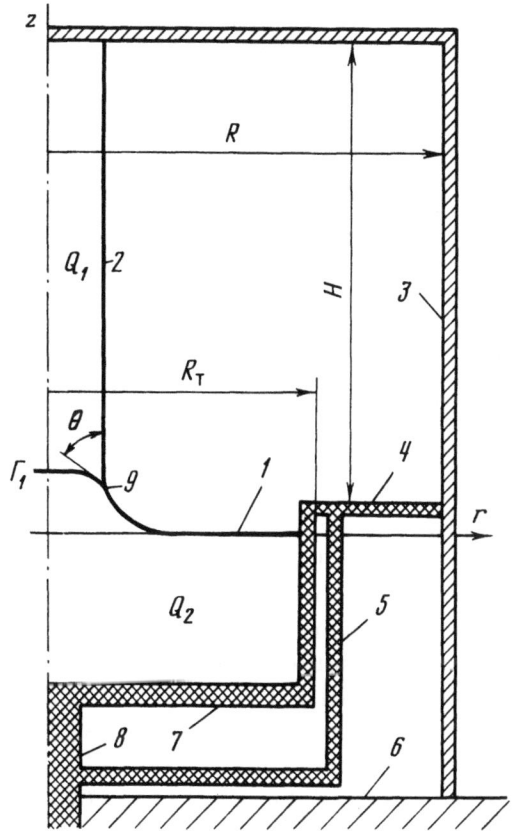

Fig. 1. Schematic representation of the Czochralski growth of crystals. (1) free surface of the melt, (2) crystal, (3) wall of a water-cooled housing, (4) graphite screen, (5) heater, (6) support plate, (7) crucible, (8) crucible support, (9) triple junction.

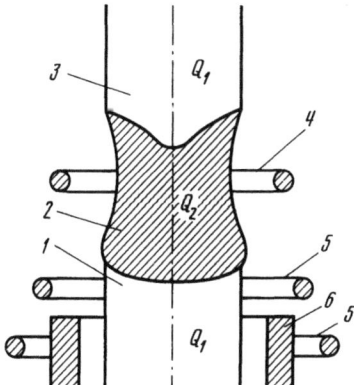

Fig. 2. Schematic of floating zone melting. (1) single crystal, (2) liquid zone, (3) melted ingot, (4, 5) main and pre-heating coils, (6) heat screen.

respectively; and q_k is the density of the effective radiation flux from the k-th elementary segment of crystal surface; q_k is found from the integral equation

$$q_k = \varepsilon_k \sigma T_k^4 + (1 - \varepsilon_k) \int q_j dF_{k-j}, \tag{12}$$

where q_j is the density of effective radiative flux from the j-th elementary segment of the surrounding inner surface, and dF_{k-j} is the angular coefficient between the k-th and j-th elementary segments.

In the case of the floating zone melting

$$-\lambda(T)\frac{\partial T}{\partial n} = \varepsilon_k \sigma T_k^4 - I_k^2 R_k,$$

where R_k is the resistance of the k-th segment and I_k is the distribution of currents in this segment (it is found from a system of linear equations defined by the laws of mutual inductance and self-inductance of the cells [10]).

Another aspect of interest is the study of a quasistationary state which occurs in the vicinity of the crystallization interface for low rates of crystal growth. In this case the terms containing derivatives with respect to time in system (3)-(8) must be set equal to zero. The problem (3)-(12) was studied stage-by-stage. Reliable methods have been suggested for the solution of the thermal and thermodiffusional problems. When hydrodynamics of the melt was taken into account, the problem was solved for regions with simplified geometry.

The crystallization interface shape, an unknown variable in the problem as formulated (defined by conditions (9), (10)), is one of the principal parameters of the process.

Two methods were successfully applied at the Computer Center of the Latvian University to the numerical solution of the Stefan conditions (9), (10) in different methods of single crystal growth. One of them is the method of "spreading" of the phase transition heat in the vicinity of the crystallization interface in a temperature interval δT; this method is based on a well-developed theory of the generalized solution of the Stefan problem [11]. Despite the error introduced by "spreading" of the latent heat, an optimal choice of the spreading interval δT ensures that an accurate shape of the crystallization interface is obtained. Calculated shapes of the interface for a number of spreading intervals are shown in Fig. 3 for the uniaxial crystallization of ingots. The calculated shape is in good agreement with the experimental data obtained by anodic etching [12] in a 4°C interval in the solid and 1.25°C in the liquid phase. This simulation approach was successfully applied not only to uniaxial crystallization but also to float-zone melting; standard programs were developed, and a large number of versions were run through a computer to optimize the induction coil power for different materials [13]. Figure 4 shows phase transition isotherms calculated for the floating zone melting process [13].

The second approach based on the idea of "straightening" the crystallization interface [14] enables one to obtain high-accuracy approximations of the Stefan conditions (9). This method also makes it possible to determine supercooling at a plane face for given growth kinetics, or to solve the reverse problem of finding kinetic coefficients for a face width given by the experiment [15].

Severe difficulties are encountered in determining the interface shape in Czochralski growth, because of a liquid column with a free surface. The main characteristic of the interface, namely, its inclination with respect to the cylinder generatrix or the triple junction (see Fig. 1), was found on the assumption that the interface is a paraboloid of revolution. This angle, θ, was first found without taking the molten column into account [16], and then considering the heat transfer conditions at the triple junction [17, 18]. Unfortunately, this

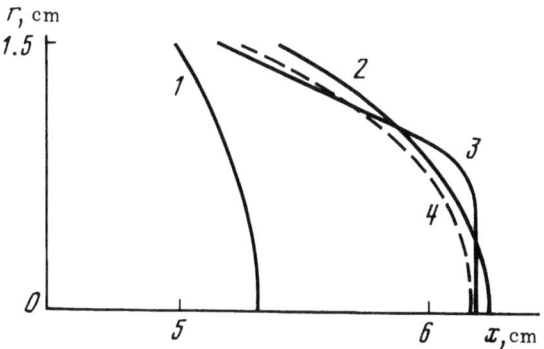

Fig. 3. Interface shape in the case of uniaxial crystallization (crystal radius R = 1.5 cm, V_0 = 0.4 mm min^{-1}). (1-3) for δT_S = 12, 4.2°C and δT_L = 5, 1.25, 0.6°C, respectively; (4) experimental curve.

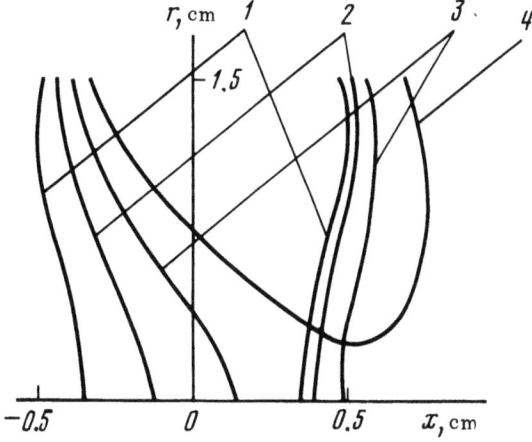

Fig. 4. Interface shape at different growth rates of silicon crystal (R = 1.5 cm, heater coil power P = 1770 W). (1-4) V_0 = 0, 2, 5, and 10 mm min^{-1}, respectively.

approach is limited to the growth of single crystals with relatively all diameters (up to 30 mm) in which the crystallization interface fits the paraboloid of revolution. Considerable difficulties are also met with respect to the boundary conditions (11), (12) for Czochralski growth because it is necessary to take into account the heat exchange between numerous surfaces and multiple re-radiation. Even by considering only single radiation processes in Czochralski growth, it was possible to reveal a non-monotonic pattern of thermal load on the crystal surface due to radiation from the melt surface [19] (Fig. 5). Generally the thermal load maximum is found at a distance from the interface not exceeding the ingot radius. The bright spot on the melt surface [20] used to control crystal diameter in current equipment for Czochralski growth of silicon is explained by the influence of the crystal on the radiative heat transfer from the melt surface. Figure 6 illustrates the distribution of heat fluxes on the melt surface (the origin is chosen as the crystal surface at the triple junction). We see that the heat flux from the surface of the crystal is produced by a non-monotonic distribution of radiative power along the melt surface.

It is becoming increasingly apparent that hydrodynamics in the liquid melt must also be taken into account. Convective flows in the liquid critically affect the interface shape, crystallization rate, and other characteristics of the crystal growth processes. However, numerical solution of the hydrodynamics equations in rotating melts is very difficult. Only a few papers are known in this field devoted to calculating the velocities and temperature distributions in the melt contained in a crucible [21-23], and to calculating velocities, temperature, and concentration in the case of vertical zone melting using a resistance heater (ampoule version) [24]. Even at the Reynolds number Re > 500, crucible rotation produces instability of the difference solution, so that the authors of refs. 21-23 had to introduce into calculations for germanium, a fictitious viscosity exceeding the actual one by two orders of magnitude. New programs have recently been developed in our Computer Center and the Institute of Rare Metals to calculate flow rates in the melt in the case of Czochralski growth and floating zone melting, by using more stable difference algorithms suggested in refs. 25, 26. The application of these methods to an analysis of hydrodynamics in melts of silicon

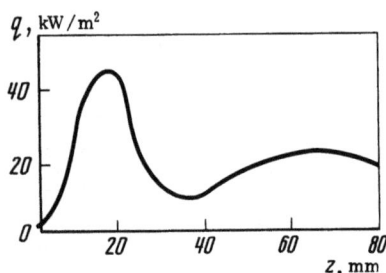

Fig. 5. Distribution of heat flux along the surface of a silicon ingot. Ingot diameter 40 mm. z = 0 corresponds to the interface.

Fig. 6. Distribution of heat fluxes on the melt surface. (1-3) reflected fluxes incident from the crystal, upper graphite annular screen, and the heater overhang, respectively; (4) the flux emitted by the melt surface; (5) the total flux emitted by the melt surface.

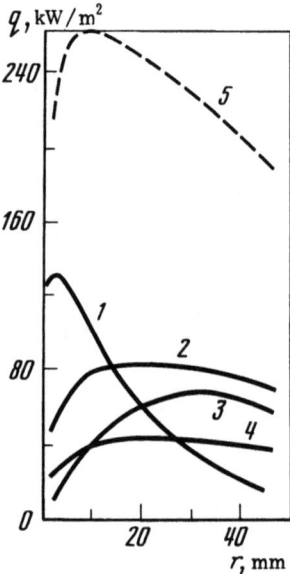

and other materials was reported at the Vth USSR Conference on Crystal Growth [27]. The method suggested in ref. 25 was also successfully applied to an analysis of hydrodynamics in a flow of vapor-gas mixture in a reactor designed for the homoepitaxial growth of silicon-based structures [28]. The same publication conclusively demonstrated that the boundary layer approximation cannot be used to analyze processes of this type. Application of the results of ref. 29 to an analysis of non-stationary modes of crystal growth opens up the possibility of numerical investigation of convective turbulence in melts and, hence, temperature fluctuations in them, as well as microscopic inhomogeneous distributions in the crystals grown.

The results of numerical analysis of temperature distribution in Czochralski growth and float-zone melting of single crystals were used to analyze the thermoelastic stress produced in boules grown from the melt. In 1972/5 a numerical solution of the thermoelasticity equations in displacements was developed in the Latvian University Computer Center [30]. The results obtained by analyzing thermoelastic stress generated in the course of the growth of large single crystals are reported in refs. 31, 32. The stress generated in different temperature regions of the boules was compared with the experimentally observed stress for dislocation generation [33]. This comparison made it possible to specify the conditions required for the growth of dislocation-free single crystals; by taking into account the relaxation of thermoelastic stress it was possible to carry out calculations of dislocation density in single crystals [34].

Literature Cited

1. N. A. Avdonin. Supercooling for growth front motion limited by the heat transfer rate. In: Growth of Crystals, 11, Consultants Bureau, New York, 274-277 (1979).
2. E. Billig. Some defects in crystals grown from melt. Proc. Roy. Soc. A, 235, 37-55 (1956).
3. A. A. Uglov. Temperature field of a single crystal in conditions of partial screening. Izv. AN SSSR, Ser. Metallurgiya i Gornoye Delo, 4, 139-145 (in Russian) (1964).
4. G. P. Boikov and V. A. Kuchin. On the temperature field in the growing crystal. Izv. Vyssh. Uchebn. Zavedenii, Fizika, 2, 15-23 (in Russian) (1959).
5. D. A. Burton, R. S. Prim, and W. P. Slichter. Distribution of impurities in crystals grown from the melt. J. Chem. Phys., 21, 1987-1991 (1953).
6. W. R. Wilcox. Heat and mass transfer in the process of crystallization. In: Fractional Solidification. New York. Marcel Dekker, 1, 47-111 (1967).
7. G. P. Ivantsov. "Diffusional" supercooling in the process of crystallization of a binary melt. Dokl. AN SSSR, 8, 179-181 (in Russian) (1951).
8. N. A. Avdonin. Description of solidification in binary systems taking account of bulk crystallization kinetics. In: Problems in Crystallization Theory. Latvian Univ. Publishing House, Riga, 2, 56-75 (in Russian) (1974).
9. V. T. Borisov, B. Ya. Lyubov, and D. E. Temkin. On kinetics of solidification in a metal ingot at different temperature conditions on its surface. Dokl. AN SSSR, 104, 223-226 (in Russian) (1955).
10. V. I. Dobrovolskaya, B. Ya. Martuzan, E. N. Martuzan, and D. G. Ratnikov. Temperature fields of floating zone melting calculated on the basis of inductor operation and surface profile of floating zone. Fiz. i Khim. Obrab. Materialov, 6, 42-46 (in Russian) (1973).
11. O. A. Oleinik. On one method of solving the general Stefan equation. Dokl. AN SSSR, 135, 1054-1057 (in Russian) (1960).
12. N. A. Avdonin, M. F. Globin, V. A. Smirnov, and V. E. Shniger. Analysis of thermal conditions in the growth of gallium arsenide ingots by uniaxial crystallization. Fiz. i Khim. Obrab. Materialov, 5, 50-55 (in Russian) (1971).
13. B. Ya. Martuzan. Investigation of thermal effects in growth of single crystals. Thesis, Latv. Univ., Riga (in Russian) (1975).
14. B. M. Budak, N. L. Gol'dman, A. T. Yegorova, and A. B. Uspensky. A method of straightening crystallization interface for solving multidimensional Stefan-type problems. In: Numerical Calculation Methods and Programming, Moscow Univ. Publishing House, Moscow, 8, 103-108 (in Russian) (1967).
15. N. A. Avdonin, M. V. Koyalo, and A. I. Pogodin. Analysis of supercooling in the growth of dislocation-free single crystals. In: Vth USSR Conference on Crystal Growth, Tbilisi. Abstracts Inst. of Cybernetics of AN GSSR, Tbilisi, 1, 56-57 (in Russian) (1977).
16. W. R. Wilcox and R. L. Duty. Macroscopic interface shape during solidification. Heat Transf., Trans. ASME, C, 88, No. 45 (1965).
17. Yu. F. Shchelkin. Determination of shape of liquid column in Czochralski growth of single crystals with free melt surface. Fiz. i Khim. Obrab. Materialov, 3, 29-33 (in Russian) (1971).
18. Yu. F. Shchelkin. On the effect of heat exchange at the interface on the shape of crystallization interface in the Czochralski growth of single crystals. Fiz. i Khim. Obrab. Materialov, 4, 36-42 (in Russian) (1971).
19. Yu. F. Shchelkin, V. A. Smirnov, I. V. Starshinova, and A. A. Kholodovskaya. Temperature field in the crystal and the melt at nonlinear boundary conditions. In: Physico-Chemical Methods in Semiconductor Materials Research. Metallurgiya, Moscow, 29-42 (Giredmet Proceedings, 44) (in Russian) (1974).

20. M. G. Mil'vidsky, V. A. Smirnov, I. V. Starshinova, and Yu. F. Shchelkin. On the analysis of thermal conditions in the Czochralski growth of single crystals. Izv. AN SSSR, Ser. Khim., 40, 1444-1451 (in Russian) (1976).
21. N. Kobayashi and T. Arizumi. The numerical analyses of the solid-liquid interface shapes during crystal growth by the Czochralski method. Jap. J. Appl. Phys. 9, 361-367 (1970).
22. N. Kobayashi and T. Arisumi. The numerical analyses of the solid-liquid interface shapes during crystal growth by the Czochralski method. Part II. Effects of the crucible rotation. Jap. J. Appl. Phys., 9, 1255-1259 (1970).
23. T. Arizumi and N. Kobayashi. Theoretical studies of the temperature distribution in a crystal being grown by the Czochralski method. J. Cryst. Growth, 13/14, 615-618 (1972).
24. C. E. Chang and W. R. Wilcox. Inhomogeneities due to thermocapillary flow in floating zone melting. J. Cryst. Growth, 28, 8-12 (1975).
25. A. J. Gosman, W. M. Pun, A. K. Runchel, D. E. B. Spalding, and M. Wolfstein. Heat and Mass Transfer in Recirculating Flows. Academic Press, London-New York, (1969).
26. V. L. Gryaznov and V. I. Polezhayev. An analysis of several different procedures and approximations of boundary conditions for numerical solution of thermal convection equations. Preprint of the Institute of Problems in Mechanics, AN SSSR, No. 4, Moscow (in Russian) (1974).
27. I. A. Remizov, I. V. Starshinova, and Yu. F. Shchelkin. Analysis of melt flow in the Czochralski growth of single crystals. In: Vth USSR Conference on Crystal Growth, Tbilisi. Abstracts, Inst. of Cybernetics of AN GSSR, 2, 173-176 (in Russian) (1977).
28. N. A. Kravchenko, A. S. Kuznetsov, B. Ya. Martuzan, and N. L. Ulanova. Calculation of heating and transfer of gas in epitaxial growth, taking into account rotation of support and temperature dependence of coefficients. In: Applied Problems of Theoretical and Mathematical Physics. Latvian Univ. Publishing House, Riga, 42-51 (in Russian) (1977).
29. V. L. Gryaznov and V. I. Polezhayev. Numerical solution of the nonstationary Navier-Stokes equations for the turbulent mode of natural convection. Preprint of the Institute of Problems in Mechanics, AN SSSR, No. 81, Moscow, (in Russian) (1977).
30. S. S. Vakrameyev. Calculation of thermal stress in crystals grown from the melt. In: Problems in Crystallization Theory. Latvian Univ. Publishing House, Riga, 2, 101-122 (in Russian) (1975).
31. S. S. Vakhrameyev, V. V. Osvensky, and V. A. Smirnov. Relation between dislocation structure of a single crystal and the thermal stress field calculated by taking into account the radial and axial variation of ingot temperature during growth from the melt. In: IVth USSR Conference on Crystal Growth, Yerevan, AN ArmSSR Publishing House, Yerevan, 2, part 2, 81-82 (in Russian) (1972).
32. S. S. Vakhrameyev, M. G. Mil'vidsky, V. A. Smirnov, and Yu. F. Shchelkin. An analysis of temperature and thermal stress fields in crystal growth from the melt. In: Growth and Doping of Semiconductor Crystals and Films. Nauka, Novosibirsk, part 1, 162-168 (in Russian) (1977).
33. V. B. Osvensky, S. S. Shifrin, and M. G. Mil'vidsky. Multiplication of dislocations in semiconductor crystals at high temperatures. Izv. AN SSSR, Ser. Fiz., 37, 2357-2361 (in Russian) (1973).
34. S. S. Vakhrameyev, V. B. Osvensky, and S. S. Shifrin. Calculation of dislocation density distribution on the basis of thermal stress fields in real semiconductor single crystals. In: Vth USSR Conference on Crystal Growth, Tbilisi. Abstracts, Inst. of Cybernetics, AN GSSR Publishing House, Tbilisi, 2, 230-231 (in Russian) (1977).

PHASE DIAGRAMS OF BINARY SYSTEMS FORMED BY RARE EARTH TRIFLUORIDES

B. P. Sobolev and P. P. Fedorov

The Institute of Crystallography of the USSR Academy of Sciences, Moscow

A. K. Galkin, V. S. Sidorov, and D. D. Ikrami

The Institute of Chemistry of the Tadzhik SSR Academy of Sciences, Dushambe

The present paper is a follow-up on an earlier investigation [1-4] of phase diagrams of a large group of RF_3-$R'F_3$ systems where R denotes rare earth elements. In recent years single crystals of single- or multicomponent fluorides with structure belonging to tysonite (LaF_3) and rhombic yttrium fluoride (YF_3) types became attractive as laser materials [5, 6], infra-red-to-visible light converters [7], and ionic conductors with high fluorine-ion conductivity [8, 9]. Detailed knowledge of the phase diagrams of the respective systems is necessary for successfully growing crystals of rare earth fluorides and for activating them with lanthanide ions. The polymorphism and morphotropic alternation of structure types in the sequence of rare earth trifluorides [10] limit the possibility of growing single-component matrices with particular structures. These restrictions are partially removed by going to more complex compositions [3]. We have analyzed phase diagrams of RF_3-$R'F_3$ systems and mutual isomorphous substitutions of rare earth ions in tysonite, rhombic, and hexagonal yttrium fluoride structural types in order to accumulate additional information on the possibility of growing multicomponent materials.

The total number of binary systems composed of the trifluorides of yytrium, lanthanum, and the lanthanides (with an exception of promethium) is 105. It is thus reasonable to assume that the trial and error approach to the selection of compositions for crystal growth should prove ineffective. At present we have analyzed 34 systems (33 for the first time; the phase diagram of the LaF_3-YF_3 system was reported earlier [11]), partly discussed in ref. 2. The experimental data obtained make it possible to discuss the main features of phase diagrams of RF_3-$R'F_3$ systems.

Experimental

Rare earth fluorides of "chemically pure" grade (purity at least 99.99% with respect to the principal element), melted in a fluorinating atmosphere produced by Teflon pyrolysis, were used as the starting materials. The oxygen content in the trifluorides, measured by vacuum melting [12], was between 0.005 and 0.05 percent by weight. Phase diagrams were derived from differential thermal analysis data (DTA) [13]. The mean oxygen concentration in the samples after DTA was 0.025 wt.%. The phase transition points observed in LnF_3 were in good agreement with earlier measurements [10]. The phase equilibria in the subsolidus of a number of systems were studied by X-ray methods for samples annealed in a fluorinating atmosphere [3].

Experimental Results and Their Analysis

Polymorphic and morphotropic transitions divide the sequence of lanthanide trifluorides into four subgroups [10, 13]. In subgroup A (La-Nd) fluorides retain the tysonite-type structure up to the melting point. In subgroup B (Sm-Gd), the tysonite structure type is retained in a high-temperature modification which transforms to the rhombic modification of β-YF$_3$ type as the temperature is lowered. In subgroup C (Tb-Ho), the fluorides crystallize from the melt into the β-YF$_3$ structure. Fluorides in subgroup D have a high-temperature modification of the hexagonal α-YF$_3$ type. The polymorphism of lanthanide trifluorides is schematically shown in Fig. 1. It defines the subdivision of RF$_3$-R'F$_3$ systems into ten subgroups which are bounded by solid lines in Fig. 2. Phase diagrams of systems belonging to different subgroups are differently organized, and to a first approximation are determined by phase transitions in the free components. The total number of systems in each subgroup is specific to the subgroup, and varies from three (B-B' combinations) to twenty (A-D). Phase diagrams obtained in the present study include representatives of each subgroup. Let us analyze these subgroups.

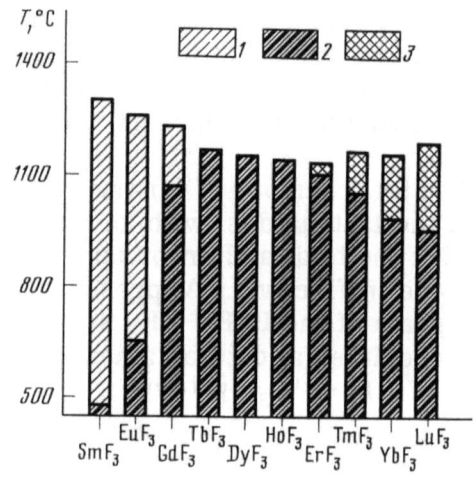

Fig. 1. Polymorphism in lanthanide trifluorides of different types (1) LaF$_3$; (2) β-YF$_3$; (3) α-YF$_3$ (α-UO$_3$).

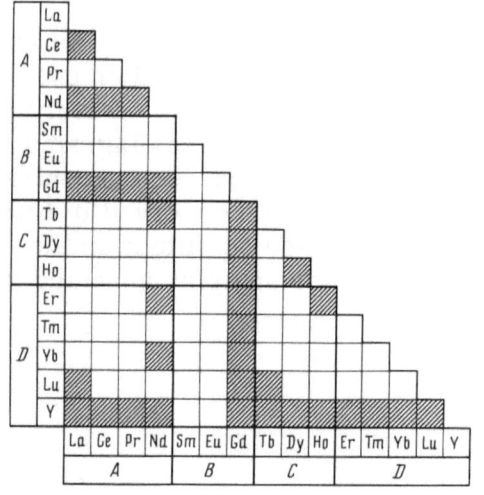

Fig. 2. Subdivision of RF$_3$-R'F$_3$-type systems into subgroups (shaded squares represent systems studied in this work).

Pair Combinations of Fluorides from the Same Subgroups:
A-A', B-B', C-C', D-D'. The first of these (six systems) is of practical significance because it characterizes the behavior in systems often grown as single crystals (pure or doped). The results obtained for diagrams of this type were discussed in ref. 4. The components of the systems form continuous solid solutions with phase diagrams showing no maxima or minima. Small positive deviations from Vegard's law were found on the concentration dependences of the unit cell parameters.

The B-B' combination (three systems) is the only study that was unsuccessful, because of partial reduction of samarium and europium. Note that phase diagrams of these systems are easily predictable, and their compositions cannot be regarded as promising for the growth of single crystals.

In the C-C' combination (three systems), the system formed by dysprosium and holmium fluorides was studied. Continuous solid solutions with a structure of rhombic yttrium fluoride type were observed. These systems are similar to those of the A-A' subgroup, and can be utilized for growing single crystals of complex compositions with this type of structure (the corresponding phase diagrams have a simple form and are not shown here).

The combination D-D' (ten systems) was analyzed in detail for typical systems composed of fluorides of yttrium and heavier lanthanides (Fig. 3a). Continuous solid solutions are formed between high-temperature (α-YF$_3$ type) and low-temperature (rhombic yttrium fluoride type) forms of the appropriate fluorides. On cooling, the high-temperature solutions decomposed almost completely. This decomposition is easily detected by DTA, with the two-phase regions clearly revealed. This behavior excludes the subgroup in question from the list of potential sources of crystals, because for all compositions a drastic phase transition in the solid phase far from the melting point occurred. The variable-composition phases in the system with erbium fluoride, in which the polymorphous transition temperature and the melting point are close [2], may constitute a partial exemption from this rule.

Combination A-B. Four of the twelve systems have been studied here (Fig. 4). Continuous solid solutions of LnF$_3$ and the high-temperature modification α-GdF$_3$ with tysonite-type structure are formed in all investigated systems. A polymorphic transition which is clearly detected on thermograms for pure GdF$_3$ is suppressed by an admixture of LnF$_3$ as small as 3 mol.%. If the LaF$_3$-GdF$_3$ system is quenched from the melt, or annealed at 440 and 800°C for 24 and 72 h, a solid solution with tysonite structure survives, containing 60 mol.% GdF$_3$. The lattice parameters are linear functions of the composition.

Compositions adjacent to component A have been crystallized from the melt as single crystals [14].

Combinations A-C and A-D. The systems belonging to this group are very varied and possess phase diagrams of great complexity. In the case of the LaF$_3$-LuF$_3$ system (Fig. 5a), solid solutions of limited range are formed (limiting concentrations 8 mol.% LuF$_3$ and 7 mol.% LaF$_3$). The eutectic coordinates are 1038°C and 69 mol.% LuF$_3$.

Phase diagrams for the NdF$_3$-(Tb, Er, Yb)F$_3$ systems are also complicated [15]. They include regions of tysonite solid solutions with maximum content of 35 mol.% TbF$_3$, 22 mol.% ErF$_3$, and 18 mol.% YbF$_3$. Systems composed of NdF$_3$ with ErF$_3$ and YbF$_3$ form solid solutions based on α-ErF$_3$ and α-YbF$_3$ with up to 4 and 12 mol.% NdF$_3$, respectively. Minima are observed on the equilibrium curves of solid solutions formed by high- and low-temperature modifications of ErF$_3$ and YbF$_3$ with 2 and 6 mol.% NdF$_3$. Variable composition phases based on rhombic TbF$_3$ (the peritectic melting point corresponds to 61 mol.% NdF$_3$) and β-ErF$_3$ (up to 72 mol.%) show extended homogeneous regions. Solid-state annealing

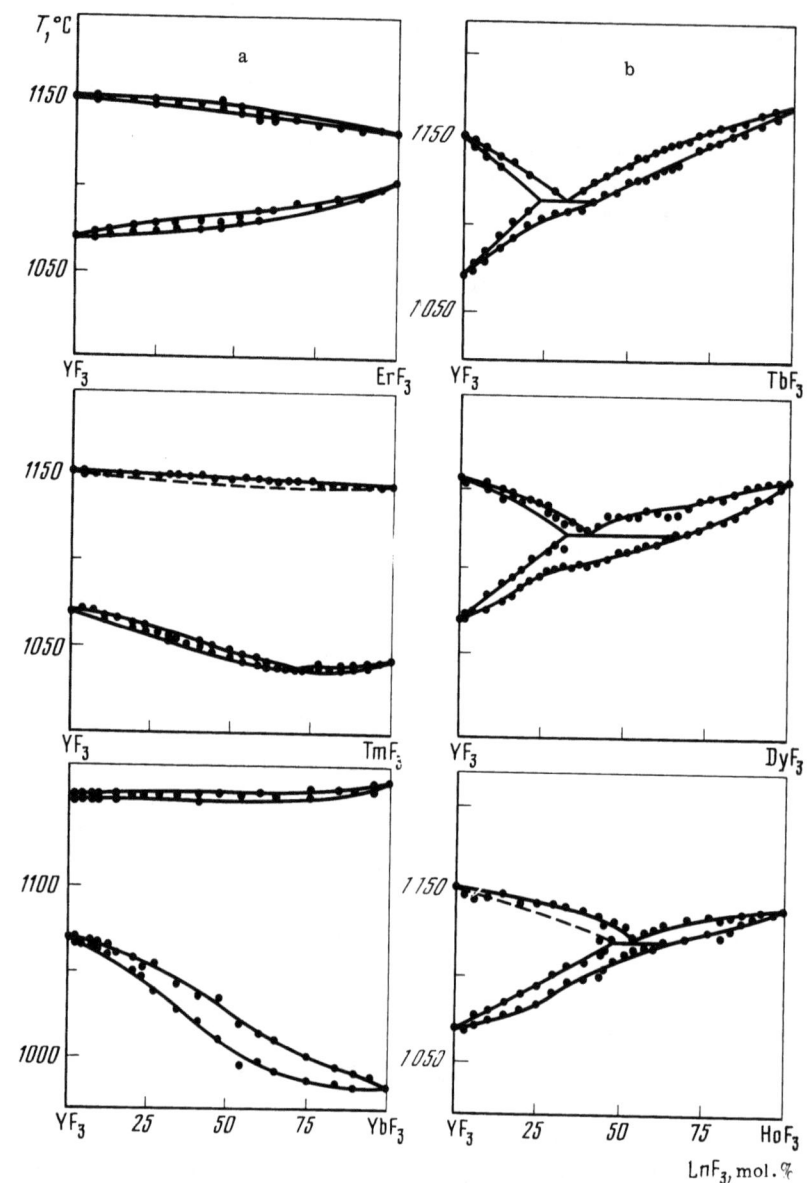

Fig. 3. Phase diagrams of YF_3-LnF_3.
(a) Ln = Er, Tm, Yb; (b) Ln = Tb, Dy, Ho.

revealed dome-shaped curves of decomposition in rhombic phases into two isostructural phases of different composition. The concentrational dependences of unit cell parameters are governed by Vegard's law and, for both phases of the decomposed solid solution, fall on the same straight lines. These results agree with the description [16] of a "compound" $0.3\,ErF_3 \cdot 0.7\,NdF_3$ with a structure similar to rhombic β-YF_3, and point to the physical-chemical nature of this phase as a component of an extensive solid solution based on β-ErF_3. In ref. 16 this phase was mistakenly interpreted as an individual compound because it is separated by a wide two-phase region from the right-hand branch of rhombic phases on the isothermal 1000°C section studied earlier. The temperature at the critical point on the decomposition curve increases as we follow the Tb-Er-Yb sequence, and the decomposition curves of the low- and high-temperature phases in the NdF_3-YbF_3 system intersect. This transforms the left-hand branch of solid solutions into an independent phase with 25±2 mol.% content of YbF_3. This individuality of a part of a decomposed solid solution was earlier

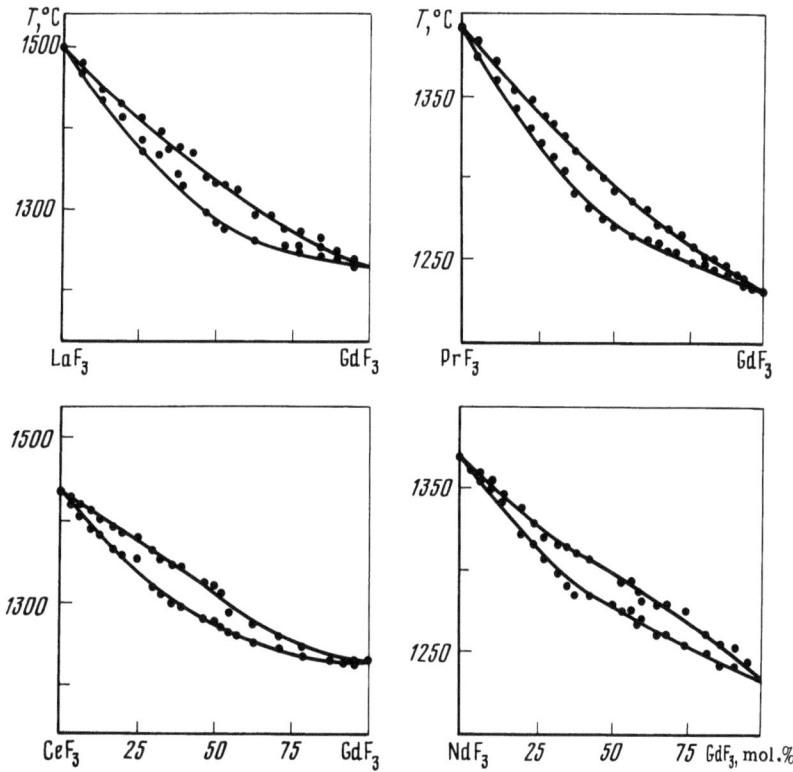

Fig. 4. Phase diagrams of GdF_3-$(La-Nd)F_3$ systems.

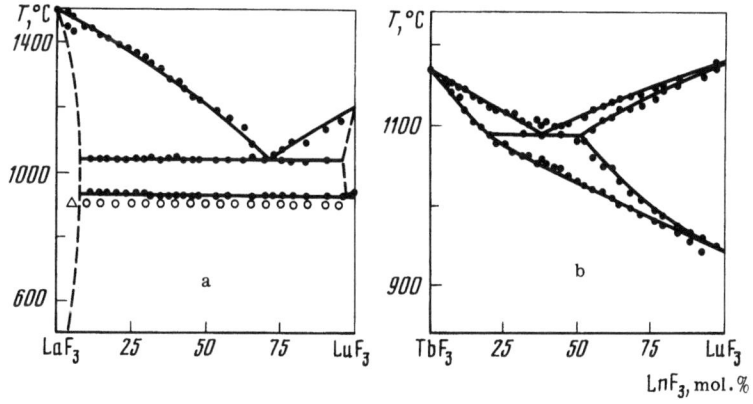

Fig. 5. Phase diagrams of the systems: (a) LaF_3-LuF_3, and (b) TbF_3-LuF_3. Black circles represent DTA data; triangles indicate single-phase regions; open circles indicate two-phase regions.

regarded as one of the theoretically possible ways for the formation of berthollide phases [17]. The phase diagrams of systems from the A-D combination are the most complex and variable, the predictions are the least certain, and further research is required.

Combinations B-C and B-D. Three out of the nine systems from the first of these combinations were studied. Systems composed of GdF_3 with Tb and Ho were discussed in ref. 3; the phase diagram of the GdF_3-DyF_3 system is given in Fig. 6. Among the fifteen systems of the second combinations, five were studied (see Fig. 6 and ref. 3).

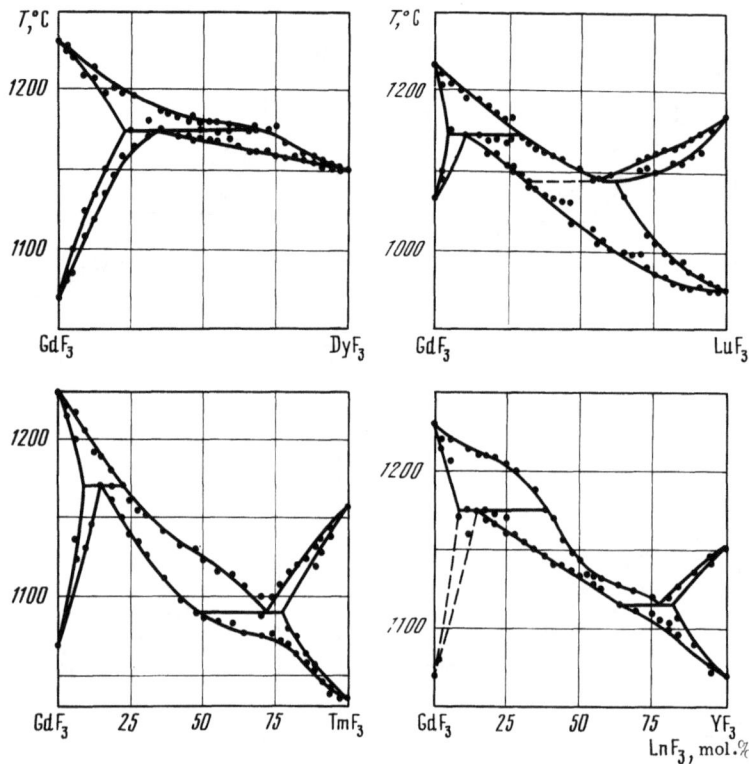

Fig. 6. Phase diagrams of GdF_3-(Dy, Tm, Lu, Y)F_3 systems.

All systems form continuous solid solutions of rhombic modifications of the β-YF_3 type and limited solutions composed of high-temperature modifications of the α-GdF_3 and α-YF_3 types. In the systems of the second group the melting of the rhombic phases is incongruent. Unit cell parameters exhibit small positive deviations from Vegard's law (annealed at 900°C). Systems comprising these two combinations are undoubtedly of great interest as potential sources for multicomponent materials having the structure of rhombic YF_3.

Combination C-D. Five of the fifteen systems of this type were studied (see Fig. 3b and ref. 2). Rhombic forms of trifluorides form continuous solid solutions. Variable-composition phases with the α-YF_3 type structure "taper out" on phase diagrams of all systems as the content of the fluoride belonging to subgroup C increases. Eutectic temperatures in the systems YF_3-(Tb, Dy, Ho)F_3 and TbF_3-LuF_3 (see Fig. 5b) are 1120±5 and 1090±5°C, respectively. The phase diagrams shown in the figures indicate that crystals with the β-YF_3 type structure can be grown over a moderately wide range of compositions.

Phase diagrams of some RF_3-$R'F_3$ systems taken together with earlier published diagrams [1-4] make it possible to draw conclusions on the basic features of the phase diagrams for nearly all of the ten types. Phase diagrams of systems not yet investigated can be predicted with satisfactory accuracy for all combinations, excepting trifluorides very distant in the sequence (A-C and A-D). An earlier attempt [14] to expand the number of crystalline substances in the rare earth fluoride class by growth experiments pointed to very limited possibilities of growing $Ln_{1-x}Ln'_xF_3$ solid solutions with tysonite type structure. Recommendations for the use of high-temperature solutions, allowing crystallization in the stability range of the rhombic modifications (β-YF$_3$ type) of trifluorides [14] and thus enabling one to avoid, at least partially, the limitations imposed by polymorphism, were indeed justified [18].

The observed thermal stabilization of the structure of the rhombic YF$_3$ type in combinations B-C and B-D [3], confirmed in the present study for other systems in these combinations, opens new routes for growing crystals directly from the melts. The incongruent melting of rhombic solid solutions creates additional difficulties, but these are easier to overcome than limitations imposed on crystal size by specific features of crystallization from high-temperature solution as for BeF$_2$ [18], which, moreover, is highly toxic.

Crystals with compositions Gd$_{0.5}$Er$_{0.5}$F$_3$ and Gd$_{0.5}$Y$_{0.5}$F$_3$ were grown by the Czochralski technique in equipment using induction heating. The pulling rate was 10-14 mm h^{-1}. Cooling of single crystals leads to cracking by thermal stress into blocks of up to 100 mm^3 in volume. Solid-solution single crystals in the above systems were also grown by the Bridgman-Stockbarger method (crucible lowered at a rate of 7 and 15 mm h^{-1}), but in this case the single crystal blocks were smaller (up to 20 mm^3).

As in the case of growing single-component crystals, crystallization in composite systems requires high purity with respect to oxygen. At oxygen contents as low as 0.035 wt. %, crystallization in the GdF$_3$-YF$_3$ system exhibits an additional thermal effect similar to that in the YF$_3$-Y$_2$O$_3$ system at 1090°C [13]. The optical quality of the crystals is sharply degraded by the segregation of finely dispersed oxygen-containing phases.

Literature Cited

1. O. V. Kudryavtseva, L. S. Garashina, K. Rivkina, and B. P. Sobolev. On the solubility of LnF$_3$ in lanthanum fluoride. Kristallografiya, 18, 843-844 (in Russian) (1973).
2. B. P. Sobolev, I. D. Ratnikova, P. P. Fedorov, B. V. Sinitsyn, and G. S. Shankalamian. Polymorphism of ErF$_3$ and position of a third morphotropic transition in the lanthanide series. Mater. Res. Bull., 11, 999-1004 (1976).
3. B. P. Sobolev, V. S. Sidorov, P. P. Fedorov, and D. D. Ikrami. Stabilization of the structure of the rhombic β-YF$_3$ type in GdF$_3$-LnF$_3$ systems. Kristallografiya, 22, 1009-1014 (in Russian) (1977).
4. B. P. Sobolev, P. P. Fedorov, N. L. Tkachenko, K. B. Seiranyan, A. M. Kevorkov, Kh. S. Bagdasarov, V. S. Sidorov, and D. D. Ikrami. Refractory fluoride-containing materials in (Ca, Sr, Ba)F$_2$-LnF$_3$ systems. In: Research in Optical Coating materials. VNII Luminophore Publishing House, Stavropol', No. 15, 73-78 (in Russian) (1977).
5. A. A. Kaminsky. Laser Crystals. Nauka, Moscow, (in Russian) (1975).
6. R. C. Pastor, M. Robinson, A. C. Pastor, and K. T. Miller. Impurity conditioned solid-solid transformation in simple halides. In: 2nd Amer. Nat. Conf. on Cryst. Growth. Princeton Univ. Press, 1, 26, (1972).
7. L. C. van Uitert, L. Pictorski, and W. H. Grodkiewitz. Preparation of efficient infra-red stimulable rare earth fluoride phosphors. Mater. Res. Bul., 4, 777-780 (1969).
8. K. Lee and A. Sher. F^{19} nuclear magnetic resonance line narrowing in LaF$_3$ at 300°K. Phys. Rev. Lett., 14, 1027-1029 (1965).
9. L. E. Nagel and M. O'Keeffe. Highly-conducting fluorides related to fluorite and tysonite. In: Fast Ion Transport in Solids. Solid State Batteries and Devices. Amsterdam: North-Holland Publ. Co., 165-172 (1973).

10. B. P. Sobolev, P. P. Fedorov, K. B. Seiranian, and N. L. Tkachenko. On the problem of polymorphism and fusion of lanthanide trifluorides. Part II. J. Solid State Chem., $\underline{17}$, 201-212 (1976).
11. R. H. Nafziger, R. L. Lincoln, and N. Riazance. Alkaline-earth fluoride — LaF_3 systems with implications for electroslag melting. J. Inorg. and Nucl. Chem. $\underline{35}$, 421-426 (1973).
12. G. G. Glavin and Yu. A. Karpov. Measurement of oxygen content in metallic rare earth elements and their fluorides. Zavodskaya Laboratoriya, $\underline{30}$, 306-308 (in Russian) (1964).
13. B. P. Sobolev, P. P. Fedorov, D. B. Shteynberg, B. V. Sinitsyn, and G. S. Shankalamyan. On the problem of polymorphism and fusion of lanthanide trifluorides. Part I. J. Solid State Chem. $\underline{17}$, 191-199 (1976).
14. D. A. Jones and W. A. Shand. Crystal growth of fluorides in the lanthanide series. J. Cryst. Growth, $\underline{2}$, 361-368 (1968).
15. V. S. Sidorov, P. P. Fedorov, V. P. Sobolev, and D. D. Ikrami. On the rhombic "compound" $Er_{0.7}Nd_{0.7}F_3$ and similar phases in NdF_3-LnF_3 systems. In: Vth USSR Symposium on Chemistry of Inorganic Fluorides, Dnepropetrovsk, Abstracts. Nauka, Moscow, 258-259 (in Russian) (1978).
16. K. Okamura and S. A. Yajima. A new lanthanoid double compound $(Er_{0.3}Ns_{0.7})F_3$. Bull. Chem. Soc. Jap., $\underline{46}$, 1531-1533 (1974).
17. P. P. Fedorov, P. I. Fedorov, and B. P. Sobolev. On some possible cases of berthollide formation in binary systems. Zh. Neorg. Khim. $\underline{18}$, 3319-3322 (in Russian) (1973).
18. W. H. Grodkiewicz and L. G. van Uitert. Pat. 3667921 (USA). Flux growth of rare earth fluorides. June 6 (1972).

Part VI
Growth of Crystals from Solutions

PROGRESS IN FLUX GROWTH OF LARGE CRYSTALS

V. A. Timofeeva

The Institute of Crystallography of the USSR Academy of Sciences, Moscow

This survey outlines the basic trends in the flux growth of crystals (using as examples iron garnets and other incongruently melting compounds).

During the last 25-30 years this method progressed substantially from spontaneous [1-15] to controlled [16-51] crystallization, and from small (1-3 mm) to large (5-8 cm) crystals. Several stages are distinguishable in flux growth history.

Spontaneous Crystallization in Small Volumes (below 500 ml). This restriction typical of the 50's was caused by a scarcity of platinum crucibles and rare earth oxides. Nowadays work on this scale is only carried out in pilot research.

Spontaneous Crystallization in Large Volumes (from 500 ml to 3 liters). The work on iron garnet growth in large volumes was initiated in the 60's by Neilsen [37] and continued by a number of researchers [6, 8, 38]. An increase in the volume of the crystallizing flux produces larger crystals (up to 5 cm); however, as a rule, the larger the crystal is, the poorer structural quality it has. The quality is the highest in small (1-1.5 cm) crystals, so that the yield of "useful" crystals is rather low. It was apparent that large defect-free crystals cannot be grown without research into the thermodynamic state of the system and without studying the crystallization kinetics. All subsequent stages in the history of flux growth are associated with taking into account physical-chemical properties of high-temperature solutions and, primarily, solubility data for crystal-forming oxides [5, 10, 18, 22].

Localization of Crystallization Centers. Local regions of supercooling at the crucible base are generated by apertures of different shape and different diameter made in the lining of the furnace base. The inverted temperature gradient (the base is cooler than the top of the melt) ensures spontaneous local nucleation of crystals (in accordance with the solubility curve). Crystal growth is then controlled by appropriate temperature programs. Crystals are grown in the static or dynamic modes [39] (Table 1).

This technique makes it possible to grow sufficiently large crystals (exceeding 20 mm maximum size) in comparatively small volumes (0.5-1 l), but the number of crystallization centers remains arbitrary because the nucleation stage remains uncontrollable. Consequently, both large and small crystals grow simultaneously.

Flux Growth of Large Crystals by Accelerated Crucible Rotation Technique. The method is based on cyclic growth ⇌ dissolution of crystals in the vicinity of the equilibrium solubility curve. The oscillation amplitude is approximately

20°C, with oscillation periods of several minutes. Multiple cycles of the system serve to enlarge favored nuclei at the base, so that only a few (ideally, only one) crystals "survive" in the course of the subsequent programmed cooling of the system (Table 2).

Crystals successfully grown by the oscillating temperature technique in relatively small volumes (below 1 liter) reached 250 g in the case of $Y_3Fe_5O_{12}$ and 210 g in the case of $Gd\,AlO_3$.

The morphology of slowly cooled crystals (0.2 - 0.5°C h^{-1}) is determined by layer growth of equilibrium faces, so that reasonably uniform crystals are produced (the best of them containing only a few dislocations).

Controlled Crystallization. We shall consider several versions of this method [46-58]: crystal growth in temperature gradient mass transfer conditions; the seeded pulling method (modified Czochralski growth); crystal growth on seeds introduced into the bulk of the high-temperature solution. All of these methods involve growth on seed crystals. The reported techniques differ in a number of technological features (introduction and positioning of seeds, melt stirring, specifics of the crystallization mode).

In the case of gradient mass transfer (Table 3), crystals grow either on rigidly fixed seeds introduced into the melt prior to the experiment (Fig. 1) [23], or on floating seeds dipped into the melt after the saturation temperature is reached [55]. The crystals used as seeds are mostly those grown under spontaneous crystallization conditions. As in the former cases (see Tables 1 and 2), crystal growth is governed by the growth of smooth faces via the layer growth mechanism. It should be mentioned here that the rate of the gradient mass transfer growth must be a maximum for seeds with rough faces perpendicular to the crucible axis (by analogy with the hydrothermal growth of quartz along $\langle 0001 \rangle$).

Experiments of this type are rare, for a number of technical reasons, such as aggressively volatile media, rather high temperatures (1200 - 1300°C), and scarcity of large homogeneous seed crystals from which various sections could be cut.

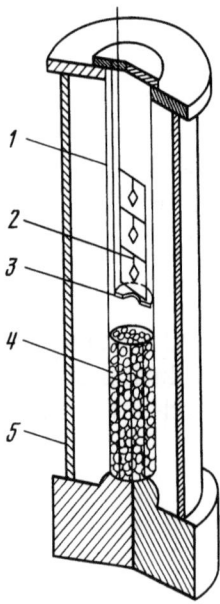

Fig. 1. Schematic of a crucible for gradient growth. (1) quartz tube, (2) seed, (3) diaphragm, (4) nutrient solution, (5) housing.

TABLE 1. Crystal Growth with Localized Crystallization Centers.

Crystal	Initial batch composition, wt %	γ, %	m, %	Number of crystals with sizes		
				>20 mm	10 mm	5 mm
$Y_3Al_5O_{12}$	Y_2O_3—12.55; PbO—39.9; Al_2O_3—9.45; PbF_2—34.9; B_2O_3—3.6	4.0	11.0	3	8	5
$Y_3Ga_5O_{12}$	Y_2O_3—5.8; PbO—39.9; Ga_2O_3—14.0; PbF_2—36.8; B_2O_3—3.5	4.6	13.0	1	11	12
$Y_3Fe_5O_{12}$	Y_2O_3—11.1; PbO—39.2; Fe_2O_3—15.6; PbF_2—32.2; B_2O_3—1.8	3.0	12.0	7	2	10

Note: γ = evaporation losses; m = weight of crystals as a percentage of initial batch.

TABLE 2. Crystal Growth by the Accelerated Crucible Rotation Technique.

Crystal	Solvent	Growth conditions		Size, mm (weight, g)	Source
		T, °C	B, °C h^{-1}		
RPO_4	$Pb_2P_2O_7$	—	—	1.5×1×15	[42]
$RAsO_4$				1×1×6.5	
$GdAlO_3$	PbO—PbF_2—B_2O_3	1300—900	0.3—0.5	35×30×25 (200)	[43]
$Y_3Fe_5O_{12}$	PbO—PbF_2—B_2O_3	1300—1050	—	(250)	[44]
$BaFe_{12}O_{19}$	BaO—B_2O_3—PbO	1300—1200	—	15×2.5	[45]
$Y_3Fe_5O_{12}$	PbO—$0.2B_2O_3$	1250—1200	0.4	10×10×10	[13]

Note: B = mean cooling rate; R = rare earth element.

TABLE 3. Gradient Crystal Growth.

Crystal	Solvent	Growth conditions			Size mm	Source
		T, °C	ΔT, °C	B, °C h^{-1}		
Al_2O_3	PbO—B_2O_3	1000	50	—	10×10	[52]
	Na_2O—WO_3	1200	50	—	6×8	[53]
	PbF_2	1100	50	0.5	40×40×12	[54]
	Na_3AlF_6	1200—1000	—	—	20×20×5	[28]
$Y_3Al_5O_{12}$	PbO—PbF_2—B_2O_3	1375	20—30	0.2—0.5	10×10×25	[55]
$Zn_2Ba_2Fe_{12}O_{22}$	$Ba_2B_2O_5$	1000	10	0.5	20×16×4	[56]

Note: ΔT = temperature difference.

Modified Czochralski growth. (Table 4) is used with non-volatile molten salts (BaO-B_2O_3, Na_2WO_4, $K_2Mo_2O_7$, Bi_2O_3, TiO_2, etc.) on the basis of conventional equipment for single-component systems. It produces high-quality crystals with different crystallographic orientations and with different mechanisms of growth on smooth and rough faces. Figure 2 shows a $KY(MoO_4)_2$ crystal grown by this method from a solution of about 60 wt.% in molten $K_2Mo_2O_7$ [25].

Possible progress in this technique of flux growth lies with a search for non-aggressive non-volatile solvents and for lower growth temperatures.

Growth of Crystals on Seeds Introduced into the Bulk of the Solution (Table 5). This is analogous to growing water-soluble crystals when zero-gradient conditions are created in the crystallization ambient and the seed is introduced

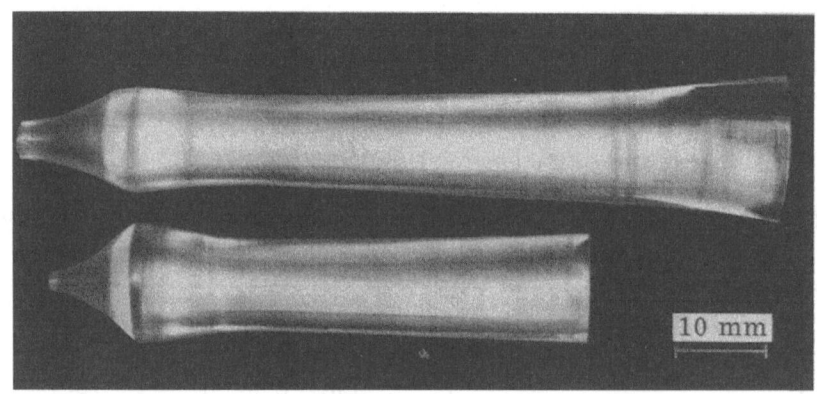

Fig. 2. $KY(MoO_4)_2$ crystals flux-grown by the Czochralski techniques (photograph by A. A. Pavlyuk).

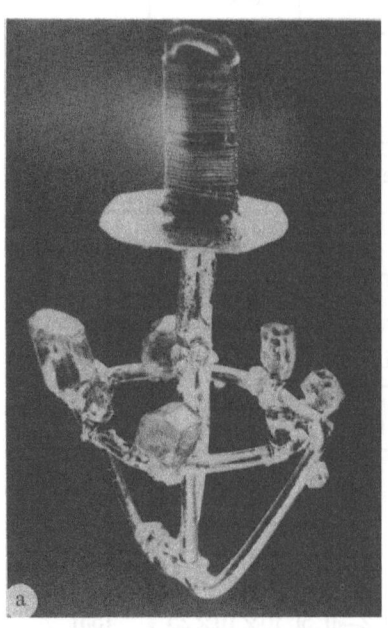 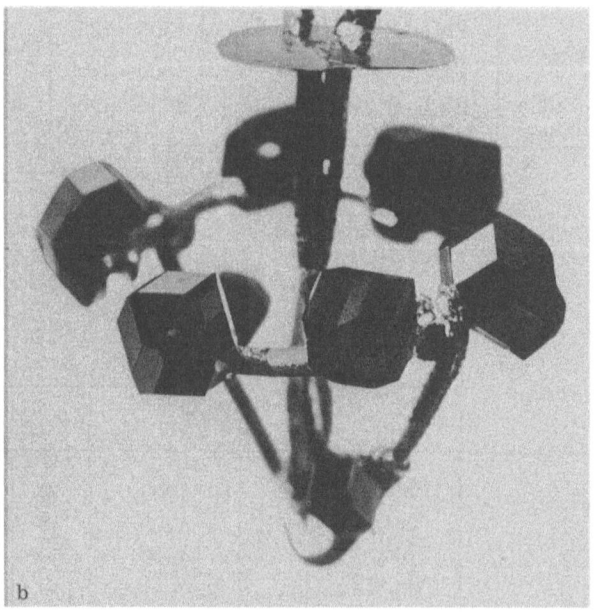

Fig. 3. Seeded growth of large crystals (photographs by L. N. Bezmaternykh and G. I. Shvartzman. (a) annular holder with seeds, (b) $Y_3Fe_5O_{12}$ crystals obtained by seeded growth.

TABLE 4. Growth of Crystals by Modified Czochralski Method.

Crystal	Solvent	c, wt.%	T, °C	ΔT, °C	ω, rpm	V, mm h^{-1}	Size, mm	Source
$Y_3Fe_5O_{12}$	$1BaO - 0,6 B_2O_3$	30	1100	50	400	0.04--0.25	20×20×40	[49]
$Y_3Fe_5O_{12}$	$BaCO_3 - B_2O_3$	30	1100--1080	25	—	—	10	[21]
$Y_3Al_5O_{12}$	$BaO - B_2O_3$	—	1320--1250	10	60	0.2	8× 8×10	[19]
$Y_3Fe_5O_{12}$	$BaO - B_2O_3$	33.5	1250--1100	—	—	—	10×10×20	[22]
$KY(MoO_4)_2$, $KLa(MoO_4)_2$	$K_2Mo_2O_7$	—	1100--1000	—	—	—	10×10×60	[25]
$BaTiO_3$	TiO_2	—	1400--1300	—	—	0.3	10×10×30	[57]
RVO_4, (R — Tb, Dy, Tm)	$Pb_2V_2O_7$	—	1200	—	—	—	4× 4×19	[58]
$Bi_2Mo_2O_{12}$	MoO_3	—	650	—	—	—	50×20	[32]

Note: C = solution concentration, ω = crystal rotation rate, V = pulling rate.

TABLE 5. Growth on Seed Crystals.

Crystal	Solvent	Growth conditions		Size, mm (wt. g)	Source
		T, °C	B, °C h^{-1}		
$CoFe_2O_4$	$Na_2B_4O_7$	1300--1100	1--2	10×10×15	[26]
$Y_3Fe_5O_{12}$	$PbO - PbF_2$	1250--1000	0.2--0.5	(100)	[20]
$(BaPb)TiO_3$	$BaB_2O_4 - PbB_2O_4$	1290--950	0.2	15×8	[46]
$NiFe_2O_4$	$BaO - B_2O_3$	1230		10× 8× 8	[19]
$BaZn_2Fe_{12}O_{19}$	$BaO - B_2O_3$	1250		(15)	[48]
$BaY_2MgGe_2O_{12}$	$GeO_2 - BaO$	1570	0.1	(48)	[31]
$BaTiO_3$	TiO_2	1377	0.1	(24)	[31]
$Ba_2MgGe_2O_7$	$BaO - GeO_2$	1345	0.1	(72)	[31]
$Y_3Fe_5O_{12}$	$BaO - B_2O_7$	1200--1050	0.1	10×10×10	[34]
$Y_3Fe_5O_{12}$	$PbO - PbF_2$	1280--1000	0.1	(42)	[51]

into the bulk of the moltent salt solution after preliminary heating of the solution and its subsequent cooling to the saturation temperature (Fig. 3) [18, 31, 48].

The seeds used in closed systems (sealed crucibles) are spontaneously formed $Y_3Fe_5O_{12}$ crystals whose growth proceeds by the development of smooth faces [20]. These seeds are first fixed to the inner part of the lid (Fig. 4a), then are dipped into the prepared saturated flux (Fig. 4b), and then grow to form a large crystal in the course of slow cooling (2 - 0.5°C h^{-1}) (see Fig. 4b, dashed line). The crystals are removed from the melt by rotating the crucible by 180°C (Fig. 4c) and are annealed by slowly cooling the furnace to room temperature. Large crystals are grown on oriented seeds [21, 36] in similar solutions to those in which liquid phase epitaxy of thin films is carried out [41, 50].

However, there is not an exact correspondence between the growth of thin films and bulk crystals: the imperfection of a film sharply increases with the film thickness. The main factor here is that films (up to 10 μm thick) grow in the kinetic mode (Fig. 5, low-slope segments of curves 1-5), while thicker layers grow in the diffusion mode (steeper segments of curves 1-5) [50].

An analysis of crystallization kinetics and growth mechanisms of $Y_3Al_5O_{12}$ crystals on 2×30×60 mm flat seeds oriented on (111) and (100) faces (Table 6) demonstrated that bulk crystals develop from dilute solutions (3-10 wt.%) at 785-1000°C (Fig. 6) owing to a difference in growth rates of smooth and rough faces [36]. Under these conditions the growth rate anisotropy reaches one order of magnitude.

Large $Y_3Fe_5O_{12}$ crystals can be grown under similar conditions, but this results in mismatch effects associated with the difference in lattice parameters between $Y_3Al_5O_{12}$ (12.01 Å) and $Y_3Fe_5O_{12}$ (12.37 Å). This misfit can be overcome by a successive alteration of $Y_3Al_5O_{12}$ and $Y_3Fe_5O_{12}$ compositions with isomorphous substituted interlayers of $Y_3Al_{5-x}Fe_xO_{12}$. For $x \approx 0.1-0.2$ mole the mismatch effects are completely suppressed.

A morphological analysis of iron garnet films grown on (111) substrates shows that liquid phase epitaxy proceeds by the normal growth on atomically rough surfaces. The morphology of large garnet crystals grown on seeds with the same (111) orientation differs from that of crystalline films. Figure 6b shows how smooth {110} faces completely replace rough {111} faces.

Experiments demonstrated that a mechanically polished surface of the seed becomes macroscopically rough during growth (see Fig. 6a). This roughness corresponds to an equilibrium profile of the crystal surface whose mean orientation deviates from that of the smooth face [40].

The roughness is caused by instabilities of the growth surface. Davies and White have shown [41] that first individual crystals are formed when the overgrowth reaches a thickness of about 100 μm. Thicker layers (above 100 μm) grow by the development of the already formed individual overgrowth crystals. The local orientations of the growth surface become varied, with segments of smooth faces appearing on it (Fig. 7). However, the growth rate of the rough surfaces remains relatively high.

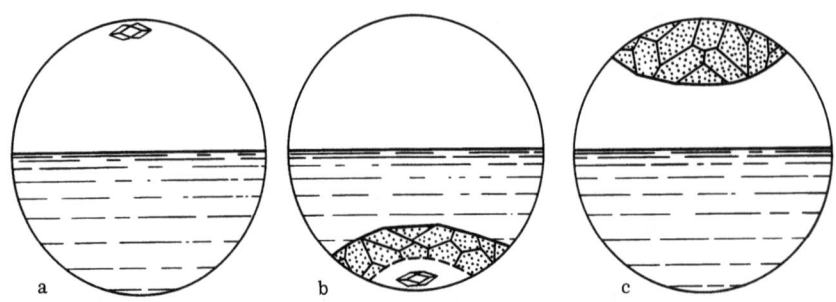

Fig. 4. Growth of $Y_3Fe_5O_{12}$ crystals (after Tolksdorf). (a) initial state, (b) intermediate stage, (c) final stage.

TABLE 6. Crystallization of $Y_2Al_5O_{12}$ on Flat Seed Crystals.

Solvent	c, wt.%	v, liters	T_r, °C	B, °C h^{-1}	{hkl}	d, mm	V_1, μm h^{-1}	V_2, μm h^{-1}	V_1/V_2
PbO — 0.85 PbF$_2$ — 0.16 B$_2$O$_3$	10—7	2.7	990—875	0.22	{120}	3.0	6.4	0.8	8.0
			975—870	0.28	{120}	0.8—0.45	2.1—1.2	0.25—0.15	8.4*
			930—905	0.17	{100}	1.15	7.7	1.0	7.7
			900—800	0.29	{100}	6.5	19.7	3.2	6.2
			885—770	0.28	{120}	2.8	6.5	0.9	7.2
PbO — 0.64 PbF$_2$ — 0.22 B$_2$O$_3$	8—7	1.5	920—835	0.20	{100}	6.0	14.0	3.5	4.0
			895—820	0.13	{100}	3.5	6.1	0.7	8.7
PbO — 0.60 PbF$_2$ — 0.03 B$_2$O$_3$	8—6	1.3	1000—945	0.62	{100}	1.8	20.2	4.5	4.5
			965—940	0.10	{111}	1.25	5.0	0.8	6.3
			955—870	0.86	{100}	3.4	34.0	8.0	4.2
PbO — 0.54 PbF$_2$	3.4	0.2	865—785	0.48	{120}	1.5	10.2	1.2	8.5

*) Static Mode.
Note: v = solution volume; T_r = temperature range of crystallization; {hkl} = seed orientation; d = mean thickness of overgrowth in fast-growth direction; V_1 and V_2 = mean growth rate of fast-growing and smooth faces, respectively.

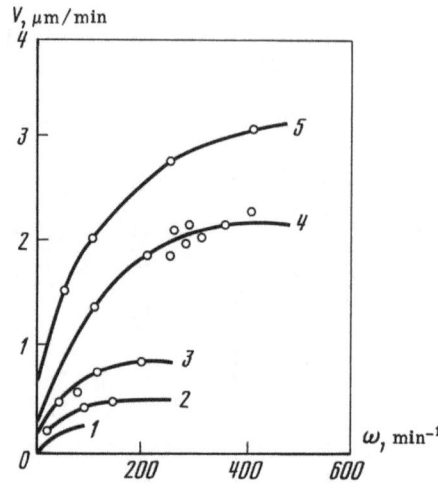

Fig. 5. Growth rate V as a function of seed rotation rate for $Y_{1.7}Eu_{0.65}Tm_{0.65}Fe_{4.3}Ga_{0.7}O_{12}$ film. (1-5) saturation temperature 866, 925, 925, 1015, 933°C and ΔT = 19, 9, 22, 22, and 28°C, respectively.

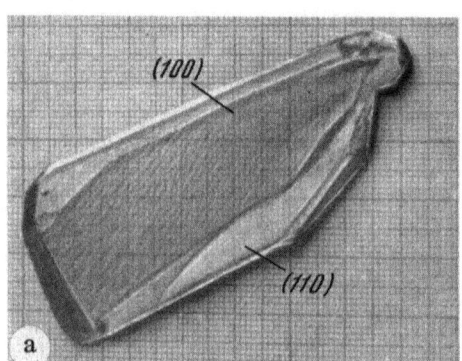 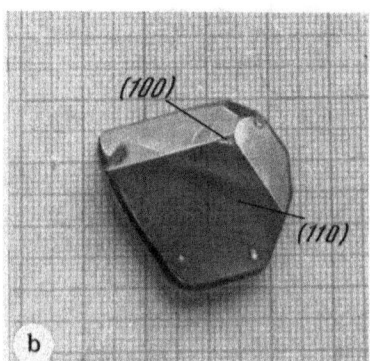

Fig. 6. $Y_3Al_5O_{12}$ crystals grown on seeds with orientation (a) (100), and (b) (111) (photographs by A. B. Bykov). (110) smooth face, (100) rough face.

At increased cooling rates the degree of roughness of the surface increases; that is, the individual overgrowth crystals grow bigger and segments of smooth faces on each of them become more pronounced. At high cooling rates (higher than 1°C h^{-1}) only a few large individual crystals, completely faceted by smooth faces, are formed on the growth surface.

Rough faces without macroscopic inclusions of the solution appear at growth rates below 10 μm h^{-1} (seed rotation rate 30 rpm). At higher growth rates the solution is captured in the re-entrant polyhedral angles of the rough surface.

The following conclusions can be drawn from the above exposition.

In recent years the method of flux growth of large crystals advanced from spontaneous to controlled crystallization.

In conditions of thermodynamic instability of the flux (liquation and volatility of solutions at high temperatures, formation of co-crystallizing phases, etc.), other techniques, such as accelerated crucible rotation, localization of crystallization centers, and some others appear more promising than spontaneous crystallization. In these cases measures are taken to limit the number of spontaneous nucleation sites, so that it becomes possible to grow one large single crystal or several large crystals of different sizes in the course of controlled cooling.

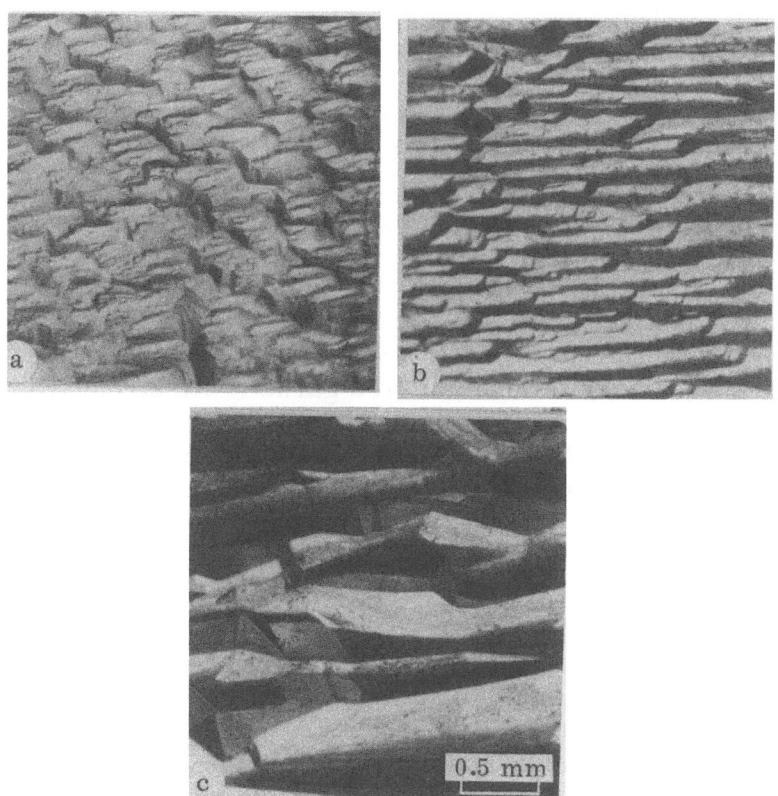

Fig. 7. Morphology of $Y_3Al_5O_{12}$ crystals grown on oriented seeds (photographs by A. B. Bykov). (a-c) enhancement of roughness caused by increased cooling rate of the solution.

Considerable progress has been achieved in various approaches to controlled crystallization in which seed crystals grow either in the static mode (in gradient mass transfer conditions) or in the dynamic mode (in zero-gradient bulk of the solution), or by the Czochralski growth of crystals pulled from the molten salt solutions. In these methods crystals are grown taking into account the thermodynamic changes in the system, with kinetic parameters ensuring a correlation between the growth rate of the seed and the changes in supersaturation of the melt.

Literature Cited

1. J. P. Remeika, J. Galt, and B. Matthias. Properties of single crystals of nickel ferrite. Phys. Rev., 79, 391-394 (1950).
2. J. W. Nielsen. The growth of magnetic garnet crystals. J. Appl. Phys., 29, 21-26 (1958).
3. A. G. Titova. Growth of yttrium ferrite single crystals. Fiz. Tverd. Tela, 1, 1971-1974 (in Russian) (1959).
4. E. A. D. White. A new technique for the production of synthetic corundum. Nature, 191, 901-902 (1961).
5. R. C. Linares. Growth of yttrium aluminium garnet single crystals. J. Amer. Ceram. Soc., 45, 119-121 (1962).
6. L. G. van Uitert, W. H. Grodkewicz, and E. F. Dearborn. Growth of large optical quality yttrium and rare-earth aluminum garnets. J. Amer. Ceram. Soc., 48, 105-112 (1965).

7. V. A. Timofeeva. Conditions of growth of Al_2O_3, ZnO, Ga_2O_3 crystals. In: Growth of Crystals, vol. 6, ed. N. N. Sheftal', Consultants Bureau, New York (1968).
8. B. V. Zaitsev, S. Sh. Gendelev, A. G. Titova, and V. G. Kurilenko. Growth of yttrium iron garnet crystals in large crucibles. Elektronnaya Teknika. Ser. 7. Ferritovaya Tekhnika, No. 4, 10-16 (in Russian) (1968).
9. B. M. Wanklyn. The flux growth of single crystals of rare-earth perovskites (orthoferrites, orthochromites and aluminates), J. Cryst. Growth, 5, 323-328 (1969).
10. O. M. Konovalov and L. L. Nagornaya. Investigation of melts in growth of yttrium iron garnets. Izv. AN SSSR, Ser. Fiz., 37, 1240-1243 (in Russian) (1970).
11. B. V. Mill', A. G. Mikhal'chenkov, N. D. Ursulyak, and N. I. Utkin. Synthesis of single crystals of scandium-neodymium iron garnet. Elektronnaya Tekhnika, Ser. 14, Materialy, No. 3, 90-93 (in Russian) (1970).
12. J. M. Robertson and B. V. Neate. Some observations on the growth of YIG under oxygen pressure by the flux melt technique. J. Cryst. Growth, 13/14, 576-578 (1972).
13. D. Jonker. Investigation of phase diagram of the system $PbO-B_2O_3-Fe_2O_3-Y_2O_3$ for growth of single crystals of $Y_3Fe_5O_{12}$. J. Cryst. Growth, 28, 231-239 (1975).
14. V. K. Yanovsky and V. I. Voronkova. Growth and main properties of crystals of rare-earth and yttrium oxytungstates with Ln_2WO_6 composition. Izv. AN SSSR. Neorg. Materialy, 11, 91-94 (in Russian) (1975).
15. Z. N. Zonn, V. A. Ioffe, and P. P. Gorokhov. Czochralski growth of single crystals of alkali and alkaline-earth metal metavanadates. In: Vth USSR Conference on Crystal Growth, Tbilisi, Abstracts, Inst. of Cybernetics AN GSSR, Tbilisi, 2, 154-155 (in Russian) (1977).
16. V. A. Timofeeva and A. V. Zalessky. Crystallization of ferrites from vapor and melt. In: Growth of Crystals, Nauka, Moscow, 2, 88-94 (in Russian) (1958).
17. M. Kestigian. Growth of single crystals. J. Amer. Ceram. Soc., 46, 563-566 (1963).
18. V. A. Timofeeva and J. Kvapil. On solubility and crystallization of $Y_3Al_5O_{12}$ from $PbO-B_2O_3$ and $PbO-B_2O_3-PbF_2$ fluxes. Kristallografiya, 11, 289-294 (in Russian) (1966).
19. S. H. Smith and D. Elwell. Growth of nickel ferrite crystals from barium borate by a pulling method. J. Cryst. Growth, 3/4, 471-474 (1968).
20. W. Tolksdorf. Growth of yttrium iron garnet single crystals. J. Cryst. Growth, 3/4, 463-466 (1968).
21. G. A. Bennet. Seeded growth of garnet from molten salts. J. Cryst. Growth, 3/4, 458-462, (1968).
22. L. N. Mezmaternykh, G. I. Shvartsman, D. V. Tsynchik, V. G. Mashchenko, and A. P. Afanas'ev. Investigation of seeded growth of YIG single crystals from high-temperature solutions. Izv. AN SSSR, Ser. Fiz., 34, 1246-1249 (in Russian) (1970).
23. L. I. Potkin. Flux growth of scheelite group tungstate and molybdate single crystals. In: Growth of Crystals, vol. 9, ed. N. N. Sheftal' and E. I. Givargizov, Consultants Bureau, New York (1975).
24. H. J. Scheel and D. Elwell. Stable growth rates and temperature programming in flux growth. J. Cryst. Growth, 12, 153-161 (1972).
25. P. V. Klevtsov, L. P. Kozeeva, and A. A. Pavlyuk. Polymorphism and crystallization of potassium-rare earth molybdates $KLn(MoO_4)_2$ (Ln=La, Ce, Pr, Nd). Kristallografiya, 20, 1216-1220 (in Russian) (1975).
26. N. I. Lenyuk. Solubility of $YAl(BO_3)_4$ in molten potassium trimolybdate and seeded growth of single crystals. Izv. AN SSR, Neorg. Materialy, 12, 554-556 (in Russian) (1976).

27. T. Fukuda, Y. Uematsu, and T. Ito. Kyropoulos growth and properties of $KnBO_3$ single crystal. J. Cryst. Growth, 24/25, 450-454 (1974).
28. K. Watanabe and Y. Sumiyoshi. Relationship between habit and etch figures of corundum crystals grown from molten cryolite flux. J. Cryst. Growth, 32, 316-328 (1976).
29. F. W. Perry. Single-seed growth of $(Ba, Pb)TiO_3$ from borate melts. J. Phys. and Chem. Solids, Suppl. 1, 483-487 (1967).
30. T. M. Bruton, J. C. Brice, O. F. Hill, and P. A. Whiffin. The flux growth of some γ-Bi_2O_3 crystals by the top seeded technique. J. Cryst. Growth, 23, 21-24 (1974).
31. V. Belruss, J. Kalnajs, A. Linz, and R. C. Folweiler. Top-seeded solution growth of oxide crystals from non-stoichiometric melts. Mater. Res. Bull, 6, 899-906 (1971).
32. T. Chen. Crystal Growth of $BaMoO_4$, $Bi_2O_3 \cdot 3MoO_3$ and $Bi_2O_3 \cdot 2MoO_3$ from molten salt solution by pulling seed method. J. Cryst. Growth, 20, 29-34 (1973).
33. V. A. Timofeeva, A. B. Bykov, and O. V. Kazakevich. Seeded growth of garnet and orthoferrite crystals from flux solvents. Vth USSR Conference on Crystal Growth, Tbilisi. Abstracts. Inst. of Cybernetics AN GSSR, Tbilisi, 2, 123-124 (in Russian) (1977).
34. O. M. Konovalov, E. N. Sablin, and V. I. Salo. Seeded growth of YIG single crystals from flux solvents. In: Proceed. of Conf. on Electronic Devices. TsNII Elektronica Publ., Moscow, 9, 154-155 (in Russian) (1970).
35. O. V. Kazakevich and V. A. Timofeeva. Seeded growth of orthoferrite single crystals. Kristallografiya, 22, 1115-1117 (in Russian) (1977).
36. A. B. Bykov and V. A. Timofeeva. Phase equilibria in solvent-garnet systems. Kristallografiya, 23, 180-186 (in Russian) (1978).
37. J. W. Nielsen. Improved method for the growth of yttrium iron and yttrium gallium garnets. J. Appl. Phys. 31, 518-525 (1960).
38. A. G. Titova and R. A. Petrov. Dynamic growth of yttrium-gallium-iron and bismuth-vanadium-calcium garnets. Elektronnaya Tekhnika, Ser. 7. Ferritovaya Tekhnika, No. 4, 3-10 (in Russian) (1968).
39. L. N. Averina, B. N. Birman, O. M. Konovalov, and T. R. Mnatskanova. Growth of YIG single crystals by spontaneous crystallization with a fixed number of crystallization centers. In: Growth of Crystals, vol. 9, ed. N. N. Sheftal' and E. I. Givargizov, Consultants Bureau, New York (1975).
40. A. A. Chernov. The Spiral Growth of Crystals. Soviet Physics Uspekhi, 4, 115-148 (1961).
41. J. E. Davies and E. A. D. White. Interface breakdown in garnet liquid phase epitaxy. J. Cryst. Growth, 27, 261-265 (1974).
42. W. Hintzmann and L. Muller-Vogt. Crystal growth and lattice parameters of rare-earth doped yttrium phosphate, arsenate and vanadate prepared by the oscillating temperature flux technique. J. Cryst. Growth, 5, 274-279 (1969).
43. H. J. Scheel and F. O. Schulz-Dubois. Flux growth of large crystals by accelerated crucible rotation technique. J. Cryst. Growth, 8, 304-306 (1971).
44. W. Tolksdorf. New experimental development in flux growth. In: 1976. Crystal Growth and Materials. New York - Oxford: North-Holland Publ. Co., 2, 639-659 (1977).
45. J. Aidelberg, J. Flicstein, and M. Schieber. Cellular growth in $BaFe_{12}O_{19}$ crystals solidified from flux solvents. J. Cryst. Growth, 21, 195-202 (1974).
46. F. W. Perry, G. A. Hutchins, and E. L. Cross. Compositional inhomogeneity of $(Ba, Pb)TiO_3$ crystals. Mater. Res. Bull., 2, 409-418 (1967).
47. W. A. Bonner, E. F. Dearborn, and L. G. van Uitert. Growth of $K(Ta, Nb)O_3$, single crystals for optical and semiconductor studies. J. Phys. and Chem. Solids, Suppl. 1, 437-440 (1967).

48. S. Sh. Gendelev and R. I. Zvereva. Morphological features of seeded crystallization of $Ba_2Zn_2Fe_{12}O_{22}(Zn_2Y)$. In: Proceedings of Conference on Electronic Devices. TsNII Elektronika, 9, 186-187 (in Russian) (1977).
49. R. A. Laudise, R. C. Linares and E. F. Dearborn. Growth of yttrium iron garnet on seed, from molten salt solution. J. Appl. Phys., 33, 1362-1363 (1962).
50. W. A. Bonner. Reproducible garnet films preparation using fast growth rates. Mater Res. Bull., 9, 885-894 (1974).
51. A. G. Titova, Yu. L. Sapozhnikov, and R. A. Petrov. Dynamic growth of YIG single crystals. Elektronnaya Tekhnika. Ser. 7. Ferritovaya Tekhnika, No. 1, 70-71 (in Russian) (1968).
52. R. C. Linares. Properties and growth of flux ruby. J. Phys. and Chem. Solids, 26, 1817-1820 (1965).
53. V. I. Voronkova, V. K. Yanovsky, and V. A. Koptsik. Growth of corundum single crystals from molten tungstate flux. Izv. AN SSR, Neorg. Materialy, 4, 1727-1731 (in Russian) (1968).
54. E. A. D. White and J. W. Brightwell. The growth of ruby crystals from solution in molten lead fluoride. Chem. and Ind., No. 9, 1662-1663 (1965).
55. V. A. Timofeeva. Isomorphic impurity distribution in garnets as a function of the solubility of the trivalent oxides in molten salts and of the method of crystal growth. J. Cryst. Growth, 3/4, 496-499 (1968).
56. W. Tolksdorf. Herstellung Ferrit-Einkristall mit Y-structure aus schmelzflussiger Losung. J. Cryst. Growth, 18, 57-61 (1973).
57. J. Albers. Top-seeded solution growth of $BaTiO_3$ single crystal from the TiO_2-rich melt. In: Abstract-book ECCG-1. Zurich, Sept. 12-18, p. 150-154 (1976).
58. S. H. Smith and B. M. Wanklyn. Flux growth of rare-earth vanadates and phosphates. J. Cryst. Growth, 21, 23-28 (1974).

GROWTH OF EMERALD SINGLE CRYSTALS

G. V. Bukin, A. A. Godovikov, V. A. Klyakhin, and V. S. Sobolev

The Institute of Geology and Geophysics of the USSR Academy of Sciences, Siberian branch, Novosibirsk

Transparent crystals of the green chrome-containing variety of beryl, known as emerald, are first-grade gemstones. In addition to its use for jewelry, emerald can also be used in low-noise microwave amplifiers [1-3]. Much effort has therefore been devoted to emerald synthesis and emerald crystal growth.

The methods of emerald crystal growth must be those using temperatures below the melting point (flux growth, hydrothermal growth, gas transport reactions), which for emerald is incongruent [4, 5].

Flux Crystallization of Emerald

When emerald is grown by this method, it is desirable to use solvents which dissolve emerald with minimum deviations from congruence but at the same time with sufficiently high efficiency.

Studies on emerald synthesis and flux growth [2, 3, 6-21] reported crystallization either by slowly cooling a saturated solution (from 1250 to 700°C at a cooling rate gradually reduced from 10 to 3°C h^{-1}), or by seeded growth in temperature gradient conditions. In both cases the temperature of the experiment was dictated by the solution composition and by the melting point and volatility of the solvent.

Usually emerald is synthesized in platinum crucibles inert to molten fluxes. Usually BeO, $BeCO_3$, Al_2O_3, SiO_2, or natural beryl are used as initial components. Either Cr_2O_3 or Li_2CrO_4, as well as Fe_2O_3, are chosen as chromophores. As a rule, these reactants are mixed in a ratio corresponding to the beryl stoichiometry. Typical solvents used in the synthesis are alkali metal salts of molybdenum, tungsten, and vanadium, and oxides of molybdenum, vanadium, and boron.

In complete agreement with earlier publications [3, 9, 10, 15, 18, 21], our investigation of the above-listed solutions confirmed that the emerald growth rates are very low. Thus, growing a crystal with volume above 1 cm^3 by the temperature gradient method at a temperature about 1000°C takes at least a year. If emerald is crystallized by the temperature lowering technique, then numerous small crystals are obtained in the range 1250-800°C even at low cooling rates (1-2°C h^{-1}). The crystals thus grown have a high density of defects (cracks, and inclusions of solvent chrysoberyl, phenacite, etc.).

Selection of Solvent. A number of solvents suitable for emerald growth were chosen by qualitatively testing the solubility of beryl, analyzing the phase composition of the final products, and evaluating the behavior of the solvents and their dissolving capacity as functions of temperature. The solvents were considered promising if they dissolved emerald congruently in amounts not less than 5-10 wt.%. High-quality crystals obtained for molten solutions based on V_2O_5 [2, 3, 16, 18, 19] made us select this type of flux despite a partial decomposition of emerald into phenacite and quartz. In particular, a low melting-point (480°C) eutectic E_1 in the PbO-V_2O_5 system proved especially convenient (Fig. 1) [22]. This mixture does not form solid solutions with emerald, has a high solubility for emerald and a relatively low evaporation rate, and is dissolved by weak solutions of inorganic acids. In addition, it has a favorably low viscosity (4-6 centipoise).

The solubility data for beryl in molten PbO-V_2O_5 allowed part of the liquidus curve (Fig. 2) of a pseudobinary system $Al_2\{Be_3[Si_6O_{18}]\}$-PbO-V_2O_5 to be plotted. As the eutectic is very close to the PbO-V_2O_5 composition (0.2 mol%), emerald is crystallized out almost completely when the saturated molten solution is cooled. Although the liquidus line is rather steep, the difference in saturation concentrations in the temperature range 800-1250°C reaches 0.8-1.4 wt.% even for a temperature difference of 50°C; thus, it is possible to achieve crystallization in temperature gradient conditions.

A specific growth technique must be chosen taking into account the dependence of the density, ρ, of the flux on its composition in the PbO-V_2O_5 system (Fig. 3) [23]. Thus, the low density of beryl (2.6-2.8 g cm^{-3}) and negligible changes in solution densities with temperature (Fig. 4) [23] are arguments in favor of recommending compositions based on V_2O_5 with a density of 2.3-2.6 g cm^{-3} for growing emerald crystals by cooling saturated solutions (in the normal temperature gradient modification) and possibly, by pulling crystals from the flux by the Czochralski technique. On the other hand compositions with density above 3 g cm^{-3} can be used for crystal growth only if a reversed temperature gradient is utilized. (in the case of the normal temperature gradient the dissolution zone is in the bottom portion of the crucible, and crystallization takes place in the upper (cooler) zone).

High-quality crystals can be grown at reasonably high growth rates only from low viscosity solutions (4-8 centipoise) (Fig. 5) [23], so that it is advisable to crystallize from PbO-V_2O_5 at temperatures above 900°C and with compositions 20 PbO + 80 V_2O_5 to 60 PbO + 40 V_2O_5.

Emerald synthesis and crystal growth from slowly cooled solutions is possible owing to the high solubility of emerald in molten PbO-V_2O_5 (11 wt.% at 1250°C). The upper limit of the cooling temperature range 1250-800°C is defined by the fact that beryl is stable in PbO-V_2O_5 up to 1250°C, and the lower limit — by a reduction in solubility and increase of viscosity to 20 centipoise at 800°C.

Two to three days at a temperature 50°C above the liquidus temperature of the initial composition is sufficient for saturation of the PbO-V_2O_5, dissolution of all the components, and homogenization (in accordance with the kinetic solubility curves for beryl).

Solutions were cooled at 0.5-5°C h^{-1} rates down to 800°C, and later left to cool in the furnace with the power turned off.

The sizes of emerald crystals grown by spontaneous crystallization depend on the cooling rate, temperature stability, and crucible volume, but do not exceed 5 to 7 mm. The experiments were made either in the static mode or in the crucible-rotation mode. Well-faceted crystals showed the hexagonal prism (10$\bar{1}$0) and the pinacoid (0001) faces (Fig. 6).

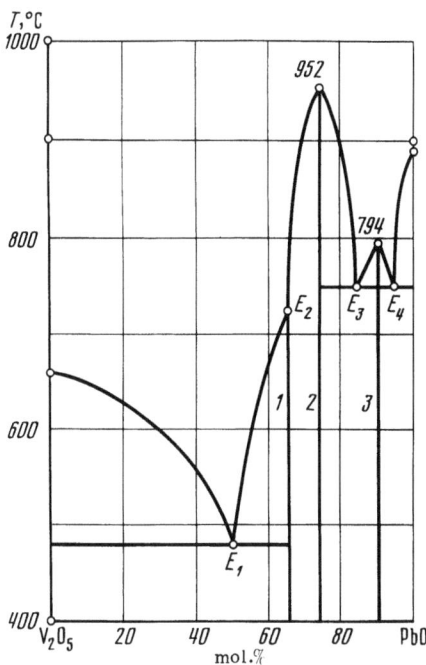

Fig. 1. Phase diagram of the PbO-V_2O_5 system. E_1(480°C), E_2(722°C), E_3, E_4 are eutectics; (1) 2PbO·V_2O_5; (2) 3PbO·V_2O_5; (3) 8PbO·V_2O_5.

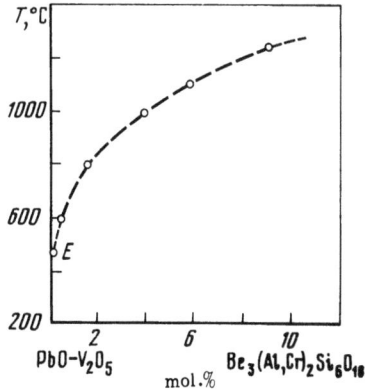

Fig. 2. Liquidus line of the system $Al_2\{Be_3[Si_6O_{18}]\}$-PbO-V_2O_5 in the range 0-12 mol % $Al_2\{Be_3[Si_6O_{18}]\}$.

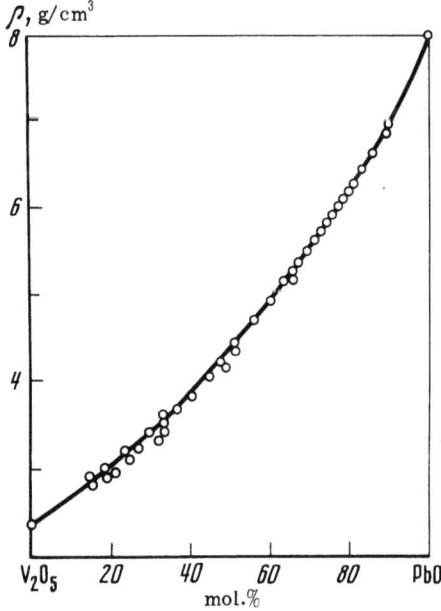

Fig. 3. Density of the PbO-V_2O_5 solvent as a function of its composition at 1000°C.

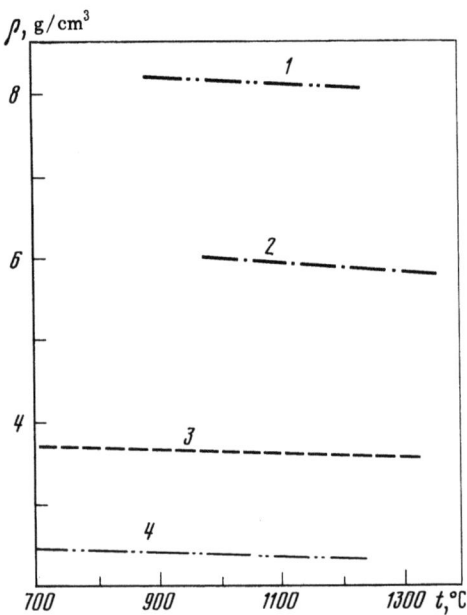

Fig. 4. Polytherms of density in solutions of the PbO-V_2O_5 system. (1) PbO; (2) 74.6 PbO + 25.4 V_2O_5; (3) 35PbO + 65V_2O_5; (4) 7.9PbO + 92.1V_2O_5.

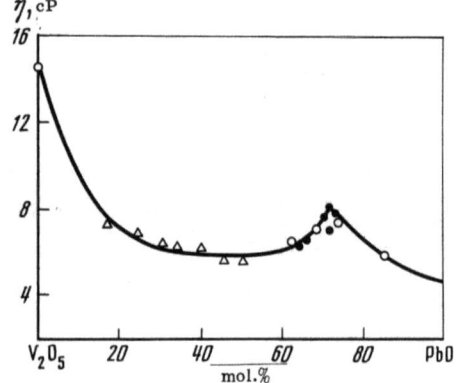

Fig. 5. Viscosity of solutions as a function of composition for the PbO-V_2O_5 system at 1000°C.

Fig. 6. Emerald crystals formed spontaneously in the course of cooling the saturated flux.

Seeded growth of emerald was also investigated. This shortens the duration of the process since it eliminates the slow cooling stage necessary for nucleation. On the whole, the cooling of saturated $PbO-V_2O_5$-based solutions is advisable for synthesis of the charge material (later used for recrystallization by the temperature gradient technique), as well as for the growth of seed crystals.

Growth of emerald crystals by the temperature gradient technique enables one to maintain constant crystallization conditions and thus grow more uniform crystals. Here supersaturation is produced by the temperature difference between the dissolution zone and the cooler growth zone. Both the normal and reversed vertical temperature gradients were used in the experiments (Fig. 7).

Extensive decomposition of emerald at temperatures above 1200°C into phenacite, chrysoberyl, and cristobalite makes it imperative to conduct crystallization at lower temperatures. On the other hand, an increase in viscosity of the melt to 6 centipoise below 900°C causes imperfect growth and the formation of a large number of inclusions. The optimal temperature interval is from 1200 to 1000°C. Under these conditions crystals form as druses of gem-quality emeralds (Fig. 8). If the temperature gradient exceeds $3°C\ mm^{-1}$, a large number of nuclei are generated, and dense intergrown aggregates of small crystals are formed (Fig. 9). Rotation with periodical acceleration makes it possible to increase the growth rate on oriented seeds and to obtain large crystals (Fig. 10).

Hydrothermal Growth of Emerald

Hydrothermal crystallization of emerald was carried out in two stages: first the conditions of synthesis and crystallization of beryl were studied, and then the conditions of chromophore ion incorporation in beryl, and the synthesis and growth of emerald.

Hydrothermal synthesis of beryl was conducted over a wide temperature range. At present most information has been accumulated on growth in the range 450-650°C, in which beryl was successfully synthesized both in pure water and in most of the available solvents with acid or slightly alkaline pH: chlorides and fluorides of alkali metals and ammonium, hydrochloric, and boric acids, sodium tetraborate, carbonate, and bicarbonate, and in carbonate-fluoride solutions [24-30]. Seeded growth of beryl in these solutions was also achieved.

Seeded crystallization of beryl was studied by synthesis in an autoclave (Fig. 11) of 165 cm^3 volume made of stainless steel 12Cr18Ni10Ti. Quartz, as well as beryllium and aluminum hydroxides, were used as the charge components; these components were spatially separated: quartz was at the top, over the seed, and the source of beryllium and aluminum was at the autoclave bottom.

It was found that the rate of beryl crystallization at the seed increases in the sequence of inorganic acids $CO_3^{2-} \to BO_3^{3-} \to NO_3^{1-} \to SO_4^{2-} \to Cl^{1-} \to F^{1-}$.

Beryl is easily synthesized from solutions of salts of the above acids, provided the activity of alkaline and alkali-earth metals is moderate, since a high activity leads to the formation of various silicates (albite, adular, petalite, anorthite, etc.). High activity of fluorine ion is also undesirable, as topaz is extensively crystallized. However, if fluorine is present in strongly bound complexes, the beryl synthesis is successful. Thus, seeded growth of beryl was achieved in the presence of the carbonate ion and aluminum-fluoride complexes of the type of $N_m Al_n F_{3n+m}$. These complexes act as buffer systems controlling the activity of the alkaline cation, aluminum, and fluorine.

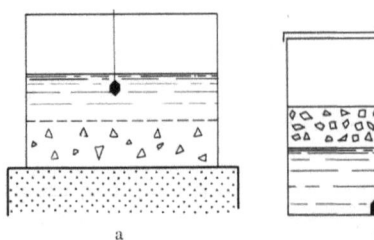

Fig. 7. Schematic representation of the experiments on flux growth of emerald crystals in the (a) standard and (b) reversed temperature gradients.

Fig. 8. A druse of emerald crystals.

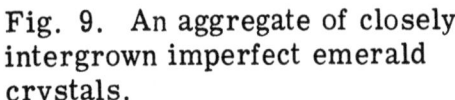

Fig. 9. An aggregate of closely intergrown imperfect emerald crystals.

Fig. 10. Emerald crystals grown by the reversed temperature gradient method.

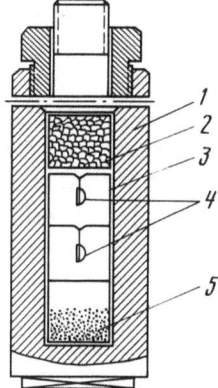

Fig. 11. Schematic of an apparatus for the hydrothermal synthesis of emerald. (1) autoclave, (2) charge (source of SiO_2), (3) seed holder, (4) seeds, (5) charge (source of Al_2O_3, BeO, Cr_2O_3).

Fig. 12. Emerald crystals grown on seeds by hydrothermal techniques.

Fig. 13. As-grown emerald crystals and cut gem crystals.

A controlled partial pressure of oxygen is of principal importance in the synthesis and growth of colored varieties of beryl, whose coloring is determined by **variable-valence** elements [31]. Oxygen affects the incorporation of **variable-valence** elements in two ways: first, it provides the necessary partial pressure PO_2, at which the element in the solution has the required **valence**; second, oxygen may influence the solubility of compounds comprising **variable-valence** elements. The oxygen partial pressure was controlled by introducing into the system selected salts of oxygen containing acids or other compounds with variable **valence**. This enabled us to grow beryl crystals colored with Fe, Ni, Mn, Co, and Cu impurities.

The greatest difficulties are encountered in synthesizing the chromium-containing emerald, but both the chromium- and vanadium-containing emeralds have now been synthesized [26-28, 32]. The best results were obtained by synthesizing the chromium-doped emerald in high-concentration solutions of HCl and NH_4Cl [26, 27]; the growth rate achieved was up to 0.3 mm/day. One disadvantage of these processes is the rapid reduction of growth rate with time due to phenacite formation in the charge. Consequently, the desired thickness of overgrowth (5-6 mm) was obtained [33] by repeating the process two or three times. Gold lining had to be used to resist the highly aggressive solvents.

The difficulties met in emerald growth are caused by the necessity to synthesize it only in acid media, since in the presence of OH^- chromium salts are hydrolyzed and a poorly soluble chromium oxide is formed. Also the Cr^{3+} ion is easily oxidized in alkaline media to Cr^{6+} which, however, is not incorporated into the crystal lattice. Most of the typical alkaline solvents are therefore excluded. Introduction of solvents containing fluorine ions and producing high rates of transfer of beryl components leads to chromium deposition in the form of a poorly dissolved chromium fluoride [26], since the concentration of Cr^{3+} in the solution is insufficient for its incorporation into the lattice in the required amounts (0.2-0.5 wt.%). Our investigations have shown that these difficulties can be overcome if complex-composition acid solutions are employed. With this approach crystalline layers of deep-green emerald up to 10 mm thick and up to 10 g by weight were grown on seeds of natural beryl or synthetic emerald (Fig. 12).

Vapor Synthesis and Growth of Beryl Crystals

No information is available in the literature on the synthesis and growth of emerald crystals by gas transport techniques. However, the method of transferring the components of emerald (beryllium, silicon, aluminum) as volatile components [34] is, in principle, possible by gas transport reactions. In the present paper we have studied experimentally the seeded CVD crystallization of beryl from BeO, Al_2O_3 and SiO_2 oxides, by using complex compounds as carrier agents.

The color of the crystals of emerald and other varieties of beryl is determined by selective light absorption in the visible part of the spectrum and depends on the density and structural positions of chromophore ions: chromium, iron, vanadium, nickel, manganese and cobalt. A selection of chromophore ions and their relative concentrations made it possible to grow synthetic emeralds with coloring very similar to natural stones from the Ural emerald mines (507-519 nm range) and to obtain gem-quality stones (Fig. 13).

Literature Cited

1. F. E. Goodwin. Maser action in emerald. J. Appl. Phys., 32, 1624-1625 (1961).
2. R. C. Linares, A. A. Ballman, and L. G. van Uitert. Growth of beryl crystals for microwave applications. J. Appl. Phys., 33, 3209-3210 (1962).
3. A. A. Ballman, R. C. Linares, and L. G. van Uitert. Pat. 3,234,135 (USA) (1966).
4. A. van Valkenburg and C. E. Weir. Beryl studies: $BeO \cdot Al_2O_3 \cdot 6SiO_2$. Bull. Geol. Soc. Amer., 8, 1808-1811 (1967).
5. R. P. Miller and R. A. Mercer. The high-temperature behavior of beryl melts and glasses. Miner. Mag. 35, 250-276 (1965).
6. I. I. Ebelmen. Memoire sur une nouvelle methode pour obtenir der combinasions cristallisees par la voie seche, et sur des applications de la reproduction des especes minerals. Ann. chim. phys. 22, 213-244 (1848).
7. H. Traube. Uber die künstliche Darstellung des Berylls. Neues Jahrb. Mineral., S. 9-16 (1894).
8. P. Hautefeuille and A. Perrey. Sur la reproduction de la phenacite et de l'emeralde. C. R. Acad. Sci., 1, 1800-1803 (1988).
9. H. Espig. Die Synthese des Smaragds. Chem. Techn., 12, 327-331 (1960).
10. H. Espig. Die Synthese des Smaragds. Zpravy Vyr kumn. ustavu monokryst., 8, 84-88 (1962).
11. A. Armstutz and A. Borloz. Synthesis of emerald. Arch. Sci. Phys. Natur., 17, 39-41 (1935).
12. A. L. Gentile, D. M. Cripe, and F. N. Andres. The flame fusion synthesis of emerald. Amer. Miner., 48, 940-943 (1963).
13. F. N. Andres, D. M. Gripe, and A. D. Gentile. Synthesis of emerald, alexandrite and beryllium oxide single crystals. Hughes Res. Lab. Exp. Res. Project Rept. October, 1648-1655 (1962).
14. M. Ushio and Y. Symioshi. Flux-growth of emerald single crystals. Nippon Kagaku Kaishi, 9, 1648-1655 (1972).
15. M. Ushio and Y. Sumiyoshi. Growth rate of each plane of synthetic emerald single crystal by the V_2O_5 flux method. Department of Applied Chemistry, Faculty of Engineering, Gunma University: Kiryi-shi, No. 3, 506-513 (1973).
16. M. Ushio and Y. Sumiyoshi. J. Chem. Soc. Jap., Chem. and Ind. Chem., No. 5, 941-947 (in Japanese) (1973).
17. R. A. Lefever, A. B. Chase, and H. B. Sobon. Synthetic emerald. Amer. Miner. 47, 1450-1453 (1952).
18. R. C. Linares. Growth of beryl from molten salt solutions. Amer. Miner., 52, 1554-1559 (1967).
19. E. M. Flanigen, D. W. Breck, N. R. Mumbach, and A. M. Taylor. Characteristics of synthetic emeralds. Amer. Miner., 52, 744-772 (1967).
20. C. Sacamoto. Pat. 73-23278 (Jap.) Growing emerald single crystals. 25.02.70.
21. E. M. Flanigen and A. M. Taylor. Pat. 3,341,302 (USA). Flux-melt method for growing single crystals having the structure of beryl. Sept. 12 (1967).
22. Gmelin Kraut. Handbuch der anorganische Chemie. Vanadium. Berlin: Faris Chemie – G. M. G. H., 48, 567-568 (1967).
23. A. F. Mundus-Tabakayev, G. V. Bukin, and A. B. Kaplun. Viscosity and density of vanadium-containing fluxes V_2O_5-PbO, V_2O_5-Li_2O, and $LiVO_3$-MoO_3. In: Thermal Properties of Liquid High-Temperature Solutions and Alloys. Inst. of Thermal Physics SO AN SSSR, Novosibirsk, 135-143 (in Russian) (1977).
24. D. Ganguli and P. Saha. A reconnaissance of system BeO-Al_2O_3-SiO_2-H_2O. Trans. Indian Ceram. Soc., 26, 102-110 (1967).
25. E. M. Flanigen. Pat. 3,567,642 (USA). Hydrothermal process for growing crystals having the structure of beryl in an alkaline halide melium. March 2 (1971).

26. E. M. Flanigen and N. R. Mumbach. Pat. 3,567,643 (USA). Hydrothermal process for growing crystals having the structure of beryl in an acid halide medium. March 2 (1971).
27. P. G. Yancey. Pat. 3,723,337 (USA). Hydrothermal process for growing crystals having the structure of beryl in highly acid chloride medium. March 27 (1973).
28. E. H. Yemelyanova, S. V. Grum-Grzhimailo, O. N. Boksha, and A. M. Varina. Artificial V-, Mn-, Co-, Ni-containing beryl. Kristallografiya, 10, 59-62 (in Russian) (1965).
29. A. A. Beus and Yu. P. Dikov. Geochemistry of Beryllium in the Processes of Endogeneous Mineralization. Nedra, Moscow, (in Russian) (1967).
30. V. A. Klyakhin, A. S. Lebedev, T. P. Ragozina, and A. Ya. Rodionov. Behavior of beryl in hydrothermal solutions. In: Physico-Chemical Processes of Mineralization: Theoretical and Experimental Data. Inst. of Geol. and Geophysics SO AN SSSR, Novosibirsk, 82-106 (in Russian) (1976).
31. V. A. Klyakhin, A. S. Lebedev, and D. A. Fursenko. Experimental evaluation of PO_2 of hydrothermal solutions in autoclaves. In: Physico-Chemical Processes of Mineralization: Theoretical and Experimental Data. Inst. of Geol. and Geophysics SO AN SSSR, Novosibirsk, 156-157 (in Russian) (1976).
32. A. M. Taylor. Synthetic vanadium emerald. J. Gemmology. 10, 211-217 (1967).
33. K. Nassau. Synthetic emerald: the confusing history and the current technologies. J. Cryst. Growth, 35, 211-222 (1976).
34. A. V. Novoselova and B. P. Sobolev. On the role of fluorine-containing compounds in beryllium transfer and phenacite formation. Geokhimiya, 1, 20-28 (in Russian) (1959).

SOME TECHNOLOGICAL PROCEDURES AND EQUIPMENT FOR HYDROTHERMAL GROWTH OF SINGLE CRYSTALS

V. I. Popolitov, A. N. Lobachev and A. Ya. Shapiro

The Institute of Crystallography of the USSR Academy of Sciences, Moscow

We shall discuss several new growth procedures and new equipment based on them, for the hydrothermal synthesis and growth of single crystals.

Quartz Reactor

A well-known bottleneck impeding the progress in hydrothermal growth of single crystals is the difficulty of extracting reliable information on physical chemical processes taking place inside steel autoclaves, and the ensuing impossibility of an accurate analysis of the interrelations between factors limiting crystallization.

We have designed, installed, and tested a quartz reactor with 200 cm^3 volume in order to directly observe the process of dissolution of solid ingredients, and the synthesis and seeded growth of crystals. The reactor (Fig. 1) is a cylindrical container 1 made out of transparent fused quartz. Mounted inside the container are the basket for charge 2, diaphragm 3 separating the dissolution and growth zones, and the crystal holder 4 with seed crystals 5 suspended from it. The reactor is sealed by a corrugated Teflon plug 6 (using the principle of noncompensated area) which is rigidly fixed by a support cylinder 7 and a support slab 8. The quartz reactor is heated with two heater coils 9 mounted on plates 10. All subunits are fixed to columns 11 and the foundation 12. The vertical distribution of internal temperature was measured in the designed reactor in different operation modes of the upper and lower heaters, with and without the diaphragm.

Two C-A thermocouples were used during the experiments to measure the temperature distribution along the outer wall of the reactor in the upper and lower zones. In addition, a metal capillary, sealed at the bottom, was introduced into the reactor through a seal. The temperature distribution inside the reactor was measured by sliding a thermocouple through this capillary (experiments were conducted with the reactor temperature $T \leq 200°C$). This yielded the vertical temperature distribution inside the reactor as a function of the temperature distribution on the outer walls. The following conclusions have been drawn from the studies.

Experiments without a Diaphragm between the Growth and Dissolution Zones. When the temperature of the lower heater is higher than that of the upper one, intense stirring of the solution is observed, produced by gas bubbles propagating from the reactor bottom upward to the seal. The velocity of motion of the gas bubbles is so high that the temperature inside the reactor is essentially uniform. With a filling factor 0.7

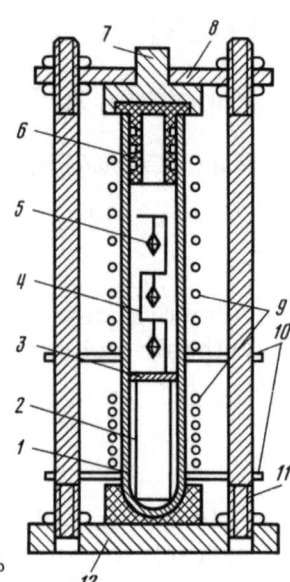

Fig. 1. Quartz reactor. (a) general view; (b) sectional view; (1) cylindrical quartz container; (2) basket for charge; (3) partition; (4) crystal holder; (5) seed crystals; (6) Teflon plug; (7, 8) support cylinder and plate; (9) heaters; (10) plates; (11) columns; (12) foundation.

and temperatures inside the reactor 110, 150 and 200°C, the gas bubble velocity W was 23, 26, and 28 cm s^{-1}, respectively. The gas bubble velocity clearly increases at higher temperatures, possibly owing to reduced viscosity of the liquid and increased difference between the densities of the liquid and vapor phases.

If the lower heater is kept at a lower temperature than the upper one, no gas bubbles are generated and a negative temperature gradient sets up along the reactor axis.

Experiments with a Diaphragm between the Dissolution and Growth Zones. When the temperature of the lower heater exceeds that of the upper heater, a large number of gas bubbles are generated and the solution is vigorously stirred. The temperature difference between the dissolution and growth zones inside the reactor is determined by the external temperature difference and by the velocity of the gas bubbles. As the external temperature difference rises, the velocity of the bubbles increases, which results in a reduced internal temperature difference. At constant external temperature difference and different diameters of holes in the diaphragm, the internal temperature difference ΔT is smaller the higher is the velocity of the gas bubbles (Fig. 2).

The reactor was tested by studying the growth of paratellurite crystals as a function of several parameters. The experiments were run at temperature differences of 1, 2, 3, 4, and 5°C, and with constant area of holes in the diaphragm (7% of the total diaphragm area). The temperature in the dissolution zone was invariably kept at 150°C.

Preliminary experiments on the recrystallization of TeO_2 in aqueous solutions of hydrohalic acids of different concentrations, have shown that the most effective solvents are aqueous solutions of HCl. The concentration of HCl in experiments on the growth of TeO_2

was 1-4 wt.%. TeO_2 single crystals, obtained initially by recrystallization of TeO_2 from aqueous HCl solutions of different concentrations, were used as the charge. The charge mass was constant, and in all experiments the TeO_2 particle size was 1-1.5 mm. The HCl solution was saturated in advance with tellurium dioxide at 150°C, that is, at the crystal growth temperature.

The seeds were hydrothermally grown crystals suspended by quartz or Teflon crystal holders.

Mass Transfer in Seeded Crystallization of TeO_2. The mass transfer of the solute from the dissolution zone to the growth zone is known to be a complicated function of a number of variables. In some cases the solute mass transfer may constitute the limiting stage of crystal growth. Two situations are usually observed: excessive mass transfer and insufficient mass transfer. Consequently, one needs to know the mass transfer kinetics of the crystalline medium in order to control single crystal growth. Control of mass transfer in steel autoclaves meets with severe difficulties because of the impossibility of monitoring the process. In the experiments of the present study mass transfer takes place not in the homogeneous region but in a two-phase system (gas-liquid), so that one has to find the effects of hydrodynamic and other parameters on the mass transfer kinetics. The effects of various factors on mass transfer in paratellurite were studied by determining the mass of material transferred to TeO_2 seeds placed in the growth zone, as a function of the velocity and size of gas bubbles, temperature difference, and concentration of the aqueous solution of hydrochloric acid, C_{HCl}.

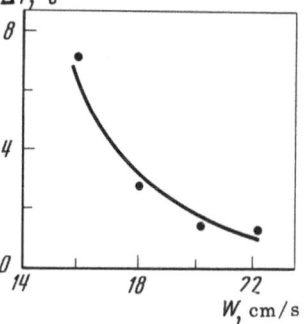

Fig. 2. Temperature difference as a function of the velocity of gas bubbles.

Fig. 3. Gas bubbles.

Effect of Velocity of Gas Bubbles on Paratellurite Mass Transfer. The experiments were run at $\Delta T = 3°C$, with filling coefficient 0.7, dissolution zone temperature $T_1 = 150°C$, and $C_{HCl} = 4\%$. The holes in the diaphragm occupied 7 percent of its total area, although the number and diameter of the holes were varied.

The velocity and size of gas bubbles were recorded by filming (Fig. 3), for diameters of holes in the diaphragm 3.5, 3.0, 2.7 and 2.4 mm and relative total area of the holes 7%. A bubble formed at a hole first grows and then, when the buoyancy, $\pi d^3 g(\rho_L - \rho_G)/6$, becomes equal to the bubble separation force (depending on the surface tension, $k = \pi d_0 \sigma$), it breaks off. The bubble diameter at the break-off moment is given by the relation

$$d = [6 d_0 \sigma / g (\rho_L - \rho_G)]^{1/3}, \qquad (1)$$

where d_0 is the hole diameter; σ is the surface tension; ρ_L, ρ_G are the solution and bubble densities, respectively; and g is the gravitational constant.

Bubble velocities depend on their size, the difference between the liquid and gas densities, and the coefficient of drag r [1]:

$$W = [6 g d (\rho_L)/r\rho_G - \rho_L]^{1/2}. \qquad (2)$$

These relations provide a good fit to the experimental data:

Number of holes in the diaphragm	2	3	4	5
d_0, mm	3.5	3.0	2.7	2.4
d, mm	7.1	6.4	5.9	5.1
W, cm s^{-1}	24.0	22.0	20.0	18.0

The following values of mass transfer m were obtained at different values of gas bubble velocities:

W, cm s^{-1}	18	19	20	22	23	25
m, g day^{-1}	0.074	0.078	0.084	0.086	0.085	0.086

The mass transfer of TeO_2 is thus practically independent of the gas bubble velocity; that is, the limiting stage of the mass transfer process is the heterogeneous reaction at the interface between the TeO_2 charge and the solvent.

Mass Transfer as a Function of Gas Bubble Size. Experiments conducted with the same parameters of the unit have shown that mass transfer is independent of the size of gas bubbles. This fact is theoretically supported by equations (1) and (2) (the gas bubble velocity and size are interrelated) and by the independence of mass transfer on the bubble velocity (see the experimental data given above).

Effect of Temperature Gradient on Mass Transfer of Paratellurite. The experiments were conducted at ΔT equal to 1, 2, 3, 4, and 5°C, filling coefficient 0.7, $T_1 = 150°C$, relative hole area of a diaphragm 7%, and $C_{HCl} = 4$ wt.%. The data on mass transfer were obtained only for seeded crystallization of TeO_2. When the temperature difference between the dissolution and growth zones reached $\Delta T = 6°C$, parasitic seeding was observed on the surface of the main seeds, so that mass transfer was studied only at $\Delta T < 5°C$. The results obtained are given below:

ΔT, °C	1	2	3	4	5
m, g day^{-1}	0.038	0.060	0.081	0.087	0.120

The amount of TeO_2 dissolved per unit time is obviously in good agreement with the equation

$$m = k\Delta T,$$

where k is the proportionality factor with dimension g·°C^{-1}·day^{-1}.

Mass Transfer as a Function of HCl Concentration. The experiments were conducted at C_{HCl} equal to 1, 2, 3, and 4 wt.% and $\Delta T = 4°C$. The remaining parameters were kept constant. The following dependence of TeO_2 mass transfer to the growth zone on HCl concentration was obtained:

C_{HCl}, wt.%	1	2	3	4
m, g day^{-1}	0.025	0.030	0.062	0.097

The apparent nonlinear dependence of mass transfer on concentration can be tentatively explained as follows. As a rule, in hydrothermal systems, mass transfer involves chemical reactions with complex formation, with active participation of the solvent. The composition and structure of the carrier complexes may alter with changing concentration of the solvent, and this must influence the kinetics of mass transfer to the growth zone. Presumably, this situation applies to the case in question. The fact that at $C_{HCl} > 4$ wt.% a secondary phase is formed with composition $Te_6O_{11}Cl_2$ supports this interpretation.

Growth Kinetics of TeO_2 Crystals. In the conditions of coexisting liquid and vapor phases the growth of TeO_2 crystals proceeds with vigorous stirring of the saturated solution by gas bubbles; consequently, bulk diffusion from solution cannot limit the growth process; that is, the limiting factor of growth kinetics lies in surface processes.

A direct experimental confirmation of the conclusion eliminating bulk diffusion as the limiting stage in the growth of TeO_2 crystals is the dependence of TeO_2 crystal growth rate on the velocity of gas bubbles.

The following dependence of growth rate on (100) faces of paratellurite single crystals, $V_{(110)}$, on W was observed experimentally at $\Delta T = 3°C$, $C_{HCl} = 4$ wt.%, $T_1 = 150°C$:

W, cm s^{-1}	18	19	20	22	23	25
$V_{(110)}$, day^{-1}	0.107	0.114	0.120	0.118	0.121	0.121

We find that in the initial stages of growth, the growth rate rises as a result of the increase in the velocity of gas bubbles with respect to the growing crystal; however, the situation is soon reached in which any further increase in bubble velocity cannot enhance the rate of crystal growth.

We have also measured the growth rate of (100) faces of TeO_2 single crystals for different values of internal temperature difference, for $C_{HCl} = 4$ wt.%.

ΔT, °C	1	2	3	4	5
$V_{(110)}$, mm day^{-1}	0.041	0.077	0.120	0.161	0.196

With other parameters constant ($T_1 = 150°C$, $W = 20$ cm s^{-1}), $V_{(110)}$ is practically a linear function of ΔT and can be approximated by the equation

$$V_{(110)} = k(\Delta T - \Delta T_{crit}),$$

where k is the proportionality coefficient, mm·°C^{-1}·day^{-1}; ΔT_{crit} is the critical temperature difference below which $V_{(110)} = 0$.

The data obtained for $\Delta T = 3°C$, $T_1 = 150°C$, $W = 20$ cm s^{-1} confirm the nonlinear dependence of growth rate on the concentration of HCl:

C_{HCl}, wt.%	1	2	3	4
$V_{(110)}$, mm day^{-1}	0.020	0.038	0.072	0.120

As in the case of mass transfer, this relation can be explained by a change in the mechanism of TeO$_2$ transfer as the HCl concentration increases, and consequently, by a change in the mechanism of surface reaction at the crystal-solution interface. Therefore, under the conditions of our experiments the growth rate of a face (hkl) is a function of a number of variables, that is,

$$V_{hkl} = f[(\Delta T - \Delta T_{crit}), W, C_{HCl}],$$

and is thus determined by the surface process kinetics and, probably, by adsorption of the solute at the growth surface with subsequent migration of particles to those parts of the growing crystal where they are incorporated into the lattice.

It is shown, therefore, that a hydrothermal experiment in a quartz reactor, allowing visual observation, allows a rapid correction of the growth mode (by varying temperature difference within the solution and hole diameter in the diaphragm and by selecting the concentration of the solvent, etc.). To a large extent this should facilitate the search for optimal conditions of mass transfer and crystal growth and accelerate very substantially this time-consuming process.

Autoclave for High-Rate Synthesis and Growth of Single Crystals

The process of hydrothermal synthesis and growth of single crystals is usually carried out with vertical temperature gradients of the order of 1°C cm^{-1}. This is not always sufficient to achieve a required supersaturation, so that the rate of synthesis and growth of a number of single crystals in practice remains rather low. We have designed and tested a special autoclave developed to accelerate the synthesis and growth of single crystals by hydrothermal techniques. The design of this autoclave (Fig. 4) is based on an idea suggested by N. N. Sheftal', namely, the use of a space of only 0.2-10 mm between the initial charge and the seed crystal. In this configuration the seed crystal is in direct contact with the external surface of the seal, while the charge dissolution zone and growth zone are separated with a special annulus 0.2-10 mm thick. The vertical temperature gradient is produced by cooling an internal cavity in the seal by a liquid or a gas coolant, circulated at a controlled rate. Direct contact between the cooled seal and the seed crystal minimizes the heat inflow to the growth zone, so that the vertical temperature gradient can be very substantially increased. The growth cycle duration is determined by the physical-chemical conditions of crystallization and by the composition of the single crystals to be grown.

A working model of this autoclave was used for experiments on quartz growth. An initial SiO$_2$ charge, shaped as a plate, was placed at the autoclave bottom; the separating ring was installed, and on this ring a quartz single-crystal seed (a disc cut normal to [0001]) was lowered. The working space between the charge and the seed crystal was filled with an aqueous solution of K$_2$CO$_3$ with concentration of 30 wt.%. The dissolution zone temperature was 450°C, with the vertical temperature gradient 40°C cm^{-1}; the growth duration was seven days. When the autoclave was opened the extracted single crystal was found to have grown in the direction [0001] at a rate of 1.65 mm day^{-1}.

This autoclave design can also be used for the synthesis and recrystallization of various inorganic compounds. In particular, we experimented with the recrystallization of Te and TeO$_2$, and with the synthesis of SbSI and SbSBr in various solvents under vertical

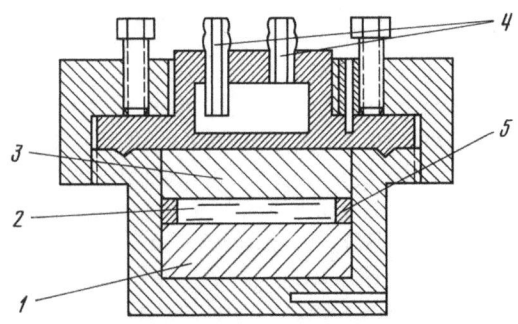

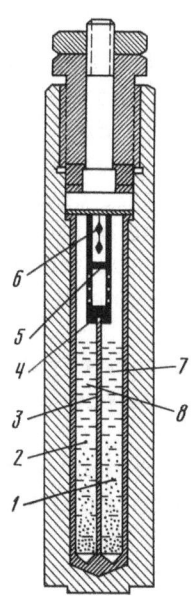

Fig. 4. Autoclave for high-rate synthesis and growth of single crystals. (1) charge; (2) solvent; (3) single crystal seed; (4) union nipple for coolant inflow; (5) separating ring.

Fig. 5. Schematic of the separation of initial charge components and solvents into isolated dissolution zones. (1, 2) dissolution zones; (3) vertical partition; (4) cup; (5) horizontal partition; (6) seed single crystals; (7, 8) solvents.

temperature gradients between 30 and 50°C cm^{-1}. These experiments demonstrated, similarly to those for crystal growth, certain attractive potentials of this autoclave design and of the investigated growth technique.

Separation of Initial Charge Components and Required Solvents into Isolated Dissolution Zones

One of the principal difficulties in hydrothermal synthesis of compounds in multicomponent systems (comprising simple elements, oxides, salts) is the unequal solubilities of the components and their chemical reactions with aqueous solutions of acids, salts and alkalis used as solvents. This often leads to the formation of a number of separate phases and therefore reduces the yield of the required single-crystal product.

Another problem which sometimes limits the synthesis and growth of inorganic compounds is the creation and maintenance of the redox potential of the hydrothermal medium. A change in this parameter may transform the initial components into a different valence state and thus modify the mechanism of synthesis and growth of the desired compound. We have developed and tested a new method of separating the initial components and solvents into independent dissolution zones, in order to optimize the synthesis and growth of single crystals. The technique is illustrated in Fig. 5. An autoclave with a volume of 500 cm^3 has a lining forming two independent dissolution zones 1, 2 separated by a vertical partition 3. A common zone of synthesis and growth, whose area is bounded by a cup 4 with holes, is formed above the partition. Depending on the purpose of the experiment, this cup may be of two types:

(a) with holes of different diameters covering its lateral surface;
(b) with holes covering only a part of the lateral surface of the cup; and a horizontal partition 5 separating the undisturbed and drilled parts of the cup; this partition serves to change the temperature gradient, that is, the supersaturation in the upper zone of the cup.

Seed crystals are suspended either in arbitrary parts of the cup (case (a)), or in its upper part above the partition (case (b)). If some specific compounds are to be synthesized, no seed crystals are used. The use of a cup prevents the solvents from mixing (if they are in a supercritical condition), and also helps in controlling the bulk diffusion of the dissolved components to the growth zone; that is, to control the growth rate of the crystals (by altering the number of holes in the cup). The horizontal partition in the cup provides an additional method of controlling the supersaturation in the upper growth zone. This method was used to grow pyro- and ferroelectric single crystals from the ABO_4 group ($A-Sb^{3+}$, Bi^{3+}; $B-Nb^{5+}$, Ta^{5+}, Sb^{5+}) and in particular, $SbSbO_4$, $SbNbO_4$, $SbTaO_4$ [2, 3], whose initial components A_2O_3 and B_2O_5 have unequal solubilities. A necessary condition of growing ABO_4-type single crystals on seeds is the selection of solvents which would ensure identical solubility of the oxides A_2O_3 and B_2O_5 in isolated zones, so that their dissolved species in which elements A and B are in the tri- and pentavalent states, respectively, are transferred to the growth zone at similar rates.

Departures from this last condition led either to oxidation or to reduction of A^{3+} and B^{5+} to other valences, thus disturbing the growth mechanism of the crystals.

The optimum conditions for the growth of such crystals were determined directly by physical-chemical analysis; it was found that their growth kinetics is determined by temperature, mass transfer rate of the species A_2O_3 and B_2O_5, dissolved in various solvents, to the common growth zone (aqueous solutions of KF were used for A_2O_3, and mixed aqueous solutions of KHF_2 and H_2O_2 for B_2O_5), the value of the Eh-pH of the medium, and temperature gradient.

The process of single crystal growth involves several stages:

— dissolution of the initial components A_2O_3 and B_2O_5 in different solvents within individual zones;
— synchronized mass transfer of the dissolved species A_2O_3 and B_2O_5 to the common growth zone;
— bulk diffusion of the dissolved species to the growth zone;
— seeded growth of single crystals using a temperature gradient along the autoclave axis (for example, growth of $SbSbO_4$ within a cup without a partition), or due to the additional gradient produced by the horizontal partition (growth of $SbNbO_4$).

The method described is efficient, simple to implement in actual autoclaves, and can be used for synthesis and crystal growth in multicomponent systems. This last problem can be solved by using three or more independent dissolution zones with one common zone for synthesis and growth.

Literature Cited

1. A. G. Kasatkin. Basic Processes and Equipment in Chemical Technology. Khimiya, Moscow (in Russian) (1960).
2. V. I. Popolitov, A. N. Lobachev, and M. N. Tseitlin. Hydrothermal crystallization of antimony orthoantimonate. In: Crystal Growth from High-Temperature Aqueous Solutions. Nauka, Moscow, 198-216 (in Russian) (1977).
3. V. I. Popolitov, V. F. Peskin, and A. N. Lobachev. On crystallization of antimony orthoniobate and orthotantalate. In: Crystal Growth from High-Temperature Aqueous Solutions. Nauka, Moscow, 217-227 (in Russian) (1977).

THE ROLE PLAYED BY Me^{2+} IN HYDROTHERMAL CRYSTALLIZATION OF GERMANATES OF DIVALENT METALS

L. N. Demianets, N. G. Duderov, and V. A. Kuznetsov

The Institute of Crystallography of the USSR Academy of Sciences, Moscow

T. N. Nadezhina

Moscow State University

Chemical reactions between oxides of divalent metals and germanium in high-temperature alkaline solutions produce a large number of germanates differing in their composition and crystal structure. The specific compounds formed are determined by the physical-chemical conditions of synthesis (temperature, pH of the crystallization medium, solvent concentration, etc.) and by crystal-chemical factors which are most clearly revealed by the effects of the metals in the reactant oxides on the processes of the crystallization of germanates. This aspect is discussed in the present paper. The analysis of the phase diagrams given below takes into account structural crystal-chemical characteristics of individual components and makes it possible to formulate basic relationships for the formation of germanates of divalent metals, to find the roles played by mono- and divalent cations in the formation of final products of crystallization, and to reveal the general features characteristic of germanate systems.

The experimental bases for the outlined analysis are the phase diagrams for crystallization in three-component systems MeO(M)-GeO$_2$(G)-Na$_2$O(S)-H$_2$O(H) (Me = Pb, Ba, Sr, Cd, Mn, Zn, Fe, Co and Ni). In systems containing GeO$_2$ and MeO where Me = Zn or Mn we used the data of refs. 1-3. The phase diagrams chosen for the analysis correspond to a constant temperature (500°C), with the type of the divalent metal oxide (M) and the ratios M/G, S/H being variable parameters. The interval of variation of the molar ratio M/G was (6:1)-(1:6), and the ratio S/H was varied in the range corresponding to the variation of NaOH from 0 to 50 wt.%. Crystalline reaction products were obtained by the method of direct temperature difference.

The diversity in the germanates of divalent metals stems from the possibility of varying widely the coordination of the cations (those of sodium, divalent ions, or germanium) with respect to oxygen. Germanium can enter the compounds with coordination numbers 6 or 4, depending on the crystallization conditions. Even in the simplest oxygen-containing compound, GeO$_2$, polymorphic modifications are possible with Ge coordination with respect to oxygen equal to 4 (hexagonal modification, α-quartz structure type) and to 6 (tetragonal modification, rutile structure type). In water the hexagonal modification is transformed to the tetragonal when a GeO$_2$(hex) + H$_2$O mixture is heated to 185°C. Above this temperature the tetragonal modification is stable. Consequently, with no mono- or divalent ions present, germanium is characterized by its maximum possible coordination 6. By introducing monovalent, divalent, or higher-valence cations into the simplest system GeO$_2$-H$_2$O we suppress the more "cationic" function of germanium (germanium with coordination number 6) and ultimately obtain compounds with Ge in four-coordination with oxygen ions [4-7]. This process is complete at different temperatures and different concentrations of mineralizers, depending on the cation type, its charge, and its concentration in the system, or it stops at an intermediate stage. Consider the basic factors which determine this process in systems which contain, as principal components, an oxide of a divalent metal and germanium oxide: the phase composition of the products of hydrothermal crystallizations in GeO_2- and MeO-containing systems in NaOH solutions, at the dissolution zone temperature 500°C, is presented in Table 1. Germanate compositions are written in the oxide form: M, G, S stand for oxides of a divalent metal, germanium, and monovalent metal, respectively, and H denotes water.

The paragenesis of crystal phases in alkaline aqueous solutions at 500°C can be described on the basis of data shown in paragenesis tetrahedra (see Fig. 1). The lines which connect the points corresponding to a given composition of crystal phases represent the paragenesis and the sequence of crystallization of the phases; the lines connecting specific compositions with the H_2O apex represent phases which crystallize in the absence of the mineralizer (NaOH); and the lines leading from the Na_2O apex of the tetrahedron reach the phases which are formed at the maximum concentration of alkali (50-60 wt.%).

Two very different types of four-component systems are established by a comparative analysis of crystallization phase diagrams, paragenesis tetrahedra, and composition and structure of the final crystal phases; the criterion is the radius of the divalent cation Me^{2+} (or its ratio to the germanium cation radius). The boundary between the two groups is the radius of the divalent cation $R \approx 1$ Å (Cd^{2+}). Systems with larger cations (Ba, Pb, Sr) are characterized by one type of phase diagram, and systems with smaller cations (Mn, Fe, Co, Zn, Ni) by another.

These two types of systems differ in the following features: the phase composition of the final crystallization products; the position of boundaries for mixed compounds (containing mono- and divalent cations); the germanium coordination in the crystallized compounds; the type and degree of polymerization of the Ge radical in the solid phases.

Systems Containing MeO Oxides with $R_{Me^{2+}} < 1$ Å

The formation of Me_2GeO_4-type germanates is typical in these systems. In Ni-containing systems this phase is the only compound of NiO with GeO_2 observed in the experimental range; in N vs. C_{NaOH} coordinates (where $N = NiO:GeO_2$ is the molar ratio, and C_{NaOH} is the mineralizer concentration) this phase occupies the whole range of values of N from 4:1 to 1:4 and C_{NaOH} from 0 to 50 wt.%. In the case of Co^{2+} and Fe^{2+}, whose radii are, respectively, 0.78 and 0.80 Å (after Belov and Bokii), we typically observe (along with Me_2GeO_4) double germanates of the type A_2MeGeO_4 (for example, Na_2CoGeO_4, Na_2FeGeO_4) containing an alkaline cation. These phases logically replace the alkali-free germanate Me_2GeO_4 as the mineralizer concentration increases. The larger the radius of the divalent cation, the lower is the NaOH concentration at which, in the general case, the transition to double germanates is observed; one exception is the Fe-containing system in which the double germanate is formed at a high concentration of NaOH [2, 3, 7]. In the case of the largest cations in this sequence (Zn, Mn), metastable crystallization of diorthogermanate $Na_2Zn_2Ge_2O_7$ and $Na_2Zn_3[GeO_4]_2$ for zinc is observed, and stable compounds $Na_2Mn_2Ge_2O_7$ and $Na_2MnGe_2O_6$ are found for manganese (at lower concentrations of solvent).

We see, therefore, that as $R_{Me^{2+}}$ increases, a successive transition is observed to the formation of double germanates and to increased stability of crystallization of more complicated germanates; the germanium radical also becomes more complicated: diorthogroups $[GeO_7]$ and island radicals (six-member rings composed of two diorthogroups $[Ge_2O_7]$ and two tetrahedra $[GeO_4]$) appear. More complex radicals in Mn-germanates appear because of specific properties of Mn-cations compared with the Zn-cation for which the structure is

TABLE 1. Germanates of Divalent Metals Synthesized in Systems $MeO-GeO_2-Na_2O-H_2O$

Compound	Symbol	Coordination number Ge	Coordination number Me	Ge-radical	Reference
		$PbO-GeO_2-Na_2O-H_2O$			
$Pb_3GeO_4(OH)_2$	M_3GH	4	—	$[GeO_4]$	*
Pb_3GeO_5	M_3G	4	6	$[GeO_4]$	[8]
$Pb_5Ge_3O_{11}$	M_5G_3	4	6	$[Ge_2O_7][GeO_4]$	[9]
$PbGeO_3$	MG	4	6(8)	$[Ge_3O_9]$	[10,11]
$NaPb_3Ge_3O_8OH$	SM_4G_6H	4	—	$[Ge_3O_9]_\infty$	*
$Pb_2Ge_7O_{16} \cdot 7H_2O$	$M_2G_7H_7$	4	—	$[Ge_4O_{16}][GeO_4]$	[12]
$PbGe_4O_9$	MG_4	6; 4	8	$[GeO_6][Ge_3O_9]$	[13]
		$BaO-GeO_2-Na_2O-H_2O$			
BaH_2GeO_4	MGH	4	9	$[GeO_4]$	[14]
$Ba_5[(Ge, C)(O, OH)_4]_3(OH)$	$M_5G_3H_2$	4	7; 9	$[GeO_4]$	[15]
$BaGeO_3 \cdot 0,15 H_2O$	$MGH_{0,15}$	4	—	—	[14]
$BaGeO_3$	MG	4	—	$[Ge_3O_9]$	[12]
$Ba_2Ge_7O_{16} \cdot 7 H_2O$	$M_2G_7H_7$	6; 4	—	$[Ge_4O_{16}][GeO_4]$	[12]
$BaGe_4O_9$	MG_4	6; 4	10	$[Ge_3O_9][GeO_6]$	[16]
$Ba_3Ge_9O_{20}(OH)_2$	M_3G_9H	6; 5; 4	8; 9	$[GeO_6][GeO_5][GeO_4]$	[17]
$Ba_8Ge_7O_{22} \cdot 10 H_2O$	$M_8G_7H_{10}$	—	—	—	[18]
$NaBa_3Ge_2[GeO_4]_2(OH)_6 \cdot [H_2O, OH]_2$	$SM_6G_8H_{10}$	6; 4	9(12), 7(10)	$[GeO_6][GeO_4]$	[18,19]
$Na_3Ba_6Ge_8O_{23} \cdot 4 H_2O$	$S_3M_6G_8H_4$	6; 4	—	$[GeO_6][GeO_4]$	[18]
		$SrO-GeO_2-Na_2O-H_2O$			
SrH_2GeO_4	MGH	4	—	$[GeO_4]$	[20]
$SrGeO_3$	MG	4	6(8)	$[Ge_3O_9]$	[12]
$Sr_5Ge_6O_{17} \cdot 1,3 H_2O$	$MG_5G_6H_{1,3}$	4	8; 6	$[Ge_3O_9]$	[21]
$SrGe_4O_9$	MG_4	6; 4	—	$[GeO_6][Ge_3O_9]$	[12]
$Na_4SrGe_3O_3[GeO_4]_3$	S_2MG_6	6; 4	9	$[GeO_6][GeO_4]$	[22]
		$CdO-GeO_2-Na_2O-H_2O$			
$Cd_9Ge_4O_{16}(OH)_2$	M_9G_2H	4	6	$[GeO_4]$	[23]
Cd_2GeO_4	M_2G	4	6	$[GeO_4]$	[24]
$CdGe_2O_5$	MG_2	4; 6	6(8)	$[GeO_6][GeO_4]$	[25,26]
$Cd_2Ge_2O_6$	MG	4	6	$[GeO_3]_\infty$	[27]
$NaCd_2Ge_3O_8(OH)$	SM_2G_6H	4	6	$[Ge_3O_9]_\infty$	[25]
		$MnO-GeO_2-Na_2O-H_2O$			
Mn_2GeO_4	M_2G	4	6	$[GeO_4]$	[28]
$Na_2Mn_7[GeO_3]_{10}$	SM_7G_{10}	—	—	—	[1]
$Na_2MnGe_6O_6$	SMG_2	4	6	$[Ge_6O_{18}]$	[29]
$Na_2Mn_2Ge_2O_7$	SM_2G_2	4	6; 4	$[Ge_2O_7]$	[30]
Na_2MnGeO_4	SMG	4	4	$[GeO_4]$	*
		$ZnO-GeO_2-Na_2O-H_2O$			
Zn_2GeO_4	M_2G	4	4	$[GeO_4]$	[31]
$Na_2Zn_3[GeO_4]_2$	SM_3G_2	4	4	$[GeO_4]$	[32]
$Na_2Zn_2Ge_2O_7$	SM_2G_2	4	5; 4	$[Ge_2O_7]$	[3]
Na_2ZnGeO_4	SMG	4	4	$[GeO_4]$	[33]
		$CoO-GeO_2-Na_2O-H_2O$			
Co_2GeO_4	M_2G	4	6	$[GeO_4]$	[28]
Na_2CoGeO_4	SMG	4	4	$[GeO_4]$	*

TABLE 1, Continued.

Compound	Symbol	Coordination number Ge	Coordination number Me	Ge-radical	Reference
FeO—GeO$_2$—Na$_2$O—H$_2$O					
Fe$_2$GeO$_4$	M$_2$G	4	6	[GeO$_4$]	[28]
Na$_2$FeGeO$_4$	SMG	4	4	[GeO$_4$]	*
NiO—GeO$_2$—Na$_2$O—H$_2$O					
Ni$_2$GeO$_4$	M$_2$G	4	6	[GeO$_4$]	[28]

*The form of the germanium radical was determined with infra-red spectroscopic data.

dictated by quantum chemistry relations (which demand four-coordination for zinc regardless of the ion size) and not by the closest packing principle (as for smaller cations).

TABLE 1 and Fig. 1e-h show that the absence of compounds with germanium in six-coordination or in two coordinations (6, 4) is a common feature for the group of systems with $R_{Me}2+ < 1$ Å. In all of these germanates, germanium is found only in a tetrahedral environment. Among the alkali-free germanates (Ni$_2$GeO$_4$, Co$_2$GeO$_4$, Fe$_2$GeO$_4$), the most frequent structure is spinel in which large oxygen atoms form a close-packed lattice, and the voids are filled with cations with appropriate radii: the small germanium ion (R = 0.44 Å) fits into the smaller tetrahedral sites, thus achieving a natural (from the standpoint of quantum chemistry) tetrahedral environment of oxygen ions. The larger Me ions occupy octahedral sites. Another structural type obtained under hydrothermal conditions, namely phenacite Zn$_2$GeO$_4$, is characterized by close packing of oxygen ions with tetrahedral sites occupied by zinc and germanium ions.

Finally, manganese orthogermanate Mn$_2$GeO$_4$ has the olivine structure typical for larger cations with $R_{Me}2+ < 1$ Å. The structure is composed of serrated [MeO$_6$] bands linked with tetrahedrally coordinated ions of germanium and divalent metal.

In double germanates of the A$_2$MeGeO$_4$ type germanium retains its positions in tetrahedral sites, and cations "give up" the larger sites to the larger sodium ions, moving to tetrahedrally coordinated sites according to their sizes. The tetrahedral coordination of Me^{2+} is also retained in more complicated germanates: Na$_2$Zn$_3$[GeO$_3$]$_2$, Na$_2$Zn$_2$Ge$_2$O$_7$, Na$_2$Mn$_2$Ge$_2$O$_7$. In the last of these compounds some of Mn^{2+} ions also occupy coordination polyhedra with coordination number 5.

In this group of systems, therefore, the degree of polymerization of Ge-radicals is quite low: no ribbon, laminated, or skeleton compounds are formed.

Systems containing MeO Oxides with $R_{Me}2+ > 1$ Å

A wide variety of alkali-free germanates with very different values of ratio M/G is typical of systems containing oxides of the larger divalent metals (see Fig. 1a-c): 3 for Pb$_3$GeO$_5$, Pb$_3$GeO$_4$(OH)$_2$; 1.7 — Ba$_5$[(Ge, C)(O, OH)$_4$]OH; 1.66 — Pb$_5$Ge$_3$O$_{11}$; 1 — PbGeO$_3$, BaGeO$_3$, SrGeO$_3$; 0.8 — Sr$_5$Ge$_6$O$_{17}$·1.3H$_2$O; 0.33 — Ba$_3$Ge$_9$O$_{21}$; 0.3 — Pb$_3$Ge$_7$O$_{16}$·7H$_2$O; 0.25 — SrGe$_4$O$_9$, BaGe$_4$O$_9$, PbGe$_4$O$_9$.

As in most systems with smaller cations, double germanates are formed in all of these systems. However, one essential feature of the phase diagrams for larger cations is that the regions of crystallization of double germanates are associated with low- and medium-concentration solutions of the mineralizer. This means concentrations of the order of 5-15 and 12 wt.% for Sr- and Ba-containing systems, respectively, and a very wide field of crystallization of double germanate from 10 to 40 wt.% alkali concentration for the Pb-containing system. Sr- and Ba-containing systems are typically formed in regions with a substantial excess of the Ge-component (M:G $\approx$ 1:6), while for Pb this excess is slightly lower (M:G in the range from 1:2 to 1:5).

In the ranges of low and medium concentrations of alkali and high concentration of GeO_2, double germanates "invade" the fields in which alkali-free germanates crystallize in the regions of low and medium concentration of alkali and high concentration of GeO_2. This sequence of phase alternation cannot be understood in terms of the physical chemistry of the phases, and lies in the crystal chemistry of these compounds. Here, as in the case above, we use the principle of closest packing of large atoms, illustrating it with the clearest example of the Sr-containing system. All compounds within this system are governed by the closest-packing principle. The Goldschmidt rule states that the closest packing is slightly strained, without essentially deviating from the packing of oxygen ions, by departures from the critical cation-to-anion radius ratio. Strontium ions in such structures may reveal a duality of behavior: Ba-like (Ba: atomic number Z = 56 and $R_{Ba}^{2+} = 1.38$ Å), since Ba often acts as a geometrically similar partner in the close-packed lattices, and Ca-like (Ca: Z = 20 and $R_{Ca}^{2+} = 1.04$ Å), coordination number 6 being fairly common for Ca. As a result, in the close-packed lattice Sr atoms occupy octahedral sites. The mean spacing between two oxygen atoms in Sr-polyhedra, for a typical packing parameter 2.7, is nearly equal to the edge of the Ge-octahedron, so that joint crystal structures become possible. If Ge occupies octahedral sites (this is possible if the Ge concentration in the solution is high), strontium ions move into the lattice at the expense of available vacancies. Easily polarized sodium ions, with possible coordination number from 4 to 8, fill smaller voids of the strained close-packed lattice and form very specific structures. As the content of SrO in the system and the alkalinity of the solution are increased, the solubility of SrO is greatly enhanced. As a result, large cationic complexes, forming massive cationic blocks in the solid phase, can appear in the crystal-forming medium, with only tetrahedrally coordinated Ge being able to occupy the voids in these blocks.

Crystal chemical arguments thus make it possible to explain the location of crystallization regions associated with double germanates in systems with large divalent cations.

Another important distinction of the systems discussed compared with systems with smaller cations is the coordination of Ge in the crystal phases. Even in alkali-free germanates both phases with the tetrahedral and phases with the double coordination of Ge with respect to oxygen (6 and 4) are formed. Thus, benitoite-type germanates are typical for Sr- and Pb-containing systems ($SrGe[Ge_3O_9]$ and $PbGe[Ge_3O_9]$), with Ge occupying octahedral and tetrahedral coordination polyhedra and forming from them a mixed skeleton. In barium germanate $Ba_3Ge_9O_{21} \cdot H_2O$ Ge occupies polyhedra with coordination numbers 6, 5, and 4.

Systems containing oxides with larger divalent cations typically form germanates with a higher degree of polymerization of germanium-oxygen radicals. Ge may form rings ($[Ge_4O_{16}]$ in $Pb_2Ge_7O_{16} \cdot 7H_2O$ and $BaGe_4O_9$) or chains ($[Ge_3O_8OH]_\infty$ in $NaPb_2[Ge_3O_8OH]$, and $[GeO_3]_\infty$ in $BaGeO_3$ and in strontium germanate in phase K [21]).

A boundary situation of a system possessing the features of both above-discussed types is that of the Cd-containing system (see Fig. 1d). As for systems typical of smaller cations, it exhibits the formation of orthogermanate Cd_2GeO_4 (spinel structure type), and has no compounds with Ge in coordinations 4 and 6, or 6. On the other hand, the formation

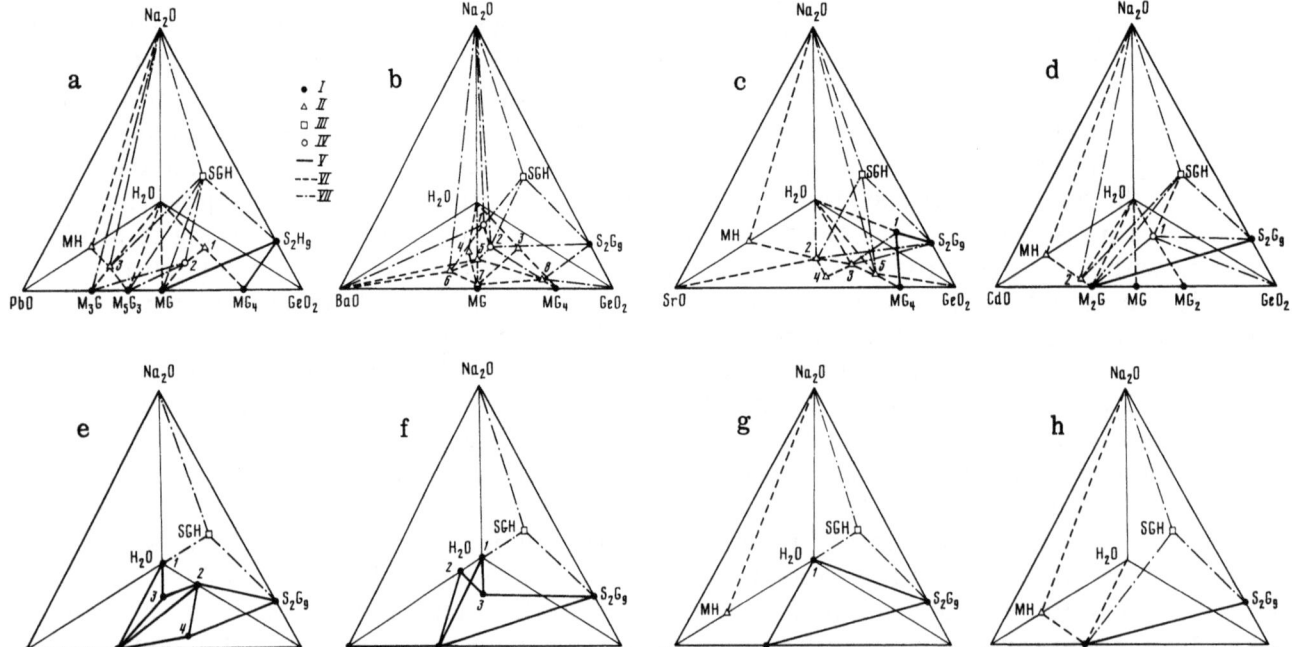

Fig. 1. Paragenesis tetrahedra in MeO-GeO$_2$-Na$_2$O-H$_2$O systems for the crystallization temperature 500°C. I. points on the surface of triangle MeO-GeO$_2$-Na$_2$O; II. points on the surface of triangle MeO-GeO$_2$-H$_2$O; III. points on the surface of triangle Na$_2$O-GeO$_2$-H$_2$O; IV. points inside the tetrahedron; V. paragenesis lines on the surface of triangle MeO-GeO$_2$-Na$_2$O; VI. paragenesis lines on the surface of triangles MeO-GeO$_2$-H$_2$O and Na$_2$O-GeO$_2$-H$_2$O; VII. paragenesis lines within the tetrahedron. [a] PbO: (1) M$_2$G$_7$H$_7$, (2) SM$_4$G$_6$H, (3) M$_3$GH; [b] BaO: (1) SM$_6$G$_8$H$_{10}$, (2) S$_3$M$_6$G$_8$H$_4$, (3) M$_2$G$_7$H$_7$, (4) M$_8$G$_7$H$_{10}$, (5) MGH, (6) M$_5$G$_3$H$_2$, (7) MGH$_{0.15}$, (8) M$_3$G$_9$H; [c] SrO: (1) S$_2$MG$_6$, (2) MGH, (3) phase L, (4) M$_5$G$_6$H$_{1.3}$, (5) MG$_3$H; [d] CdO: (1) SMG$_6$H, (2) M$_9$G$_4$H; [e] MnO: (1) SMG (2) SMG$_2$, (3) SM$_2$G$_2$, (4) SM$_7$G$_{10}$; [f] ZnO: (1) SMG, (2) S$_2$M$_3$G$_2$; [g] FeO, CoO: (1) SMG; [h] NiO.

of more complicated alkali-free germanates (e.g., $Cd_9Ge_4O_{16}(OH)_2$), of double germanates in regions with low solvent concentrations, of polymerized Ge-radicals (chain $[Ge_3O_8OH]_\infty$ in $NaCd_2Ge_3O_8OH$) are features which are typical for systems with large divalent cations.

Conclusions

Divalent cations play a decisive role in the synthesis of germanates in highly alkaline hydrothermal media. The chemical and crystal chemical characteristics of divalent metal ions are among the most important parameters determining the type of the phase diagram.

In accordance with the steric factor, two basic types of crystallization diagrams can be distinguished in four-component systems $MeO-GeO_2-Na_2O-H_2O$ with $R_{Me^{2+}} > 1$ Å and $R_{Me^{2+}} < 1$ Å; these diagrams differ in the composition of the crystallization products, the position of phase boundaries, the germanium coordination, and the form and degree of polymerization of the Ge-radical.

Literature Cited

1. I. M. Ismail-Zade and B. N. Litvin. Hydrothermal synthesis of single crystals of Na-Mn silicates and germanates. Azerb. Khim. Zhurn., 6, 135-140 (in Russian) (1969).
2. I. P. Kuz'mina. Hydrothermal synthesis of Na(K) germanates and zinc germanates. Kristallografiya, 13, 854-858 (in Russian) (1968).
3. I. P. Kuz'mina, B. N. Litvin, and V. S. Kurazhkovskaya. Hydrothermal synthesis of germanates. In: Processes of Crystallization in Hydrothermal Growth. Nauka, Moscow, 164-186 (in Russian) (1970).
4. A. N. Lobachev, I. P. Kuzmina, L. N. Demianets, V. V. Iljuchin, and N. V. Belov. Über der Rolle der Alkalimetal beim Aufbau der Silikate und Germanate under hydrothermallen Bedingungen. Krist. und Tech., 5, 425-432 (1970).
5. L. N. Demianets, V. V. Il'yukhin, I. P. Kuz'mina, A. N. Lobachev, and N. V. Belov. On the role of alkali metals and germanium in hydrothermal crystallization of germanates. In: Growth of Crystals, 9, ed. N. N. Sheftal' and E. T. Givargizov, Consultants Bureau, New York (1975).
6. L. N. Demianets and A. N. Lobachev. Synthesis of germanates in high-temperature alkaline solutions. In: Crystal Growth in High-Temperature Aqueous Solutions, Nauka, Moscow, 88-101 (in Russian) (1977).
7. N. G. Duderov, L. N. Demianets, and A. N. Lobachev. Hydrothermal crystallization of germanates of transition metals. In: Crystal Growth in High Temperature Aqueous Solutions, Nauka, Moscow, 102-119 (in Russian) (1977).
8. R. R. Neurgonkar, R. W. Wolfe, and R. E. Newnham. Crystal data and crystal growth of ferroelastic Pb_3GeO_5. J. Appl. Cryst., 7, 307-309 (1974).
9. R. E. Newnham, R. W. Wolfe, and C. N. W. Darlington. Prototype structure of $Pb_5Ge_3O_{11}$. J. Solid State Chem., 6, 378-383 (1973).
10. K. Sugii, H. Iwasaki, and S. Mayasawa. Crystal growth and optical properties of lead germanium oxide single crystals. J. Cryst. Growth, 10, 127-132 (1971).
11. A. N. Lazarev and V. A. Kolesova. IR absorption spectra of lead metasilicate and lead metagermanate. Izv. AN SSSR, Neorg. Mater, 6, 1445-1449 (in Russian) (1970).
12. G. Eulenberger, A. Wittman, and H. Nowotny. Germanate mit zweiwertigen Metallionen. Monatsh. Chem., 93, 1046-1054 (1962).
13. C. R. Robbins, A. Perloff, and S. Block. Crystal structure of $BaGe[Ge_3O_9]$ and its relation to benitoite. J. Res. Nat. Bur. Stand., A70, 385 (1966).

14. V. I. Ponomarev, O. S. Filipenko, E. A. Pobedimskaya, and N. V. Belov. Crystal structure of $BeGeO_3 \cdot H_2O$. Dokl. AN SSSR, 215, 584-587 (in Russian) (1974).
15. Yu. A. Malinovsky, E. A. Pobedimskaya, and N. V. Belov. Germanium carbonate-apatite $Ba_5[(Ge, C)(O, OH)_4]_3(OH)$. Synthesis and crystal structure. Kristallografiya 20, 644-646 (in Russian) (1975).
16. Yu. I. Smolin. Crystal structure of barium tetragermanate. Dokl. AN SSSR, 181, 595-598 (in Russian) (1968).
17. Yu. A. Malinovsky, E. A. Pobedimskaya, and N. A. Belov. Crystal structure of $Ba_2Ge_9O_{20}(OH)_2$. Dokl. AN SSSR, 227, 1350-1353 (in Russian) (1976).
18. Yu. A. Malinovsky, V. A. Kuznetsov, E. A. Pobedimskaya, and N. V. Belov. Hydrothermal synthesis of barium silicates and germanates. In: Crystal Growth in High-Temperature Aqueous Solutions, Nauka, Moscow, 136-157 (in Russian) (1977).
19. Yu. A. Malinovsky. Synthesis and crystallographic analysis of a group of natural and synthesized barium silicates and germanates. Thesis, Moscow State University (1976).
20. H. Nowotny and G. Szekely. Über einige Germanate. — Monatsh. Chem., 83, 568-572 (1952).
21. T. N. Madezhina, E. A. Pobedimskaya, V. V. Ilyukhin, N. N. Kinishina, and N. V. Belov. Crystal structure of the second modification of strontium metagermanate. Dokl. AN SSSR, 233, 1086-1089 (in Russian) (1977).
22. T. N. Nedezhina, E. A. Pobedimskaya, and N. V. Belov. Crystal structure of alkaline strontium germanate $Na_4SrGe_3[GeO_4]_3O_3$. Kristallografiya, 19, 867-869 (in Russian) (1974).
23. E. L. Belokoneva, M. A. Simonov, N. G. Duberov, A. N. Lobachev, and N. V. Belov. Group of synthetic cadmium germanates and their X-ray analysis. Kristallografiya, 18, 973-977 (in Russian) (1973).
24. E. L. Belokoneva, Yu. A. Ivanov, M. A. Simonov, and N. V. Belov. Crystal structure of cadmium orthogermanate Cd_2GeO_4. Kristallografiya, 17, 217-219 (in Russian) (1972).
25. M. A. Simonov, E. L. Belokoneva, Yu. K. Yegorov-Tismenko, and N. V. Belov. Crystallochemical features of the structures of the cadmium group of germanates. Vestn. MGU, Ser. Geol. No. 3, 41-49 (in Russian) (1977).
26. E. L. Belokoneva, A. V. Arakcheeva, M. A. Simonov, and N. V. Belov. Crystal structure of $CdGe_2O_5$. Kristallografiya, 21, 303-306 (in Russian) (1976).
27. A. N. Kornev, L. N. Demianets, B. A. Maksimov, V. V. Ilyukhin, and N. V. Belov. Hydrothermal synthesis in the $KCl-CdO-GeO_2-H_2O$ system and crystal structure of $Cd_2[Ge_2O_6]$. Kristallografiya, 17, 289-291 (in Russian) (1972).
28. A. Novrotsky. Thermodynamic relations among olivine, spinel and phenacite structures in silicates and germanates: I. Volume relations in the system $NiO-MgO-GeO_2$ and $CoO-MgO-GeO_2$. J. Solid State Chem., 6, 21-41 (1973).
29. L. P. Otroshchenko and V. I. Simonov. Crystal structure of synthetic sodium-manganese metasilicate $Na_5(Mn, Na)_3Mn[Si_6O_{18}]$. Dokl. AN SSSR, 208, 845-848 (in Russian) (1973).
30. L. P. Astakhova and V. I. Simonov. Determination of crystal structure of $Na_2Mn_2Si_2O_7$ by the method of superposition. Kristallografiya, 14, 3-8 (in Russian) (1969).
31. Ch'in Hang, M. A. Simonov, and N. V. Belov. Crystal structure of willemite $Zn_2[SiO_4]$ and its germanium analog $Zn_2[GeO_4]$. Kristallografiya, 15, 457-460 (in Russian) (1970).
32. G. F. Plakhov, M. A. Simonov, and N. V. Belov. Crystal structure of zinc orthosilicate $Na_2Zn_3[SiO_4]_2$, phase F. Kristallografiya, 20, 46-51 (in Russian) (1975).
33. E. A. Kuz'min, V. V. Ilyukhin, and N. V. Belov. Crystal structure of synthetic Na-Zn germanate Na_2ZnGeO_4. Kristallografiya, 6, 978-979 (in Russian) (1968).

Part VII
Defect Structure in Crystals: Relation to Growth Conditions

THEORETICAL AND EXPERIMENTAL STUDIES OF GENERATION OF STRESS AND DISLOCATIONS IN GROWING CRYSTALS

V. L. Indenbom

The Institute of Crystallography of the USSR Academy of Sciences, Moscow

V. B. Osvensky

The Institute of Rare Metals, Moscow

The application of the theory of dislocations and the theory of internal stress to the problem of growing perfect crystals and controlling the dislocation structure of the growing crystal has led to significant advances which have affected the progress in solid state physics as a whole and in its technical applications. The experience accumulated to date enables one to draw some general conclusions on the role played by dislocations and internal stress in crystal growth, including certain aspects of the technology of growing crystals with prescribed structural and physical parameters.

The present paper is mostly based on review lectures presented by V. L. Indenbom to the 2nd International School on Crystal Growth (Varna, 1975) and on the recent results of the studies of semiconductor crystals grown by different growth techniques.

Ignoring specific features of different types of crystals, one can roughly evaluate the levels of stress corresponding to the basic effects of stress in the growth of crystals. In terms of strain ε, that is, normal stress σ divided by Young's modulus E, or tangential stress τ by the shear modulus G, we can say that at the level of $\varepsilon \approx 10^{-3}$ stresses may fracture the crystal spontaneously or during mechanical, thermal, or chemical treatment. At the level of 10^{-4}, stresses may cause plastic flow. The level of 10^{-5} is typical for the growth of dislocation-free crystals. Of course, the indicated levels vary in specific cases, but not by one or two orders of magnitude, as is often claimed in the literature.

The most important sources of stress in growing crystals are a nonuniform temperature distribution causing thermoelastic stress, as well as dislocations and impurity inhomogeneities causing residual stress.

Calculation of the thermoelastic stress provides a foundation for studying stress and dislocation generation in crystal growth and is of independent value when a safe mode of crystal cooling is sought. Assuming a typical mean value of the thermal expansion coefficient $\alpha = 5 \cdot 10^{-6}$ °C^{-1}, one can rephrase the above-given stress levels in terms of an equivalent temperature difference δT. A crystal may fracture at $\delta T \approx 200$°C, plastic flow sets in at $\delta T \approx 20$°C, and dislocation-free crystals can be grown if $\delta T \lesssim 2$°C. In practical situations the problem of maximum importance is to prevent formation of excessive thermoelastic stresses and strains $\varepsilon = \alpha \delta T$. For example, it was found feasible to maintain the thermoelastic strain in semiconductor growth at a level smaller than the thermal strain αT by three orders of magnitude.

In the case of a general arbitrary temperature distribution and arbitrary shape of crystals the thermoelastic stress field σ_{ij} is found by solving a system of equations

$$\sigma_{ij} = C_{ijkl}\varepsilon_{kl}; \tag{1}$$

$$e_{ikl}e_{jmn}\frac{\partial^2}{\partial x_k \partial x_m}(\varepsilon_{ln} + \alpha_{ln}T) = 0; \tag{2}$$

$$\frac{\partial}{\partial x_j}\sigma_{ij} = 0 \tag{3}$$

with boundary conditions at the free surface of the crystal

$$n_j \sigma_{ij} = 0. \tag{4}$$

Here C_{ijkl} is the elastic moduli tensor; ε_{kl} denotes thermoelastic strain; $\alpha_{ln}T$ represents temperature strain; e_{ijk} is a completely antisymmetric unit tensor of rank 3; n_j is the normal to the crystal surface. Equation (1) expresses Hooke's law, equation (2) corresponds to the Saint-Venant compatibility conditions, and equations (3) and (4) to equilibrium conditions.

For isotropic cylindrical crystals of radius R equilibrium equations in cylindrical coordinates transform to

$$\frac{1}{r}\frac{\partial}{\partial r}(r\sigma_r) + \frac{\partial}{\partial z}\sigma_{rz} = \frac{1}{r}\sigma_\varphi; \quad \frac{1}{r}\frac{\partial}{\partial r}(r\sigma_{rz}) + \frac{\partial}{\partial z}\sigma_z = 0, \tag{5}$$

where the components of the elastic stress tensor are related to strains by Hooke's law:

$$\sigma_r = Ga\left[\frac{\partial U}{\partial r} + K\left(\frac{U}{r} + \frac{\partial W}{\partial z}\right) - cT\right]; \quad \sigma_\varphi = Ga\left[\frac{U}{r} + K\left(\frac{\partial U}{\partial r} + \frac{\partial W}{\partial z}\right) - cT\right];$$

$$\sigma_z = Ga\left[\frac{\partial W}{\partial z} + K\left(\frac{U}{r} + \frac{\partial U}{\partial r}\right) - cT\right]; \quad \sigma_{rz} = Ga\left[\frac{\partial U}{\partial z} + \frac{\partial W}{\partial r}\right], \tag{6}$$

where $a = 2(1-\mu)/(1-2\mu)$; $K = \mu/(1-\mu)$; $c = \alpha(1+\mu)/(1-\mu)$; G is the shear modulus; μ is Poisson's ratio; T is the known temperature distribution which in the case of crystal growth is found by numerical solution of the Stefan problem; and U and W are the radial and axial displacements. The boundary conditions are fixed under the assumption that the surfaces, including the interface, are free of external forces, and symmetry conditions hold along the axis:

$$\sigma_r = \sigma_{rz} = 0 \quad \text{for} \quad r = R;$$
$$\sigma_z = \sigma_{rz} = 0 \quad \text{for} \quad z = 0, l; \tag{7}$$
$$U = 0, \frac{\partial W}{\partial r} = 0 \quad \text{for} \quad r = 0.$$

In the general form for an axisymmetrical temperature field T(r, z), problem (5) - (7) can be solved numerically by computer [1, 2].

In some practically important cases, in which the system of thermoelasticity equations cannot be solved rigorously, a semiqualitative analysis of thermoelastic fields is carried out, based on finding effective temperature differences δT which, in fact, determine the level of thermoelastic strains and stresses ($\sigma \approx E\alpha\delta T$). In many cases a semiqualitative estimate can be obtained from the curvature of the thermal field [3]:

$$\sigma = \alpha E L^2 \frac{\partial^2 T}{\partial z^2},$$

where L is a characteristic length which usually approximates to 0.2 of the crystal diameter.

In the simplest cases δT can be evaluated directly from the profiles of the temperature field. For example, in the case of plates and cylinders grown on the end face, stresses at this end face are practically determined by the temperature field at a distance of about one half of the sample diameter D. If the field is linear over this distance, stresses are not generated, and if the temperature profile is appreciably curved, then stress is proportional to this curvature and changes sign together with the curvature of T(z). The corresponding method of evaluating the temperature difference δT, which determines the level of thermoelastic stress, is shown in Fig. 1.

In the case of a wafer, in which temperature varies only with depth, and distribution T(z) is a symmetrical parabola (Fig. 2a), the maximum magnitude of stress, twice as large as the stress in the half-thickness plane, is reached at the wafer surface, and the effective difference δT is equal to 2/3 of the total temperature difference between the surface and the half-thickness plane. For the case of a radial parabolic distribution of T in a cylinder, δT comes to one half of the maximum temperature difference between the axis and the cylinder surface.

It should be emphasized that even large temperature differences induce no thermoelastic stress if the temperature field is linear in coordinates. This effect is known as the free thermal bending (FTB). As an example, consider Fig. 2b in which the wafer is again

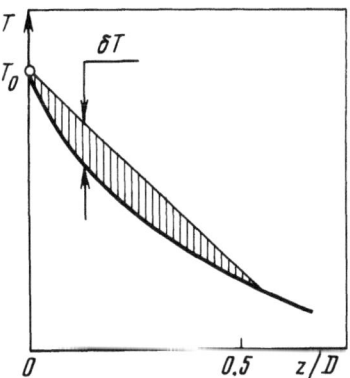

Fig. 1. Schematic method of determining effective temperature difference δT for a plate or a cylinder growing on the end face. T_0 marks the crystallization temperature.

Fig. 2. Effective temperature differences δT determining thermoelastic stress in (a) symmetrical and (b) nonsymmetrical temperature distribution over the wafer thickness. OO' marks the level from which temperature drops are measured.

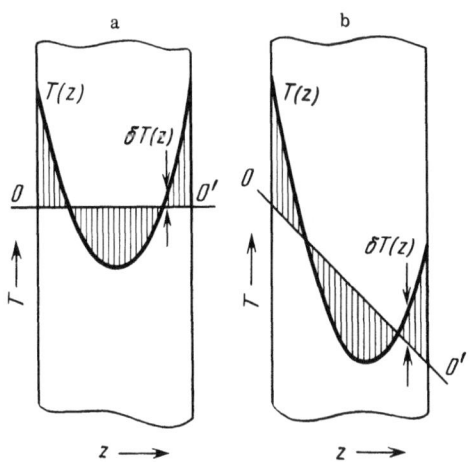

described by a parabolic distribution of T but with a maximum not at the half-thickness plane. Although the temperature difference has increased for the same curvature of the parabola, the stress field remains unaltered; indeed, the compared temperature distributions differ by a linear temperature field inducing no stress (see Fig. 2). The FTB effect is extremely important in all cases when thermoelastic stress is generated. In fact, the fundamental problem of lowering thermoelastic stresses to the level ensuring safe crystal growth is always solved by creating the FTB condition. Note that in platelet crystals FTB is achieved for any temperature field $T(x, y)$ satisfying the Laplace equation $\Delta T(x, y) = 0$.

In complicated problems, for example, in growing large low-dislocation crystals when the level of thermoelastic strain ε must be reduced substantially compared with the level of the thermal strain ε^T, semiqualitative methods of analyzing thermoelastic fields are not valid. In these cases it is necessary to find exactly the temperature field in the growing ingot (Stefan problem) and to solve the system of thermoelasticity equations (1) - (4) (or (5) - (7)) numerically. This problem has been solved, for example, for the growth of semiconductor single crystals from the melt [2, 4].

Calculation of thermoelastic stresses in silicon, germanium, and gallium arsenide show that in actual conditions of Czochralski growth the stress in semiconductor crystals usually varies from tens to hundreds of grams per square millimeter, which corresponds to $\sigma/E \approx 10^{-5}$-10^{-4}. In the conditions of a steady-state temperature field

$$\Delta T = \frac{1}{r}\frac{\partial}{\partial r}\left(r\frac{\partial T}{\partial r}\right) + \frac{\partial^2 T}{\partial z^2} = 0.$$

The distribution and level of thermoelastic stress (τ_3) along the axis of the growing crystal correlates with the corresponding distribution of radial temperature gradient at the ingot surface $(\partial T/\partial r)_s$ and curvature of the axial temperature distribution $(\partial^2 T/\partial z^2)_s$, while there is no unambiguous relation between the stress and the axial temperature gradient. This is clear, for example, for a gallium arsenide crystal grown by the flux encapsulation Czochralski technique (Fig. 3). The observed change in the sign of $\partial T/\partial r$, that is, in the direction of the radial heat flux, which in this case is associated with the conditions of encapsulation growth, changes the sign of the curvature of distribution $T(z)$ and that of the thermoelastic stress profile ($-\partial T/\partial r$ corresponds to the radial heat flux away from the ingot axis toward the lateral surface). As the radial temperature gradient $(\partial T/\partial r)_s$ characterizes the specific heat flux from the ingot surface, q [5] ($q = -\lambda(\partial T/\partial r)$, where λ is the thermal conductivity), one can say that the magnitude and distribution of thermoelastic stress in a growing boule are determined by the distribution of q, that is, by the characteristics of heat transfer between the crystal and the ambient. The distribution curve q along the ingot axis in the Czochralski growth is always characterized by one or several well-pronounced maxima [6]. Consequently, all practicable methods of lowering the level of thermal stress in fact reduce to appropriately effecting heat transfer at the lateral surface of the growing ingot in order to bring down the maximum and shift it to the area of lower temperatures where the crystal is less susceptible to plastic strain. The desirable limiting situation is a thermal insulation of the lateral ingot surface and linear axial distribution of temperature, with the ensuing FTB effect. In actual conditions of growth, the stresses in cross sections of the growing ingot reach a maximum at the surface (with the exception of regions adjacent to the crystallization interface) and grow as the crystal diameter increases.

In ingots growing under the conditions of a steady-state temperature field, when $\Delta T = 0$, the profiles of thermoelastic stress are frequently characterized by opposite signs of axial stress σ_z and tangential stress σ_φ in the central and peripheral parts of the ingot (positive stress corresponds to extension, and negative to compression). Such stress profiles (Fig. 4) are sharply different from well-known profiles [7] which are typical of nonstationary temperature fields for $\Delta T \neq 0$, and can be compared with similar profiles for

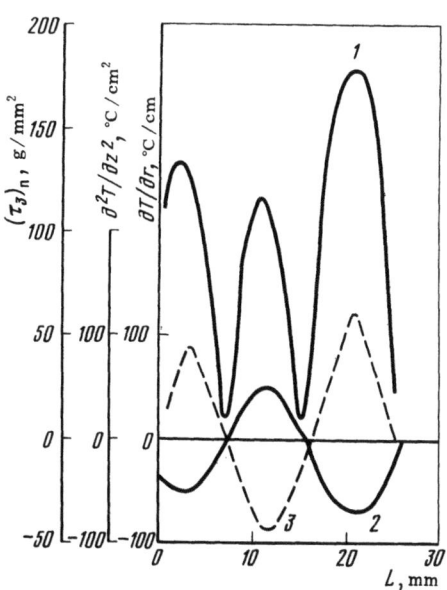

Fig. 3. Distribution of normalized thermoelastic shear stress (1) and radial temperature gradient (2) at the lateral surface of the boule, and curvature of the axial temperature distribution (3) along a gallium arsenide crystal grown by the Czochralski liquid encapsulation technique (B_2O_3 flux).

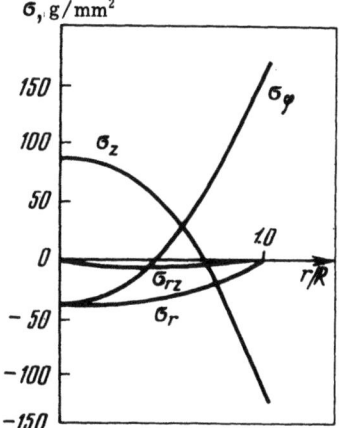

Fig. 4. Thermoelastic stress profile in a silicon crystal grown by the Czochralski technique.

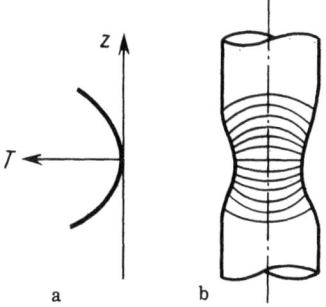

Fig. 5. (a) Effective axial temperature distribution in a boule with the linear temperature field subtracted; (b) the corresponding "barrel-like" strain (thin lines schematically trace bent lattice planes).

residual stress [3]. The distribution and sign of σ_φ (see Fig. 4) can be related to the direction of the radial heat transfer flux corresponding to cooling the crystal at its surface. The following simplified scheme (Fig. 5) is helpful for a qualitative analysis of σ_z distribution in the ingot cross section. If the temperature distribution within a part of the crystal can be written, after subtracting the linear temperature field inducing no stress, in the form (see Fig. 5a).

$$T = \frac{\partial^2 T}{\partial z^2}\left(\frac{z^2}{2} - \frac{r^2}{4}\right), \tag{8}$$

then the ingot undergoes a "barrel-like" strain (see Fig. 5b), which corresponds, for $(\partial^2 T/\partial z^2) > 0$, to axial extension at the center and to axial compression at the lateral surface. Figure 4 represents precisely this type of change in σ_z.

A qualitative calculation of thermoelastic stress in field (8) sufficiently far from the ingot end face yields

$$\sigma_z = \alpha G \frac{\partial^2 T}{\partial z^2}\left(\frac{R^2}{2} - r^2\right);$$
$$\sigma_\varphi = \alpha G \frac{\partial^2 T}{\partial z^2}\left(\frac{3r^2 - R^2}{4}\right); \qquad (9)$$
$$\sigma_r = \alpha G \frac{\partial^2 T}{\partial z^2}\left(\frac{r^2 - R^2}{4}\right),$$

that is $\sigma_z = -(\sigma_\varphi - \sigma_r)$. In the general case a steady-state temperature field can be given as the Fourier-Bessel expansion. Smoothly varying harmonics then give a stress distribution of type (9), but, as the wave number increases, the ratio of the components of stress tensor becomes more complicated and even the sign of the stress may change.

It must be noted that usually the signs of σ_z and σ_φ are identical for the growth of dislocation-free and low-dislocation semiconductor crystals, while these signs are opposite in crystals with a relatively high dislocation density. The dislocation structure of the ingot thus depends not only on the magnitudes but also on the signs of the stress components. This is understandable because plastic strain is caused by shear stress found from the difference between the corresponding principal stresses.

The thermoelastic stress profile in actual conditions of growing large crystals from the melt is thus very complicated and cannot be analyzed reliably on the basis of only simplified estimates of the temperature field.

If crystals are grown under large temperature differences, the main type of residual stress is the thermoplastic stress due to dislocations which are generated in the process of plastic strain caused by thermoelastic stress. Thermoplastic stresses are equal in magnitude (with the sign reversed) to thermoelastic stresses which relax at elevated temperatures. If this relaxation were complete, the residual stress σ_0 can be estimated by the formulas for thermoelastic stress, with the signs reversed.

Two situations can be distinguished in an analysis of the generation of residual stress [8]: (1) "rigid" crystals (like corundum) in which the plasticity zone where the stress relaxes completely is so narrow that residual stress is actually frozen in at the growth interface; (2) plastic crystals characterized by a wide plasticity zone adjacent to the crystallization interface. It has been shown [8-11] that in the first case the abrupt interface model is a good approximation; this appreciably simplifies the calculation of internal stress and thus allows the use of the standard methods of the theory of thermoelasticity. The plastic properties of the crystal are then taken into account as boundary conditions at the crystallization interface. The method of reducing the residual strain and stress problem to the standard thermoelasticity problem is based on a simple rule: the time rates of variation of the total (transient and residual) stresses are numerically equal, in the case of a steady-state crystallization mode, to thermoelastic stresses which would be produced by a virtual temperature field numerically equal to the time rates of change of the original temperature field:

$$\frac{\partial}{\partial t}\sigma_{ij}\{T(t)\} \equiv \sigma_{ij}\left\{\frac{\partial T}{\partial t}(t)\right\}.$$

In the practically important cases (for example, growth of cylindrical and platelet crystals at their end faces), calculations can only be numerical. In the usual system of thermoelasticity equations displacements are replaced with the corresponding components of the vector of displacement velocity, and boundary conditions are set for derivatives of stress with respect to time. By solving this system one obtains the distribution of displacement velocities $\dot{U}$ and $\dot{W}$. The appropriate formulas then yield time rates of changes for stress $\dot{\sigma}_{ij}$, after which integration yields total stress at each moment of time t (or, which is equivalent, at each distance from the interface):

$$\sigma_{ij}(t) = \int_{t_0}^{t} \dot{\sigma}_{ij}(t') \, dt',$$

where t' is the variable of integration over time, and residual stresses are found in the limit $t \to \infty$.

Note that a simple parabolic stress profile can be obtained not only for the radial quenching of "soft" crystals but also for cylindrical "rigid" crystals grown on end faces [9]. The sign of the residual stress is determined by the convexity or concavity of the temperature profile (Fig. 6).

Semiconductor crystals belong to the group of plastic crystals in which strain often develops in regions comparatively remote from the crystallization interface [7, 8, 12].

A rigorous solution of the problem of calculating residual thermoplastic stress and strain in plastic crystals requires a solution to the general system of equations of the theory of internal stress where the total strain ε^T comprises, in contrast to equation (2), the residual (plastic) strain ε^0:

$$\varepsilon_{ij}^T = \varepsilon_{ij} + \alpha_{ij}T + \varepsilon_{ij}^0. \tag{10}$$

Plastic strains will be found if the system of equations (10) is supplemented by a rheological equation of the type

$$\dot{\varepsilon}^0 = f(\varepsilon^0, \sigma, T),$$

which describes the kinetics of plastic strain caused by a real thermal stress (taking account of relaxation) at time t, at a given point within the sample. The magnitude and distribution of thermoplastic stress and strain thus depend not only on the thermoelastic field but also on the specific plasticity characteristics of the crystal. In addition to purely mathematical difficulties, the formulation and solution of such a problem for plastic crystals become complicated also because thermoplastic strains in the case (most important for practical applications) of growing perfect crystals with dislocation density $N_d \leq 10^3$-10^4 cm^{-2} are so low that it is impossible to obtain a rheological equation for these early stages of plastic straining from the standard macroscopic stress-strain curves. In this situation it is preferable to "construct" a rheological equation, by using the fundamental kinetic equation of the theory of dislocations and the empirical quantitative relations for motion, multiplication, and interaction

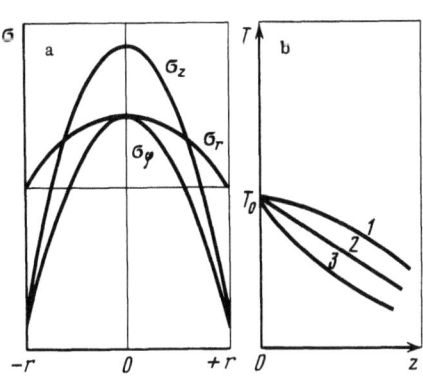

Fig. 6. (a) Typical profiles of residual stress in platelet crystals in the case of radial quenching; (b) typical cases of temperature distribution along the axis of a growing boule, corresponding to different types of residual stress profiles in rigid crystals. (1) $\partial^2 T/\partial z^2 < 0$ (stress profile similar to the case of radial quenching); (2) $\partial^2 T/\partial z^2 = 0$ (no residual stress); (3) $\partial^2 T/\partial z^2 > 0$ (quenching stress profile with reversed sign).

of dislocations obtained by independent direct experiments. For example, the rheological equation for semiconductor crystals can be written in the form

$$\frac{da}{dt} = Aa^m (\tau - Ba^{m/2})^n \exp\left(-\frac{u}{kT}\right),$$

where a and τ are the shear strain and stress referred to the appropriate slip systems, and the dislocation motion activation energy u and constants A, B, m, and n are taken from experimental data [13]. By taking account of the crystallography of slip for a known orientation of a growing crystal, we readily transform shear strain a to thermoplastic strain in a cylindrical system of coordinates. With the residual strain field known, we can also find the dislocation density distribution in the growing crystal. In the general form the problem as outlined above can be solved only by numerical methods.

Such evaluation of residual shear stress and strain was carried out to a first approximation for semiconductor crystals [13]. According to the results obtained for the growth of, for example, gallium arsenide single crystals with $N_d \approx (2-3) \cdot 10^4$ cm^{-2}, the degree of relaxation of thermoelastic stress is approximately 80 percent.

In the general case of parabolic distribution of plastic strain in a cylindrical ingot of radius R,

$$\varepsilon_z^0 = Ar^2; \quad \varepsilon_\varphi^0 = Br^2; \quad \varepsilon_r^0 = Cr^2,$$

the distribution of residual stress is given by the relations

$$\sigma_z^0 = E[A + \mu/2(3B - C)](R^2 - 2r^2)/2(1 - \mu^2);$$
$$\sigma_\varphi^0 = E[\mu A + 1/2(3B - C)](R^2 - 3r^2)/4(1 - \mu^2);$$
$$\sigma_r^0 = E[\mu A + 1/2(3B - C)](R^2 - r^2)/4(1 - \mu^2),$$

from which in the particular case A = B = C we find the relation

$$\sigma_r^0 + \sigma_\varphi^0 = \sigma_z^0,$$

which was derived earlier [7, 14] for the case of the steady-state radial heat transfer from the ingot. If, however, B = C = 0 (cylindrical crystal growing at its end face), we have

$$\sigma_r^0 + \sigma_\varphi^0 = \mu \sigma_z^0.$$

Consequently, monitoring of residual stress in different cross sections should yield important information on the processes of plastic straining in specific modes of growth and cooling of crystals. These data can be obtained experimentally, for example, by photoelasticity measurements, as reported for semiconductors in refs. 5, 12.

Several stages, corresponding to consecutive steps in the achievements of the dislocation theory, can be singled out in the progress of the concepts dealing with the formation of dislocation structure in a growing crystal. The earliest publications [15, 16] suggested a macroscopic theory of the generation of dislocations during growth, based on an assumption that dislocations completely remove elastic distortions in the growing crystal. Later [8] it was stressed that it is essential to take account of the total strain of the substrate (including the FTB effect) and the latent energy of lattice defects. If the contribution of the second factor is negligible, dislocations indeed completely remove the stress in the surface layer. If, however, the energy is not negligible, the stress can be estimated from the dependence of the latent energy on strain. The macroscopic theory of dislocations is not valid in single crystals with low dislocation density, and it is necessary to operate with stress fields of individual dislocations taking into account stress relaxation at the free surface.

This approach made it possible to relate the problem of controlling the dislocation structure in a growing crystal to the problem of interaction of a tilted dislocation with the crystal surface [3]. The basic cases for isotropic elastic crystals were treated in ref. 17. The solution of the problem of the moment of force in the interaction of a tilted dislocation and crystal surface, with crystal anisotropy being taken into account, was found by Lothe [18]; this proved to be general and fairly simple. It coincides with the solution of a simplified problem dealing with the equilibrium of a dislocation in the linear tension approximation; a linear (per unit length) energy $E(\theta)$ is assigned to the dislocation, and is assumed to depend on the dislocation orientation fixed by azimuth angle θ.

The total linear energy of a tilted dislocation in a plane-parallel plate with thickness a is $w = aE(\theta)/\cos\theta$ where θ is the angle by which the dislocation deviates from the normal to the plate surface. An equilibrium orientation of the dislocation corresponds to a minimum of function $w(\theta)$, and hence to a maximum of function $\cos\theta/E(\theta)$. Figure 7 shows that the procedure by which this maximum is found is geometrically very simple: a tangent corresponding to the orientation of the free surface is traced on the polar diagram $1/E(\theta)$ characterizing the orientational dependence of elastic energy of the dislocation. The point of tangency indicates an equilibrium orientation of the dislocation emergent at this surface. A change in surface orientation leads to a changed orientation of the dislocation. An obvious corollary is the possibility of controlling the trajectory of the dislocation, including, for example, squeezing the dislocation out to the lateral surface of the growing crystal. Another important corollary is a conclusion concerning the conditions for the generation of growth dislocations on inhomogeneities and inclusions. This generation is substantially facilitated if angled V-shaped dislocations with both rays emerging at the surface can be formed. But this dictates that two tangents of the same orientation exist on the $1/E(\theta)$ diagram (see Fig. 7). Such a situation is only possible if the diagram has a concave segment corresponding to a forbidden sector of orientations, with a negative linear tension along the dislocation axis, the dislocations then being unstable. The sector of forbidden orientations in Fig. 7 is shaded.

The general state-of-the-art in the problem of producing a desired dislocation structure in a growing crystal is discussed in review [3], on the basis of current concepts covering various types of dislocation structures, dislocation free path length, and the difference between the scalar and tensor dislocation densities. Various aspects of these problems are discussed in refs. 4, 12, 19 in relation to the melt growth of semiconductor crystals. Current achievements in the growth of single crystals with perfect structure, including dislocation-free crystals, are reviewed in ref. 4.

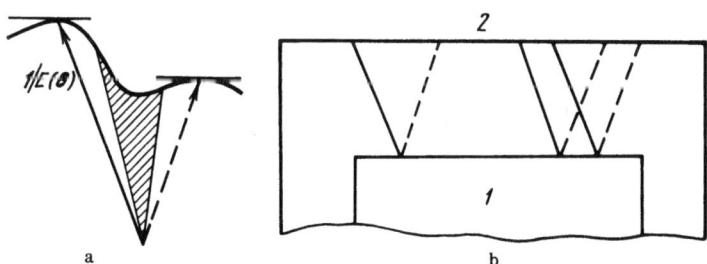

Fig. 7. Orientation of dislocations in the growth of nonplastic crystals. (a) polar diagram of dislocation energy as a function of orientation; (b) selection rule for the orientations of dislocations; (1) seed, (2) growth interface.

The calculation of dislocation density in growing crystals is a very complicated problem. At the same time, there are situations, such as the selection of technological operations for growing perfect crystals, in which a semi-qualitative analysis of dislocation structure formation proves very useful; this analysis is based on comparing calculated thermoelastic stress in an ingot with the experimentally established critical stresses for the generation of dislocations at appropriate temperatures. Strictly speaking, one is justified in calculating stresses, assuming the crystal to be elastic, only if the shear stresses are below a critical level. Therefore this approach, although not meant to yield a quantitative estimate of dislocation density, makes it possible to determine conditions for the growth of dislocation-free crystals, to carry out a comparative analysis of growth modes, and to single out the regions of the most intensive generation of dislocations within the growing crystal. Thus we obtain useful information for the adjustment of thermal conditions in crystal growth.

The approach outlined above has proved effective with semiconductor crystals [4]. We have mentioned that the stress field in a growing ingot can be found by solving the system of thermoelasticity equations by the method described in ref. 2. Examples of such numerical calculations for the growth of germanium, silicon, gallium arsenide, and indium antimonide single crystals can be found in refs. 2, 19, 20. Experimentally measured critical stress for generation of dislocations, τ_{crit}, can be used as a criterion of admissible thermoelastic stresses above which dislocations are generated in the ingot [21, 22]. The following values of τ_{crit}, in g mm^{-2}, are obtained by extrapolating to the melting point, for different semiconductors:

Si	Ge	GaAs	InSb	InAs
60–90	15	7	20	50

A very low level of critical stress in real crystals indicates that the most probable mechanism of generation of dislocations in dislocation-free and low-dislocation crystals is the heterogeneous generation mechanism caused by microinhomogeneities in the crystals. Experiments with silicon crystals [23] have demonstrated that microscopic dislocation loops (several tens of microns in diameter), formed during growth by coalescence of point defects (vacancies and interstitials), can serve as sources for the heterogeneous generation of dislocations. Estimates show that loops of about 30 μm diameter can serve as dislocation sources in silicon at stresses of about 100 g mm^{-2}, while stresses of the order of several tens of grams per square millimeter are sufficient to multiply loops with diameters around 10^2 μm. As the size of such loops grows with diminishing growth rate, both dislocation-free and dislocation-containing ($N_d \approx 10^3$ cm^{-2}) silicon crystals can grow at the same level of thermoelastic stress, depending on the growth rate.

An analysis of dislocation density and stress distributions in semiconductor crystals enables one to classify thermal stress τ_T by its effect on the formation of dislocation structures. If $\tau_T \gg \tau_{crit}$ (which usually corresponds to $N_d \approx 10^4$-10^5 cm^{-2}), we observe a clear correlation of the mean value and distribution of dislocation density with the thermoelastic stress field. There is no doubt that in these conditions generation of dislocations mostly follows from multiplication of dislocations by thermal stress. If the thermoelastic stress is nearly equal to τ_{crit}, then no intensive multiplication takes place, and most of the dislocations are the "primary" dislocations whose number and distribution are determined by the concentration, efficiency, and distribution of heterogeneous generation sources, or by specific features of the processes operative at the crystallization interface. Consequently, the correlation of mean density and distribution of dislocations with the thermoelastic stress field in low-dislocation crystals ($N_d \approx 10^2$-10^3 cm^{-2}) is usually suppressed and is not always unambiguous. If $\tau_T < \tau_{crit}$, the crystal in the general case grows dislocation-free; this has been confirmed for germanium and silicon crystals. Even in this case, however, dislocations can be generated either directly in the course of crystallization (for example, after capturing inclusions at the growth interface), or because the region of plasticity in the growing crystal comprises more effective sources of generation than those in the samples repeatedly heated in experiments conducted to measure τ_{crit}. It was mentioned above that small dislocation loops may serve as such effective sources. The well-known effect of breakdown of dislocation-free growth in silicon single crystals is explained by this factor. The dislocation density in indium antimonide crystals under the conditions $\tau_T < \tau_{crit}$ may also reach 10^3 cm^{-2} and be anomalously distributed [19]. We note in conclusion that all basic types of stress manifest themselves in epitaxial films, even more clearly than in bulk crystals: thermoelastic stress, stress generated by dislocations, and stress related to impurity inhomogeneities or to a misfit in the lattice parameters of heteroepitaxial layers. The behavior of stress generation, stress relaxation, formation of dislocation structure, and interaction between dislocations and crystal surface described above remains valid in this particular case.

Literature Cited

1. V. L. Indenbom, I. S. Zhitomirsky, N. N. Morozovskaya, and T. S. Chebanova. Numerical solution of the problem of residual stress generated by extending a semi-infinite cylinder. In: Thermal Stress in Structural Elements, Naukova Dumka, Kiev, No. 9, 136-148 (in Russian) (1970).
2. S. S. Vakhrameyev, M. G. Mil'vidsky, V. B. Osvensky, V. A. Smirnov, and Yu. F. Shchelkin. Analysis of thermoelastic stress in the Czochralski growth of semiconductor single crystals from the melt: relation to dislocation structure. In: Growth of Crystals, 12, ed. A. A. Chernov, Consultants Bureau, New York (1984).
3. Stress and dislocations in crystal growth. Izv. AN SSSR, Ser. Fiz. 37, 2258-2267 (in Russian) (1973).
4. M. G. Mil'vidsky and V. B. Osvensky. Growth of perfect single crystals. In: Problems in Modern Crystallography, Nauka, Moscow, 79-109 (in Russian) (1975).
5. Yu. F. Shchelkin, V. A. Smirnov, I. V. Starshinova, and A. A. Kholodovskaya. Calculation of temperature field in the melt and crystal for nonlinear boundary conditions. In: GIREDMET Proceedings, Metallurgiya, Moscow, 55, 29-42 (in Russian) (1974).
6. M. G. Mil'vidsky, V. A. Smirnov, I. V. Starshinova, and Yu. F. Shchelkin. On the analysis of thermal conditions in the Czochralski growth of single crystals. Izv. AN SSSR, Ser. Fiz., 40, 1444-1451 (in Russian) (1976).
7. V. L. Indenbom and V. I. Nikitenko. Analysis of stress in semiconductors by optoelectronic amplifier. In: Stress and Dislocations in Semiconductors, AN SSSR Publishing House, Moscow, 8-42 (in Russian) (1962).
8. V. L. Indenbom. On the theory of generation of stress and dislocations in crystal growth. Kristallografiya, 9, 74-83 (in Russian) (1964).
9. I. S. Zhitomirsky, M. K. Likht, and T. S. Chebanova. On the theory of stress in solids with mobile boundaries. In: Thermal Stress in Structural Elements. Naukova Dumka, Kiev, No. 6, 227-235 (in Russian) (1966).
10. V. L. Indenbom, I. S. Zhitomirsky, and T. S. Chebanova. Theoretical analysis of stress generated in growing crystals. In: Growth of Crystals, 8, ed. N. N. Sheftal', Consultants Bureau, New York (1969).
11. V. L. Indenbom, I. S. Zhitomirsky, and T. S. Chebanova. Internal stress generated in steady-state growth of crystals. Kristallografiya, 18, 39-47 (in Russian) (1973).
12. M. G. Mil'vidsky and V. B. Osvensky. Growth of perfect crystals of semiconductors from the melt. In: Growth of Crystals, 12, ed. A. A. Chernov, Consultants Bureau, New York (1984).

13. S. S. Vakhrameyev, V. B. Osvensky, and S. S. Shifrin. Calculation of dislocation density in real melt-grown crystals of semiconductors, on the basis of thermal stress fields. In: Vth USSR Conference on Crystal Growth. Tbilisi, Abstracts. Inst. Cybernetics AN GSSR Publ. House, Tbilisi, 2, 230-233 (in Russian) (1977).
14. V. I. Nikitenko and V. L. Indenbom. Joint analysis of stress and dislocations in germanium crystals. Kristallografiya, 6, 432-438 (in Russian) (1961).
15. E. Billig. Growth of monocrystals of germanium from an undercooled melt. Proc. Roy. Soc. London A, 229, 346-354 (1955).
16. V. L. Indenbom. Macroscopic theory of generation of dislocations in crystal growth. Kristallografiya, 2, 594-603 (in Russian) (1957).
17. V. L. Indenbom and G. N. Dubnova. Interaction between dislocations in lattice sites and equilibrium configurations of dislocations. Fiz. Tverd. Tela, 9, 1171-1177 (1967).
18. J. Lothe. The force of dislocations emerging at free surfaces. Phys. Norv., 2, 153-157 (1967).
19. M. G. Mil'vidsky, V. B. Osvensky, S. S. Shifrin, and S. P. Grishina. Thermal stress as a factor determining generation of dislocations in melt-grown semiconductor crystals. In: Growth and Doping of Semiconductor Crystals and Films. Nauka, Novosibirsk, Part 2, 272-279 (in Russian) (1977).
20. S. S. Vakhrameyev, M. G. Mil'vidsky, V. A. Smirnov, and Yu. F. Shchelkin. Analysis of temperature and thermoelastic stress fields in single crystals grown from the melt. In: Growth and Doping of Semiconductor Crystals and Films. Nauka, Novosibirsk, part 1, 162-168 (in Russian) (1977).
21. M. G. Mil'vidsky, V. B. Osvensky, B. A. Sakharov, and S. S. Shifrin. Stress-induced generation of dislocations in perfect single crystals. Dokl. AN SSSR, 207, 1109-1111 (in Russian) (1972).
22. V. B. Osvensky, S. S. Shifrin and M. G. Mil'vidsky. Multiplication of dislocations in semiconductors at high temperatures. Izv. AN SSSR, Ser. Fiz., 37, 2357-2361 (in Russian) (1973).
23. A. J. R. de Kock, P. I. Roksnoer, and P. G. T. Boonen. The introduction of dislocations during the growth of floating zone silicon crystals as a result of point defect condensation. J. Cryst. Growth, 30, 279-294 (1975).

GROWTH DEFECTS IN SEMICONDUCTOR CRYSTALS

A. N. Buzynin, N. I. Bletskan, Yu. N. Kuznetsov, and N. N. Sheftal'

Crystals grown from the melt contain defects classified by their origin into growth and thermal defects [1, 2]. The former are associated with peculiarities of the growth processes, and the latter with non-uniform post-growth cooling of the crystal. Approximate formulas [3, 4] and computer-based calculations [5-8] are known for the evaluation of the density of thermally generated dislocations. Growth defects have been studied in much less detail [1, 8].

We shall consider growth defects in Czochralski-grown silicon and gallium phosphide single crystals with different orientation, shape, and structure of the crystallization interface and different doping levels. We shall discuss methods of defect suppression and the effects of clusters in the substrates on the structure of epitaxial layers. The experimental methods used were X-ray topography and selective etching.

Cluster Structure and Stress in Dislocation-Free Crystals

Experiments demonstrated that the shape and structure of a crystallization interface affect the density and distribution of clusters and the parameters of residual stress in silicon crystals. Thus, in a boron-doped 35 mm diameter silicon crystal grown along [111] with a slightly convex crystallization interface (faces {111} were not present during growth, since the lateral ridges were parallel on all three pseudofaces), the cluster density is $N_c \approx 2 \cdot 10^4$ cm^{-2} within the facet-effect zone, and $N_c \approx (4-6) \cdot 10^4$ cm^{-2} outside of this zone (Fig. 1a).

The distribution of clusters was different in a similar crystal which was grown with a slight tilting (less than 1°) of the pulling direction from [111] towards [110] so that one of the three inclined lateral {111} faces participated in growth (this is apparent from the crescent-shaped lateral ridges on one of the three pseudofaces, Fig. 1b). The lowest density of clusters ($2.6 \cdot 10^4$ cm^{-2}) and the maximum uniformity of their distribution were observed within the facet-effect zone (Fig. 2a). The cluster density was found to be somewhat higher in the region formed by the curvilinear growth surface formed by steps parallel to the front face (111). It was still higher (($4-5) \cdot 10^4$ cm^{-2}) in the region formed by the curvilinear growth surface consisting of steps parallel to an inclined (111) face, with clusters within the curvilinear interface zone being distributed less uniformly than within the facet-effect zone; this is caused by impurity bands. Finally, the cluster density rose abruptly (up to $5.8 \cdot 10^5$ cm^{-2}) within the region of a diffuse interface of growth pyramids of the frontal and inclined {111} faces (Fig. 2b). This character of cluster distribution is maintained in crystal cross sections along the crystal length, if the growth interface shape is conserved. If the crys-

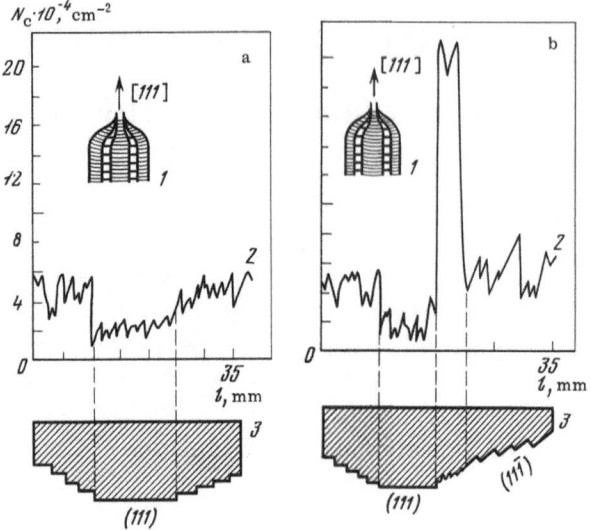

Fig. 1. Cluster structure of silicon crystals with (a) single-face crystallization interface, and (b) lateral (111) face emerging at the crystallization interface. (1) schematic morphology of the crystal, (2) cluster distribution across the sample cross section, (3) schematic structure of crystallization interface.

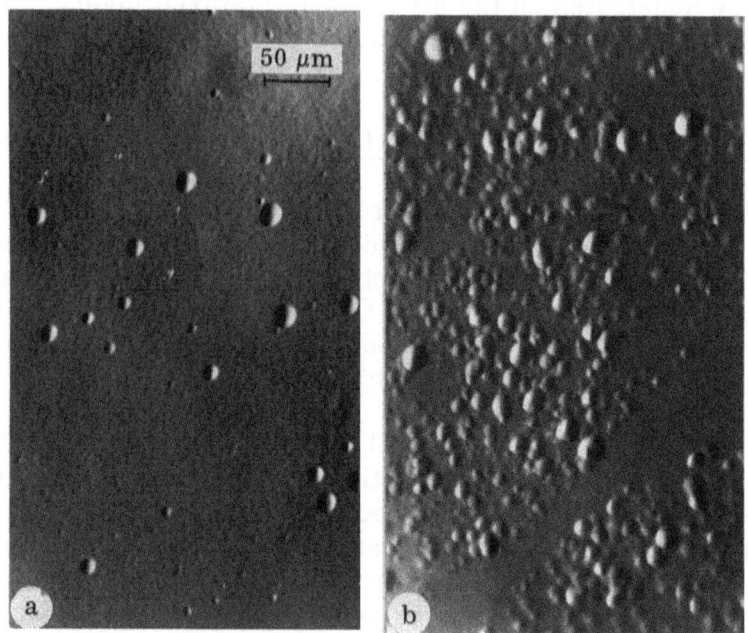

Fig. 2. Micrographs of cluster etch pits in silicon crystal within (a) facet-effect zone, and (b) diffuse boundary of growth pyramids of {111} faces.

tallization interface shape changes in the course of growth, the distribution of clusters within different cross sections changes too. In off-center grown crystals with a variable configuration of the crystallization interface, the cluster density is fairly uniform over a cross section ($N_c \approx (2-6) \cdot 10^4$ cm^{-2}).

The distribution of regions with a high level of residual stress correlates with the distribution of clusters within the crystal [1].

Dislocation Structure of Silicon Single Crystals with Low Dislocation Density

It is found that the shape of the growth interface, the number of participating {111} faces, the extent to which they are pronounced, and their misorientation affect the dislocation structure of the crystal.

The distribution of dislocations, clusters, and residual stress in crystals with a low density of dislocations ($N_d \approx 10^2$-10^3 cm^{-2}) grown in thermal fields designed to obtain dislocation-free crystals (with seed dislocations not inherited), is very similar to the distribution in dislocation-free crystals [1]. The dislocation density reaches a maximum at the diffuse interfaces of growth pyramids of {111} faces and at the segments of the curvilinear crystallization interface; it is a minimum inside the growth pyramids of {111} faces. The number of regions with a high dislocation density varies when the number of {111} faces, participating in the growth, changes [1, 9].

The degree of imperfection of the crystal also depends on the seed contour and the boule growth angle at which the crystal reaches the steady-state diameter. Thus, two methods are possible to seed a triangular contour with {112} lateral faces: seeding to one or the other of the seed bases (Fig. 3) [10]. In the first case (Fig. 3a) the lower lateral {111} faces emerge at the crystallization interface along the whole perimeter of the seed base; in the second case (Fig. 3b) they emerge only at three points on the perimeter. It has been established that in the first case the crystal is less perfect and the generation of dislocations is more difficult to suppress than in the second case. Stress in dislocation-free silicon crystals of 20 mm diameter, grown by the two seeding methods without a bottleneck stage, was measured by X-ray techniques. Wafers cut from a crystal grown by the first (unfavorable) method show three regions of stress concentration in the seeding area; these regions correspond to the site at which lateral {111} faces emerge at the growth interface. If dislocations are generated in such a crystal, they also concentrate in these three regions (Fig. 4). No stress concentration areas were detected in the crystals with the second (favorable) method of seeding. Moreover, dislocations are less frequently generated in such crystals. We notice that additional stress concentration areas appear if the upper lateral {111} faces take part in the growth at large growth angles of the boules. Also, if the conical part is sufficiently high, stresses increase towards the base and may generate dislocations. This effect is generally suppressed if the cone angle is nearly 90°, that is, if the cone height is not too large.

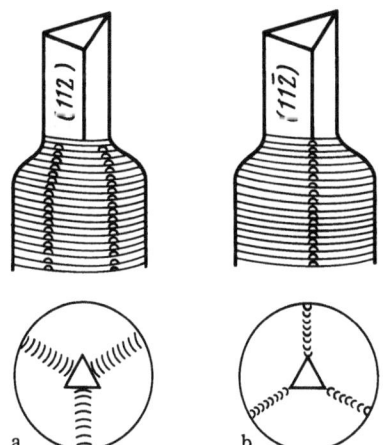

Fig. 3. Methods of seeding by a trihedral seed: (a) unfavorable and (b) favorable for crystal perfection.

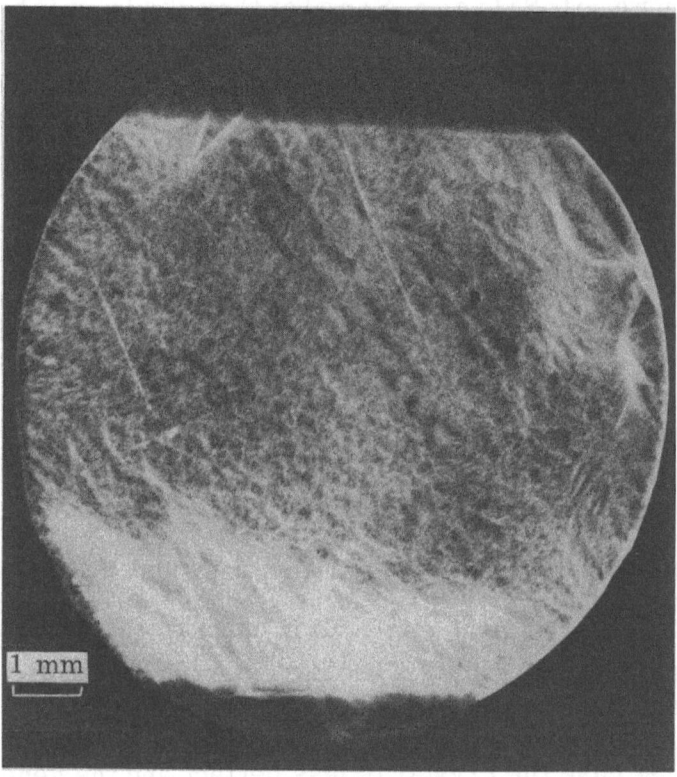

Fig. 4. Lang topograph of silicon crystal grown on a trihedral seed with unfavorable seeding configuration.

It was found that crystal perfection is affected by tilting the pulling direction with respect to the exact crystallographic direction [9]. The important factor is not so much the angle as the direction of misorientation. Thus, even a small (below 1°) misorientation of the growing crystal from [111] toward [110] is detrimental to high perfection. This misorientation leads to the melt separation from the frontal face and to formation of depressions at the lateral surface (Fig. 5a); it also facilitates the generation of dislocations and increases the probability of twinning and transformation to polycrystalline growth (Fig. 5b). The generation of dislocations in an 80 mm diam. crystal growth with this misorientation is clearly pronounced in its morphology: narrowing and smoothing of the helical thread, reduced width of pseudofaces, and a change in the form of lines on pseudofaces from rectilinear to crescent shaped (see Fig. 5a, b). Growth with misorientation towards [100] is favorable for crystal perfection. A deep helical thread on the lateral surface, a wide pseudoface along this surface, and rectilinear lines on it (Fig. 5c) indicate that these silicon crystals (80 mm diam. and more than 1 m long), grown with this misorientation, are dislocation-free.

Crystals with High Dislocation Density

An analysis of crystals with medium and high density of dislocations has shown that the dislocation structure depends on the character of the impurity distribution and the degree of roughening of the growth interface. In addition, the shape of the interface proved to be related to changes in the dislocation structure caused by annealing of the crystal [1].

The distribution inhomogeneity of impurity striations in large-diameter silicon crystals (diam. exceeding 80 mm), grown along [111], is accompanied by a similar striation

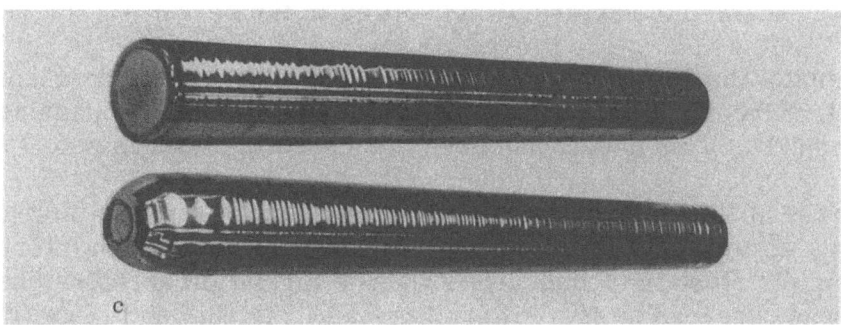

Fig. 5. Large diameter silicon crystals grown along [111] with misorientation of 30' with respect to (a.b) [110], and (c) [100]

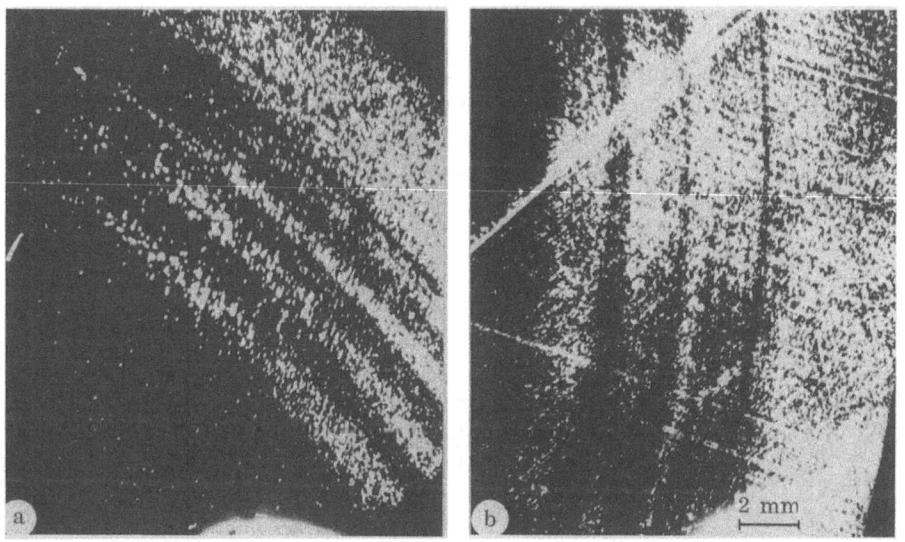

Fig. 6. Striated dislocated pattern in silicon crystal cross section produced (a) in the course of growth, and (b) by thermal shock.

inhomogeneity in dislocation distribution (Fig. 6a). The dislocation density within impurity bands exceeds the mean density by more than an order of magnitude. This distribution of dislocations was also observed when dislocations were generated by a thermal shock produced by extracting the crystal (with no as-grown dislocations) from the melt. The pattern of the distribution of dislocation sources within the impurity bands, "latent" in the dislocation-free crystal, was as if "developed" by thermoelastic stresses (Fig. 6b).

Crystals grown with a roughened structure of the crystallization interface, typically accompanying a high level of doping or large supercooling of the melt, have a substantially increased mean dislocation density and sharply pronounced local clusters over the mean level. Thus, the dislocation density N_d within a cross section of a gallium phosphide crystal (30 mm diam.), grown from a highly Te-doped melt, was more than 10^5 cm^{-2} in the central part and close to the diagonals of the square (i.e., on the diffuse boundaries of growth pyramids of $\{111\}$ A and $\{111\}$ B faces), while of the order of 10^4 cm^{-2} in other regions (Fig. 7). Figure 7 also shows band-shaped clusters of defects close to the sample edges; presumably, they were caused by chain-like capture of impurities in the re-entrant angles of macrosteps.

If roughening of the stepped structure of the growth interface is accompanied by a polycentric growth of layers, dislocation clusters and small-angle boundaries may be formed at the contact sites [11]. This aspect has been discussed in detail in ref. 1.

A high degree of interface roughening due to the large supercooling and high impurity content in the melt, as well as a large curvature of the crystallization interface, results in a very non-uniform distribution of impurities and the formation of inclusions at macrostep end faces. This was observed both in silicon crystals (Fig. 8a) and in gallium phosphide crystals; substantial impurity inhomogeneities and inclusions in transparent gallium phosphide crystals can be observed in optical transmission micrographs (Fig. 8b). Note that a similar relationship between crystal homogeneity and the degree of roughening of the stepped crystallization interface was also observed in crystals grown from solutions [12, 13]; the role played by the stepped relief in the formation of solution inclusions was discussed in ref. 14.

Clusters and the Structure of Epitaxial Layers

Thermal treatment modifies the defect structure of dislocation-free silicon crystals. The clusters are transformed and grow, and new clusters, stacking faults, and macroscopic impurity segregations (manifested by pits and hillocks in selective etching patterns) are formed on the latent growth microdefects. Dislocations can be generated if the clusters are large or if their density is high.

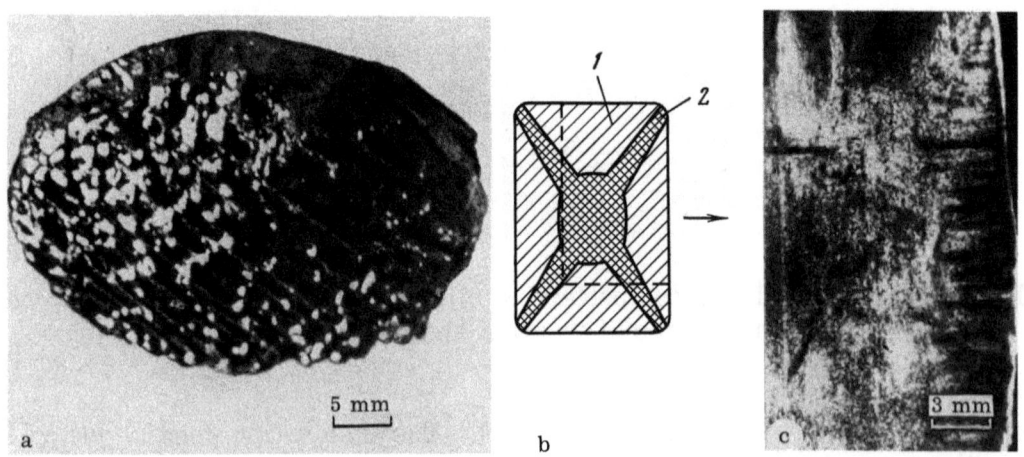

Fig. 7. Distribution of dislocations in a gallium phosphide crystal grown from heavily doped melt. (a) macroscopic structure of crystallization interface at the moment of separation from the melt; (b) schematic distribution of dislocations in crystal cross section; (c) Lang X-ray topograph; (1) $N_d \approx 1 \cdot 10^4$ cm^{-2}; (2) $N_d \approx 2 \cdot 10^5$ cm^{-2}.

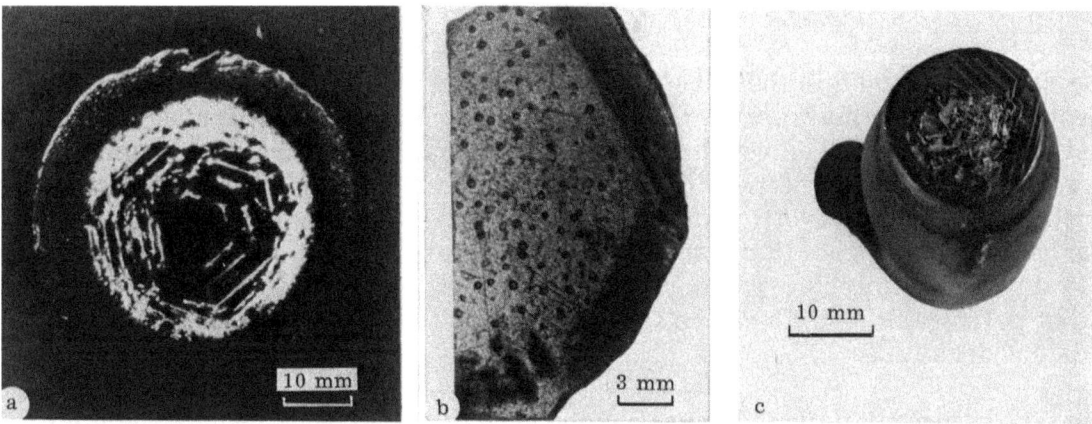

Fig. 8. Inclusions in the end faces of macroscopic steps in crystals of (a) silicon; (b) gallium phosphide; (c) crystallization interface of a gallium phosphide crystal extracted from the melt.

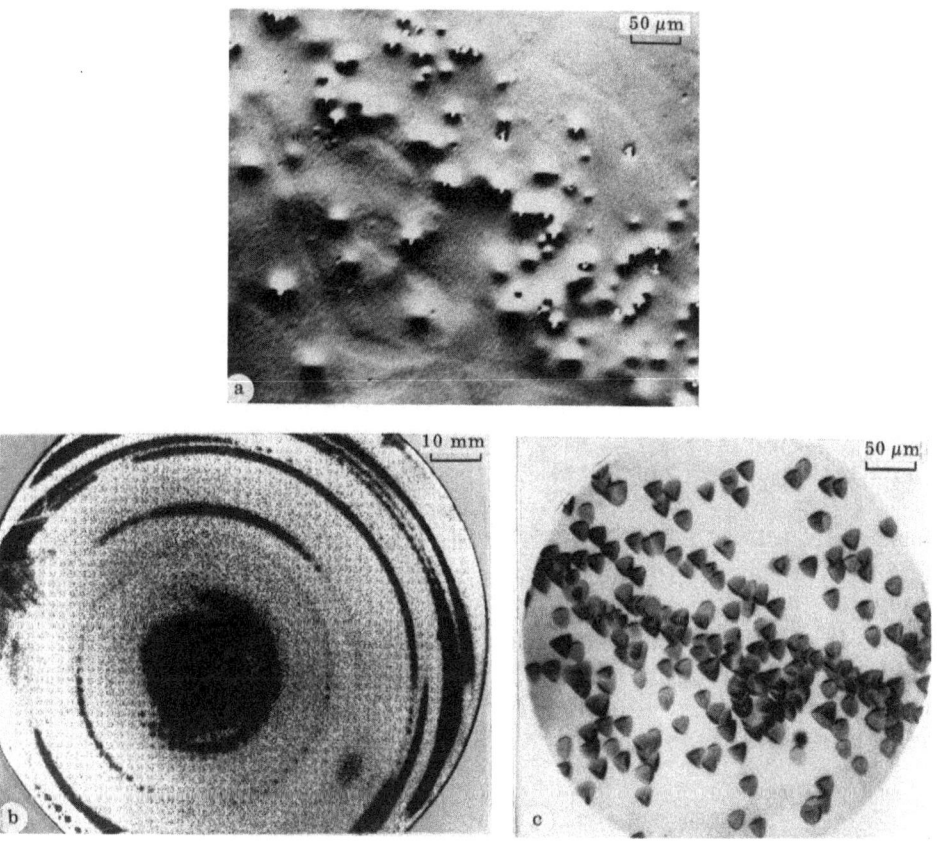

Fig. 9. (a) Clusters in dislocation-free silicon substrates revealed by etching, and (b, c) stacking faults and dislocations in epitaxial layers grown on these substrates.

Clusters in silicon substrates lead to the formation of stacking faults and dislocations, on the cluster sites, in the epitaxial layers. This is deduced from the observed correlation between the density and distribution curves (Fig. 9).

Discussion of Results

The results obtained in the present work demonstrate that various types of growth defects, from clusters to macroscopic inclusions, are related to similar structural features of the crystallization interface, which lead to inhomogeneous capture of the ambient medium into the crystal in the course of growth. By the "ambient" we mean both foreign impurities and impurities similar to the lattice atoms, including the supercooled melt [12]. It is found that diffuse boundaries of growth pyramids of $\{111\}$ faces, regions of transition from flat to rounded crystallization interfaces, curvilinear segments of the interface, and contact sites in the case of polycentric growth of one face are conducive to the formation of regions with enhanced defect density.

The type of growth defects generated depends on the extent to which the ambient is captured by the crystal. Thus, impurity- and vacancy-decorated microscopic inclusions generate clusters during growth or during the thermal treatment of the crystal. At still higher rates of capture, dislocations and their clusters, small-angle boundaries, and macroscopic inclusions are formed. The rate of capture of the ambient medium increases as roughening of the interface increases. The distribution of growth defects depends on the shape and specific macroscopic structure of the growth interface.

A comparative analysis of the sectorial structure of polyhedral crystals [15] and crystals with constrained growth forms can explain the general features of the observed distribution of growth defects. The following features are found for crystals with constrained growth forms (Fig. 10):
— growth pyramids on their faces are shaped as solid, hollow, or truncated cylinders in constrast to growth pyramids of polyhedral crystals;
— broad diffuse boundaries of growth pyramids appear, representing growth pyramids with rounded, non-crystallographic surfaces; these are absent on polyhedral crystals;
— growth pyramids of crystallographically similar faces (a and a' in Fig. 10b), being non-equivalent with respect to growth conditions, are thus non-equivalent with respect to structural perfection, in contrast to growth pyramids on analogous faces of polyhedral crystals which are practically equivalent in this respect.

The above arguments indicate that in order to improve the homogeneity of a growing crystal, it is expedient to create conditions in which the crystal will grow in the most perfect growth pyramid with a planar or nearly planar growth interface. The observed effect of seed contour (see Fig. 3, 4) and boule growth angle on crystal perfection is caused by the unequal length of the boundaries of growth pyramids. The effect produced by crystal misorientation on its perfection (see Fig. 5) can be explained by the fact that in the case of misorientation from [111] toward [100] the layers at the interface propagate in the direction of the maximum

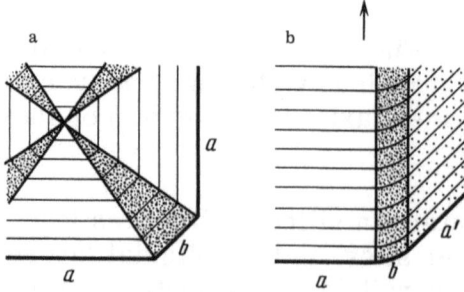

Fig. 10. Sectorial structure of crystals. (a) polyhedral, (b) with constrained growth form (sectional view).

tangential growth rate, while in the case of misorientation from [111] to [100] they propagate in the direction of minimum tangential growth rate. As the crystal perfection in the first case is higher, one may conjecture that a lower velocity of microscopic steps facilitates their coalescence into macroscopic steps with roughening of the relief. This, in turn, facilitates the incorporation of the ambient medium (degrading the crystal perfection), twinning (which, as we observe, is facilitated by the presence of macroscopic steps at the crystallization interface), and the transition to polycrystallinity (which usually starts at inclusions). The tendency for crystals to twin as the layers on the growth interface propagate in the direction of the minimum tangential growth rate was also mentioned in ref. 16.

The required direction of layer propagation is provided by a predetermined misorientation of the crystal. When a crystal grows along [111], layers nucleate at the interface at the point of maximum supersaturation, namely, at the site on the front (111) face which is the most elevated with respect to the melt. Hence, the misorientation determines the location of layer nucleation and the direction of the layer advance. In particular, the misorientation from [111] towards [100] ensures the propagation of the layers in the direction of the maximum tangential growth rate. The increased dislocation density in the impurity striae (see Fig. 6) may be caused both by microscopic inclusions in these striae and by the impurity concentration gradient in the lattice [17]. By using the known linear dependence of the lattice parameter on concentration of the substitutional impurity, the following formula is suggested for an approximate evaluation of the dislocation density of impurity origin, N_{di}, with non-zero gradient ∇C of impurity concentration in the crystal:

$$N_{di} = \alpha_C \nabla C / b,$$

where α_C is the coefficient of impurity-induced lattice expansion, and b is the Burgers vector. With the density of thermally induced dislocations [3, 18] taken into account, the general formula for estimating the dislocation density of thermal and impurity origin takes the form

$$N_d = (\alpha_C \nabla C + \alpha_T \nabla T)/b - 2\tau_{crit}/GbD,$$

where α_T is the linear thermal expansion coefficient, ∇T is the temperature gradient, τ_{crit} is the critical shear stress, G is the shear modulus, and D is the crystal diameter.

The data obtained indicate that in order to prevent the generation of growth defects it is necessary to grow crystals with the most perfect growth pyramid and with a crystallization interface as near to planar as possible, to ensure monocentric layered growth, minimum development of relief on the growth interface, and propagation of the layers in the direction favorable for perfect crystal growth. These requirements can be met by a choice of suitable growth direction, thermal conditions of growth and cooling, appropriate seed contour, predetermined misorientation of growth, supercooling, and purity of the melt.

Literature Cited

1. A. N. Buzynin, N. I. Bletskan, and N. N. Sheftal'. Linear defects and their genesis. In: Real Crystallization Processes. Nauka, Moscow, 113-130 (in Russian) (1977).
2. V. L. Indenbom and V. B. Osvensky. Theoretical and experimental analysis of generation of stress and dislocations in growing crystals. In: Vth USSR Conference on Crystal Growth, Tbilisi. Abstracts. Inst. of Cybernetics AN GSSR Publ. House, Tbilisi, 1, 15-17 (in Russian) (1977).
3. E. Billig. Growth of monocrystals of germanium from an undercooled melt. Proc. Roy. Soc. London, A 229, 346-363 (1955).
4. V. L. Indenbom. Types of lattice defects. Theory of dislocations. In: Physics of Defect-Containing Crystals. Tbilisi, Part 1, 5-98 (in Russian) (1966).

5. S. S. Vakhrameyev, M. G. Mil'vidsky, V. B. Osvensky, V. N. Smirnov, and Yu. F. Shchelkin. Analysis of thermoelastic stress: relation to dislocation structure. In: Growth of Crystals, Yerevan Univ. Publ. House, Yerevan, 12, 287-293 (in Russian) (1977).
6. S. S. Vakhrameyev. Calculation of thermal stress in melt-grown crystals. In: Problems in the Theory of Crystallization, Latv. Univ. Publ. House, Riga, 2, 108-122, (in Russian) (1976).
7. V. L. Indenbom, I. S. Zhitomirsky, and T. S. Chebanova. Internal stress generated in crystals grown in a steady-state mode. Kristallografiya, 18, 39-47 (in Russian) (1973).
8. M. G. Mil'vidsky and V. B. Osvensky. Growth of perfect crystals. In: Problems in Modern Crystallography, Nauka, Moscow, 79-109 (in Russian) (1975).
9. A. N. Buzynin, N. I. Bletskan, V. V. Zaichko, and N. N. Sheftal'. On the factors leading to generation of dislocations in silicon crystals. In: Growth and doping of semiconductor crystals and films, Nauka, Novosibirsk, 291-297 (in Russian) (1977).
10. Pat. 417857 (USA). 16.03.76, 23-273 SP.
11. I. V. Salli and E. I. Falkevich. Industrial Production of Semiconductor Silicon. Metallurgiya, Moscow, (in Russian) (1970).
12. N. N. Sheftal'. On real crystallization processes. In Growth of Crystals, AN SSSR Publishing House, Moscow, 1, 5-30 (in Russian) (1957).
13. A. V. Belyustin, G. D. Pavlova, and O. A. Gur'yev. Effect of solution supersaturation and impurities on stepped relief and homogeneity of crystals. Kristallografiya, 17, 647-650 (in Russian) (1972).
14. A. A. Chernov and S. I. Budurov. On growth forms of macroscopic steps. Faceting of step end faces. Kristallografiya, 9, 388-395 (in Russian) (1964).
15. G. G. Lemmlein. Sectorial Structure of Crystals. AN SSSR Publishing House, Moscow-Leningrad (1948).
16. P. I. Antonov, N. S. Grigoriev, and A. V. Stepanov. Specifics of growth of shaped germanium crystals with different orientations. In: Growth and Synthesis of Semiconductor Crystals and Films, Nauka, Novosibirsk, 268-273 (in Russian) (1975).
17. W. A. Giller. Production of dislocations during growth from the melt. J. Appl. Phys., 29, 611-618 (1958).
18. S. V. Tsivinsky. Generation of dislocations in crystals grown from the melt. Fiz. Metal. i Metalloved., 25, 1013-1020 (in Russian) (1968).

FORMATION OF DEFECTS IN EPITAXIAL HETEROSTRUCTURES AND MULTICOMPONENT SOLID SOLUTIONS OF SEMICONDUCTOR COMPOUNDS

M. G. Mil'vidsky and L. M. Dolginov

The Institute of Rare Metals, Moscow

The progress achieved in developing optoelectronic semiconductor devices employing heterojunctions has attracted much attention to the problem of fabrication of these epitaxial structures. The structural perfection of the heterostructures, and particularly the **minimization of defects** at the heterojunction, determine whether or not the heterostructures can be used in devices.

In the present paper we discuss several aspects of defect formation in epitaxial heterostructures based on solid solutions of A^3B^5 semiconductor compounds; these compounds are the basic materials for modern optoelectronics. Heterostructures were prepared by liquid phase epitaxy (LPE) by cooling suitable high-temperature solutions from the initial growth temperature T_{cr}.

It is well known that the basic factors affecting defect formation in such heterostructures are: misfit of lattice constants of the different materials at the interface (Δa); thermal stress; composition gradients in the solid solution across the expitaxial layer; defects inherited from the substrate.

We have discussed these factors in an earlier publication [1]; it was shown that perfect heterojunctions cannot be prepared by combining binary A^3B^5 compounds, nor by using heterostructures based on most ternary solid solutions. An analysis shows that the regions near heterojunctions are regions of increased defect density even in compositions based on $Al_xGa_{1-x}As-GaAs$, $Al_xGa_{1-x}P-GaP$, $Al_xGa_{1-x}Sb-GaSb$, where the differences in lattice parameters of the contacting materials are a minimum [2-3]. As a result, in most cases the heterojunction is a region with low efficiency of radiative recombination; this is confirmed by studying photo-, cathodo-, and electroluminiscence in these materials [4-5]. A good correlation is found between the results of structural (Fig. 1) and physical measurements. In particular, the photoluminiscence efficiency, Φ, of such structures is drastically lowered if the distance d from the investigated layer to the substrate-solid solution interface is small (Fig. 2).

In view of this fact, a promising approach is the development of heterostructures based on quaternary and more complex solid solutions in which an appropriate variation of the component ratios can result in practically independent changes in the band gap and lattice periodicity. This creates a requirement for fabricating lattice matching heterostructures for operation over a wide spectral range, using substrates of binary compounds [6].

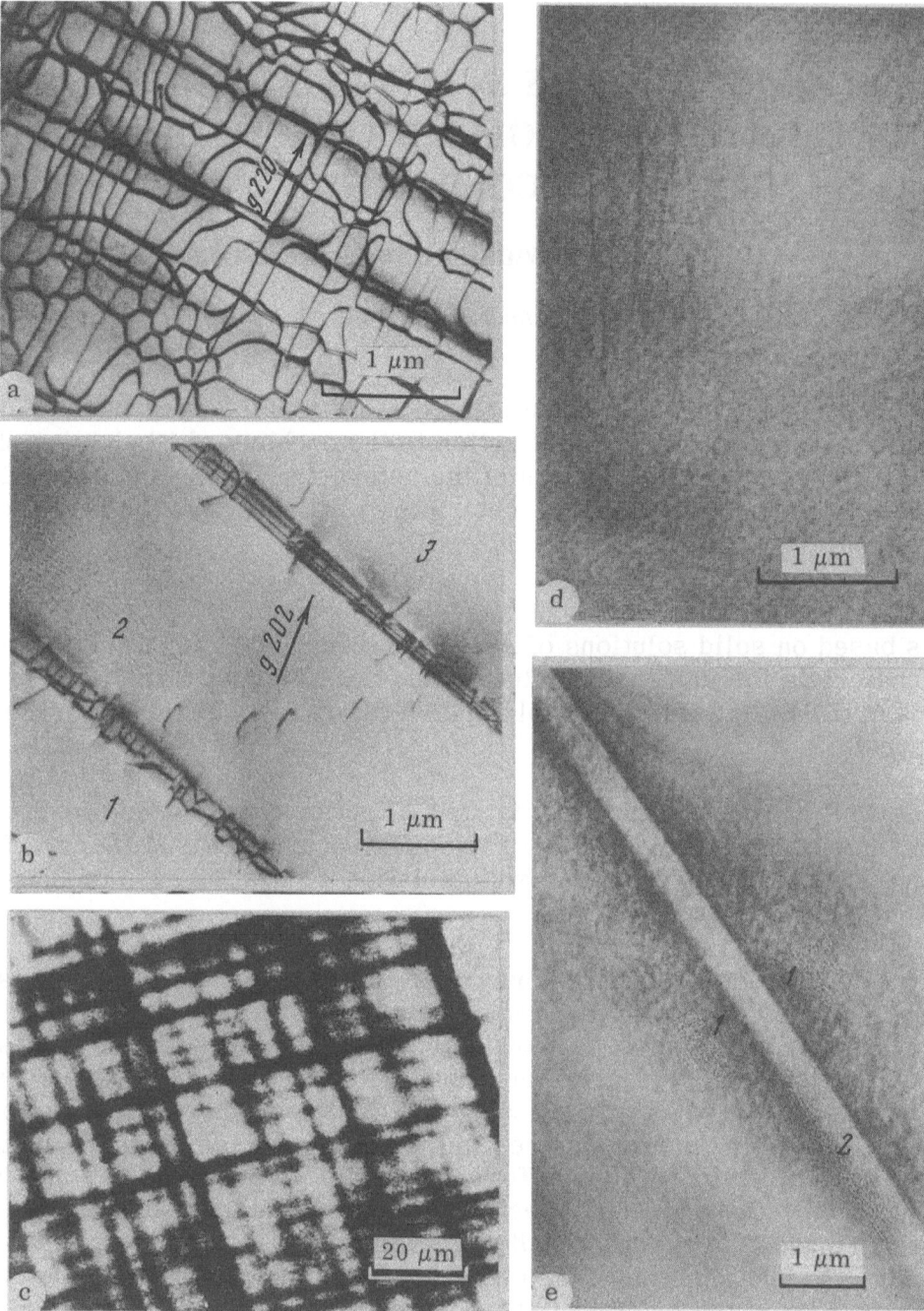

Fig. 1. Structural perfection and cathodoluminescence in heteroepitaxial compositions. (a) misfit dislocation network in the $GaAs-GaAs_{0.93}Sb_{0.07}$ heterojunction plane (transmission electron microscopy); (b) misfit dislocation network in a transverse section of the $GaAs-GaAs_{0.98}Sb_{0.02}-GaAs_{0.95}Sb_{0.05}$ heterocomposition (TEM): (1) GaAs, (2) $GaAs_{0.98}Sb_{0.02}$, (3) $GaAs_{0.95}Sb_{0.05}$; (c) SEM cathodoluminescence image of the $GaAs-GaAs_{0.93}Sb_{0.07}$ heterojunction plane; (d) heterojunction plane in $InP-Ga_xIn_{1-x}As_yP_{1-y}$ ($\Delta a \approx 1 \cdot 10^{-4}$ Å at T_{cr}, TEM); (e) transverse section of the $InP-Ga_xIn_{1-x}As_yP_{1-y}$ structure ($\Delta a \approx 3 \cdot 10^{-4}$ Å at T_{cr}, TEM): (1) InP, (2) $In_xGa_{1-x}As_yP_{1-y}$.

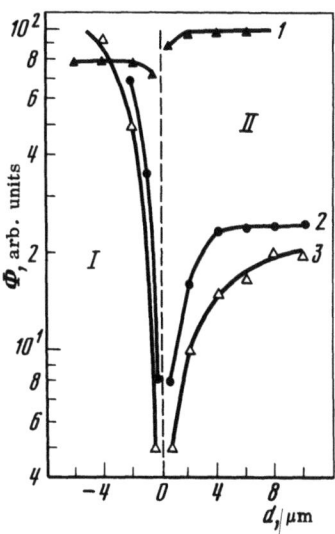

Fig. 2. Distribution of photoluminescence intensity across the heterostructure (band gap width at the junction approximately 1.2 eV). I. substrate; II. solid solution; (1) InP-$Ga_xIn_{1-x}As_yP_{1-y}$; (2) GaAs-$Ga_xIn_{1-x}As$; (3) GaAs-$GaAs_xSb_{1-x}$.

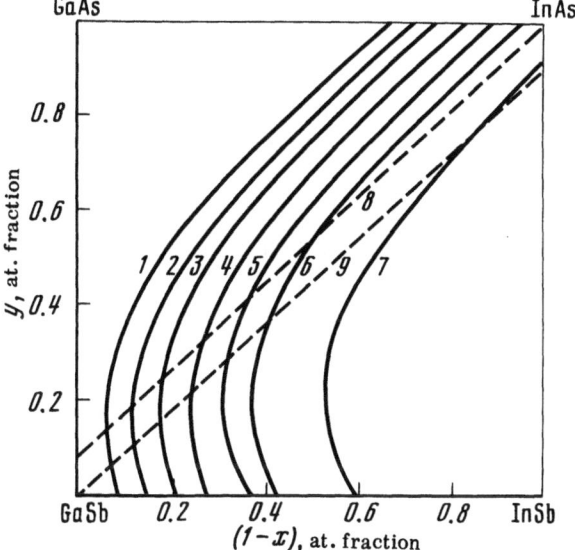

Fig. 3. Constant band-gap lines (300 K) in a quaternary solid solution $Ga_xIn_{1-x}As_ySb_{1-y}$. (1-7) 0.65, 0.60, 0.55, 0.50, 0.45, 0.40, and 0.30 eV, respectively; (8) $a = a_{InAs}$; (9) $a = a_{GaSb}$.

The conditions for lattice matching can be presented graphically on a plane if one assumes the formation of a quaternary solution from the appropriate binary compounds. Note that substitution in the sublattice of the group III or group V elements yields a triangular diagram, while simultaneous substitution in both sublattices yields a square diagram (Fig. 3). Lattice matching curves for constant energy gap can be traced on these diagrams, with the intersection points giving the compositions of quaternary solid solutions with a given band gap having lattice spacings equal to that of the binary compounds (or ternary solid solutions). One also can trace within these diagrams a curve corresponding to solid solutions with equal values of the thermal expansion coefficient. However, in a quaternary system the isoexpansion and lattice matching curves can coincide only at separate points. For example, they will coincide in heterostructures of the type "quaternary solid solution-binary compound" only at the point corresponding to the binary compound. Nevertheless, the quaternary solid solution approach is a way to solve the problem of varying stresses caused by the misfit of lattice parameters ($\sigma_{\Delta a}$) and thermal expansion coefficients ($\sigma_{\Delta \alpha}$) in a heterostructure having a given band gap of the quaternary solution in the active region of the structure (Fig. 4).

The residual stress σ_R listed in Table 1 was calculated by the formula

$$\sigma_R = Et_s^2/6R(1-\nu)t_f$$

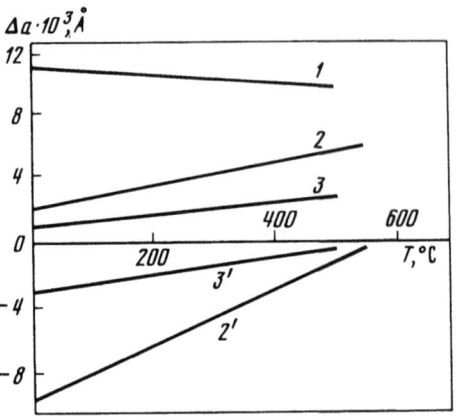

Fig. 4. Lattice parameters of quaternary solid solutions as functions of temperature. (1) GaInAsSb-GaSb; (2, 2') GaInAsSb-InAs; (3, 3') GaInAsP-InP.

where E is Young's modulus; t_s and t_f are the substrate and film thickness, respectively; R is the sample bending radius; and ν is Poisson's ratio.

To a first approximation, the parameters of a quaternary solid solution can be predicted by interpolating the data for compositions with a smaller number of degrees of freedom [6]. Constant band gap curves plotted using this method for solid solutions $Al_xGa_{1-x}As_{1-y}P_y$, $Ga_xIn_{1-x}As_yP_{1-y}$, $Ga_xIn_{1-x}As_ySb_{1-y}$, $Al_xGa_{1-x}As_ySb_{1-y}$ are in good agreement with the experiments. A preliminary analysis makes it possible to conjecture that lattice matching epitaxial heterostructures can be fabricated in a wide range of band gap values of quaternary solid solutions. The corresponding preliminary data concerning the "quaternary solid solution—A^3B^5 binary compound" systems are listed in Table 2.

By employing ternary solid solutions as substrates it is possible to widen the range of band gaps in quaternary solid solutions of A^3B^5, A^2B^4, or A^4B^6 compounds to 0.06-4 eV [6].

TABLE 1. Stress in Heteroepitaxial Layers of Quaternary Solutions

x, molar fraction	y, molar fraction	$\Delta a^*/a$, % (at T_{cr})	$\Delta \alpha^*$, °C$^{-1} \cdot 10^6$	$\sigma \Delta a$, kg mm^{-2} (at T_{cr})	$\sigma \Delta \alpha$, kg mm^{-2}	σ_R^*, kg mm^{-2}	T_{cr}, °C
\multicolumn{8}{c}{$Ga_xIn_{1-x}As_ySb_{1-y}$—InAs}
0.91	0.17	+0.16	1.86	16.4	7.7	0.7	530
0.90	0.20	−0.01	2.79	3.1	8.9	7.8	530
0.91	0.21	−0.08	3.00	9.1	6.7	13.3	530
\multicolumn{8}{c}{$Ga_xIn_{1-x}As_ySb_{1-y}$—GaSb}
0.91	0.06	+0.2	−0.012	16.6	1.9	15.2	480
0.91	0.09	−0.04	−0.097	1.6	1.5	2.7	480
0.91	0.09	−0.07	−0.36	3.4	2.0	5.3	530
\multicolumn{8}{c}{$Ga_xIn_{1-x}As_yP_{1-y}$—InP}
0.1	0,25	+0.09	0.523	8,6	2.3	1.9	670
0.18	0.36	−0.03	0.657	3.2	2.1	6.5	670
0.11	0,30	−0.16	1.015	22.9	3.9	3.4	670

* Measurements carried out in MISiS (chair of semiconductor materials science) on samples prepared in the Institute of Rare Metals.

TABLE 2. Quaternary Solid Solutions of A^3B^5 Compounds Isoperiodic with Binary Compounds

Solid solution	Binary compound	Range of energies of direct optical transitions, eV
$Ga_xIn_{1-x}As_yP_{1-y}$	InP	0.8 —1.34
	GaAs	1.45—2.1
$Al_xGa_{1-x}As_ySb_{1-y}$	GaSb	0.72—1.1
	InAs	0.36—0.75
$Ga_xIn_{1-x}As_ySb_{1-y}$	GaSb	0.30—0.72
$Al_xIn_{1-x}As_ySb_{1-y}$	InAs	0.36—0.90
$InAs_xSb_yP_{1-x-y}$	InAs	0.36—0.60

At the present time quaternary solid solutions of semiconductors are mostly prepared by LPE. Consequently, it is necessary to carry out thermodynamic calculations and experimental studies of phase equilibria in quaternary semiconductor systems, and especially for compositions enriched with the solvent metal, from which the required quaternary solid solution is crystallized.

The most widely used method of calculating phase equilibria in quaternary systems is based on treating a quaternary solution as a "quasibinary" solution of two ternary solid solutions, in one of which the substitution occurs in the group III element sublattice, and in the other in the group V element sublattice [7]. In addition, both the liquid and the solid phases are treated by the regular solution approximation.

The available experimental data on solubility in the Ga-In-As-P and Ga-In-As-Sb systems are in satisfactory agreement with the calculations carried out by the method suggested in [7]. As an example, Fig. 5 shows solubility isotherms in the Ga-In-As-Sb system for compositions of practical importance. The solidus composition was assumed to be equal to the composition of the epitaxial layer at the substrate-epitaxial layer interface. The epitaxial layers were grown on {111} surfaces of indium arsenide and gallium antimonide substrates. The experimental points on the liquidus isotherms were obtained by differential thermal analysis.

As in the case of the epitaxy of ternary solid solutions on binary substrates, thermodynamic equilibrium between the high-temperature solution and the substrate does not occur in LPE of quaternary solid solutions on binary substrates (or a ternary solid solution). This absence of equilibrium in the substrate-solution isothermal system must lead to substrate dissolution. However, a quasi-equilibrium in a substrate-solution system may set up without appreciable substrate dissolution, for example, owing to diffusional processes at the interface. Diffusion will result in a boundary layer having an intermediate composition (between the substrate composition and the equilibrium solid phase composition). The actual process of LPE growth of solid solutions is usually achieved from a high-temperature solution supercooled with respect to its liquidus point. The degree of supercooling in the melt is determined to a large extent by the kinetics of dissolution of the substrate in the melt in isothermal contact with it. Usually this is minimal for epitaxy on a substrate whose dissolution leads to an increase in the liquidus point of the high-temperature solution.

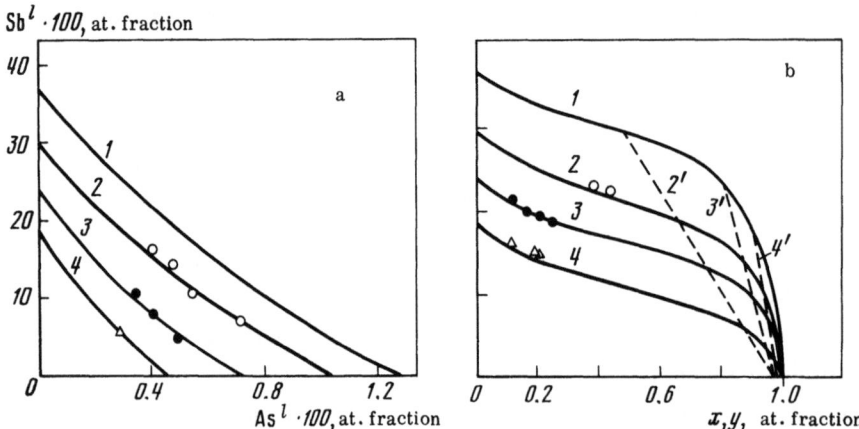

Fig. 5. (a) Liquidus isotherms and (b) arsenic content (1-4) and gallium content (2'-4') in $Ga_xIn_{1-x}As_ySb_{1-y}$ solid solution as a function of melt composition. 1, 2 and 2', 3 and 3', 4 and 4' correspond to x_{Ga}^l = 0, 10, 20, and 30 at. %, respectively.

The initial high-temperature solution used for the LPE of quaternary solutions usually contains an exact proportion of the components, in correspondence with the composition of the crystallizing solid phase lattice matching to the substrate. If the substrate when dissolved in the melt increases its liquidus point, the composition of the initial melt is sometimes chosen with the substrate dissolution in the course of epitaxial growth taken into account. For example, the substrate is placed under the solution at a temperature lower than the crystallization point, and then the substrate-solution system is heated to the prescribed temperature for epitaxial growth.

It is mandatory to take into account the characteristics of the quaternary system phase diagram and the absence of thermodynamic equilibrium between the melt and the substrate when choosing an optimum substrate material for preparing perfect heterojunctions. Uncontrolled substrate etching changes the composition of the molten solution at the interface and makes it difficult to obtain a quaternary solid-solution epitaxial layer lattice matched to the substrate.

Obviously, the specific device for which the heterostructure is intended must also be considered in choosing an optimum substrate material for heterostructure. For example, in the case of electroluminescent diodes in which a quaternary solid solution is the active medium, it is important that the radiation generated within the heterojunction be smaller in energy than the band gap of the substrate, otherwise the efficiency of the device will be much reduced by absorption of the radiation in the substrate material. For this reason an optimum substrate material for a heterostructure based on $Ga_xIn_{1-x}As_ySb_{1-y}$ which is used in electroluminescent diodes for the 2-2.5 μm spectral range, is, for example, GaSb and not InAs, although the dissolution of the substrate in the In-Ga-As-Sb molten solution is much lower in the case of InAs substrate than for GaSb substrates. Consequently, InAs must be considered as the optimum substrate material for lattice matched heterostructures based on $Ga_xIn_{1-x}As_ySb_{1-y}$ for those devices in which the situation outlined above is not of principal importance.

When quaternary layers lattice matched to a ternary solid solution are grown, the ternary is usually obtained in the process of growing the epitaxial heterostructure as a whole. In order that the ternary layer grown on a binary substrate should have a perfect structure, it is necessary to form a layer with a composition realizing the transition from the substrate to the prescribed composition of the final solid solution. Studies [4, 8] show that if the growing epitaxial layer has a lattice spacing exceeding that of the binary substrate, the distribution of the compositions in the transient region must presumably be stepwise. In this case the transient region is a series of several epitaxial layers (with the composition constant across each of them) with composition changing stepwise at a heterojunction. The density of dislocations in the layers is appreciably reduced by the "incorporation" of tilted dislocations into the misfit dislocation network at each interface. If the ternary layer has a lattice spacing smaller than that of the substrate material, the transient region presumably must have a smooth composition gradient across the layer thickness from the substrate composition to the prescribed constant-composition solid solution [8]. If these requirements are not satisfied, the structure of ternary epitaxial layers sharply deteriorates; cracks occur, and often it becomes impossible to grow single crystal layers differing from the substrate material in composition. This, of course, makes the formation of lattice matched heterostructure of the type "quaternary solid solution—ternary solid solution" very doubtful.

One of the alternatives to the formation of structurally perfect layers of ternary solid solutions, on which quaternary solid solution is to be grown, is to use as a transient layer, in growing the ternary solid solution, a graded-composition quaternary solid solution. Here the quaternary solid solution must be lattice matched with the binary substrate and must have a high value of the distribution coefficient ($K \gg 1$) of the constituent not contained in the ternary solid solution. For example, if $Ga_xIn_{1-x}As(GaAs)$, $GaAs_xSb_{1-x}(GaAs)$, and $InAs_xSb_{1-x}$· (InAs) epitaxial layers are grown, solid solutions $Ga_xIn_{1-x}As_yP_{1-y}$, $GaAs_xP_ySb_{1-x-y}$, and $InAs_xP_ySb_{1-x-y}$, respectively, can be used as the transient region materials lattice matched with the indicated binary compounds. On cooling, the quaternary solutions tend to be depleted in phosphorus, so that at a certain distance from a heterojunction between a binary compound and a quaternary solid solution we may find a structurally perfect layer of the ternary solid solution (containing no phosphorus) with the lattice spacing considerably different from that of the substrate material; such, for example, is the case for $InAs_xSb_{1-x}$ ($x \approx 0.85$) when $InAs_xP_ySb_{1-x-y}$(InAs) is used as the transient layer material [9]. A quaternary solid solution lattice matched to the substrate may then be grown on the layers of the ternary solid solution obtained. One example of the successful use of this type of structure in actual devices is the heterostructure $GaAs - GaAs_xP_ySb_{1-x-y} - GaAs_ySb_{1-y} - Al_zGa_{1-z}As_ySb_{1-y} - GaAs_ySb_{1-y} - Al_zGa_{1-z}As_ySb_{1-y}$ [10].

Thermal etching of the substrate prior to placing it under the solution sometimes results from its decomposition (e.g., the formation of metal drops on the surface due to evaporation of the group V component). To avoid this, the stage of growing heteroepitaxial compositions on a binary substrate is usually preceded by the growth of a "buffer" layer of the substrate composition, which improves the planarity of the final heterojunction.

In contrast to liquid phase homoepitaxy in which the initial stages of growth of the epitaxial layers are mostly determined by the solution supercooling, by inhomogeneity of the substrate material, and by thermal fluctuations in the molten solution, the decisive influence on the initial stages of liquid phase heteroepitaxy in actual growth conditions is exerted by the lattice misfit at the interface. Preliminary studies indicate that the value of Δa at the heterojunction interface may significantly affect the course of nucleation and layer growth in multicomponent solid solutions. One important feature of the multicomponent systems investigated is that even a very slight change in the solid solution composition (not leading in itself to substantial changes of the basic chemical characteristics) can result in appreciable changes of the lattice spacing at the interface; that is, it can change Δa, and consequently, the interfacial surface energy. This ultimately changes the energy of nucleation, and thus the characteristics of nucleation and layer growth in the initial stage of heteroepitaxy. This behavior is most pronounced in systems in which the distribution coefficients of the major components of the solid solution are much different from unity, creating considerable changes in the composition of the layers in the course of growth.

As an example, consider the growth of single crystal layers of $Ga_xIn_{1-x}As(GaAs)$ by doping the initial ternary melt with small amounts of phosphorus possessing the distribution coefficient $K \gg 1$ and serving to form an intermediate layer of the quaternary solid solution $Ga_xIn_{1-x}As_yP_{1-y}$. Segregation of phosphorus within this layer makes the lattice period vary from a_{GaAs} to the lattice spacing for $Ga_xIn_{1-x}As$ of a given composition. If the misfit at the interface $\Delta a \leq 1 \cdot 10^{-3}$ Å, the intermediate layer of the quaternary solid solution begins to grow, at low supercoolings, by three-dimensional nucleation. The density of the nuclei formed is low and nearly independent of time (Fig. 6a), and their composition remains practically constant (both through and across the layer) in the course of growth, until the first continuous layer is formed; the thickness of this layer may reach 8-10 μm. This growth is accompanied by trapping of the solution along the boundaries of the merging nuclei and by the intensive generation of dislocations close to these boundaries (Fig. 6b). The surface of these layers is characterized by a cellular structure with traces of melt incorporation along the cell boundaries. In this case sharply pronounced growth selectivity appears, most probably because the depletion of the solution in phosphorus around the growing nuclei makes the formation of fresh nuclei energetically disadvantageous, since their composition no longer provides perfect matching to the substrate. Moreover, the depletion of the melt in phosphorus reduces the equilibrium temperature.

If the starting composition of the solution is such that the initial lattice parameter misfit $\Delta a \approx 10^{-2}$ Å, considerably greater supercooling is necessary for growth. The density of the nuclei forming is thereby sharply increased, and although the growth is still three-dimensional, the first continuous layer of the quaternary solid solution is completed at a thickness of 1 μm or less. The contribution of dislocations, generated in such layers at the boundaries of merging nuclei, to the general defect density is considerably reduced. The layers grown under these conditions have mirror-like surfaces and a typical morphology characteristic of layers with a three-dimensional network of misfit dislocations.

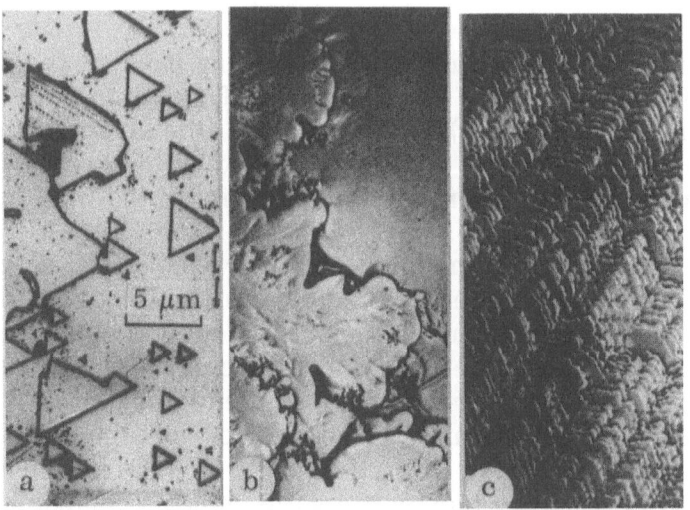

Fig. 6. Stages of growth of heteroepitaxial structure $GaAs-Ga_xIn_{1-x}As_yP_{1-y}-Ga_xIn_{1-x}As$ with layer thickness (a) 5 μm, (b) 10μm, and (c) 15 μm.

The crystallization of epitaxial layers of multicomponent solid solution thus necessitates the determination of the optimum supercooling ΔT of the solution, since in a number of systems the relation of ΔT to the composition of the crystallizing solid solution (and as a result, to Δa) may prove essential; this is especially important in growing thin (less than 0.5 μm) layers of solid solutions.

The decisive role of Δa during the initial stages of growth makes it difficult to evaluate the direct effect of the supercooling in the solution because it usually affects the composition of the crystallizing epitaxial layers and, consequently, the value of Δa. The experimental data available in our laboratory indicate that in the case of heterostructure crystallization in the system $Ga_xIn_{1-x}As_yP_{1-y}$(InP) in which, according to [11], supercooling of the solution (up to 10°C) produces practically no effect on the composition of the crystallizing solid solution, no appreciable changes in the surface morphology of layers were noticed when the supercooling of the solution was varied in the same range. However, an increase in the supercooling to 5-7°C for crystallization of $Ga_xIn_{1-x}As_ySb_{1-y}$(GaSb) results in the formation of growth pyramids resembling those observed under similar conditions in LPE of GaAs. The growing pyramids coalesce and form a terrace-shaped (and sometimes cellular) surface on the epitaxial layers. No growth pyramids were observed at lower values of supercooling, and the surface was mirror-smooth, sometimes with a pronounced "network"-like morphology.

Any discussion of problems related to the initial stages of the growth of epitaxial layers by liquid phase heteroepitaxy must also take into account the effect of epilayer composition stabilization. This effect was first observed in the crystallization of $Ga_{0.5}In_{0.5}P$ on GaAs substrates [12], and in a wider range of solid solution compositions on $GaAs_xP_{1-x}$ substrates [13].

Analysis [14] shows that substantial stabilization of composition is only possible for the case of a considerable departure of the solid solution from the ideal composition; for example, when the solution is close to the spinodal decomposition point. As for the quaternary solid solutions lattice matched to a binary compound, the above-mentioned effect may take place, for example, in the systems

$$Ga_xIn_{1-x}As_yP_{1-y} (GaAs, Ga_xIn_{1-x}P) \text{ for } x \approx 0.5 \text{ and } y \to 0$$

and

$$Ga_xIn_{1-x}As_ySb_{1-y} (GaSb, InAs) \text{ for } 0.3 < y < 0.7 \text{ and } x \to 1.$$

The effect of composition stabilization must be taken into account when constructing the phase diagrams of quaternary systems because the composition of the epitaxial layer, often assumed to be the same as the solidus composition, may in practice be very different.

In the case of LPE of quaternary solid solutions, as in the epitaxy of ternary solid solutions, there is an orientation dependence of effective distribution coefficients of the major constituents entering the solid solution. One can say that at least for the solid solutions $Ga_xIn_{1-x}As_yP_{1-y}$(InP) and $Ga_xIn_{1-x}As_ySb_{1-x}$(InAs) there is a tendency for a reduction of the distribution coefficients of the constituents to increase the lattice spacing of the epitaxial layer (arsenic and indium in $Ga_xIn_{1-x}As_yP_{1-y}$ and antimony and indium in $Ga_xIn_{1-x}As_ySb_{1-y}$), when the substrate orientation is changed from $(\bar{1}\bar{1}\bar{1})$ to (111). In the case of the $Ga_xIn_{1-x}As_ySb_{1-y}$(GaSb, InAs) it was shown that the difference in distribution coefficients on the polar (111) surface levels out as the temperature of epitaxial growth is increased (Fig. 7).

If the growth of the solid solution layer is not preceded by the formation of a "buffer" layer of the substrate material, the orientational dependence of the distribution coefficients

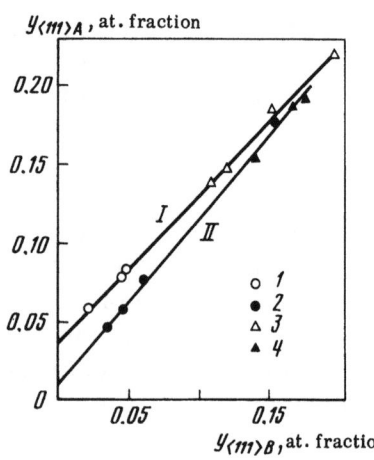

Fig. 7. Arsenic content in epitaxial layers $Ga_xIn_{1-x}As_ySb_{1-y}$ at growth temperatures 800 K (I) and 880 K (II). (1, 2) GaSb substrate with orientations (111) and ($\bar{1}\bar{1}\bar{1}$), respectively; (3, 4) InAs substrate with orientations (111) and ($\bar{1}\bar{1}\bar{1}$), respectively.

of the major constituents may lead to inhomogeneous composition in the grown epitaxial layers. The reason for this is that chemical and thermal etching of the substrate fail to achieve the ideal smoothness of the substrate surface oriented in the desired crystallographic plane. This factor, together with the undesirable dissolution of the substrate in the solution, results in the possibility of the crystallization of a quaternary solid-solution layer on substrate surface segments deviating from the required crystallographic plane. The growth of a "buffer" layer of the substrate material allows the epitaxial layer—substrate interface to be brought to the desired crystallographic plane, which makes it possible to improve the homogeneity of epitaxial layers.

It must be mentioned in conclusion that although the time spent on practical application of quaternary solid solutions in devices has been quite short, considerable success had been achieved. Thus, high-efficiency light-emitting diodes [5], photodiodes [15], IR lasers with low generation threshold [16. 17], and high-efficiency photocathodes [18] have been made utilizing InP-$Ga_xIn_{1-x}As_yP_{1-y}$ heterostructures. At present the $Ga_xIn_{1-x}As_y$·P_{1-y} $Ga_xIn_{1-x}P$ structure yielded the shortest wavelength laser [9], and $Ga_xIn_{1-x}As_ySb_{1-y}$·(GaSb) yielded the longest wavelength room-temperature lasers [20, 21]. There is no doubt that quaternary solid solutions represent very promising semiconductor materials for a number of optoelectronic devices [6].

Literature Cited

1. V. B. Osvensky and M. G. Mil'vidsky. Defect formation in heteroepitaxial structures of A^3B^5 compounds for optoelectronics. Kristallografiya, 22, 431-444 (in Russian) (1977).
2. L. V. Druzhinina, V. T. Bublik, L. M. Dolginov, P. G. Eliseev, J. Z. Pinsker, P. G. Kerbelev, V. B. Osvensky, and M. G. Shumsky. Crystal perfection of heterostructures in a system of AlAs-GaAs solid solutions and its effect on parameters of injection lasers. Zh. Tekh. Fiz., 44, 1503-1510 (in Russian) (1974).
3. L. M. Dolginov, L. V. Druzhinina, M. G. Mil'vidsky, L. M. Morgulis, E. G. Shevchenko, and T. G. Yugova. Structural defects in epitaxial layers of A^3B^5 solid solutions. In: Growth and Doping of Semiconductor Crystals and Films, Nauka, Novosibirsk, part 2, 240-245 (in Russian) (1977).
4. A. V. Govorkov, L. V. Dolginov, M. G. Mil'vidsky, L. M. Morgulis, V. B. Osvensky, Yu. M. Fisman, and T. G. Yugova. Structural defects and luminescence in $GaAs_xSb_{1-x}$ epitaxial layers. Kristallografiya, 22, 1060-1068 (in Russian) (1977).

5. L. M. Dolginov, N. Ibrakhimov, M. G. Mil'vidsky, V. Yu. Rogulin, and E. G. Shevchenko. High-efficiency electroluminescence in $Ga_xIn_{1-x}As_{1-y}P_y$. Fizika i Tekhnika Poluprovodnikov, 9, 1319-1323 (in Russian) (1975).
6. L. M. Dolginov, P. G. Eliseev, and M. G. Mil'vidsky. Multicomponent solid solutions and their application to lasers. Kvantovaya Elektronika, 3, 1381-1393 (in Russian) (1976).
7. M. Ilegems and M. B. Panish. Phase equilibria in III-V quaternary systems — Application to Al-Ga-P-As. J. Phys. Chem. Solids, 35, 409-420 (1974).
8. G. H. Olsen. Interfacial lattice mismatch effects in III-V compounds. J. Cryst. Growth, 31, 223-234 (1975).
9. R. E. Gertner, D. T. Cheung, A. M. Andrews, and J. T. Longo. Liquid phase epitaxial growth of $InAs_xSb_yP_{1-x-y}$ layers on InAs. J. Electron. Mater., 6, 163-166 (1977).
10. R. E. Nahory, M. A. Pollack, and J. C. DeWinter. Growth and continuous compositional grading of $GaAs_{1-x-z}Sb_xP_z$ by liquid phase epitaxy. J. Appl. Phys. 48, 320-324 (1977).
11. J. J. Hseih, M. C. Finn, and J. A. Rossi. Conditions for lattice matching in the LPE growth of GaInAsP layers on InP substrates. In: 6th Intern. Symp. N. American Conf. on GaAs and Relat. Comp. (Inst. Phys. Conf. Ser. 336). London, 37-44 (1977).
12. G. B. Stringfellow. The importance of lattice mismatch in the growth of $Ga_xIn_{1-x}P$ epitaxial crystals. J. Appl. Phys. 43, 3455-3460 (1972).
13. I. N. Arsentyev, D. Z. Garbuzov, S. G. Konnikov, V. D. Rumyantsev, and D. N. Tretyakov. Fabrication of $Ga_xIn_{1-x}As_yP_{1-y}$ solid solutions by liquid phase epitaxy and their study by electron microprobe. In: Growth and Doping of Semiconductor Crystals and Films, Nauka, Novosibirsk, part 2, 268-272 (in Russian) (1977).
14. V. V. Voronkov, L. M. Dolginov, A. N. Lapshin, and M. G. Mil'vidsky. Effect of stabilization of composition in epitaxial solid-solution layers. Kristallografiya, 22, 375-378 (in Russian) (1977).
15. L. M. Dolginov, P. G. Eliseev, M. G. Mil'vidsky, B. N. Sverdlov, V. M. Chupakhina, and E. G. Shevchenko. Photoelectric properties of heterostructures based on four-component GaInAsP solutions, with high quantum efficiency at 1060 nm. Pis'ma v Zh. Eksp. Teor. Fiz., 2, 631-634 (in Russian) (1976).
16. L. M. Dolginov, P. G. Eliseev, M. G. Mil'vidsky, B. N. Sverdlov, and E. G. Shevchenko. CW stripe-geometry heterolaser based on four-component GaInAsP solid solution. In: Physics Comments, Lebedev Institute of Physics, Moscow, No. 8, 38-41 (in Russian) (1976).
17. J. J. Hsieh, J. A. Rossi, and J. P. Donnely. Room-temperature CW operation of GaInAsP/InP double-heterostructure diode lasers emitting at 1.1 μm. Appl. Phys. Lett. 28, 709-711 (1976).
18. J. S. Escher, G. A. Antypas, and J. Edgecumbe. High quantum-efficiency photoemission from an InGaAsP photocathode. Appl. Phys. Lett., 29, 153-155 (1976).
19. Zh. I. Alferov, I. N. Arsentyev, D. Z. Garbuzov, V. D. Rumyantsev, and V. P. Ulin. On maximum levels of output energy in injection GaInAsP heterolasers emitting in the yellow-green range at 77 K. Pis'ma v Zh. Eksp. Teor. Fiz. 2, 481-483 (in Russian) (1976).
20. L. M. Dolginov, L. V. Druzhinina, P. G. Eliseev, M. G. Mil'vidsky, and B. N. Sverdlov. New room-temperature injection heterolaser for 1.5 - 1.8 μm range. Kvantovaya Elektronika, 3, 465 (in Russian) (1976).
21. L. M. Dolginov, L. V. Druzhinina, P. G. Eliseev, I. V. Krukova, V. I. Lescovitch, M. G. Milvidski, B. N. Sverdlov, and E. G. Schevchenko. Multicomponent solid-solution semiconductor lasers. IEEE J. Quant. Electron., QE-13, No. 8, 609-611 (1977).

GROWTH AND STRUCTURE OF SYNTHETIC AMETHYST CRYSTALS

L. I. Tsinober, V. E. Khadzhi, E. M. Tsyganov, M. I. Samoilovich and A. A. Shaposhnikov

The USSR Research Institute of Synthesis of Mineral Raw Materials, Aleksandrov, Vladimirskaya Region

The 50's was the period of a qualitatively new stage in studying quartz crystals. Hydrothermal techniques for growing large single crystals of quartz were developed, and very soon practically all of the natural and previously unknown modifications of quartz crystals were synthesized. The growth and extensive application of large quartz crystals was a stimulus to studying the structure of this mineral; at the same time, the growth of crystals under defined physical chemical conditions offered an opportunity to study the genesis of imperfections. This was also the time when such highly informative methods of studying the structure of crystals as radio-frequency, optical and Mössbauser spectroscopy, high-resolution electron microscopy, etc. were developed and widely used.

The present paper presents the main results of a program devoted to growing and studying the structure of amethyst crystals. Amethyst is one of the most valuable gem-grade modifications of quartz.

We shall begin by briefly enumerating the basic distinctive features of amethyst, established for natural gems during the "pre-synthesis" period [1, 2].

1. **The violet coloring of amethyst is "developed" in the crystal by an ionizing radiation** and undergoes bleaching on heating the samples to 400°C. Annealing at temperatures above 500°C produces clouding of many (although far from all) amethyst stones; this is usually not observed in colorless crystals and in smoke-colored quartz crystals.

2. **Coloring of amethyst crystals has a well-pronounced sectorial structure.** The ratio of coloring intensities (I) for basic growth pyramids is given by the inequalities

$$I_{\langle R \rangle} > I_{\langle r \rangle} \gg I_{\langle m \rangle},$$

where the following notation for quartz crystal faces is used: R denotes the basic positive rhombohedron; r denotes the basic negative rhombohedron; and m denotes the hexagonal prism. Growth pyramids on the faces are denoted by the corresponding letters in angle brackets.

3. **Twinning is very typical for amethyst crystals:** Dauphine twins and especially (in $\langle R \rangle$ pyramids) polysynthetic Brazil twins.

4. **Anomalous symmetry of light absorption (coloration), typical for biaxial crystals,** was observed in many amethyst crystals.

5. Mineralogical data show that amethyst crystals are as a rule formed in conditions of high iron concentration, that is, from paragenesis with oxide ferriferous minerals: hematite, goethite, limonite, etc. Consequently, it has been suggested by some researchers that the amethyst coloring is associated with iron as an impurity in quartz. However, no physically substantiated evidence of this relationship was found.

Growth of Amethyst Crystals

Reproducible hydrothermal growth of quartz crystals with amethyst coloring was first achieved in the SiO_2-K_2CO_3-H_2O system in the presence of iron impurity in 1959 [3]. Quartz crystals grown in this system on platelet seeds with pinacoidal and other growth surfaces had green and russet primary coloring [4]. The spectroscopic analysis data, the results of annealing, and electron microscopy analysis [5] have shown that these types of coloring are mostly associated with the incorporation of a nonstructural colloidal impurity "colored" by ions of trivalent (russet coloring) and divalent (green coloring) iron. In russet-colored crystals optical spectra and electron spin resonance spectra indicate that the Fe^{3+} ions also occupy interstitial positions in the quartz lattice [6]. However, growth pyramids of the basic rhombohedra of these crystals do not capture (or capture only weakly) nonstructural impurities, and remain practically colorless. Irradiation of growth pyramids <R> and <r> causes the development of the violet amethyst coloring. The same growth pyramids in sodium systems are known to absorb selectively the structural impurity of aluminum and alkali metal ions (as compensators), and irradiation then gives them a smoke coloring [7]. Both the intensity of the amethyst coloring and its purity in the first experiments on amethyst growth were far from sufficient for their use as gem material. A relatively pure violet coloring was observed only in <R> pyramids, while in <r> pyramids it had a substantial smoke hue. Subsequent experiments combined with a thorough study of the physical properties of the crystals made it possible to greatly improve the growth technique. The studies progressed in the following directions.

1. Measures were taken to remove the undesirable aluminum impurity from the system (by using especially pure charges and by introducing special components binding aluminum into poorly soluble compounds).

2. Much effort went into selecting optimum thermobaric parameters. This made it possible to achieve effective growth of amethyst on the faces of both positive and negative rhombohedra.

3. Special acidifying additives were found, ensuring the presence of iron mostly in the oxide form (a necessary condition for the formation of potential centers of amethyst coloring).

4. It is also important that the resistance of amethyst centers to thermal and irradiation treatment in crystals grown by the modified techniques proved substantially higher than the stability of smoke coloring centers which are difficult to suppress completely in the system studied. This made it possible to develop a method of heat and light treatment of the as-grown amethyst crystals which considerably improves the purity of the violet shade of coloring.

At the present moment an industrial technology of amethyst crystal grown in the potassium system is available at the Research Institute of Synthesis of Mineral Raw Materials. The gem quality of these crystals rivals that of the best natural stones and in many cases, according to the expert opinion, exceeds it.

The spectral analysis data on the content of the main impurity elements in different types of quartz crystals are listed in Table 1 which shows that the iron concentration in amethyst crystals (samples 5-7) is from several hundredths to several thousandths of one percent, while in crystals grown in the sodium system (samples 8-10) it never exceeds $1 \cdot 10^{-4}$ wt.%. There is a correlation between the coloring intensity and iron and potassium contents (cf. samples 5 and 6, 7). Especially high concentrations of iron (up to 0.081 wt.%) and potassium (up to 0.14 wt.%) are observed in crystals with green, russet, and orange coloring, obviously owing to the trapping of these elements as nonstructural impurities (samples 1-4). A similar behavior is found for aluminum and sodium impurities in crystals grown in the sodium system (cf. samples 8, 9 grown at low supersaturation, and sample 10 grown at high supersaturation and containing nonstructural impurities).

The conclusion that amethyst coloring is produced by iron impurity is also supported by reference experiments on growing quartz crystals in the K_2CO_3-SiO_2-H_2O system without iron impurity (with a platinum lining). These experiments yielded quartz crystals with slightly smoked irradiation-induced coloring in $<R>$ and $<r>$ pyramids.

Let us briefly mention the comparative properties of the sodium system in which the piezo-electric and optical quartz crystals are grown, and the potassium system in which amethyst crystals are grown. Iron in the sodium system is known to form a number of stable silicates, such as acmite $NaFe Si_2O_9$, riebeckite $Na_2Fe_3^{2+}Fe_2^{3+}Si_8O_{22}(OH)_2$, tuhualite $Na_3Fe_3Si_{12}O_{30}$, and phenaxite $Na_4Fe^{2+}Fe^{3+}Si_8O_{20}(OH)$. On one hand, this results in the self-passivation of autoclave walls, and on the other hand, this results in growing quartz crystals practically free of iron impurity which is known to considerably downgrade the piezo-electric and optical characteristics of the crystals. The same fact precludes the possibility of growing amethyst crystals in the sodium system. In the potassium system, however, such stable iron silicates are not formed, so that it becomes possible to grow quartz crystals with iron as a structural impurity (amethyst) or a nonstructural impurity (green and russet quartz).

A similar remark is valid with respect to the NH_4F-SiO_2-H_2O system in which iron can be free and quartz can be grown. This enabled the scientists of the Research Institute of Synthesis of Mineral Raw Materials to develop an original technique for growing amethyst crystals [8].

Investigation of Real Structure of Amethyst

Crystal Morphology.
Quartz crystals grown in the potassium and sodium systems have very similar morphologies with only minor differences. The shape of synthetic amethyst crystals is determined by the shape of R and r seed plates on which they are grown, by the ratio of growth rates of the basic faces, and by the duration of the growth cycle. A method of growing amethyst crystals on seeds shaped into the complete simple form (R- or r-rhombohedra) was also developed. Such crystals look very much like natural amethyst crystals.

The sectorial distribution of amethyst coloring in synthetic crystals is similar to that observed in natural crystals: coloring of the $<R>$ pyramid is as a rule more intense than that of the $<r>$ pyramid (Fig. 1). Both in the potassium and in the sodium systems the growth rate of the hexagonal prism faces is very low (by two to three orders of magnitude lower than the growth rate of R faces). However, it proved possible in some very prolonged cycles to achieve overgrowth up to 0.2 mm in thickness on m faces. The corresponding growth pyramids $<m>$ have practically no amethyst coloring centers and remain colorless after irradiation. We thus find for m faces an analogy with natural crystals. Preliminary

TABLE 1. Concentration of Impurity Elements in Synthetic Quartz Crystals, wt. %

Sample No.	Coloring initial	Coloring after γ-irradiation	Growth pyramid	Al	Fe	Na	K	Li	Growth conditions, admixtures
				System K_2CO_3–SiO_2–H_2O					
1		Green*	⟨c⟩**	$1 \cdot 10^{-3}$	$8.1 \cdot 10^{-2}$	$2.6 \cdot 10^{-3}$	$1.4 \cdot 10^{-1}$	$3.6 \cdot 10^{-4}$	Fe
2		Russet*	⟨c⟩	$1.1 \cdot 10^{-3}$	$6.8 \cdot 10^{-2}$	$2.5 \cdot 10^{-3}$	$1.3 \cdot 10^{-1}$	$8 \cdot 10^{-5}$	Fe
3		Green*	⟨n⟩**	$1.8 \cdot 10^{-3}$	$3.7 \cdot 10^{-2}$	—	—	—	Fe
4		Orange*	⟨c⟩	$1.5 \cdot 10^{-3}$	$3.6 \cdot 10^{-2}$	$8 \cdot 10^{-4}$	$1.3 \cdot 10^{-1}$	$2.2 \cdot 10^{-3}$	Fe & compounds of Li
5	Colorless	Violet	⟨r⟩	$1.1 \cdot 10^{-3}$	$6 \cdot 10^{-3}$	$1.6 \cdot 10^{-4}$	$1.7 \cdot 10^{-3}$	$9 \cdot 10^{-4}$	Fe & compounds of Li
6	Colorless	Dark Violet	⟨R⟩	$1.1 \cdot 10^{-3}$	$1.4 \cdot 10^{-2}$	$5 \cdot 10^{-4}$	$7 \cdot 10^{-3}$	$2.5 \cdot 10^{-4}$	Fe
7	Colorless	Dark Violet	⟨R⟩	$1.5 \cdot 10^{-3}$	$1.4 \cdot 10^{-2}$	$7.3 \cdot 10^{-4}$	$7.6 \cdot 10^{-3}$	$3 \cdot 10^{-4}$	Fe
				System Na_2CO_3–SiO_2–H_2O					
8	Colorless	Slightly smoked	⟨c⟩	$8 \cdot 10^{-4}$	$1 \cdot 10^{-4}$	$1 \cdot 10^{-4}$	$1 \cdot 10^{-4}$	$5 \cdot 10^{-5}$	Low supersaturation Li compounds, low supersaturation
9		Colorless	⟨c⟩	$2 \cdot 10^{-4}$	$1 \cdot 10^{-4}$	$1 \cdot 10^{-4}$	$1 \cdot 10^{-4}$	$4 \cdot 10^{-5}$	
10	Colorless	Slightly smoked	⟨c⟩	$1.7 \cdot 10^{-3}$	$1 \cdot 10^{-4}$	$1.9 \cdot 10^{-3}$	$2 \cdot 10^{-4}$	$6 \cdot 10^{-5}$	High supersaturation
				Natural morion crystal					
11		Dark smoked	⟨R⟩, ⟨r⟩	$7.4 \cdot 10^{-4}$	$1 \cdot 10^{-4}$	$1 \cdot 10^{-4}$	$1 \cdot 10^{-4}$	$8 \cdot 10^{-4}$	—

* Sample exhibits post-anneal clouding.
**⟨c⟩, ⟨n⟩ denote growth pyramids of pinacoid face and positive trigonal prism, respectively.

data show that the distribution coefficient of amethyst coloring centers on faces $<R>$ and $<r>$ becomes larger as the growth rate increases.

Twinning. The formation of Dauphiné and especially Brazil twins is quite typical in synthetic amethyst, as it is in natural stores. Figure 2 illustrates the massive formation of Dauphiné twins at the r face of synthetic amethyst (R in growths in an r crystal). In many cases Dauphiné twins grow so intensively that after some time the r face is completely replaced by the R face. Figure 3 shows a sectional view of twins, their nucleation (in the vicinity of the seed surface), and the progress in the growth of twins. As R faces capture color centers much more strongly than r faces, twins have a more intensive coloring and are seen as cones with apices at the seed. Some data indicate that although these twins were produced in the course of growth, their origin is mechanical.

Sclerometric investigation of quartz [9] demonstrated that Dauphiné twins are formed in synthetic quartz crystals, and especially in amethyst crystals, at very low indenter loads. They are formed only on r surfaces facing upward during growth, that is, they nucleate on mechanical inclusions deposited on the face surface from the solution (these inclusions are visible under a microscope at the apex of each twin cone). It can be assumed that such inclusions produce a stress sufficient to form Dauphiné twins by a strain mechanism.

Brazil twins are quite frequent both in synthetic amethysts and in natural samples. Electron microscopic study of Brazil twins shows that they are formed of R lamellae 100 to 1000 Å thick (Fig. 4). It is not clear what causes a massive formation of Brazil twins in amethyst crystals; it can be assumed that these twins facilitate relaxation of stress generated in quartz when the structural iron impurity is incorporated. X-ray topography shows that synthetic amethyst is characterized by a high degree of imperfection: the density of growth dislocations in synthetic crystals cannot be counted directly, but definitely exceed 10^6 cm^{-2}.

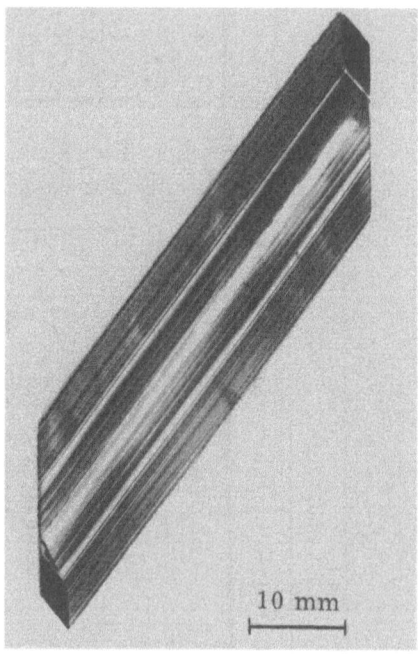

Fig. 1. Polished section of a synthetic amethyst r-crystal, parallel to yz plane (seed plate cut from an irradiation-resistant quartz).

Fig. 2. Dauphiné twins (R ingrowth on an r face) of a synthetic amethyst crystal.

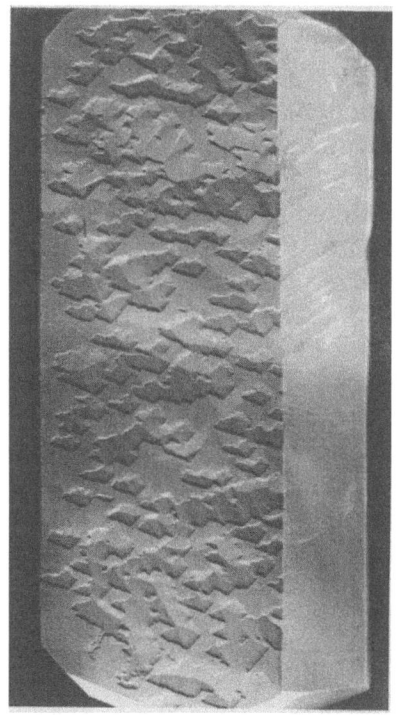

Fig. 3. Polished section of a synthetic amethyst r-crystal, parallel to yz plane.

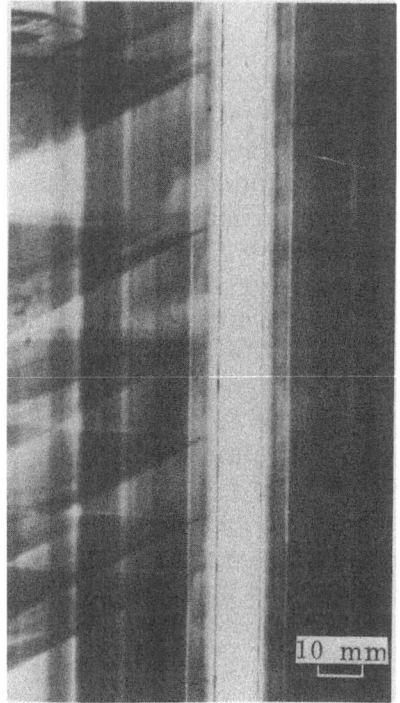

Fig. 4. Polysynthetic Brazil twins in amethyst (electron micrograph, photo by V. G. Balakirev).

On the Nature of Amethyst Coloring. It has been possible in recent years to determine the nature of amethyst coloring on the basis of ESR and optical spectroscopy data [10-12]. It was found that the initial amethyst has two types of Fe^{3+} impurity ions, namely: Fe^{3+} substituting for silicon in tetrahedra (center S_1), and Fe^{3+} in a strongly distorted tetrahedral site in the structural channels (center I_4). Compensation of the excess of negative charge, produced by substituting Si^{4+} by Fe^{3+}, is achieved (mostly nonlocally) by Li^{1+}, H^{1+}, and Fe^{3+} ions in I_4 sites. After irradiation a predominant part of the Fe^{3+} (S_1) defects are oxidized to the unusual valence state Fe^{4+} ($3d^4$). Some of the Fe^{3+} (Si_1) centers are transformed to Fe^{3+} (S_2) in which protons act as compensators. The electrons released on oxidation reduce the interstitial Fe^{3+} (I_4) ions to the divalent state with a changed coordination, Fe^{2+} (I_6). The optical spectrum typical for amethyst is mostly determined by crystal field bands of the Fe^{4+} ions. A hypothesis suggesting the presence of these ions on the basis of optical spectroscopy data was presented in a number of publications [13-15]. ESR spectroscopy data confirming the presence of the Fe^{4+} ions in amethyst were given in ref. 16. The following constants of a $\Delta M \pm 4$ transition were found for the center with effective spin $S=2$: $g_{zz} = 1.9874$, $g_{eff} = 7.9502 \pm 0.0025$, $\Delta_2 = 10.166$ (level splitting ± 2). As the observed lines in the ESR spectrum had various populations not very different from those for the initial ESR spectrum of Fe^{3+}, one can reliably assume that the centers of amethyst coloring are iron ions in the valence state (Fe^{4+}).

The bands typical for the optical absorption spectrum of amethyst (absorption coefficient D as a function of wavelength λ) lie at 340, 540, and 940 nm (Fig. 5). Certain characteristics of these bands indicate that the problem must be treated in the transient crystal-field approximation. In the case of a distorted tetrahedron (with symmetry C_2) the lower state (^5T) of the Fe^{4+} ion is split in the transient crystal field approximation into three sublevels (^5A, ^5B, ^5B), and the upper state ^5E splits into two sublevels (^5A, ^5A). The band at 540 nm may be ascribed to the $^5A(^5B) \rightarrow {}^5A$ transition, and an intensive absorption at 340 nm to the $^5A(^5B) \rightarrow {}^3B(^3A)$. The absorption band at 940 nm poses a more difficult problem. Presumably, this band is associated both with the interstitial ion Fe^{2+} ($^5E \rightarrow {}^5T$ transition) and with the $^5A(^5B) \rightarrow {}^3A(^3B)$ transition for the Fe^{4+} ions. For the case of C_2 symmetry the A→B type transitions have transverse polarization, and A→A and B→B transitions longitudinal polarization. The optical absorption spectra recorded with polarized light show that the 540, 940, and 340 nm bands of amethyst have different polarizations. It can be conjectured that the 280 nm band is partially associated with the electron capture centers. The formation of a Fe^{4+} ion is acompanied by a loss of one electron which is captured either by an interstitial Fe^{3+} ion (thus producing Fe^{2+}) or by a lattice defect of an unknown nature.

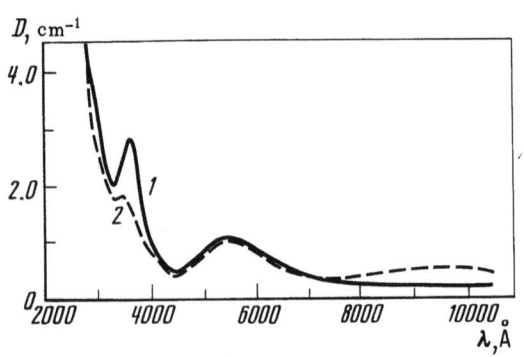

Fig. 5. Optical absorption spectra of synthetic amethyst, recorded in polarized light. (1, 2) electric vector E parallel and perpendicular to the z axis, respectively.

Anomalous Pleochroism of Amethyst Coloring. The r-crystals of synthetic amethyst (as for many natural amethyst crystals) show a light absorption symmetry different from the crystal symmetry. A similar phenomenon can be observed in <R> pyramids as well, but in this case the effect is frequently masked by Brazil twins. Figure 6 shows cross sections of absorption surfaces for the (0001) plane for the main bands of the amethyst spectrum. In normal quartz crystals these sections must be circular, while in the analyzed crystal they are elliptic. The nature of this reduction of symmetry became clear from ESR studies. It was found that the three equivalent positions of silicon in the quartz unit cell have very different populations of impurity ions (Fe^{3+} ions for amethyst) during growth on r and R faces. In natural amethysts with well-pronounced anomalous pleochroism this ratio is 30:4:1. The situation is clarified by Fig. 7, which shows a projection of the quartz structure (as tetrahedra) onto the (0001) plane. All three tetrahedra are geometrically (and therefore, physically) equivalent and can be superimposed by rotation about axis L_3. Figure 7b gives a projection of the same structure onto the rhombohedral plane (1011). The latter pattern is quite different: the tetrahedra which are equivalent in the bulk of the crystal and on the (0001) plane separate, in the projection onto the rhombohedron plane (whose intrinsic symmetry is unity), to three systems not related by symmetry elements. Tetrahedra within each such system are related by lattice translations. This nonequivalence of the systems leads to unequal distribution coefficients of the structural iron impurity. ESR data show [11, 12] that as the crystal grows via the faces of basic rhombohedra, the impurity is predominantly incorporated into one of the three sites. The difference in population of the silicon sites by

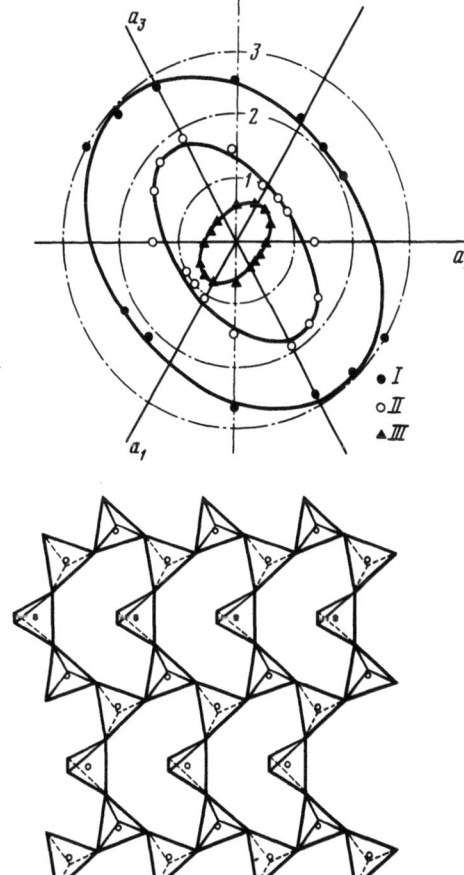

Fig. 6. Cross section of absorption surfaces in amethyst by the (0001) plane, at fixed wavelengths. (a_1-a_3) two-fold axis of quartz; (I-III) λ = 360, 550, 960 nm, respectively.

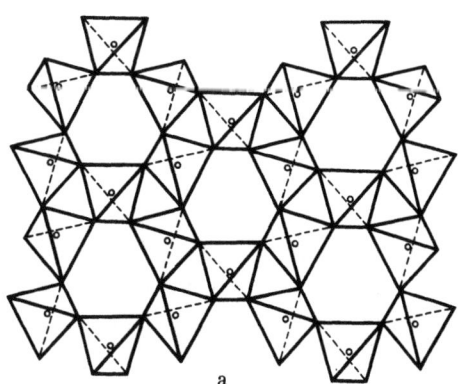

Fig. 7. Quartz structure projections onto planes (a) (0001), and (b) (1011).

the Fe^{3+} ions (potential coloring centers) reduces the symmetry of light absorption and, therefore, leads to the anomalous pleochroic effect. It has already been mentioned that the distortion in surface cross section shown in Fig. 6 (deviation from circular form typical for crystals with dichroic coloring) is associated with the fact that one of the subsystems of translationally equivalent tetrahedra is predominantly populated by the amethyst coloring centers. Therefore the difference in directions of ellipsoid axes for different bands (550 and 960 nm) indicates that these bands belong to different centers located on different L_2 axes in the quartz structure This sort of comparison of the ESR and optical data makes it possible to locate the sites which are predominantly populated by defects in the structures of various crystals. The anomalous pleochroic effect is closely related to the crystal growth conditions and depends on the state of the surface, growth temperature, and possibly, supersaturation [17].

The anomalous pleochroic effect significantly increases the gem quality of the alkaline synthetic amethyst, making it possible to fabricate from the same crystal gemstones with different shades of violet coloring (from purple-violet to lavender-blue), changing as the gem is tilted with respect to the incident light.

Conclusion

Amethyst crystals grown in the K_2CO_3-SiO_2-H_2O-Fe_2O_3 system have a structure which is strikingly similar to that of natural amethyst crystals: their optical spectra, sectorial distribution of coloring, twinning characteristics, and anomalous pleochroism are identical. One must also mention the identical IR spectra in the range of OH-defects (Fig. 8) and the necessary presence of potassium in the natural and synthetic amethyst, which is an impurity not typical for any other quartz modification. Obviously, this is a case in which the synthesis faithfully reproduces the conditions of amethyst crystallization in natural surroundings.

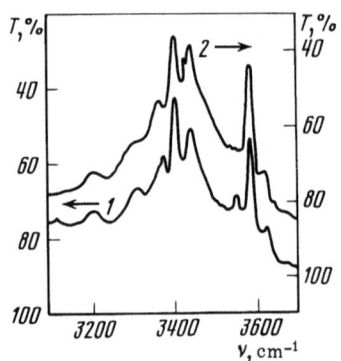

Fig. 8. Transmission spectra in the range of OH-defects of synthetic (1) and natural (2) amethyst. ν, cm^{-1}, is the wave number.

Literature Cited

1. J. D. Dana and E. S. Dana. The system of Mineralogy. 7th edition by C. Frondel. vol. 3, Silica Minerals. J. Wiley and Sons, Inc. New York and London (1962).
2. G. G. Lemmlein. Distribution of coloring in quartz crystals. Proc. of the Institute of Crystallography, No. 6, 260-288 (in Russian) (1951).
3. L. I. Tsinober and L. G. Chentsova. Synthetic quartz with amethyst coloring. Kristallografiya, 4, 633-635 (in Russian) (1959).

4. L. I. Tsinober, L. G. Chentsova, and A. A. Shternberg. On green and brown coloring of synthetic quartz crystals. In: Growth of Crystals, Vol. 2, AN SSSR Publ. House, Moscow, 61-67 (in Russian) (1959).
5. M. I. Samoilovich, L. I. Tsinober, and I. P. Khadzhi. Study of colored synthetic quartz crystals. Dokl. AN SSSR, 184, 91-93 (in Russian) (1969).
6. L. M. Matarrese, J. S. Wells, and R. L. Peterson. ESR spectrum of Fe^{3+} in synthetic quartz. Bull. Amer. Phys. Soc., 9, 502-503 (1964).
7. L. I. Tsinober. Distribution of smoke coloring in synthetic quartz crystals. In: Proceedings VNIIP, Gosgeoltekhizdat, Moscow, 3, 95-104 (in Russian) (1960).
8. V. S. Balitsky. Hydrothermal growth of large amethyst crystals in fluoride solutions. In: Vth USSR Conference on Crystal Growth, Tbilisi, Abstracts. Inst. of Cybernetics, AN GSSR Publ. House, Tbilisi, 2, 81-82 (in Russian) (1977).
9. V. G. Balakirev. Sclerometry of quartz. Izv. AN SSSR, Ser. Geol., No. 3, 83-92 (in Russian) (1976).
10. D. R. Hutton. Paramagnetic resonance of Fe^{+++} in amethyst and cytrine quartz. Phys. Lett., 12, 310-311 (1964).
11. T. Y. Barry, P. McNamara, and W. J. Moore. Paramagnetic resonance and optical properties of amethyst. J. Chem. Phys., 42, 2599-2606 (1965).
12. L. G. Chentsova, L. I. Tsinober, and M. I. Samoilovich. Investigation of amethyst-colored quartz. Kristallografiya, 11, 236-244 (in Russian) (1966).
13. G. Lehmann. Farbzentren des Eisens als Ursache der Farbe von Amethyst. Z. Naturforsch., 22a, 2080-2085 (1967).
14. E M. Tsyganov, V. E. Khadzi, A. A. Shaposhnikov, M. I. Samoilovich, and L. I. Tsinober. On specific features of amethyst crystals synthesized in the K_2O-CO_2-SiO_2-H_2O system. In: Growth of Crystals, Vol. 9. Consultants Bureau, New York (1975).
15. M. Schlesinger and A. V. Cohen. Postulated structures causing the optical color center bands in amethyst quartz. J. Chem. Phys, 44, 2599-2600 (1965).
16. R. T. Cox. ESR of an S-2 center in amethyst and its possible identification as the d^4 ion Fe^{4+}. J. Phys. C: Solid. State Phys., 9, 3355-3361 (1976).
17. L. I. Tsinober and M. I. Samoilovich. Distribution of structural defects and anomalous optical symmetry in quartz crystals. In: Problems in Crystallography, Nauka, Moscow, 207-218 (in Russian) (1975).

DEFECT STRUCTURE OF Sb_2S_3 CRYSTALS REVEALED BY ELECTRON MICROSCOPE CRYSTAL LATTICE IMAGING TECHNIQUES

A. A. Sokol, V. M. Kosevich, and A. G. Bagmut

The Polytechnical Institute, Kharkov

The electron microscope technique of crystal lattice imaging is very effective in its application for the study of defects in crystals [1]. We shall give a review of the results obtained by this technique on stibnite, Sb_2S_3, crystals [2, 3]. The studies were conducted on spherulite-type crystal films [4] so that the results obtained describe a defect structure typical for the spherulitic crystallization of amorphous films [5, 6].

Experimental. Antimony sulfide, Sb_2S_3, has an orthorhombic lattice with parameters a = 11.23, b = 11.31, and c = 3.84 Å. The lattice of Sb_2S_3 is built up of double ribbons of antimony and sulfur atoms aligned along the c axis [7]. Modeling of defects in this structure is facilitated by representing it as a "parquet pattern" (Fig. 1) [2].

Sb_2S_3 was evaporated and deposited in vacuo onto carbon substrates at room temperature. Crystallization was initiated in the microscope by electron beam heating [8]. At medium crystallization rates single-crystalline nuclei of Sb_2S_3 split into misoriented ribbon blocks which later formed spherulites [4]. The block structure of Sb_2S_3 is shown in Fig. 2; the blocks are separated by small-angle grain boundaries. The photograph inset shows an electron diffraction micrograph with the aperture encircled. In most cases the amorphous films crystallize with the c axis parallel to the film plane. With this orientation it was possible to obtain crystal lattice images for (110) and either (100), (200) or (010), (020) planes. (100) and (010) planes could not be always precisely differentiated because of the closeness of *a* and *b* lattice spacings.

In order to align the c axis of Sb_2S_3 crystals perpendicular to the substrate, the crystallization of amorphous films was conducted with a small amount of gold deposited on their surface. It should be mentioned that similar changes in the orientation of the c axis due to gold layers deposited on amorphous films were also observed in the crystallization of hexagonal selenium [9]. An Sb_2S_3 film with gold particles is shown in Fig. 3. Spherulite crystallization in Sb_2S_3 films with gold particles is suppressed because of the large number of crystallization centers; the deposit forms a polycrystalline aggregate with axial texture, as we find from the electron diffraction pattern obtained for the area shown on the micrograph. On some crystals with the c axis normal to the surface (100), (010), (110) planes were resolved simultaneously; this is shown in Fig. 4, which depicts the image of a nonsymmetrical tilted boundary AB.

Stacking Faults in the Sb_2S_3 Structure. Stacking faults were simulated by using the "parquet" model of the Sb_2S_3 structure (see Fig. 1) [2]. The following displacements, R, do not disturb the ribbon structure and provide the contact of neighboring "parquet

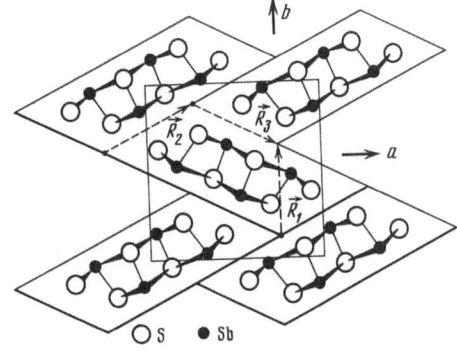

Fig. 1. Projection of Sb_2S_3 crystal structure on to the (001) plane
$\vec{R}_1 = 1/2\ [010]$; $\vec{R}_2 = 1/4\ [210]$;
$\vec{R}_3 = 1/4\ [2\bar{1}0]$

Fig. 2. Block structure of an Sb_2S_3 crystal at the spherulite growth stage.

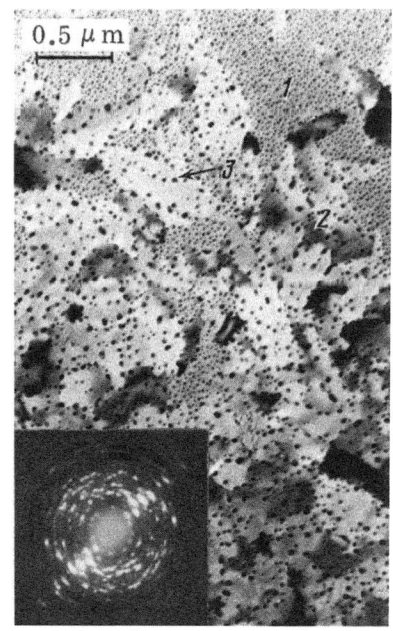

Fig. 3. Sb_2S_3 film with gold particles. (1) amorphous film and gold particles; (2) Sb_2S_3 crystals; (3) dotted line of gold particles along a boundary between blocks in Sb_2S_3.

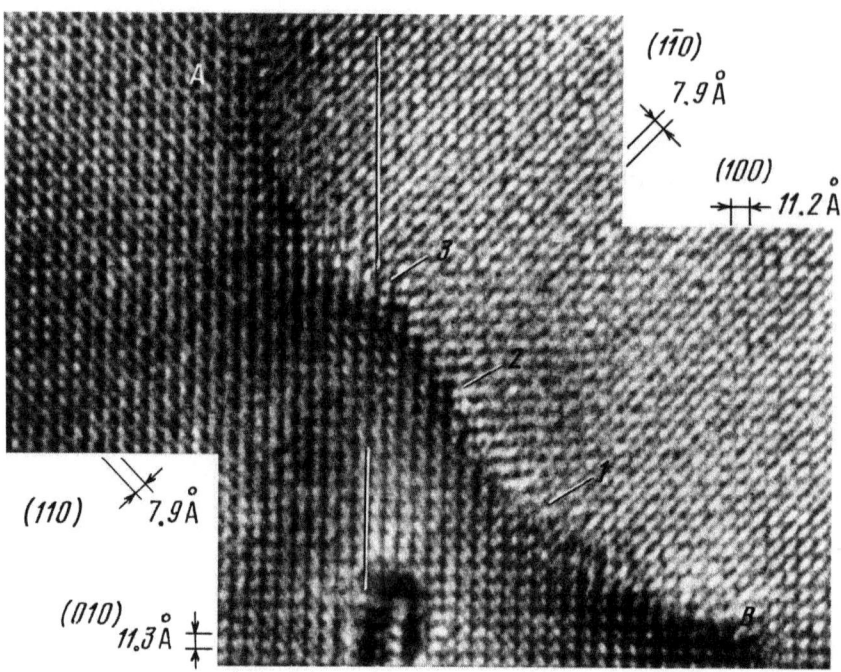

Fig. 4. Micrograph with simultaneously resolved (100), (010), (110) planes (c axis perpendicular to the plane of the photograph).

cells" without gaps: 1/2 [010]; 1/4 [210]; 1/4 [2$\bar{1}$0]. Such displacements violate the sequence of the ribbons and generate stacking faults. They are obtained most simply by adding or "subtracting" structural ribbons and, therefore, can be produced as growth faults.

In order to analyze stacking faults in the crystal lattice images we compiled a table of values of parameter n (image order). This parameter is equal to the scalar product of the diffraction vector $\vec{g}$ by the displacement vector $\vec{R}$ (or by the equivalent Burgers vector $\vec{b}$ of a partial dislocation closing the stacking fault). If n = 0, neither the dislocation nor the stacking fault appear on the image. The value of n (fractional or integral) corresponds to the shift of lines on the fault image, expressed in units of the distance between these lines.

Table of Values of n = gR

R	g					
	100	010	200	020	110	1$\bar{1}$0
1/2 [010]	0	1/2	0	1	1/2	−1/2
1/4 [210]	1/2	1/4	1	1/2	3/4	1/4
1/4 [2$\bar{1}$0]	1/2	−1/4	1	−1/2	1/4	3/4

If n = 1, a partial dislocation appears on the image, traced by an extra half-line, and the adjacent stacking fault is not visible since a shift exactly by one period of the observed interference pattern takes place along the plane of this fault.

Stacking faults in Sb_2S_3 films with the c axis perpendicular to the substrate surface are shown in Fig. 5 (c axis is perpendicular to the plane of Fig. 5). The stacking faults are planar, composed of individual segments oriented in specific crystallographic planes, and aligned normally to the film surface. Within the segment 1-2 (Fig. 5a) the fault is in the (110) plane, (010) lines are displaced by one half period, and (100) lines are not displaced; hence, this is a fault with $\vec{R} = 1/2\,[010]$ (see table). The segment 2-3-4 is the same fault in the (010) and ($\bar{1}$10) planes. The displacement within the segment 3-4 somewhat differs from one half of the period, presumably because of the changes in diffraction conditions.

Contour 1-2-3-4-5 in Fig. 5b shows the profile of a stacking fault composed of planar segments oriented almost perpendicularly to the substrate and aligned in {100} and {110} planes. The image is formed by (1$\bar{1}$0) lines which, when they cross the stacking fault planes, undergo a shift by one half period, which yields $\vec{R} = 1/2 <010>$. Stacking fault planes within segments 1-2 and 4-5 are slightly tilted, and this results in a contrast in the form of bent or interweaving interference fringes (110). Segment α-β reveals a stepwise breaking of the stacking fault plane. Generally, Fig. 5 confirms the correctness of the "parquet" simulation of the stacking fault structure.

Figure 6 gives images of stacking faults formed by (010) reflections, in a film with the c axis parallel to the surface. Figure 6a is an image of a stacking fault with a curvilinear boundary 1-3-4-2. The following result is obtained by drawing on this micrograph a contour equivalent to the Burgers contour. The number of dark fringes between the light fringes above points 1 and 2 is 25; the same number of dark fringes are located between the same light fringes around points 3 and 4. It is also clear that points 1 and 2 are the points of emergence of dislocations which displace (010) fringes by one half period; that is, dislocations with $\vec{b} = 1/2\,[010]$. These dislocations have opposite signs; therefore, the total displacement of the lattice beyond this area is zero. Possibly, either these dislocations lock into a single half-loop 1-3-4-2, or points 3 and 4 are the extreme points of two dislocation

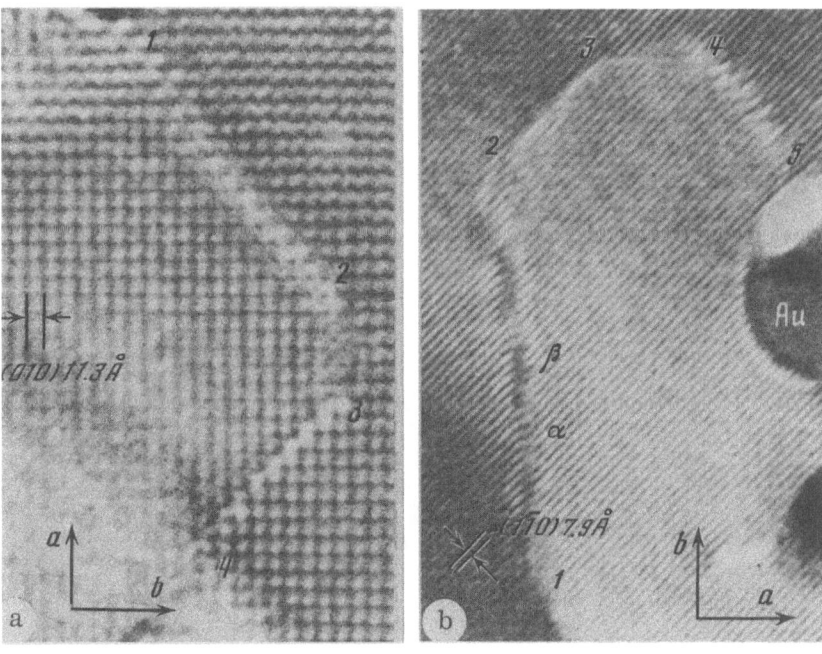

Fig. 5. Stacking faults in Sb_2S_3 films (micrographs with resolved planes: (a) (100), (010), and (b) (1$\bar{1}$0)).

segments. In either case, Fig. 6a is identified as a dislocation structure with two points of emergence (1 and 2) at one surface of the film and two (3 and 4) at the opposite surface. As each of the dislocations is partial, a tilted stacking fault with $\vec{R} = 1/2\,[010]$ must be located between them. The (010) fringes on the stacking fault image must be displaced by one half period. Indeed, this displacement is observed within the half-loop 1-3-4-2. It is interesting to note that the deformation of the lattice within the half-loop results in a transition to the (020) reflection having a period half that of (010). An important feature of Fig. 6a is that the stacking fault shown in it does not lie in a low-index plane.

One characteristic feature of stacking faults in Sb_2S_3 is that these faults easily change from one plane to another and are not necessarily planar; in this respect they are similar to antiphase boundaries.

Figure 6b is an image of a curvilinear stacking fault 1-2 formed by (010) reflections. The stacking fault surface is perpendicular to the surface plane. Points 1 and 2 are the locations of partial dislocations of the same sign; they close up this stacking fault and have $\vec{b} = 1/2\,[010]$. In some places the stacking fault has a stepped structure. Rectilinear segments are imaged by light bands parallel to (010) planes. These light bands appear because the stacking fault displaces the (010) fringes by one half period parallel to themselves; this produces a light gap between the dark fringes.

Another image of a stacking fault producing a displacement of (010) fringes by a quarter period is shown in Fig. 6c. According to the table, this is a stacking fault with $\vec{R} = 1/4 <210>$. The stacking fault profile is a broken line corresponding to the traces of $(01\bar{1})$ planes within segments 1-2 and 3-4, and to those of $(03\bar{1})$ planes within segment 2-3.

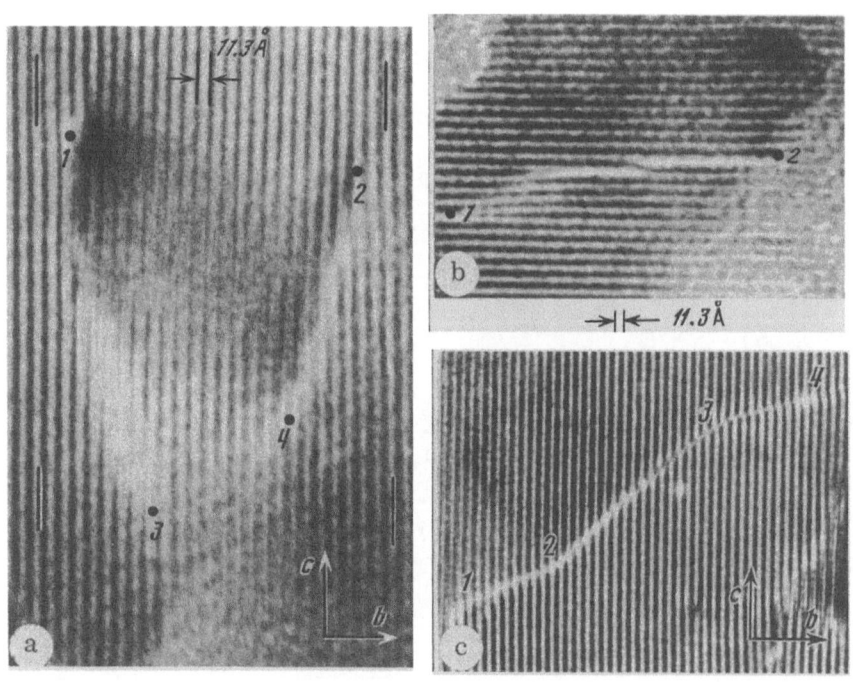

Fig. 6. Stacking faults in films with the c axis parallel to the film surface. (a) arch-shaped; (b) curvilinear; (c) with $\vec{R} = 1/4\,[210]$.

Small-Angle Boundaries. The initial stage of the spherulite growth of Sb_2S_3 crystals consists in the splitting of a single crystal nucleus into a series of misoriented ribbon-like blocks separated by small-angle boundaries (see Fig. 2).

The structure of tilted boundaries in films with the c axis parallel to the substrate is illustrated in Fig. 7. A symmetrical boundary along the (010) plane, producing a small-angle rotation by approximately 3°, is seen in Fig. 7a. The boundary is composed of partial dislocations with a Burgers vector 1/2[010]. As the image is formed by (010) planes, each dislocation gives a displacement of fringes by one half period. Dislocations are perpendicular to the plane of the photograph and are linked into pairs by stacking faults; one example is the pair of dislocations 1 and 2. Sometimes the dislocation line deviates from the normal and is projected on the micrograph plane by a segment, such as 4-4'. In this case the stacking fault linking two dislocations becomes curvilinear: the one between dislocations 3 and 4-4' is stretched like a taut sail.

Figure 7b is an image of a tilted boundary whose plane is at an angle to the surface normal. The width of the boundary projection onto the micrograph plane in its upper portion is denoted by t and is approximately equal to 50 Å; in the lower portion of the micrograph the boundary tilt increases and its projection reaches a width of 100 Å. The angle of rotation of the lattice in the upper portion of the boundary is nearly 6°. The boundary is composed of edge dislocations with $\vec{b}$ = 1/2[010] yielding n = 1 for $\vec{g}$ = (020). The images of tilted dislocations are of special interest (for segments marked by braces in Fig. 7b, these are shown magnified in Figs. 7c, d). A tilted dislocation is shown as a band along which the interference fringes (020) are shifted; this shift monotonically increases from the points of emergence of the dislocations to the crystal surface (1 and 2 in Fig. 7d) to the center. The difference between the number of (020) fringes above and below the dislocation image is unity,

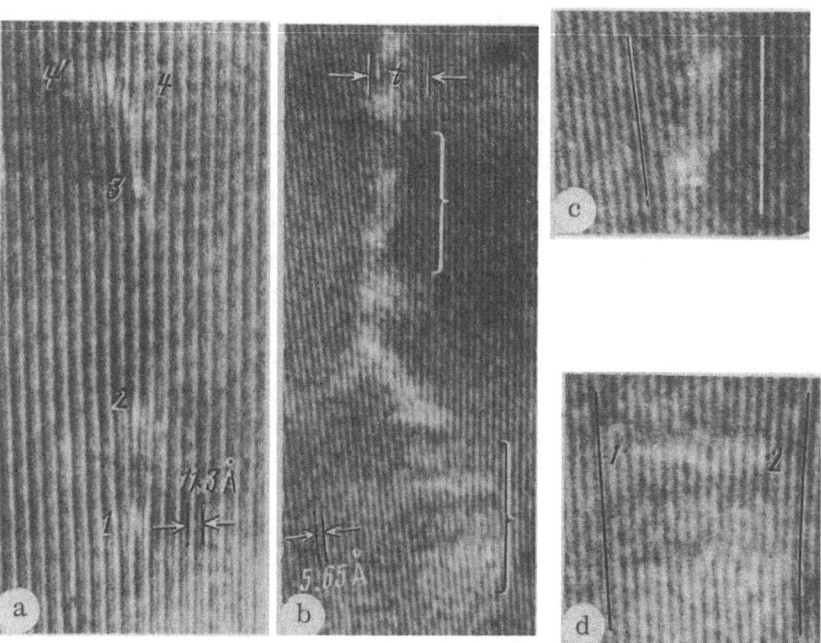

Fig. 7. Structure of tilt boundaries in films with the c axis parallel to the substrate. (a) partial dislocations bound into pairs by stacking faults (image formed by (010) reflections); (b-d) small-angle tilt boundary (image formed by (020) reflections) and its magnified segments.

as can be found by counting the number of fringes between the ends of white lines traced on the photograph. Tilted dislocations of this type were observed in silicon [1]. The second interesting feature of this micrograph relates to stacking faults. Between the white lines traced in Fig. 7c there are 12 black interference fringes (020) in the upper portion and 10 in the lower one. Hence, this area contains two partial dislocations and a stacking fault between them. As the stacking fault gives a displacement of one total period (020), it is not observable, and only the partial dislocations bordering it are seen.

A boundary with a rather peculiar geometry is shown in Fig. 8. The broken profile ABCD is a boundary across which the lattice orientation changes. This is revealed by changed diffraction conditions for the resolution of (020) planes: these planes are clearly resolved to the right of the boundary, and to the left of it they are almost out of the reflection position. Figure 8 also shows dark stitch-like branches leaving the boundary, for example, AE and BF. The contrast of these branches is definitely due to absorption. Dark spots of absorptional contrast can also be seen along the broken line ABCD, for example, in angles at A and B, and near point C. At the same time, this zigzag boundary also manifests arcs of deformation diffraction contrast, such as are CD.

Processing of this micrograph shows that (020) fringes cross the boundary ABCD without changing direction (to within the accuracy of measurements, approximately 10'). Consequently, this is a screw boundary with rotation axis [001]. As (020) planes are imaged on both sides of the boundary, the rotation at the boundary is by a small angle (which does not take (020) planes completely out of the reflection range). Apparently, the screw dislocations of which the twist boundary is composed result in arc-shaped diffraction contrast lines. As for the stitch-type branches such as AE and BF, they do not affect the positions of (020) fringes. No excess half-lines appear at the ends of branches (at points E and F), and along these branches there are no diffraction contrast effects which would indicate the presence of screw dislocations. We have to assume, therefore, that the absorption-contrast dashed lines are clusters of impurity trapped by the growth interface.

Another type of small-angle boundary in a film with the c axis perpendicular to the substrate is shown in Fig. 4 (the boundary roughly follows the diagonal AB). This is a nonsymmetrical tilted boundary with rotation axis [001]. The lattice is rotated approximately by 2°, as can be found from the orientation of white lines traced on the micrograph. The boundary is formed by two systems of dislocations. The first of them is composed of partial dislocations with Burgers vector 1/2[010]. These dislocations are connected by stacking faults with the same displacement. The second system contains dislocations with $\vec{b} = 1/2$ [210] and the corresponding stacking faults differ from those of the first system. Dislocations of the two systems alternate along the boundary, so that the stacking faults overlap. It appears that a fault with displacement $\vec{R} = 1/2[010] + 1/2[210] = 1 \cdot [110]$ is formed at the site of overlapping. Alternation of stacking faults results in a periodical variation of diffraction contrast along the boundary (see Fig. 4): a light band 1-2 transforms into a dark band 2-3, and so on.

Large-Angle Boundaries. Crystal lattice images of Sb_2S_3 spherulitic crystals reveal numerous large-angle boundaries, and in particular coincidence site boundaries [10, 11]. Figure 9 shows large-energy tilted boundaries in films with the c axis parallel to the film surface. Images are formed by (020) and (010) reflections. "Plane-to-plane" matching of images is observed on a symmetrical tilted boundary with rotation angle about 20° with respect to [$\bar{1}00$] axis (Fig. 9a). This situation occurs when the projections of reflection vectors $\vec{g}_1$ and $\vec{g}_2$ in both grains onto the boundary plane are equal. The number of planes terminating at the boundary contour is identical on both sides along the whole length of the boundary.

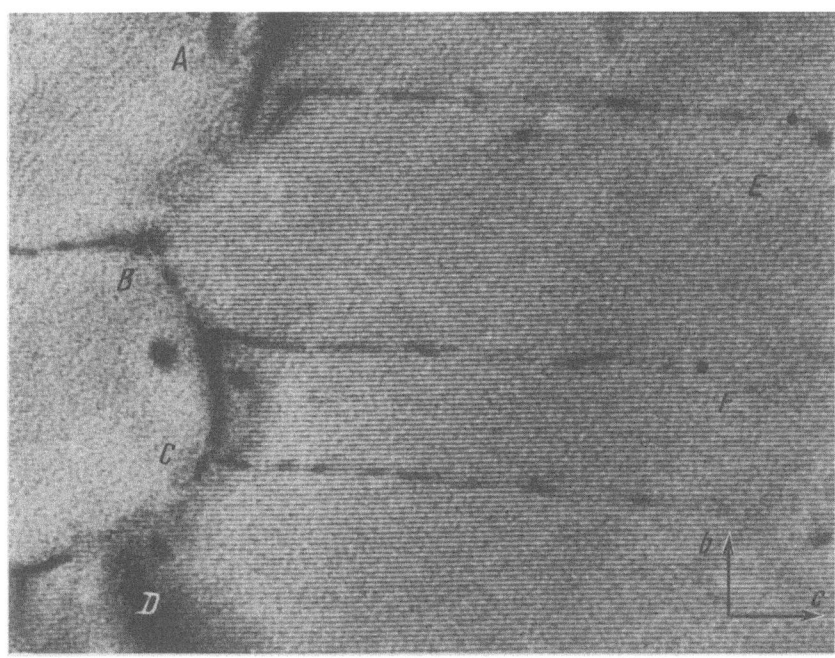

Fig. 8. Screw boundary rotation axis [001] in films with the c axis parallel to the surface (image formed by (020) planes).

Fig. 9. Large-angle boundaries in films with the c axis parallel to the surface; image formed by (a) (020) and (b) (010) reflections.

If the complex structure of Sb_2S_3 (see Fig. 1) is replaced by a geometric lattice with nodes at the apices of the orthorhombic unit cell, one can calculate rotation angles θ^* corresponding to coincidence site positions by the formula

$$\theta^* = 2\arctan(n_1 c/n_2 b)$$

where n_1 and n_2 are integers, and c and b are lattice periods [11]. For $n_1 = 1$ and $n_2 = 2$, we have $\theta^* = 19.3°$ which is in good agreement with the results of angle measurements (see Fig. 9a). This (and other micrographs not given here) indicate that the coincidence lattice site boundary is a valid concept for large-angle boundaries in Sb_2S_3.

Figure 9b shows a nonsymmetrical tilted boundary ABCD with tilt angle $\theta \simeq 44°$. Boundary segments AB and CD follow the direction of the c axis, that is, the direction of packing of Sb_2S_3 structural ribbons. These boundary planes are energetically advantageous because the bonds perpendicular to the ribbon axis are weak. The transient region BC of the boundary occupies a nonsymmetrical position in which grain boundary stacking faults must be generated [12]. It has been shown above that lattice stacking faults in Sb_2S_3 can lie in different planes; that is, the structure of Sb_2S_3 allows the formation of various configurations of stacking faults; consequently, one can assume that the grain boundary stacking faults within large-angle boundaries are also quite natural defects in Sb_2S_3.

On Defect Structure of Sb_2S_3 Spherulites. Dislocation processes play an important part in the formation of spherulite crystals [3]. The data obtained specify the types of small-angle boundaries appearing in the course of spherulitic growth of Sb_2S_3 crystals (tilted boundaries composed of partial dislocations and stacking faults; rotation boundaries; small-angle boundaries with impurity clusters). As the large-angle coincidence site boundaries are a typical element of spherulitic crystal structure, any structural model of spherulites requires that the processes of large-angle boundary formation be taken into account. One possible mechanism of their generation may be the coalescence of a series of low-angle boundaries.

Literature Cited

1. J. R. Parsons. Influence of dislocations on electron microscope crystal lattice images. In: Interatomic Potentials and Simulation of Lattice Defects. New York; London: Plenum Press, 463-473 (1972).
2. V. M. Kosevich, A. A. Sokol, and A. G. Gamut. Analysis of defect structure of Sb_2S_3 films in electron microscope crystal lattice images. Izv. AN SSSR, Ser. Fiz., 41, 2307-2309 (in Russian) (1977).
3. V. M. Kosevich, A. A. Sokol, and A. G. Bagmut. Structure of Sb_2S_3 spherulite crystals studied by crystal lattice images. Kristallografiya, 24, 143-148 (in Russian) (1979).
4. I. E. Bolotev, A. V. Kozhin, and S. B. Fisheleva. Crystallization of thin antimony films. Izv. Vyssh. Uchebn. Zavedenii Fizika, 12, 119-120 (in Russian) (1970).
5. I. E. Bolotov, V. D. Aleksandrov, A. V. Kozhin, S. B. Fisheleva, and I. N. Demidov. Dislocation structure of metal selenium crystals and its relation to spherulite formation. Fiz. Metal. i Metalloved., 37, 1191-1195 (in Russian) (1974).
6. I. E. Bolotov, A. V. Kozhin, and P. S. Mel'nikov. Electron microscope study of block formation in growing thin selenium crystals. Izv. AN SSSR, Ser. Fiz., 41, 1065-1067 (in Russian) (1977).
7. P. Bayliss and W. Nowacki. Refinement of the crystal structure of stibnite, Sb_2S_3. Z. Kristallogr., 135, 308-315 (1972).
8. M. D. Coutts and E. R. Levin. Phase transformation of As_2Se_3 and Sb_2S_3 films. J. Appl. Phys., 38, 4039-4044 (1967).

9. I. E. Bolotov and L. I. Komarova. Effect of electric field on the texture of selenium crystals growing in amorphous films. Fiz. Tverd. Tela, 17, 748-751 (in Russian) (1975).
10. H. Gleiter and B. Chalmers. High-Angle Grain Boundaries, Progress in Materials Science, Vol. 16, Pergamon Press, (1972).
11. W. Bollman. Crystal Defects and Crystalline Interfaces. Springer, Berlin, p. 255 (1970).
12. A. N. Orlov, V. N. Perevezentsev, and V. V. Rybin. Analysis of defects of crystal structure in a symmetrical tilted boundary. Fiz. Tverd. Tela, 17, 1662-1670 (in Russian) (1975).

ON THE STRUCTURE OF CRYSTALLINE GRAPHITE INTERGROWTHS

V. M. Kosevich, A. P. Lyubchenko, S. N. Grigorov and G. P. Umansky

The Polytechnical Institute, Kharkov

We shall discuss a structural model of crystalline graphite conglomerates on the basis of electron microscopy data. Conglomerates are defined here as compact intergrowths such as bicrystals, multicrystals, spherulites, noncompact dendritic forms of branched or skeletal types, and so on. We assume that the model is applicable to graphite crystallization from the melt in modified alloys based on iron, nickel, and cobalt cooled at high rates. The following features are basic for the suggested structural model.

Crystalline graphite intergrowths contain as the principal structural element large-angle boundaries of a special type, viz., coincidence lattice site boundaries (CLS boundaries) [1]. Crystallites of which the conglomerates are composed are matched along these boundaries.

A nucleus from which a graphite spherulite grows is a multicrystal faceted by basal planes and separated into sub-units by large-angle boundaries. A spherulite does not grow from one single-crystal nucleus split along the basal planes. A multicrystal forming the nucleus of a spherulite can develop from one single-crystal nucleus by multiple twinning on coincidence site planes. The simultaneous formation of several nuclei is also possible, but in this instance both the intergrowth and form changes proceed via the formation of large-angle boundaries.

Crystalline conglomerates (including spherulites) develop by incorporating fresh nuclei into the growth face either oriented parallel to this face or forming coincidence-site large-angle boundaries. Nuclei grow by the attachment of atoms to end faces of basal planes; hence, the maximum growth rate vector lies in the basal plane. Large-angle boundaries appear at the end faces, so that the crystal forms a "knee" at an appropriate angle. The grain boundary rotation vector lies in the basal plane.

Crystallites comprising a conglomerate within regions separated by large-angle boundaries may be elastically strained and consist of blocks; block boundaries may incorporate dislocations, stacking faults, and impurity clusters.

Impurities play a decisive role in the formation of regular graphite intergrowths (both compact and dendritic). Impurities are not dissolved in the graphite lattice [2], but can affect the growth of graphite by becoming attached to the basal end faces and blocking the growth surface. Impurities are either in an atomic or a molecular state (and do not form their own crystal lattice). They may form spatially ordered structures and may lead to a crystal splitting into fragments.

The degree of blocking of a growth face by impurities is determined by the physical properties of the dopant, its concentration, and crystallization rate. Only certain elements (surfactants with respect to graphite), distributed in a specific manner in the Periodic Table [2], produce a spherulizing effect in slowly cooled melts. Other impurities present in the melt may act as spherulizing agents for fast cooling.

Structurally, the following types of blocking of growth faces by impurities are important in graphite:

Weak Blocking. (Low impurity concentration at the growth surface). In this case the growing crystal breaks through the impurity barrier and forms a block boundary without any appreciable misorientation. As a result, the growing crystal consists of blocks with periodic boundaries of impurity segregation.

Medium Blocking. Crystal growth is resumed by forming a new nucleus behind the impurity barrier; however, the blocking is not total and the new nucleus connects with the initial crystal in an epitaxial large-angle orientation. This produces a large-angle coincidence lattice-site boundary along a nonbasal plane. This situation leads to the formation of a regular multicrystalline intergrowth from which a spherulite may later be formed.

Total Blocking. Crystal growth proceeds by forming a new nucleus behind the impurity barrier, but this nucleus is not matched to the initial crystal lattice.

In the first two cases the impurities are in the atomic or molecular state. Total blocking leads to the segregation of crystalline particles with separate lattices.

Let us discuss the basic experimental evidence in favor of the presented interpretation. We shall begin with electron microscopy data obtained by analyzing graphite inclusions in specimens of Ni-C alloy cooled at high rates. Two types of such inclusions were observed. One constitutes a skeletal-type conglomerate consisting of graphite platelets intergrown at various angles (Fig. 1a). The platelets perpendicular to the plane of the photograph appear as thin black needles. They are seen against a grey background of platelets parallel to the plane of photograph. The angle between intergrown graphite platelets is usually 60-70°, as, for example, in the area shown in the box. Figure 1b shows a diagram of the possible mechanism by which such intergrowths are formed. Lines in this diagram represent basal planes.

The second type of graphite inclusion is shown in Fig. 2. This is a branched dendrite with spherulitic graphite branches growing from the main "trunk." The branches are marked with arrows C. These spherulitic branches were studied in detail by electron microscopy.

The following typical features of the spherulite structure were established.

1. Spherulites are faceted by basal planes. This is shown by electron micrographs obtained for the peripheral portions of spherulite intergrowths. The micrographs always contain reflections from the basal planes (002), and vector $\vec{g}_{(002)}$ is always perpendicular to the intergrowth contour. The observed units are not necessarily planar. Analysis of the texture revealed by electron micrographs shows that in the plane perpendicular to the photograph plane, the vector $\vec{g}_{(002)}$ changes direction in the range 90±30° with respect to the electron beam. We conclude from this that spherulitic intergrowths observed in rapidly cooled Ni-C alloy have a structure similar to that of massive spherulites in modified alloys based on Ni, Co, and Fe.

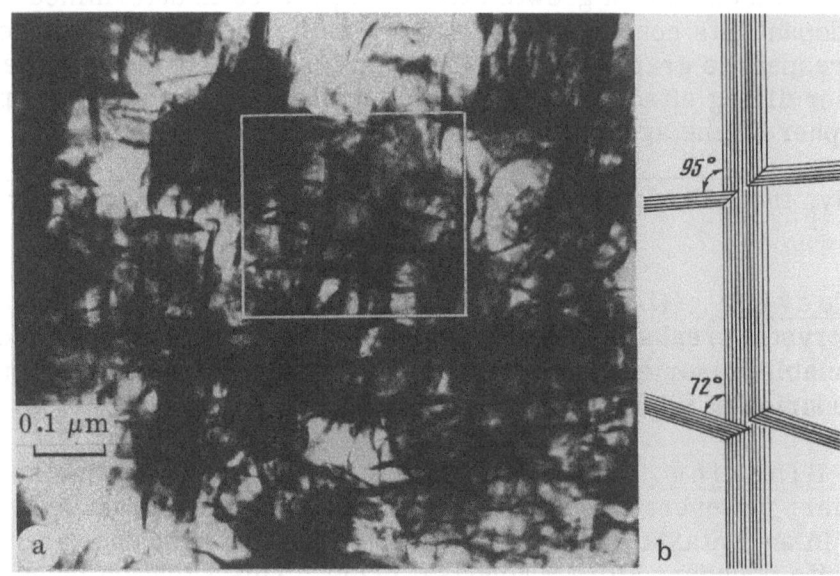

Fig. 1. (a) skeletal-type conglomerate, and (b) schematic structure of the conglomerate.

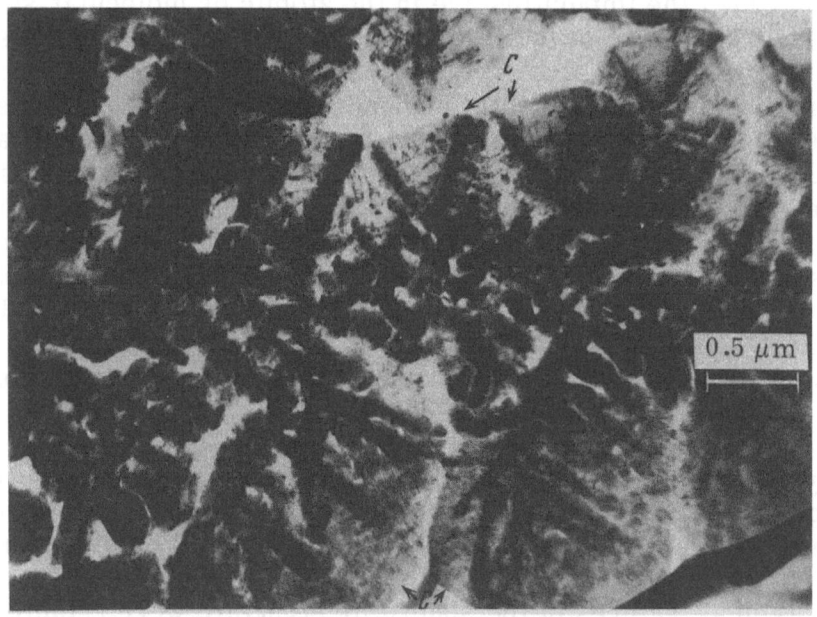

Fig. 2. Branched graphite dendrite.

2. An analysis of a large number of electron micrographs obtained from graphite spherulites and the corresponding diffraction patterns demonstrated that the basic structural element of spherulites is the large-angle tilt boundary formed by rotation about an axis lying in the basal plane. This is illustrated in Figs. 3 and 4. Spherulitic intergrowths shown in the micrographs clearly show pronounced sectorial regions (A, B, C) within which vector [002] has an almost constant direction but changes in steps by an appreciable angle if neighboring sectors are compared. This is revealed both by the faceting of spherulites and by the geometry of diffraction patterns of individual spherulites. The angle between the

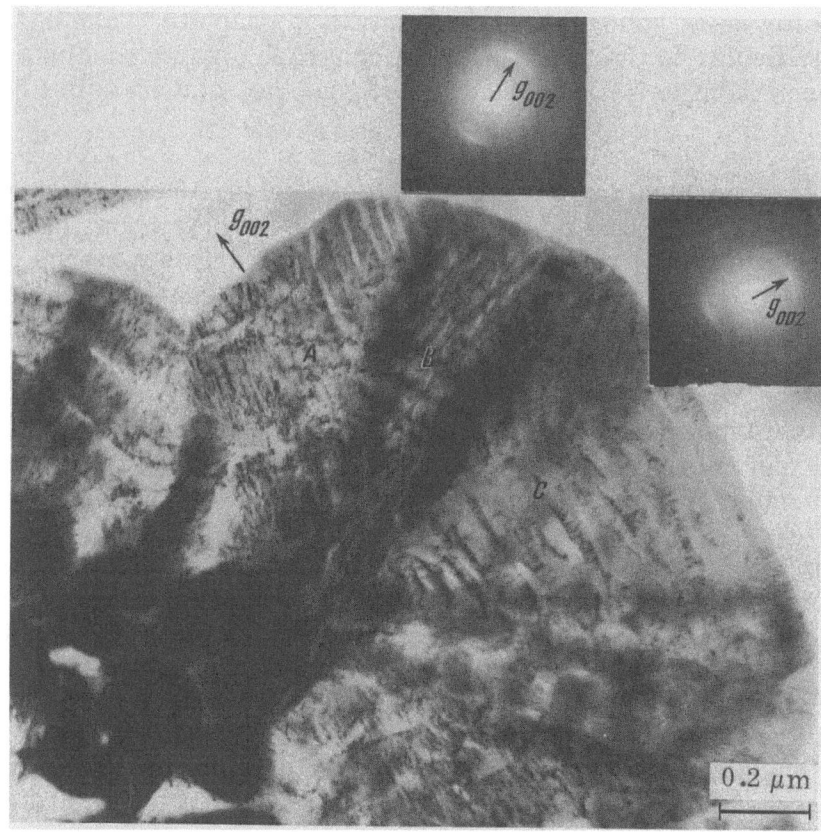

Fig. 3. Multifaceted graphite intergrowth and electron diffraction patterns of its selected areas shown in insets.

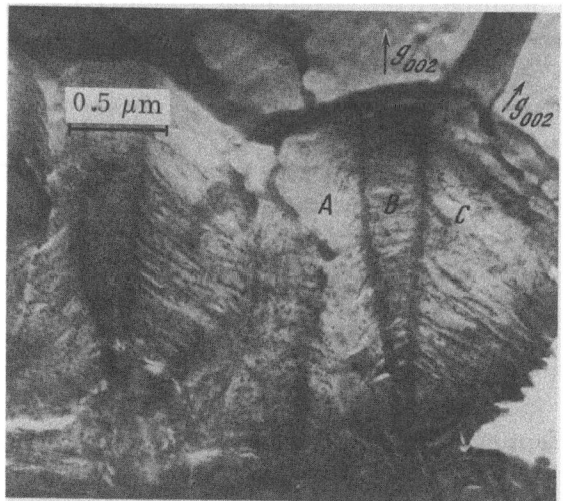

Fig. 4. A fragment of a dendritic graphite conglomerate.

adjacent sectorial areas B and C in the spherulite of Fig. 3 is approximately 40°; this misorientation angle is typical for many spherulitic intergrowths in graphite. The basal planes in areas B and C are perpendicular to the plane of the micrograph, so we conclude that the large-angle boundary between them is a tilt boundary, and that the rotation axis lies in the basal plane.

3. Spherulites in Ni-C alloy have a ribbon-like structure. Figure 5 clearly shows boundaries and pores between ribbons (indicated by arrows C). Each of the ribbons has a block structure, which follows from the (002) reflection of electron diffraction patterns of ribbons being spread into arcs, and from the appearance of dark strokes D separating the regions with uniform contrast on the electron micrograph. These dark strokes are evidently the images of small-angle boundaries.

4. Spherulites and other types of compact intergrowth in Ni-C alloy contain impurities. It should be noted that we experimented with a technical-grade alloy which contained, in addition to 1.8 wt.% carbon, about 0.5 wt.% copper.

Clusters of impurity atoms produce, in electron micrographs, a characteristic pattern of black dots and lines. These impurities are not apparent in diffraction patterns. Possibly, this is because the impurity is in the atomic or molecular state. Impurity atoms may decorate dislocations which build up small-angle boundaries in graphite ribbons. In this case the spherulite micrograph shows a series of radial dark lines similar to those in region A in Fig. 3. Impurity atoms may concentrate in pores and at boundaries between the graphite ribbons of which the spherulite is built. These lines then provide clear evidence of the laminated structure of spherulites, as in Fig. 4, sectors A, B, C.

Figure 6 shows conclusively that the listed contrast features are related to impurity clusters. Here impurity clusters outline with dots the central part of the graphite intergrowth, with graphite ribbons piling up around it. The ribbon-like structure of the intergrowth is not apparent on the micrograph; however, it was found that this intergrowth, as were all others, was faceted by basal planes, and in area A we find a dashed pattern repeating the geometry of the intergrowth. A striking feature of this micrograph is the periodic (with spacing $P \approx 250$ Å) distribution of impurities producing a layered structure in tangential directions. This typical periodic distribution of impurity is frequently observed in the growth of bulk crystals [2, 3].

The observed effect of impurity distribution within graphite segregations in rapidly cooled Ni-C alloy was compared with published experimental data on the distribution of specific spherulizing impurities in iron- and nickel-based modified alloys. As shown by radioisotope [2], X-ray microprobe analysis [4], and other local techniques of chemical analysis, the spherulizing elements are always localized in graphite at sub-optimum and optimum concentrations (from the standpoint of their spherulizing effect). X-ray studies demonstrated that the graphite lattice spacing remained unaltered. Consequently, impurity atoms must be localized in intercrystallite transition zones, such as grain boundaries. This conclusion is indirectly confirmed by X-ray microprobe data which indicate that, as the boundary density rises, the dopant concentration always increases toward the graphite spherulite center [4, 5].

The outlined experimental data provide conclusive evidence that intercrystalline boundaries are the sites of impurity concentration. The role of the impurity is then the blocking of tangential growth of the graphite basal faces, thus providing stimulation for boundary formation.

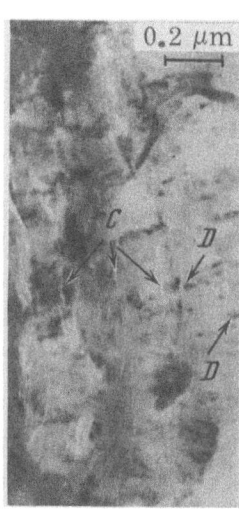

Fig. 5. A fragment of graphite segregation with a clearly pronounced ribbon-like structure.

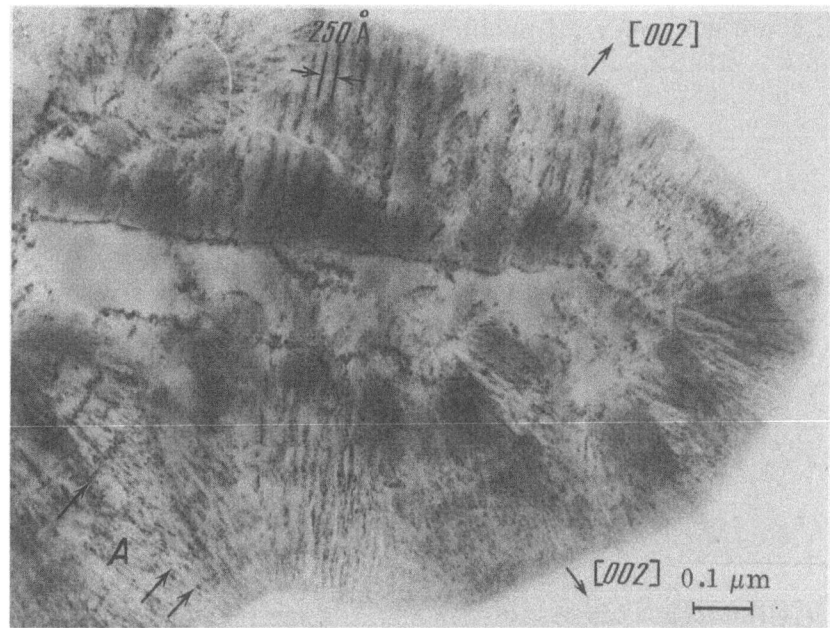

Fig. 6. A fragment of a graphite segregation with well-defined localization of impurities along boundaries.

Referring to the possible large-angle tilt boundaries in the graphite lattice, according to [6] and the calculations that the authors carried out using the theory of O-lattices [7], the main type of tilt boundaries in graphite must be twinning boundaries, because only this type of interface matches the basal planes sufficiently well. Twin boundaries for the [1$\bar{1}$0] axis are formed with the rotation angle θ = 40.30, 57.65, 62.90, 72.55, and 95.49°. For [110], such angles as, for example, 64.88 and 103.62° are possible. Boundaries formed with these misorientations can be considered as coincidence lattice site boundaries with a different site density on each side of the boundary. An analysis based on the O-lattice theory shows that the most advantageous rotation is that of 40.30° around the [1$\bar{1}$0] axis, since only in this case are planes O-elements of the O-lattice. These O-planes correspond to the crystallographic (112) planes. Consequently, a large-angle boundary in the (112) plane must have a minimum energy. In all other listed misorientations the O-elements are lines.

A comparison of the experimental data with calculations shows that graphite intergrowths comprise large-angle boundaries belonging to the group of CLS boundaries, the most frequent being the boundary with 40° misorientation.

A model illustrating the principle of impurity blocking of the growth face and the accompanying formation of a large-angle boundary is shown in Fig. 7, for a graphite lattice cross section on the $(1\bar{1}0)$ plane. The $[1\bar{1}0]$ axis is perpendicular to the plane of the figure, and the (002) basal planes are horizontal. One of the basal planes is blocked by an impurity shown by a shaded area. The subsequent basal planes turn at an angle and produce a large-angle boundary AB. A diagram showing a boundary with a misorientation angle of 40.30° (see Fig. 7a) may serve as a model for the growth of multicrystals stimulating subsequent spherulitic growth of graphite. The diagram for a misorientation angle of 72.55° (see Fig. 7b) outlines the formation of the skeletal growth of graphite, such as observed in Fig. 1.

We note in conclusion that the processes discussed in the present communication deal exclusively with structural aspects of the growth of graphite conglomerates. Consequently the results obtained could not cover all problems of spherulitic growth in graphite, which include thermodynamic, kinetic, diffusional, and other problems of comparable importance.

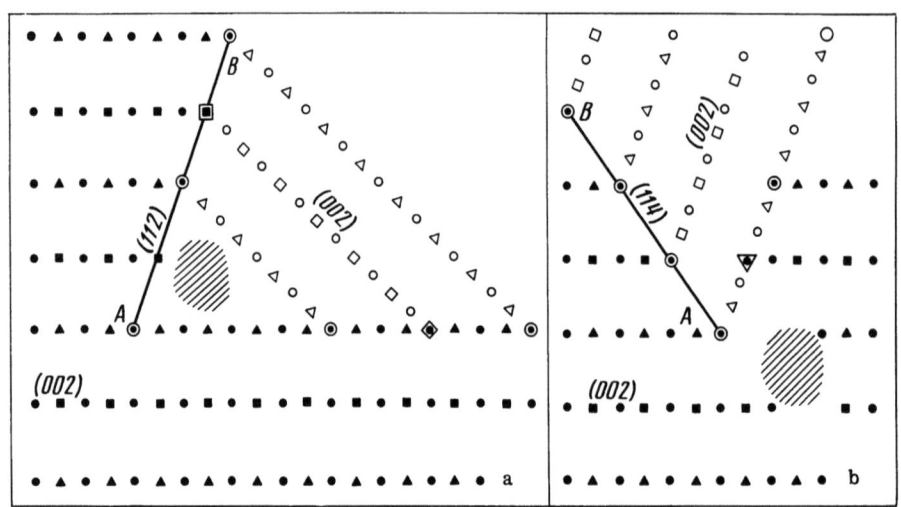

Fig. 7. Schematic model of a possible effect of impurity on the crystallography of graphite growth with misorientation angles (a) 40.30° and (b) 72.55°.

Literature Cited

1. H. Gleiter and B. Chalmers. High-Angle Grain Boundaries, Progress in Materials Science, vol. 16, Pergamon Press, (1972).
2. Utilization of Radioactive Labelling Methods in Industry. Ed. A. P. Lyubchenko. Atomizdat, Moscow, (in Russian) (1975).
3. V. M. Kosevich, A. A. Sokol, E. T. Mogilko, and V. N. Maslov. Laminated structures in gallium arsenide and garnet crystals revealed by decoration. Fiz. Tverd. Tela, 18, 1877-1882 (in Russian) (1976).
4. D. P. Ivanov and I. A. Vashukov. Experimental investigation of elements' distribution in cast iron structure. Liteinoye Proizvodstvo, No. 7, 32-35 (in Russian) (1974).

5. A. P. Lyubchenko, G. P. Umansky, and S. M. Kosmachev. On the character of adsorption of modifiers on graphite. Izv. Vyssh. Uchebn. Zavedenii, Ser. Chernaya Metallurgiya, No. 10, 156-159 (in Russian) (1977).
6. W. Bollman and B. Lux. Grain boundaries in graphite. In: Proc. 2nd Intern. Symp. Met. Cast. Iron, Geneva, St. Saphorin, 461-471 (1975).
7. W. Bollman. Crystal Defects and Crystalline Interfaces. Berlin, Springer, 255 (1970).

Part VIII
New Materials: Equipment for Crystal Growth

CRYSTAL GROWTH AND PROPERTIES OF SOME NEW IONIC CONDUCTORS

A. Rabenau

Max-Planck-Institut für Festkörperforschung, 7 Stuttgart, FRG

Introduction

Solid state electrolytes, where the electric current is due exclusively to the movement of ions, have been known for a long time. Until recently they played only a minor role in solid state science. The situation changed, however, drastically in the last two decades for two reasons: (1) the discovery of new ionic conductors with enhanced conductivity even at moderate temperatures (optimized ionic conductors; so-called "superionic conductors"), (2) the growing interest in energy storage and conversion systems in industry. New ionic conductors are required to meet the demands of solid state batteries: small rechargeable primary cells with longevity, and high energy density (secondary) batteries. On the other hand, galvanic cells with solid electrolytes are also of great importance for scientific applications in thermodynamics as well as for kinetic investigations [1].

Usually these applications do not specifically require single crystals, and ceramic materials or even pressed pellets may suffice. On the other hand, an improved understanding of enhanced ionic conductivity — an indispensable prerequisite for the development of new ionic conductors for applications in science and technology — will only arise from careful and systematic investigations of ionic transport in suitable classes of materials. Of special importance in this respect is the anisotropy of conduction, e.g., one- and two-dimensional conductors. Here single crystals of reasonable size and quality are required. Examples will be given from different groups of materials which have been the subject of recent studies at the Max-Planck Institute for Solid State Research.

Copper Telluride Halides CuTeX (X = Cl, Br, I)

The well-known classical ionic conductor α-AgI exhibits conductivities comparable with liquid electrolytes. A phase transition at 147°C into the low-temperature phase β-AgI, which is a poor conductor, is an obstacle for low temperature applications.

It can be shown that by substitution in the anionic or cationic sublattice, compounds with high conductivities could be obtained. The first example was the silver chalcogenide halide Ag_3SI, obtained by Reuter and Hardel in 1960 [2]. Others, such as $RbAg_4I_5$, followed. No inorganic copper compounds with comparable properties were known until the discovery of the copper chalcogenide halides [3]. Although they are not homologues of the respective silver compounds (Fig. 1), structure determinations of the CuTeX (X = Cl, Br, I) group [4, 5] showed that here the copper positions are only partially — statistically — occupied, a prerequisite for ionic conduction. Subsequent determinations of the partial cationic and electronic conductivities by transference, polarization and a.c. conductivity methods proved these compounds to be optimized as ionic conductors at temperatures higher than 250°C [6]. The partial electronic conductivity was found to be negligible compared with the partial ionic conductivity. For the measurements, pressed pellets of the respective polycrystalline compounds were used. In the case of CuTeBr, a phase transition is indicated by a discontinuity in the conductivity versus temperature plot (Fig. 2). DTA measurements showed that the transition is accompanied by only a small thermal effect. From calorimetric data a first-order phase transition can be concluded with $\Delta H_u = 76.8$ cal mol^{-1} and $\Delta S_u = 0.222$ cal mol^{-1} K. Structure determinations of the high temperature phase show a rearrangement of the statistical population of the Cu sites compared with the room temperature phase. The results can be explained on the basis of a displacive transition [7]. With this type of transition it has been found in other compounds that single crystals can retain their crystallinity during the transition, in contrast to AgI which undergoes a destructive transition between the α and β phases. This phenomenon is of interest in that it allows the investigation of the ionic conductivity during the transition in a single crystal specimen.

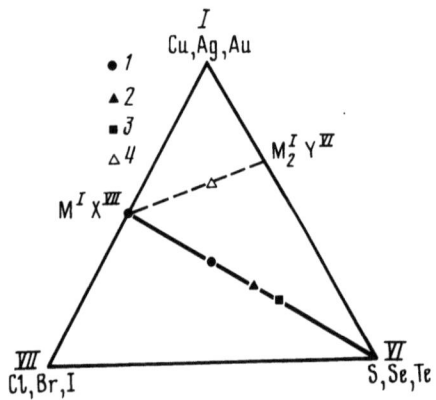

Fig. 1. Chalcogenide halides in ternary systems with Group IB elements Cu, Ag, and Au, the chalcogens S, Se, and Te, and the halogens Cl, Br, and I as components. They are indicated as M^I, Y^{VI} and X^{VII}, respectively. (1) CuClTe, CuBrTe, CuITe, AuITe; (2) CuClTe$_2$, CuBrTe$_2$, CuClSe$_2$, CuITe$_2$, AuClTe$_2$, AuBrTe$_2$, AuITe$_2$; (3) CuBrSe$_2$, CuISe$_2$; (4) Ag$_3$SBr, Ag$_3$SI.

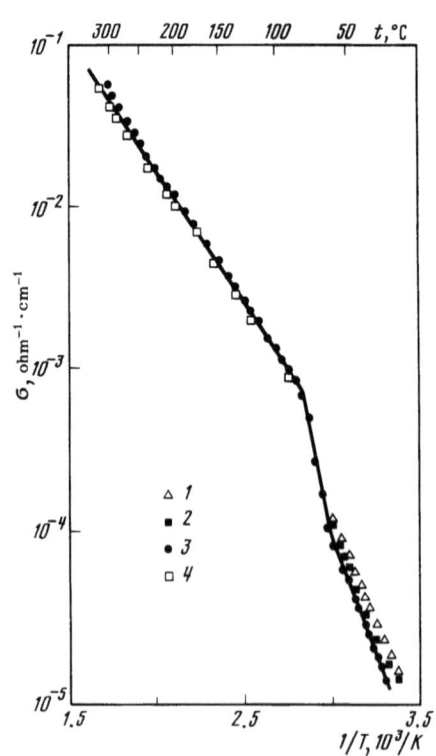

Fig. 2. Total conductivity of CuTeBr as a function of temperature. (1) 1 MHz; (2) 1 kHz; (3) extrapolated to infinite frequency; (4) with direct current.

The Growth of CuTeBr Single Crystals.

Copper tellurium bromide, CuTeBr, is a congruent melting phase in the pseudo-binary system CuBr - Te (Fig. 3). The Bridgman technique, starting from stoichiometric mixtures of CuBr and Te, has been applied. To obtain crystals of good quality, the components had to be carefully purified: CuBr was synthesized from freshly reduced copper powder (99.95) and bromine (AR grade) and distilled in streaming bromine vapor at 900°C. Tellurium (99.9995) was fused at 600°C in streaming hydrogen to remove oxide. The mixture was heated in a closed Bridgman ampoule to 550°C and thoroughly shaken. The ampoule was introduced into a Bridgman furnace (Fig. 4) and lowered by 7 mm/day and 10^{-4} rad/min, a process which had to be repeated in order to get material of high quality. From the resulting boule single crystal plates of more than 1×1 cm size could be cut.

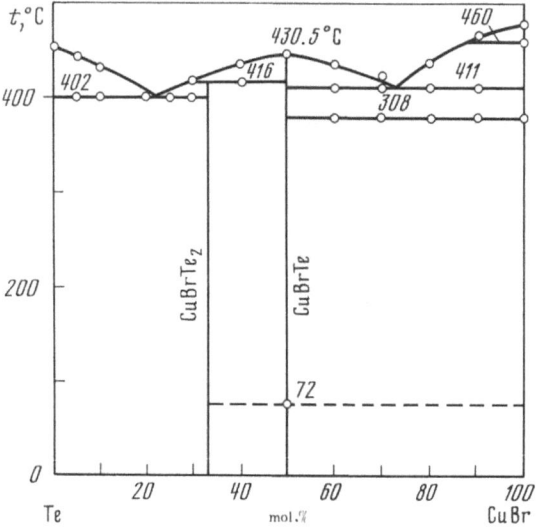

Fig. 3. Pseudobinary system CuBr - Te.

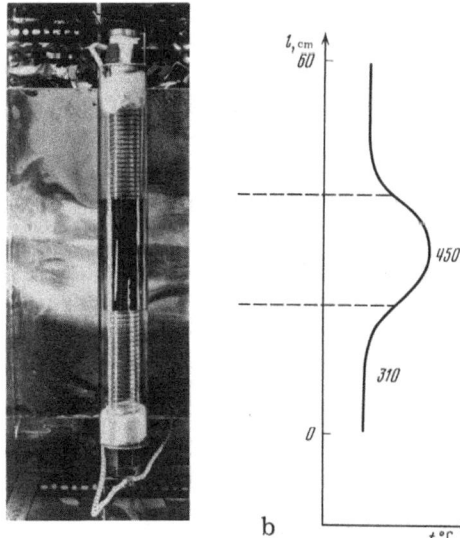

Fig. 4. Bridgman arrangement for CuTeBr (a) and temperature distribution (b).

Lithium Nitride, Li$_3$N

Highly conducting solid state Li-electrolytes are of considerable interest for a number of applications such as energy storage systems. Lithium seems to be particularly suited for this application because of its low equivalent weight combined with a strong electropositive character. Lithium nitride, Li$_3$N, attracted interest because of its high Li-content and its open structure. The ionic conductivity of pressed and sintered Li$_3$N powder proved to be quite high [8]. The crystal structure of Li$_3$N was initially proposed by Zintl and Brauer [9] and re-evaluated by Rabenau and Schulz [10].

The unusual coordination of Li$^+$ is related to its very small size compared with N^{3-}, so that its preferred tetrahedral coordination is not geometrically possible. Li$_3$N can be considered as having a typical ionic structure, N^{3-} being coordinated in a regular way by 8 Li$^+$. Li$_3$N can also be considered as a layer structure, with Li$_2$N layers perpendicular to the c-axis, these layers being connected with each other by only one Li atom per elementary cell, which occupies a central site between these layers (Fig. 5). Within each layer, each N and Li atom forms six and three bonds, respectively, whereas only two N-Li bonds per elementary cell are responsible for the connection of the layers. From this structural arrangement it can be deduced that thermal motion takes place preferentially between the Li-layers, perpendicular to the Li-N bonds. The Li-ion conductivity parallel and perpendicular to the hexagonal c-axis was therefore measured [11]. Transference measurements with the electrochemical cell $\ominus$|Mo|Li$_3$N|Li|Mo|$\oplus$ proved that only Li-transference occurs and that Faraday's law is obeyed. Results of the conductivity measurements are presented in Fig. 6 as a log σ T versus 1/T plot. The ambient conductivity perpendicular to the c-axis is nearly $10^{-3} \Omega^{-1}$cm^{-1}.

TABLE 1. Activation Enthalpies H and Conductivities E$\perp$c and E$\parallel$c of Li$_3$N

	H(eV)	$\sigma_{300K}(\Omega^{-1}cm^{-1})$	$\sigma_{400K}(\Omega^{-1}cm^{-1})$	$\sigma_{500K}(\Omega^{-1}cm^{-1})$
$\perp$c	0.290	1.2×10^{-3}	8×10^{-3}	4×10^{-2}
$\parallel$c	0.490	1×10^{-5}	6×10^{-4}	8×10^{-3}

In Table 1 the activation enthalpies and the conductivity values obtained for Li$_3$N single crystals are given. The electronic conductivity was found to be negligible, being less than $10^{-12} \Omega^{-1}$cm^{-1}. The investigations clearly show Li$_3$N single crystals to be pure Li-ion conductors. These data exceed the conductivities reported so far on polycrystalline material [8].

The Growth of Li$_3$N Single Crystals.

The system Li-Li$_3$N has a simple eutectic diagram, the eutectic composition being very close to lithium (Fig. 7). The melt in equilibrium with Li$_3$N (m.p. 815°C) has a finite, temperature-dependent solubility for Li$_3$N.

Using carefully purified nitrogen (pressure 10 atm) the reaction with Li ingots starts at about 180°C, the melting temperature of Li, resulting in a glassy, brown-red product. On fracture, small hexagonal platelets are seen (Fig. 8). For the growth of single crystals the Czochralski technique was used [12]. In order to avoid contamination, lithium nitride was synthesized in the crystal growth apparatus under reduced nitrogen pressure of 500 torr at 500°C. After the reaction the N$_2$ pressure was raised to 1 atm and the temperature to about 860°C. The crystals were grown in the equipment shown in Fig. 9 which consists of a stainless-steel vessel (1) and a stainless-steel heater (2), mounted on water-cooled electrodes of nickel (3). The heat loss is reduced by two molybdenum cylinders (4). The

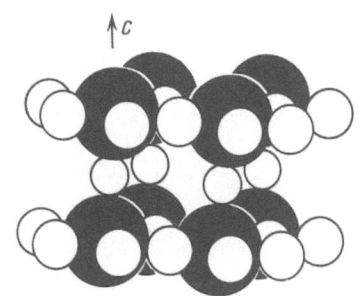

Fig. 5. Li₃N structure, schematic. Black circles: N^{3-}, open circles: Li^+. Two Li_2N layers are separated by a Li layer.

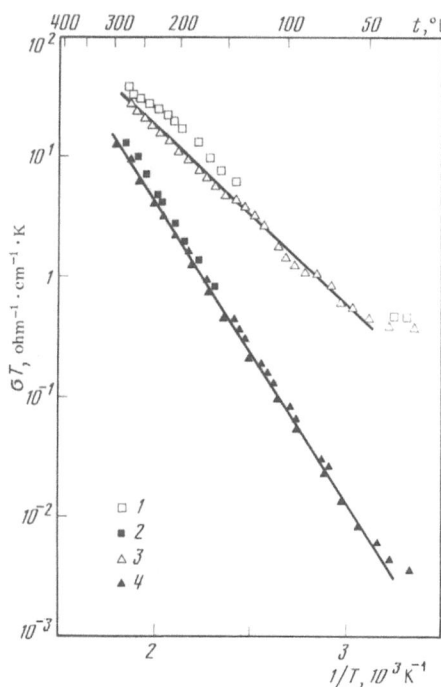

Fig. 6. Ionic bulk conductivity of Li_3N. Log σT versus $1/T$.
□ dc conductivity $E \perp c$ ■ dc conductivity $E \parallel c$
△ ac " $E \perp c$ ▲ ac " $E \parallel c$

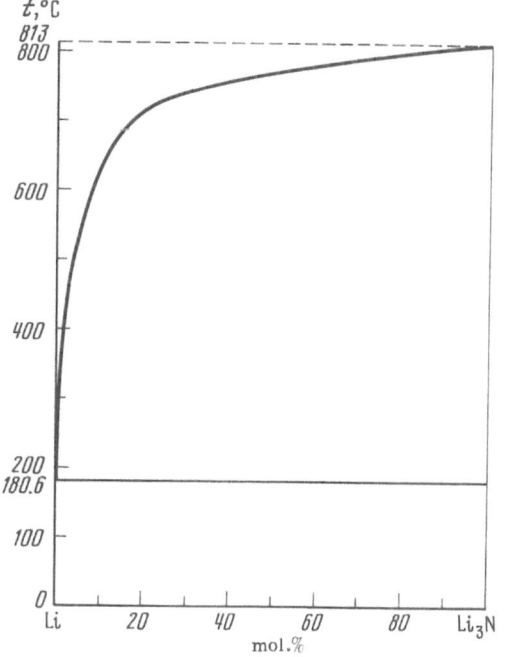

Fig. 7. Phase diagram Li – Li₃N.

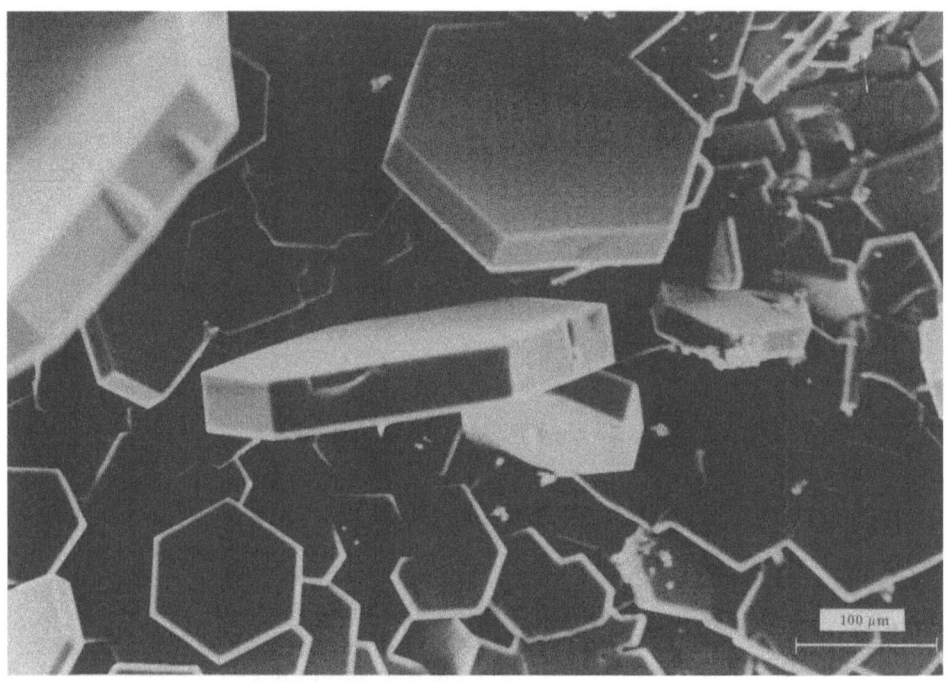

Fig. 8. Scanning electron microscope photograph of Li$_3$N.

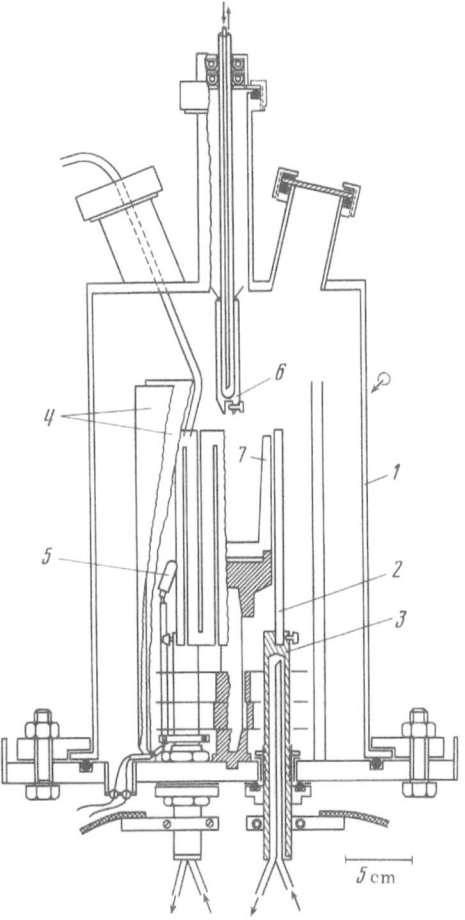

Fig. 9. Czochralski apparatus for the growth of Li$_3$N single crystals. (See text for description.)

temperature of the heater is controlled by a Pt/Pt-Rh thermocouple (5), protected by a corundum tube. The seed holder is made of nickel (6) and water cooled to dissipate the heat of solidification. Tungsten crucibles (7) were used as containers. Li_3N crystals were pulled with a rate of 5 mm h^{-1} and 5 rpm. Pieces up to 3 cm diameter and 5 cm length were obtained. High quality Li_3N single crystals are of ruby red color and are inert to the atmosphere over a long period of time.

β-Eucryptite, β-LiAlSiO$_4$

Ion diffusion in ionic conductors is of considerable interest in testing theoretical models. Recently, attention was drawn to the idea of ionic conductors with one-dimensional diffusion path. One-dimensional ionic conductivity in β-Eucryptite was proposed to analyze structural data as well as dielectric and nuclear magnetic resonance measurements. β-Eucryptite has a hexagonal quartz-like structure. Compared with quartz, half of the Si^{4+} are replaced by Al^{3+} and the charge is balanced by the incorporation of Li^+. At room temperature the Li-atoms occupy sites coordinated by four oxygens in channels running along the c-axis, each occupied site being followed by an unoccupied site. There are two sorts of structural main channels: sites occupied by Li^+ in the central channel are empty in the secondary channel (Fig. 10). With increasing temperature the Li-atoms exhibit an increasing mobility along the direction of the channels.

High quality single crystals were used for the electrochemical measurements [13]. Transference measurements were carried out with the electrochemical cell ⊖ Mo|β-Eucryptite|Li|Mo ⊕ in both directions E ∥ c and E ⊥ c. Li-transference which obeyed Faraday's law was observed only in the orientation parallel to the c-axis. No transference was observed E ⊥ c. This means that Li-diffusion can take place only in the direction of the Li-channels. An important parameter for successful transference proved to be the quality of the single crystals. Linear defects such as inclusions, grain boundaries or dislocations give rise to diffusion along energetically preferred paths, causing cracking of the samples. The results of the conductivity measurements are summarized in Fig. 11. The conductivity data can be interpreted in terms of a high partial Li-conductivity which dominates the total ac- and dc-conductivities. The anisotropy of the Li-ion conductivity is nearly 3 orders of magnitude in the temperature range from 200 °C to 600 °C. The Li-conductivity is thermally activated, $\sigma = \sigma_0 \exp(-E/kT)$, with an activation energy of 0.74 eV and $\sigma_0 = 3.4 \cdot 10^3 \, \Omega^{-1} cm^{-1}$. It should be emphasized that the Li-conductivity parallel to the structure channels above 500°C reaches values characteristic for optimized ionic conductors. High temperature single crystal X-ray investigations show that with increasing temperature the three-dimensional ordering disappears but a one-dimensional order of Li-atoms still remains, indicating a cooperative motion of Li^+ in β-Eucryptite [13].

The Growth of β-LiAlSiO$_4$ from High-Temperature Solution in LiF-AlF$_3$ Mixtures.

The growth of β-LiAlSiO$_4$ single crystals from solutions containing LiF and AlF_3 was first described by Winkler [14] and later by Tscherry and Schmid [15] and Joffe and Zonn [16]. The information provided proved, however, to be insufficient for growing crystals of the quality required. The growth process had therefore to be investigated in more detail [17].

The solubility of $LiAlSiO_4$ in the LiF-AlF$_3$ solvent was determined by analyzing quenched solutions. $LiAlSiO_4$ powder was dissolved in the solvent at temperatures between 1100 and 1150°C in Pt crucibles and the temperature was slowly lowered until the first precipitation became visible. The melt was then poured out and analyzed. The solubility of $LiAlSiO_4$ was determined for three LiF/AlF$_3$ compositions in the range of 1060 to 1100°C.

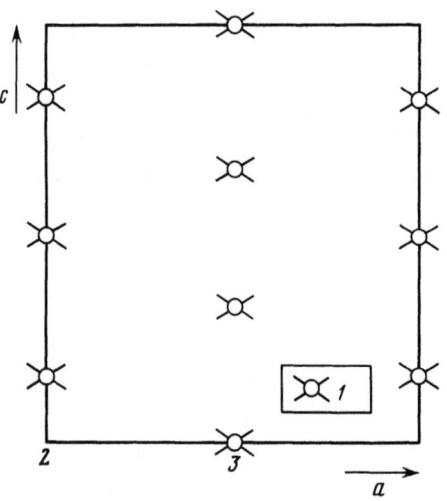

Fig. 10. Ordered distribution of the Li-atoms at room temperature. (1) Li ions coordinated to four oxygen atoms; (2, 3) central and secondary channels.

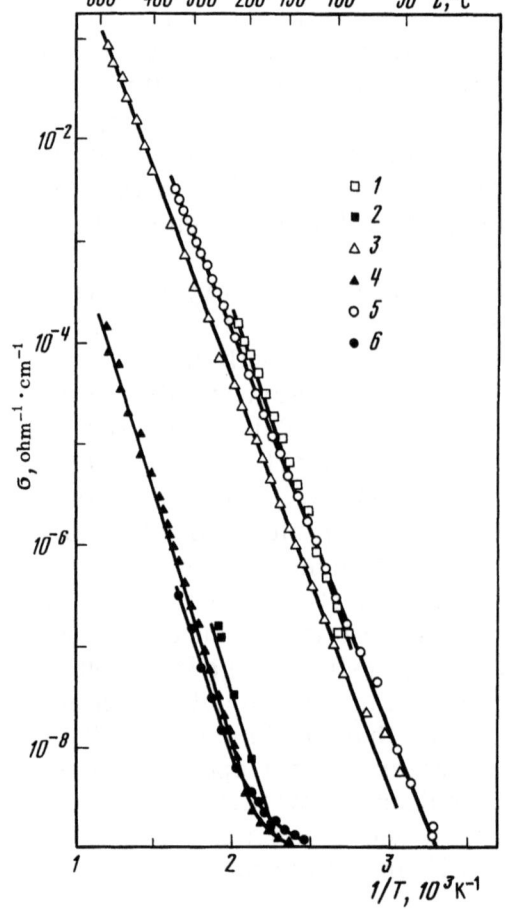

Fig. 11. Conductivity of β-Eucryptite.

△ ▲ Mo-electrodes
○ ● Li-electrodes
□ ■ dc-Li-ionic conductivity
(1, 3, 5) E ∥ c; (2, 4, 6) E ⊥ c

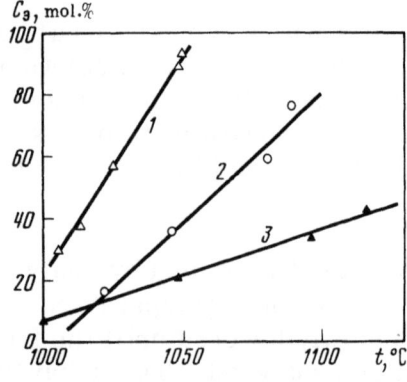

Fig. 12. Solubility of β-LiAlSiO$_4$ in
(1) 84.7 mol % LiF, 15.3 mol % AlF$_3$
(2) 90. mol % LiF, 10. mol % AlF$_3$
(3) 91.9 mol % LiF, 8.1 mol % AlF$_3$

The dependence of solubility on temperature is shown in Fig. 12. The solubility of LiAlSiO$_4$ and the slope of the solubility curves increase with increasing AlF$_3$ content.

Growth experiments were made with LiF-AlF$_3$ mixtures with 8 to 15 mol.% AlF$_3$ in the temperature range from 1100 to 1000°C. The LiAlSiO$_4$ concentration was taken from the solubility curves of Fig. 12. The β-LiAlSiO$_4$ crystals start to grow after spontaneous nucleation depending on the temperature distribution within the melt, the composition of the solvent, the composition of the solution, and the range of growth temperatures. When the temperature of the crucible is uniform, the crystals are formed preferentially at the base. In the case of low (8 mol.%) AlF$_3$ content the rate of nucleation seems to be greatly increased. Larger crystals (2-8 mm diam.) are embedded in fine crystalline eucryptite. Most of the large crystals appear opaque due to the presence of included particles. It is assumed that nuclei are mainly formed homogeneously and sink to the base. Besides eucryptite, lithium spinel (LiAl$_5$O$_8$) single crystals are frequently formed, and grow mainly on the wall of the crucible. The collection of eucryptite on the base was reduced and crystals grew preferentially along the surface of the solution when the temperature of the base was raised with respect to the surface of the melt. Most of them, however, display numerous hopper characteristics, solvent inclusions, or show initial dendritic growth. Small temperature differences of approximately 10°C lead occasionally to crystals which nucleated on the wall below the surface of the melt. In general, only these crystals display a high degree of perfection. Examples are shown in Fig. 13.

Fig. 13. β-LiAlSiO$_4$ crystals grown at crucible wall (1 mm grid).

Literature Cited

1. D. H. Whitmore. J. Cryst. Growth, 39, 160-179 (1977).
2. B. Reuter and J. Hardel. J. Angew. Chem., 72, 138 (1960).
3. A. Rabenau, H. Rau, and G. Rosenstein. Z. Anorg. und Allg. Chem., 374, 43-53 (1970).
4. J. Fenner and A. Rabenau. Z. Anorg. und Allg. Chem., 426, 7-14 (1976).
5. P. M. Carkner and H. M. Haendler. J. Solid State Chem., 18, 183-189 (1976).
6. U. V. Alpen, J. Fenner, J. D. Marcoll, and A. Rabenau. Electrochim. Acta, 22, 801 (1977).
7. U. V. Alpen, J. Fenner, B. Predel, A. Rabenau, and G. Schluckebier. A. Anorg. und Allg. Chem, 438, 5-14 (1978).
8. B. A. Boucamp and R. A. Huggins. Phys. Lett., 58, 231 (1976).
9. E. Zintl and G. Brauer. Z. Electrochem., 41, 102 (1935).
10. A. Rabenau and H. Schulz. J. Less-Common Metals, 50, 155 (1976).
11. U. V. Alpen, A. Rabenau, and G. H. Talat. Appl. Phys. Lett., 30, 621-623 (1977).
12. U. V. Alpen, E. Schönherr, H. Schulz, and G. H. Talat. Electrochim. Acta, 22, 805 (1977).
13. H. G. F. Winkler. Acta Crystallogr. 1, 27-34 (1948).
14. V. Tscherry and F. Laves. Naturwissenschaften, 57, 194 (1970).
15. V. Tscherry and R. Schmid. Z. Kristallogr., 133, 110-113 (1971).
16. V. A. Joffe and Z. N. Zonn. In: Growth of Crystals, vol. 9, ed. N. N. Sheftal', and E. I. Givargizov, Consultants Bureau, New York (1975).
17. E. Schönherr and E. Schedler. J. Cryst. Growth, 42, 289-292 (1977).

MODERN METHODS OF MONITORING AND CONTROL IN CRYSTAL GROWTH

É. L. Lube

The Institute of Crystallography of the USSR Academy of Sciences, Moscow

Crystal growth is a complex multifactor process. Success in growing crystals meeting prescribed criteria depends essentially on the relationship between the monitored and unmonitored, controlled and uncontrolled parameters of the process.

The progress in methods and means of monitoring stems from the tendency to increase the volume of information on the response of a crystal to changing growth conditions. This is necessary to find unambiguous relationships between growth conditions and crystal properties, to achieve optimization of the controllable parameters, and to restrict or minimize uncontrolled factors. The methods of crystallization monitoring can be classified into four groups by the character of the information obtained: (1) measurement of energy parameters, such as voltage, current, heater power, etc.; (2) measurement of equipment parameters, such as heater temperature, crucible temperature, crystal rotation and pulling rates, crucible displacement velocity, etc.; (3) monitoring of complex parameters of heat and mass transfer, such as position and shape of the growth interface, surface tension and viscosity of the melt, mass and cross section of the crystal, solution concentration, etc.; (4) monitoring the characteristics of the crystal, such as its composition, structure, and defects. The volume of control increases from the first group to the fourth. The methods developed in recent years to determine composition, structure, and perfection of growing crystals in the course of growth provide continuous information on the response of the system to practically all factors. Monitoring schemes of the third and fourth groups are determined to a large extent by the arrangements and conditions of the crystallization method and in their turn essentially affect the design of crystallization equipment.

In the method of horizontal uniaxial crystallization from the melt, the most informative integrated parameter of heat and mass transfer is the position of the melt-crystal interface. The sensitivity of an optical sensor of interface position depends on the relative contrast of melt and crystal images at the growth interface. In high-temperature crystallization of oxides the contrast is very low because the crystallized material is transparent and the intensity of illumination at the observation plane by the radiation emitted from the heater is very high. The method of laser-beam detection of the growth interface developed for horizontal uniaxial crystallization [1, 2] uses the difference in the spatial orientation of the melt and crystal surfaces: the former is always horizontal, while the latter deviates from horizontal because of the mass transfer due to the difference in densities of the faces and the effect of surface tension. With only the ratio of the crystal and melt densities, $\varkappa$, taken into account, the beam deflection Θ is given by the relation [3]

$$\theta = \arctan[2h_0(1-\varkappa)/l],$$

where h_0 is the initial level of the melt surface, and l is the zone length. A schematic of the method is shown in Fig. 1. The light beam of laser 1, modulated by light chopper 2, is deflected by mirror 3 to window 11 of the crystallization chamber 7 at a fixed angle to melt surface 9 in a boat crucible 10. Photocell 4 is mounted in a position adjusted to recording the beam reflected from the melt surface. The crystallization zone is scanned by the coupled reciprocating motion of the mirror and photocell driven by unit 5 along the crystallization direction. When the incident beam is reflected by the crystal surface 8, the reflected beam is deflected from the photocell window. The melt-crystal interface coordinate is the coordinate of the mirror and photocell, automatically recorded by the sensor unit 6 at the moment the output signal of the photocell drops to zero. The characteristics obtained experimentally are: scan range 30 mm, interface location accuracy 0.5 mm, sensitivity to surface deviation from horizontal 0.1°. The method makes it possible to monitor the interface position for parameters maintained at the above-mentioned levels even with the window appreciably obscured by evaporating materials.

The integrated parameter in Czochralski growth is the shape of the meniscus close to a crystal growing from the melt (Fig. 2a). The relation between the geometrical and physical characteristics of the meniscus is given by the equation [4]

$$h = [2\sigma (1 - \cos \alpha)/\rho g + (\sigma/2R \rho g)^2]^{1/2} - \sigma/2Rg,$$

where $R = 2r/\sin \alpha$ is the curvature of the molten column at the crystallization interface; r is the crystal radius; α is the angle between the tangent to the meniscus at the interface

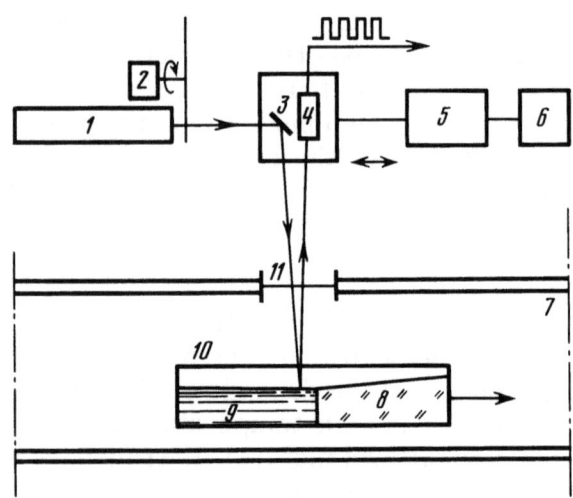

Fig. 1. Schematic of monitoring the position of melt-crystal interface in horizontal uniaxial crystallization. (1) laser, (2) modulator, (3) mirror, (4) photocell, (5) drive motor, (6) coordinate transducer, (7) crystallization chamber, (8) crystal, (9) melt, (10) boat crucible, (11) window.

Fig. 2. Schematic of monitoring the Czochralski crystal growth. (a) meniscus shape at the crystal, (b) pyrometer monitoring, (c) crystal weight monitoring; (1) retractable rod, (2) crystal, (3) pyrometer, (4) balance cell.

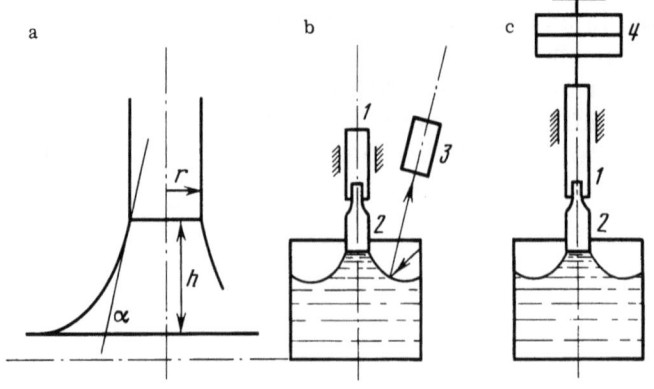

and horizontal plane; σ is the surface tension of the melt; ρ is the melt density; g is the gravitational constant; and h is the meniscus height.

A TV method of monitoring crystal diameter and meniscus cross section [5, 6] is based on the dependence of amplitude and duration of a video-signal on brightness and dimensions, respectively, of the observed object. The monitoring TV camera is mounted in front of a window in the crystallization chamber so that the meniscus and part of the crystal close to the growth interface are constantly in the view of the camera. The operator, using a marker on the monitor screen, chooses a scan line on the image as a reference, thus fixing the ordinate of a monitored cross section of the crystal or meniscus. In the case of noncircular cross sections, the rotating crystal is coupled to an angle sensor, and the varying projection of the cross section is synchronized with a prescribed angular position [6]. The accuracy of diameter measurement by the TV method is ±0.3%, and depends on the linearity of the TV scan and the accuracy of measuring the video-signal duration [5]. One of the basic difficulties in TV monitoring is caused by a low contrast of the observed brightness pattern. It has been suggested that an increase in the contrast of the images of transparent materials could be achieved by using a monochromatic illumination of the crystal through the seed and then filtering the image by selective filters [7]. A TV camera with high sensitivity in the IR spectrum has made it possible to supplement the diameter monitoring by obtaining a pattern of temperature distribution on the crystal and melt surfaces [6]. The TV monitor sensitivity is $4\,\mathrm{mV\,°C^{-1}}$; the range of linear response is 300 °C. A selection of filters in front of the camera lens provided the working range of temperature measurements of 1200-1500°C at the melt temperature 1400°C.

TV monitoring of liquid encapsulation growth employs X-rays to image the crucible with the melt and the encapsulating flux [8, 9]. The X-ray image is projected onto a fluorescent screen and then amplified by an image intensifier mounted in front of the TV camera. The image contrast depends on the angular characteristics of the X-ray focus and the ratio of the absorption coefficients of the crystallizing material, flux, and the equipment components.

Another optical monitoring technique employs a pyrometer focussed on the meniscus base (Fig. 2b) [10, 11]. The pyrometer measures radiation which is composed of the radiation from the melt itself and the background radiation reflected by the meniscus surface. The pyrometer thus records not only temperature fluctuations but also the variation of meniscus profile. If the monitoring operates only on the reflected component, an external light source with amplitude modulation is used [12]. The accuracy of crystal diameter measurements is 1%.

In contrast to optical methods, weighing techniques do not require the observation of the interface. A retractable rod 1 with crystal 2 is suspended to the weight sensor 4 (Fig. 2c) whose output signal F(t) is the sum of the rod weight and the weight of the crystal formed at time t, crystal weight increment over time dt, meniscus weight, and surface tension force [13]:

$$F(t) = m_0 g + \int_0^t \pi r^2 \rho_c g v dt + \pi r_M^2 \rho_m g h + 2\pi r \sigma_m \sin \alpha,$$

where m_0 is the mass of the rod and crystal; ρ_c and ρ_m are the crystal and melt densities, respectively; r is the crystal radius; r_M is the meniscus height; and σ_m is the surface tension of the melt. A number of perturbations affect the output signal as well as the forces listed above: electromagnetic force of the inductor, crystal buoyancy, etc. The diameter monitoring accuracy in this method is 2%. Two variants of the method are known in which the weight sensor is connected to the crystal [14, 15], or to the crucible [16, 17].

If a crystal is grown by float zoning, the complex parameter of heat and mass transfer is the width of the molten zone. Monitoring may be implemented with a TV system [18, 19], as in Czochralski growth. If access to the zone is difficult, the zone cross section is monitored by measuring the absorption of radiation from a radioactive isotope by the melt (Fig. 3a) [20]. The radiation flux from the main source 1 passes through the windows of the growth chamber 3 and the monitored portion of the molten zone, and is partially absorbed in them and recorded by counter 4. The same counter is periodically switched to recording the flux of radiation from a reference source 5. The fluxes are alternated by a rotating lead half-cylinder 7. The imbalance signal activates a compensating wedge 6 whose motion, converted into an electric signal, gives a measure of the variation in the zone cross section. An absorbing wedge 2 serves to shift the scale when the crystal diameter has to be changed. The accuracy of diameter monitoring and control is ±0.2 mm for (silicon) single crystals of 25-30 mm diameter.

A method based on measuring the torque of the viscous drag [21] eliminates the necessity to reduce the zone screening for the purpose of monitoring. The method is schematically shown in Fig. 3b. The torque M transferred from the lower rod 2 through the molten zone 3 to the upper rod 4 is a function of dynamic viscosity of the melt μ, radius r and height h of the zone, and the relative angular velocity of the rods [4]:

$$M = k\mu\omega\, r^4/h,$$

where k is a coefficient. If a motor 1 maintains constant rotation rate of the lower rod, the monitoring variable is the rotation rate of the upper rod measured by a tachometer generator 5.

In the Verneuil method, fluctuations of crystallization rate are detected by fluctuations in the height of the top of the growing crystal with respect to a fixed level. The top coordinate is monitored directly by a photocell [22] or indirectly by a thermocouple introduced into a chosen point within the muffle [23]. Sensitivity of the indirect control is ±6°C mm^{-1}, for a crystal 20 mm in diameter and muffle diameter 40 mm.

In the Kyropoulos method the crystal diameter is monitored optically, the crystal being illuminated by an external source [24]. Mass transfer due to the difference between the crystal and melt densities is found from changes of melt level in the crucible recorded by the electrocontact method [25]. This method of monitoring the melt level in the crucible was also successfully applied to mass transfer monitoring in Czochralski growth [26].

In the case of vertical uniaxial crystallization from the melt in an opaque crucible, the crystallization interface level and crystallization rate may be measured by a technique utilizing reflection of ultrasonic pulses from the crystal-melt interface [27]. An emitter 2 and receiver 3 (piezotransducers) are fixed in the seed portion of the crucible (Fig. 4). High-frequency pulses are formed by generator 1, transformed to ultrasonic pulses, propagate through the crystal 4, are partially reflected from the interface 5, and partially transmitted into the melt 6. The reflected pulses are detected by the receiver transducer and fed through an amplifier 7 to the plates of a cathode ray tube 8. The image on the screen is filmed by a camera 9. A synchronizer 11 and single-sweep 10 and time base 12 generators provide synchronization of the pattern on the screen: vertical waveforms represent the probe and reflected pulses, and a dashed line the time base. The distance from the emitter to the crystallization interface is thus found from the time interval between the pulses and the pulse propagation rate in the medium.

The ultrasonic probe technique developed for Czochralski growth [28] makes it possible to monitor both the advancement rate and changes in curvature of the crystal interface;

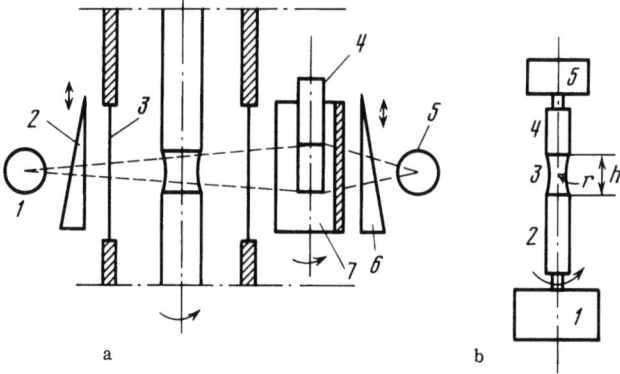

Fig. 3. Schematic of monitoring floating zone melting. (a) by absorption of radioactive radiation: (1) main emitter, (2) absorption wedge, (3) chamber window, (4) counter, (5) reference emitter, (6) compensation wedge, (7) lead half-cylinder; (b) by measuring the viscous friction torque: (1) motor, (2) lower rod, (3) molten zone, (4) upper rod, (5) tachometer generator.

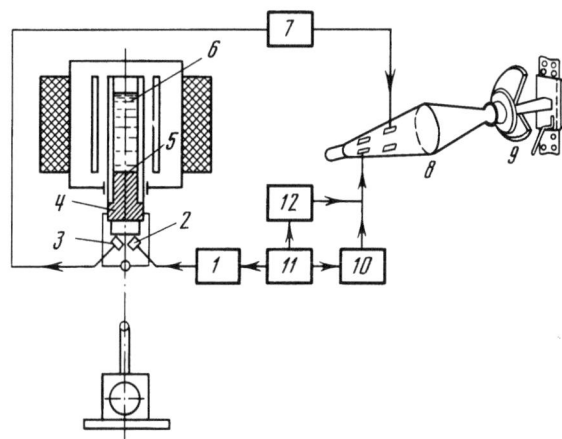

Fig. 4. Schematic of monitoring the position of the melt-crystal interface in vertical uniaxial crystallization. (1) generator; (2, 3) emitter and receiver transducers; (4) crystal; (5) crystal-melt interface; (6) melt; (7) amplifier; (8) cathode ray rube; (9) movie camera; (10) sweep generator; (11) synchronizer; (12) time base generator.

the latter is found from the attenuation of the reflected pulse amplitude due to a reduced curvature radius. The error in the measured interface advancement rate is ±10%, and the minimal curvature radius recorded is 40 mm.

A method for the determination of the sign of interface curvature by alternating irradiation of the interface with two pulsed beams having different divergence angles, and by comparing the amplitudes of reflected pulses, has been suggested [29].

Because of the temperature gradient along the crystal axis, the ultrasonic signals propagate through an acoustically inhomogeneous medium. It is necessary, therefore, to combine the ultrasonic probing with measurement of the temperature distribution along the crystal [27] or with averaging sound velocities in the crystal within the range of temperature variation; it is assumed that the mean value is valid because of only a weak dependence of Young's modulus on temperature [28].

In the case of hydrothermal synthesis of crystals, mass transfer between the dissolution and growth zones in a horizontal autoclave is monitored by a balance shown in Fig. 5 [30]. Autoclave 1 is coupled to a prism support 5 by a rod 4 passing through the walls of the furnace 2 and explosion-proof housing 3. Before the experiment the system is balanced by moving the loads 9 along the balance arms 10. Departures from equilibrium due to mass transfer are compensated by annular balance weights 6 on brackets 8. Vibrations of the system are suppressed by a damper 7. The temperatures within the zones of four sectional heaters 11 are measured by thermocouples 12. The mass of the crystals grown, m, is found by the formula

$$m/m_1 = \rho/(\rho - \rho_1),$$

where m_1 is the mass of transferred balance weights, ρ is the crystal density, and ρ_1 is the solution density. Although the balancing operation may be very accurate, the error in mass transfer evaluation is 10-20%, owing to an uncertainty in the position of the dissolving part of the charge with respect to the pivot of the balance.

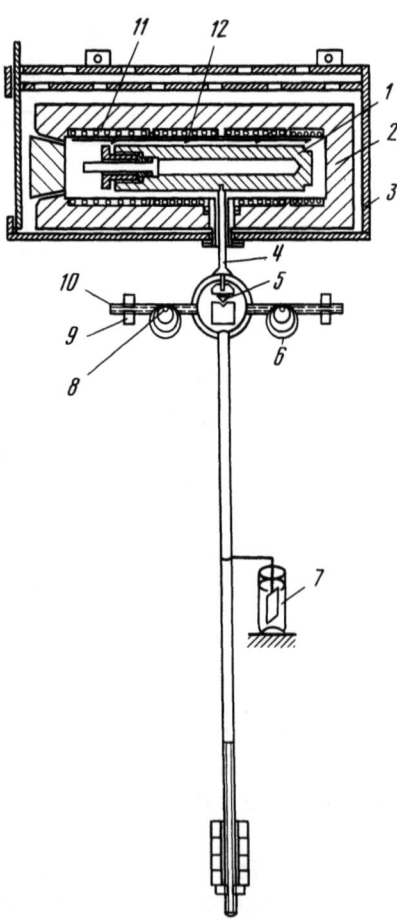

Fig. 5. Schematic of monitoring the progress of hydrothermal synthesis of crystals. (1) autoclave, (2) furnace, (3) protective explosion-proof housing, (4) rod, (5) support, (6) balance weights, (7) damper, (8) bracket, (9) adjusting load, (10) balance arm, (11) sectional heaters, (12) thermocouples.

It has been suggested that the displacement of the center of mass in a vertical autoclave could be monitored by a pendulum arrangement [30]; the displacement would be found from the variation of the oscillation period of the autoclave using a counterweight with respect to the horizontal axis.

The growth rate of crystals grown from low-temperature solutions is directly monitored optically by measuring increments to individual crystal faces under a microscope [31] or with a cathetometer [32], or by an integral weighing of the growing crystal to determine the mass increment. Optical methods provide accuracy of 1% [32]. Weighing is carried out with an analytical balance (accuracy 0.1 mg) [33] or with an electromagnetic-balancing system (accuracy 0.02 mg) [34].

Changes in solution concentration, caused by mass transfer during crystallization, are monitored by measuring the electric conductivity of the solution by a conductometric transformer-type transducer with liquid coupling coils [35]. In the case of lithium iodate growth, the transducer characteristics remained linear in the concentration range from 49.0 to 52.0 mg per 100 g of solution, and the total error of concentration measurements (including the calibration error) was 0.1%. The use of a float densitometer with electrical output for concentration measurement has also been suggested [36]. Both the electric conductivity and density of the solution depend not only on concentration but on temperature as well, so that both of these methods require that the measurements be made at constant temperature or that the results be corrected for the measurement temperature.

In vapor deposition experiments the thickness of growing films is monitored by ellipsometric techniques [37].

The amount of monitoring information reaches a maximum in the case of molecular beam epitaxy [38, 39]. A mass spectrometer measures the intensity of beams of the evaporated elements and the impurity content by computer controlled scanning. Calibration of the system for the absolute value of growth rate is made by the automatic replacement of the substrate by a quartz resonator plate, whose vibration frequency depends on the thickness of the deposited film. The elemental composition of the surface is obtained by Auger spectroscopy. The system monitors the degree of cleaning of the substrate surface, and the ratio of the deposited elements in the course of growth. High-energy electron diffraction (HEED) provides information on surface topology and crystallographic structure [38]. In another version of the monitoring system a LEED unit is used, capable of distinguishing between, for example, sphalerite and wurtzite structural types [39].

An *in situ* observation of defects generated in growing crystals has proved to be feasible for bulk crystals [40]. The X-ray source was a tube with rotating molybdenum anode, focus dimensions 0.5×10 mm, voltage 60 kV, current 0.5 A. The topographic camera is similar to the conventional Lang system. A topograph 9×13 mm in size is recorded with 10 μm resolution by an X-ray sensitive vidicon. The image is stored during the time of translation of the crystal carriage (from 3 to 10 s). A dislocation-free silicon crystal 0.8 to 1 mm in thickness was placed in a miniature furnace and the top layer was partially melted prior to crystallization. Figure 6 shows topographs obtained with 10 s (a-e) and 50 s (e-f) intervals. We see how the dislocation density, indicated by dark areas on the topographs, increases as the crystal grows.

In many cases it is not sufficiently effective to control crystallization by rigid programs of stabilization and variation of energy parameters at the input of the crystallizer. Adequate mathematical models are difficult to construct and utilize because of the complexities of the process and unavoidable unmonitored and uncontrollable perturbations.

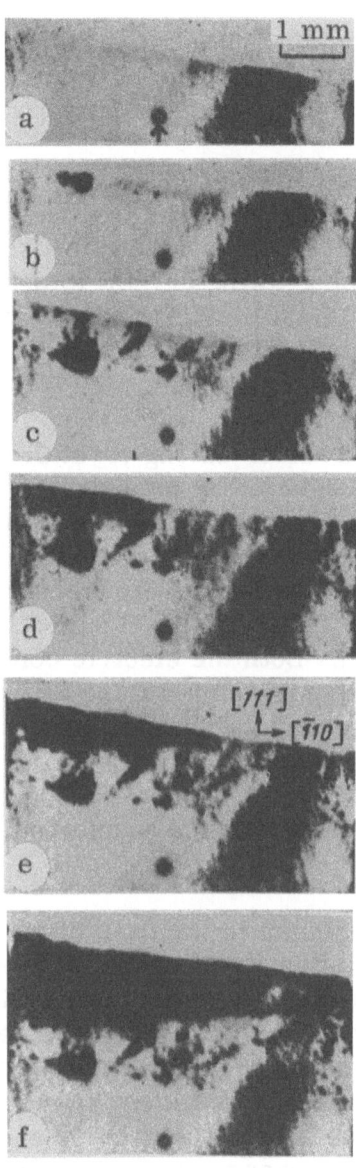

Fig. 6. Frames of a TV X-ray topographic image of a crystal growing from the melt; time intervals 10 s (a-e) and 50 s (e-f).

This has resulted in the current trend to base the control techniques on the introduction and proliferation of feedback channels.

The efficiency of feedback is determined first of all by the sensitivity of monitoring, the amount of information, and by the available parameters affecting the process. Feedback efficiency is best attained using the parameters of heat and mass exchange for control, or complex parameters reflecting the combined effects of the controlled and uncontrollable factors of these processes. As the growing crystal is highly sensitive to practically all factors, an approach that seems ideal is the one with feedback through crystal characteristics measured in the course of growth. It should be emphasized that when feedback of higher information level parameters is introduced feedback parameters of lower levels is retained. This increases the stability of the process for the case where an appreciable delay time exists between the control signals and the resultant responses of the parameters. In Czochralski growth, feedback utilizes rotation rates and the crystal and crucible vertical velocities, heater and melt temperatures, and crystal diameter (a complex parameter measured by a pyrometer focussed at the growth interface) [41]. The first two constraints serve to stabilize the prescribed crystal and crucible rotation rates. The crucible vertical velocity must ensure constant position of the melt level with respect to the heater (this is dictated by the heat transfer conditions), and also stabilize the distance between the pyrometer and the observation point. This velocity is a function of the ratio of crystal-to-crucible diameters and the crystal pulling velocity, that is, of the parameters determining fluctuations of the melt surface level in the crucible due to mass transfer. The control function of heater power output is based on a mathematical model taking account of the heat balance in the chamber, including the heat removed by the coolant water and inert gas. The model is corrected by measuring the heater temperature and the crystal diameter. The feedback on crystal diameter also affects the crystal cooling rate, to rapidly compensate for changes in the meniscus profile due to uncontrollable perturbations.

In the case of floating zone growth the crystal cross section feedback is used to control the heater power output and the position of the zone contraction-extension device [42].

In the Verneuil technique the crystal pulling rate and charge feed are controlled in response to changes of the level of the top crystal surface [22].

In hydrothermal synthesis the temperature feedback from the two autoclave zones affects the power output of two heaters [43, 44].

In molecular beam epitaxy the beam fluxes are controlled by the evaporation temperature, and the film thickness is determined by the time during which the shutter opens the cell containing the evaporating material, and by the beam flux [38, 39].

An increase in the number of monitored and controlled parameters and the complicated interrelations between these parameters necessitate a considerable functional flexibility of the required control devices. It is becoming general practice to use digital computer input coupled to monitoring devices, and output to the actuators of the crystal growth equipment [16, 38, 41, 44, 45].

Improved methods of monitoring and controlling crystal growth processes enable researchers to improve the characteristics of the crystal products. Thus, for example, crystal diameter stability in the Czochralski and floating zone techniques has reached 1-2% [13, 20], the optical homogeneity of crystals in the Verneuil technique has been improved [23], the number of inclusions in crystals in hydrothermal synthesis has been reduced [44], and a strictly periodic structure with 50 Å period has been fabricated by molecular beam epitaxy [38].

Literature Cited

1. E. L. Lube and Kh. S. Bagdasarov. Inventor's certificate No. 569320 (USSR). Photoelectric method of locating a crystallization interface. Bulletin of Inventions, No. 31, 8 (in Russian) (1977).
2. E. L. Lube. Automatic control of the melt-crystal interface in horizontal uniaxial crystallization. In: Vth USSR Conference on Crystal Growth, Tbilisi, Abstracts. Inst. of Cybernetics AN GSSR Publ. House, Tbilisi, 2, 208-209 (in Russian) (1977).
3. W. G. Pfann. Zone Melting. J. Wiley and Sons (1966).
4. S. V. Tsivinsky. Application of the theory of capillary effects to growing shaped crystals from the melt by the Stepanov technique. Inzh.-Fiz. Zhurnal, 5, 59-65 (in Russian) (1962).
5. K. J. Gäartner, K. F. Rittinghaus, A. Seeger, and W. Uelhoff. An electronic device including a TV-system for controlling the crystal diameter during Czochralski growth. J. Cryst. Growth, 13/14, 619-623 (1972).
6. D. F. C'Kane, T. W. Kwap, L. Gulitz, and A. L. Bednowitz. Infrared TV-system of computer-controlled Czochralski crystal growth. J. Cryst. Growth, 13/14, 624-628 (1972).

7. S. M. Vishnevsky, V. D. Tsitsugin, and A. F. Lavrov. Inventor's certificate No. 485763 (USSR). Method of automatic control of the Czochralski growth of single crystals. Bulletin of Inventions, No. 36, 13 (in Russian) (1975).
8. H. J. A. van Dijk, C. M. G. Yochem, G. J. Scholl, and P. van der Werf. Diameter control of LEC grown GaP crystals. J. Cryst. Growth, 21, 310-312 (1974).
9. H. D. Pruett and S. I. Lien. X-ray imaging technique for observing liquid encapsulation Czochralski crystal growth. J. Electrochem. Soc., 121, 822-826 (1974).
10. E. J. Patzner, R. G. Dessauer, and M. R. Poponiak. Automatic diameter control of Czochralski crystals. Solid State Technol., No. 10, 25-30 (1967).
11. T. G. Digges, R. H. Hopkins and R. H. Seidensticker. The basis of automatic diameter control utilizing "bright ring" meniscus reflections. J. Cryst. Growth, 29, 326-328 (1975).
12. U. Gross and R. Kersten. Automatic crystal pulling with optical diameter control using a laser beam. J. Cryst. Growth, 15, 85-88 (1972).
13. W. Bardsley, B. Cockayne, G. W. Green, D. T. Hurle, G. C. Joyce, J. M. Roslington, P. I. Tufton, H. C. Webber, and M. Healey. Developments in the weighing method of automatic crystal pulling. J. Cryst. Growth, 24/25, 369-373 (1974).
14. W. Bardsley, G. W. Green, C. H. Holliday, and D. T. J. Hurle. Automatic control of Czochralski crystal growth. J. Cryst. Growth, 16, 277-279 (1972).
15. V. S. Leibovich, V. A. Sukharev, I. M. Gol'd, V. A. Fedorov, A. I. Sereda, V. M. Shushkov, and A. V. Zhadan. Inventor's certificate. Device for automatic control of growth of single crystals from the melt. Bulletin of Inventions, No. 37, 17 (in Russian) (1975).
16. A. G. Zinnes, B. E. Newis, and C. D. Brandle. Automatic diameter control of Czochralski grown crystals. J. Cryst. Growth, 19, 187-192 (1973).
17. T. R. Kyle and G. Zydzik. Automated crystal puller. Mater. Res. Bull., 8, 443-450 (1973).
18. V. M. Byndin. Automatic TV controller of diameter of ingots grown by floating zone melting. In: Industrial Application of HF Currents. Mashinostroyeniye, Leningrad, 86-93 (in Russian) (1969).
19. H. Stut. Pat. 3757071 (USA) Method for crucible-free zone melting. Filed Mar. 2, Ser. No. 231. 182, Inst. Cl. H05b 5/00. US Cl. 219-10.43 (1972).
20. N. I. Autenshlyus, Yu. P. Betin, V. V. Dobrovensky, A. Ya. Zbarsky, U. I. Kotika, and I. L. Shenderovich. Control system for diameter of crystals grown from the melt. Pribory i Tekhnika Éksperimenta, No. 2, 226-228 (1975).
21. J. M. Quenisset and R. Naslain. Automatisation d'un appareil de fusion de zone vertical. J. Cryst. Growth, 30, 169-176 (1975).
22. F. A. Reiss. Growth problems of sapphire and ruby of optical quality. Appl. Opt. 5, 1902-1905 (1966).
23. Ken-Ichi Shiroki. Automatic control system for crystal growth by the flame fusion method. Rev. Sci. Instrum. 38, 1541-1542 (1967).
24. Ya. I. Ioffe, I. V. Smushkov, and Yu. N. Sherman. Device for monitoring crystal diameter during growth. Zavodskaya Laboratoriya, 37, 239-240 (in Russian) (1971).
25. Yu. P. Belogurov, V. A. L'vovich, A. V. Radkevich, and V. A. Sukhostat. Control of dimensions of LiF crystals grown by the Kyropoulos technique in vacuum. In: Single Crystals and Growth Equipment. VNII Monokristallov Publ. House, Kharkov, No. 2, 240-243 (in Russian) (1970).
26. B. Perner, V. Stranski, I. Gorak, E. Alekseev, V. L'vovich, and A. Radkevich. Automated unit for growth of corundum single crystals by pulling from the melt. In: Vth USSR Conference on Crystal Growth, Tbilisi, Abstrasts. Inst. of Cybernetics AN GSSR Publ. House, Tbilisi, 2, 212-213 (in Russian) (1977).
27. S. I. Yatsyk and E. B. Glotov. A setup for ultrasonic investigation of crystallization interface in temperature-resistant alloys. Zavodskaya Laboratoriya, 40, 192-193 (in Russian) (1974).

28. V. I. Il'chenko. Ultrasonic monitoring of curvature and advance rate of crystallization interface. In: Dielectrics and Semiconductors, Vishcha Shkola, Kiev, No. 9, 67-70 (in Russian) (1976).
29. D. A. Romanchenko, A. A. Snegirev, S. N. Grachev, E. V. Fillipchuk, and V. I. Il'chenko. Inventor's certificate 479486 (USSR). Method of measuring shape and curvature of crystallization interface. Bulletin of Inventions, No. 29, 11 (in Russian) (1975).
30. A. A. Shternberg. Control of crystal growth in autoclaves. In: Analysis of Crystallization in Hydrothermal Systems. Nauka, Moscow, 199-211 (in Russian) (1970).
31. T. G. Petrov, E. B. Treivus, and A. P. Kasatkin. Growth of Crystals from Aqueous Solutions. Nedra, Leningrad (in Russian) (1967).
32. Z. I. Zhmurova and V. Ya. Khaimov-Mal'kov. On the dependence of impurity distribution on growth rate in crystallization of isomorphous systems from solutions. Kristallografiya, 15, 136-141 (in Russian) (1970).
33. P. Bennema. Technique for measuring the rate of growth of crystals from solutions in dependence on the degree of supersaturation at low supersaturations. Phys. Status Solidi 17, 555-562 (1966).
34. J. C. Noisier, L. Soukiassian, and D. Defives. Apparatus for measuring crystal growth and dissolution rate. J. Phys. E: Sci. Instrum., 4, 603-607 (1971).
35. Yu. G. Agabalyan, L. S. Tatevosyan, and R. Sharkhatunyan. On the possibility of determining solution supersaturation in the growth of lithium iodate crystals. Kristallografiya, 20, 883-885 (in Russian) (1975).
36. T. G. Petrov and A. P. Kasatkin. Inventor's certificate 424042 (USSR). Automatic controller of solution supersaturation. Bulletin of Inventions, No. 14, 130 (in Russian) (1974).
37. R. Hilton and C. J. Jones. Measurement of epitaxial film thickness using an infrared ellipsometer. J. Electrochem. Soc., 113, 472-478, (1966).
38. L. Esaki and L. L. Chang. Semiconductor superfine structures by computer-controlled molecular beam epitaxy. Thin Solid Films, 36, 285-298 (1976).
39. D. L. Smith and V. Y. Pickard. Molecular beam epitaxy of II-VI compounds. J. Appl. Phys. 46, 2366-2374 (1975).
40. Jun-Ichi Chikawa. Technique for video display of X-ray topographic images and its application to the study of crystal growth. J. Cryst. Growth, 24/25, 61-68 (1974).
41. D. P. Yen, R. A. Slocum, and C. R. Valentino. Pat. 3621213 (USA). Programmed digital-computer-controlled system for automatic growth of semiconductor crystals. Filed Nov. 26, Ser. N 880273. Int. Cl. G06f 15/46, 9/06. US Cl. 235-150.
42. N. I. Autenshlus, V. V. Dobrovensky, A. Ya. Zbarsky, and I. L. Shenderovich. Inventor's certificate. Method for automatic control of crystallization in floating zone melting. Bulletin of Inventions, No. 6, 19 (in Russian) (1972).
43. N. Yu. Ikornikova, A. N. Lobachev, A. R. Vasenin, V. M. Yegorov, and A. V. Antoshin. Equipment for precision analysis in hydrothermal experiments. In: Investigation of Hydrothermal Crystallization. Nauka, Moscow, 212-223 (in Russian) (1970).
44. D. W. Rudd, A. R. Fiore, and N. S. Lias. Computerized process control for hydrothermal growth of quartz. J. Cryst. Growth 7, 29-36 (1970).
45. V. S. Leibovich, V. A. Sukharev, V. M. Shushkov, and V. A. Fedorov. Automatic control of single crystal growth. Pribory i Sistemy Upravleniya, No. 5, 7-9 (in Russian) (1975).

APPLICATION OF LASER HEATING TO CRYSTAL GROWTH

Kh. S. Bagdasarov, V. V. Dyachenko, A. M. Kevorkov, and A. Kholov

The Institute of Crystallography of the USSR Academy of Sciences, Moscow

Progress in quantum electronics has made it possible to apply laser heating to the development of new crystal growth techniques [1-6]. Compared to the ohmic, high-frequency, flame, plasma, electron beam, and radiation heating, this method of heating offers considerable advantages:
— the energy is fed to the molten zone by coherent electromagnetic radiation; the temperature distribution can therefore be adjusted by standard optical elements;
— the small divergence of the light beam makes it possible to site the heating source (laser) outside the reaction chamber;
— the energy distribution across the beam is uniform in the single-mode emission, and maximum temperature gradients are obtainable over distances comparable with the radiation wavelength;
— the inertia of the heater is low;
— the system is technically simple.

The high-power, continuous radiation makes the melting of a well-defined volume of the material possible — a *sine qua non* in crucible-free growth techniques. This characteristic, together with the possibility of producing complicated configurations of the laser beam [7], substantiates the potential of this type of heating for zone melting. Simple control of the laser beam facilitates the formation of temperature fields and helps to control the crystal-melt interface at each stage of single crystal growth. The removal of the heater from the reaction chamber is of principal importance for crystal growth under contamination-free conditions, as well as under conditions of high vacuum and high pressure. The technical simplicity of the system and its low thermal inertia are necessary for achieving high precision of heater stabilization and complete automation of growth with effective feedback. However, laser heating introduces a peculiarity related to the absorption characteristics of the stimulated emission, which is determined by the properties of the starting material, in its molten and crystal states, namely: absorption coefficients, latent heat of fusion, thermal conductivity, melting point, and geometrical configuration of the crystal-melt system. This imposes certain requirements on wavelength and radiation energy, as well as on the laser mode (CW, quasi-CW, pulsed modes).

The CW mode is the most convenient from the standpoint of the crystal grower since this is best suited to produce highly stable growth conditions. Consequently the CO_2 and CO_2-N_2-He gas lasers, operating in the CW mode and having high efficiencies (up to 20%) at high levels of power output, have been used almost exclusively in crystallization studies [8]. However, these two types of laser substantially restrict the potential of laser heating, since the wavelength of CO_2 lasers is in the 10 μm range in which most refractory materials are opaque. Moreover, the CO_2 laser beam is absorbed almost identically by the crystal and the melt, making it difficult to control temperature gradients at the growth interface. It is therefore imperative to resort to shorter-wavelength lasers emitting in the range in which most refractory crystals are transparent, for example, in the 1 μm range. This need is met with lasers based on yttrium aluminum garnet doped with neodymium ions (YAG-Nd^{3+}) operating at 1.06 μm in the CW and quasi-CW (at 50 Hz) modes. In this spectral range the melts of refractory materials are practically opaque, in contrast to the corresponding crystals, so that it is possible to generate temperature fields with the required axial and radial gradients. Furthermore, YAG-Nd^{3+} solid state lasers, in contrast to gas lasers, are small in size which is of considerable importance in designing crystal growth equipemtn and in utilizing several lasers simultaneously to form large hot zones of prescribed configuration. Solid-state lasers make it possible to generate an illuminated spot with a prescribed energy distribution; this is necessary, for example, to grow crystals with a predetermined refractive index gradient caused by an inhomogeneous distribution of dopant in the crystal.

Figure 1 shows a diagram of an experimental model of a laser-heated crystal growth unit. The laser head 1 is a system comprising a water-cooled housing with silver-coated inner walls, pump lamp, YAG-Nd^{3+} laser elements, and two mirrors. Lasing efficiency is increased by feeding the supply voltage (380 V) through a half-wave rectifier 2. The pump lamp consumes 25 kW at a current of 120±10 A. The pump supply circuit includes an arc 3 providing a current of 30 A during intervals of no emission. The lamp is started by the starter unit 4. The emission power is monitored by the instrument block 5 comprising a semitransparent beam splitter plate, a spherical-cavity calorimeter, and a microvoltmeter. The accuracy of the power output monitoring was 5%. The laser was cooled by a closed water system, with coolant flowrate 25 liters m^{-1} at 2 atm.

The beam focussing system 6 consists of a quartz lens 40 mm in diameter and focal length 100 m, and a lens displacement mechanism for moving the lens both along and normal to the beam. These displacements are necessary for focussing the beam and creating the required temperature gradients.

The addition of a second laser head substantially increased the experimental capabilities of the system and made it possible to achieve a number of modifications to the crucible-free growth of refractory single crystals. Moreover, two laser heads allow a temperature distribution control over a wider range and facilitate the melting of larger volumes of material. The temperature of the melt and crystal was measured by a disappearing-filament pyrometer with red filter KS-1.5 (effective wavelength λ = 0.65±0.01 μm).

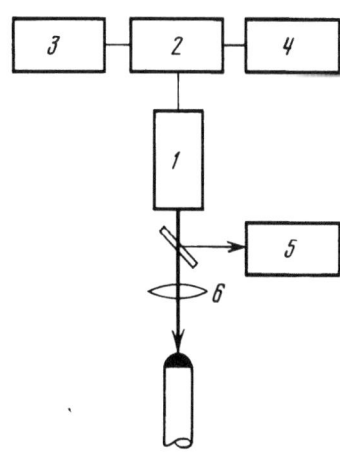

Fig. 1. Experimental model of a laser-heated crystal growth unit. (1) laser head, (2) half-wave rectifier, (3) arc, (4) starter unit, (5) block for measuring emission power, (6) focusing system.

Crystal growth experiments were mostly conducted in the Y_2O_3-Al_2O_3 system, which is complex, having two stable ($Y_3Al_5O_{12}$ and $Y_4Al_2O_9$) and one metastable ($YAlO_3$) compounds. However, it is of considerable practical importance. The main results were obtained with yttrium aluminum garnet $Y_3Al_5O_{12}$, since this compound has a very high melting point (1930°C). The melt was obtained on top of a compacted powder rod of 7-8 mm diameter. Two lasers with total power output of 250-300 W melted the end portion of the rod to a depth of 3-4 mm. If a single crystal rod was used, this power was sufficient only to melt the surface layer of a rod 4-5 mm diameter. Moreover, the size of the molten droplet was practically independent of whether the radiation was focussed into a spot 1 mm in diameter or defocussed over the whole end face of the rod. Nevertheless, focussing served to achieve a large radial temperature gradient in the melt, the temperature reaching the boiling point in the center of the focal spot. These gradients led to extremely intense convection within the melt.

Verneuil growth with laser heating is shown schematically in Fig. 2. The laser beam 1 is incident at the seed surface 2 from above, and forms a melt 3. The powder charge is fed from a screw-type feeder 4 made of a quartz tube with a metal helix forming a screw within the feeder. When the tube is rotated by an electric motor 5, the powder charge is moved to the open end of the tube and falls on the seed. The feed rate is controlled by the motor rotation rate and the tilt angle of the tube. The growth surface of the crystal is kept constant by a special lowering device 6. The method was used to grow yttrium iron garnet single crystals in air (Fig. 3).

Compared with the gas burner version of the Verneuil technique, the method described above produces much thicker molten layers, thus alleviating restrictions on the particle size distribution of the charge and allowing the use of coarser fractions (up to 0.5 mm).

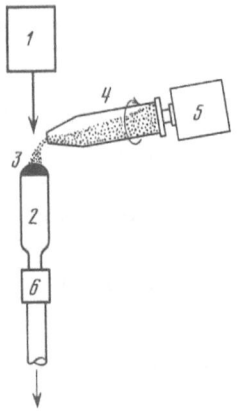

Fig. 2. Verneuil growth with laser heating. (1) laser head, (2) seed surface, (3) melt, (4) feeder, (5) electric motor, (6) lowering device.

Fig. 3. Yttrium - iron garnet single crystals with laser heating.

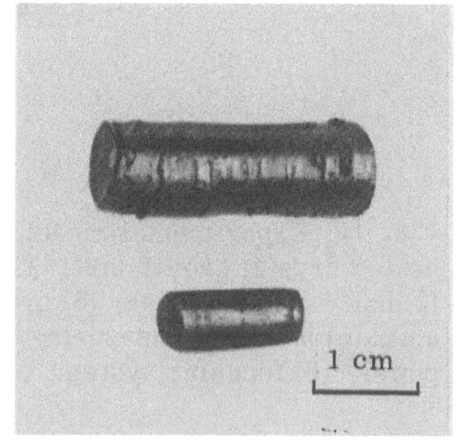

The melted-rod method is shown schematically in Fig. 4. The beams 2 of two laser units 1 are focussed by lenses 3 onto the end face of the compacted-powder rod 4 fixed in the upper lowering device 5. A similar device 6 lowers the seed 7 and the growing single crystal 8. This technique allows the growth of crystals with diameters much larger than that of the initial rod. Figure 5a shows yttrium aluminum garnet single crystals with diameter exceeding that of the powder rod by a factor of 1.5-2.

Crucible-free Czochralski Growth [9] is achieved by reversing the motion of the lowering mechanisms compared with the melted rod version in which the powder rod and the seed move downward at different velocities. Some crystals obtained by this crucible-free method are shown in Fig. 5b. In this case the diameter of the crystal pulled is smaller than that of the powder rod.

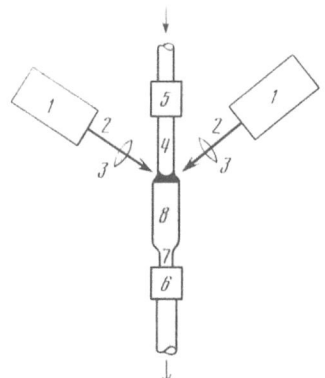

Fig. 4. Melted rod method. (1) laser head, (2) beams, (3) focusing system, (4) compacted powder rod, (5, 6) upper and lower devices for lowering, (7) seed, (8) growing single crystal.

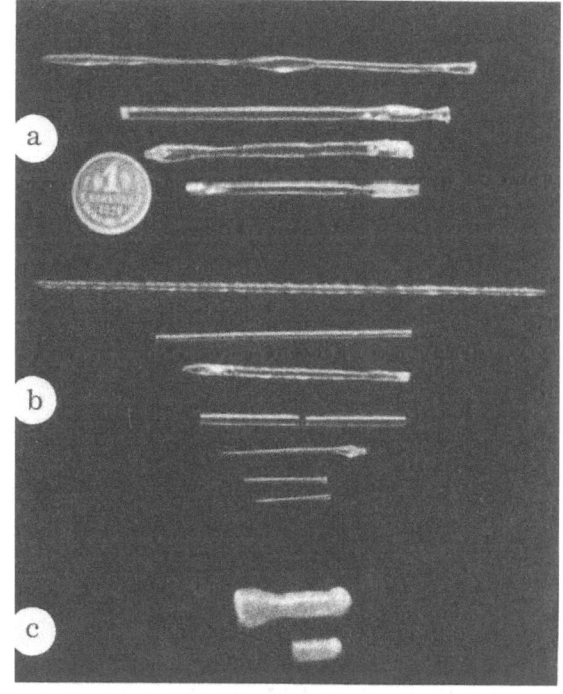

Fig. 5. Yttrium aluminum garnet single crystals activated by neodymium ions with laser heating by: (a) the melted rod method; (b) the crucible - free method; and (c) changing the absorption coefficient of the laser beam.

Crystal growth induced by changing the absorption coefficient is based on varying the material transparency by the evaporation of impurities. The method is accomplished by introducing into the initial material a volatile additive causing intense absorption of the laser emission. As the charge melts, this additive gradually evaporates and thus reduces the absorption of radiation. At the same time, the crystal grows on the seed. The necessary growth rates can be achieved by selecting the additive and the melt temperature. The method was used to grow yttrium aluminum garnet single crystals. Chromium served as the additive. Its evaporation enabled us to grow the crystals shown in Fig. 5c. This technique does not require that the seed, the initial rod, or the focussing system be moved. It is especially promising for crystal growth under extreme conditions, such as superhigh pressures, when the volume available is severely limited and mechanical displacements are hardly feasible.

Laser-heated zone melting differs from all other techniques, including the electron beam version, in that it allows the use of a complicated molten zone profile and enables the process to be carried out in any ambient, without mechanically moving the starting material. Figure 6 shows a diagram of an apparatus in which the molten zone 1 is moved by tilting the focusing lens 2. This scheme is useful because multiple rocking of the lens allows multiple recrystallization, which is very important in zone refinement methods and in growing highly perfect single crystals.

The application of laser heating thus extends the scope of crucible-free methods of crystal growth of refractory oxides and allows the direct observation of the growth processes.

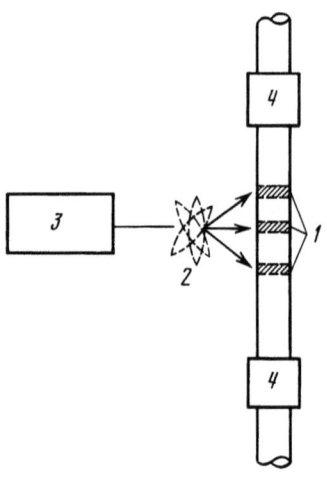

Fig. 6. Laser-heated zone melting. (1) molten zone, (2) focusing lens, (3) laser head, (4) holder for the starting material.

Literature Cited

1. D. B. Gasson and B. Cockayne. Oxide crystal growth using gas laser. J. Mater. Sci., **5**, 100-104 (1970).
2. C. A. Burrus and J. Stone. Single crystal fiber optical devices: A Nd:YAG fiber laser. Appl. Phys. Lett., **26**, 318-320 (1975).
3. C. A. Burrus and L. A. Coldren. Growth of single-crystal sapphire-clad ruby fibers. Appl. Phys. Lett., 31, 383-384 (1977).
4. K. Takagi and M. Isaii. Crystal growth of sapphire filaments by a laser-heated floating zone technique. J. Mater, Sci., **12**, 517-521 (1977).
5. K. Takagi and M. Isaii. Growth of LaB_6 single crystals by a laser-heated floating zone method. J. Cryst. Growth, **40**, 1-5 (1977).
6. U. C. Pack. Laser drawing of optical fibers. J. Appl. Opt., **13**, 1383-1386 (1974).

7. A. L. Mikaelyan and V. V. Dyachenko. Lasers with waveguide cavities. Kvantovaya Elektronika, No. 4, 937-938 (in Russian) (1974).
8. W. Volinsky and Z. Puzewicz. Application of optical quantum generators for the production of electronic parts. In: World Electrical Engineering Congress, June 21-25, Moscow, Abstracts, Sec. 4B, Paper 63, (1977).
9. R. A. Laudise. The Growth of Single Crystals. Prentice-Hall Inc., New Jersey, (1970).

SUBJECT INDEX

Acicular structure, 147-8
Adsorption energy, total, 21, 22, 39
Adsorption layer, 34, 35, 40, 43, 158-60, 167
  equilibrium, 160
Amethyst coloring, nature of, 312, 318
Autoclave, design of, 266-7

Binary systems, 129, 153, 164, 166
  lattice spacing, 303
  non-isothermal, 132-4
  phase diagrams of, 229-35
Blocking by impurity atoms, 39-43, 60, 333
  weak, medium, total, 333
Bridgman ampoule, 345
Bridgman furnace, 345
Bridgman-Stockbarger growth technique, 234
Bulk diffusion, 25
Burton-Cabrera-Frank equation, 54
Burton-Cabrera-Frank theory, 24, 27, 38

Capillary forces, 93
Capillary phenomena, 192-3
Capillary stability, 193-4, 203
CAST technology, 185
Close-packed faces, 39-43, 201
  monolayer, 111
Closest packing, principle of, 272, 273
Clusters, 291-2, 296-7, 332, 336,
  distribution in crystals, 292
Compound crystals, synthesis of, 64
Condensational restructuring, 84
Conglomerates, graphite, 332
Coordination number, 269, 272, 273
Coverage, 20, 22, 23, 42
Crystal shape stability
  applications of, 214-19
  criteria for, 213-14
Crystal-melt system, stability of, 193
Crystallization
  chloride-hydrogen, 30
  controlled, 240-247
  flux encapsulation, 282, 355
  by gas transport chemical reaction, 63-4
  heterogeneous, 97, 101, 103, 105

Crystallization (continued)
  homogeneous, 95, 99, 101, 103, 105
  microrelief of, 36
  of samples, 95
  from solution, 91
  spontaneous, 239, 246, 252
  uniaxial, 222
Crystallization centers, localization of, 239
Crystallization equipment, 353
Crystallization interface, see Interface
Crystallization pressure, 92
Czochralski growth technique
  control of impurity distribution, 200
  crucible-free, 367
  edge-defined growth, 185-6
  growth defects, 291
  interface shape, 224-6
  mathematical model, 222-3
  modified, 242
  monitoring of, 354, 356-7, 360, 361
  shape control in, 210
  shape of meniscus in, 191, 354
  shape stability in, 214-16
  solid state electrolytes, 346
  thermoelastic stress, 282

Dendritic growth, 147-8
Diamond
  autoepitaxial growth of, 62
  polycrystalline layers of, 62
Diamond-like carbon, 63
Die shaper, see Edge-definer
Diffusion
  ion, 349
  one-dimensional, 349
  surface, see Surface diffusion
Diffusion, bulk, 25
Diffusion activation energy, 24
Diffusion coefficients, 309-10
  of Ge in liquid Ge-Au, 180
Diffusion-driven transport
  in the melt, 168-9
  through a boundary layer, 40
Diffusion-limited processes, 157

Dislocations, 286-89
  critical stress of generation, 226
  density of, 288, 293-6, 307
  generation of, 145, 204, 279, 286, 293, 295, 308
  multiplication of, 288
  orientation of, 287
  partial, 330
  screw, 121, 176
  thermally induced, 299
  tilted, 287
Dissolution mechanism
  dislocation, 176
  kinetic coefficient for, 176
  normal, 175-6

Echelon (train) of steps, 35, 36, 39, 54, 55
Edge-definer (die shaper), 198, 200, 202, 209, 216, 219
Edge-defined film-fed growth (EFG), 185, 187-93, 200, 208, 218, 219
Energy barrier, height of, 96
Epitaxial layer, 30-1, 296-7
Epitaxy
  chemical vapor deposition (CVD), 27, 46, 359
  liquid-phase (LPE), 244, 301, 305, 306
  molecular-beam (MBE), 359, 361
  planar selective, 46, 47, 49
Etching, thermal, 307
Evaporation structure, 54
Excitation of vapor phase, 63

'Facet effect,' 201, 204
Field erosion, 77
Film growth
  diffusion mode, 244
  kinetic mode, 244
Flux growth, 246

Gas transport epitaxy, 46-56
Generation of nuclei, 42
Grain boundaries
  large angle, 328-30, 332, 334, 336, 337, 338
  small angle, 327-28, 330, 336
  tilted, 337
  twinning, 337
Growth
  barrier-free, 36
  crucible-free, 364
  by CVD, 27, 46, 361
  floating zone, (see also Zone melting), 210, 218-19
  from high-temperature solution, 349-51

Growth (continued)
  from the melt: Czochralski, see Czochralski growth; hydrothermal, see Hydrothermal growth; Stepanov, see Stepanov growth
  seeded, 240, 242, 255, 268
    by temperature gradient, 255,
    from low-temperature solutions, 359
  of shaped crystals, see Shaped single crystals
  web-dendritic, 208
  by zone melting, see Zone melting
Growth defects, 298
  distribution of, 298
Growth forms, 138-40
  constrained, 298
  unstable, 146-7
Growth interface, see Interface
Growth mechanism
  dislocation, 140, 145
  kinetic, 28, 34
  layer, 34-5, 43, 138, 143
  normal, 38, 138, 143, 144, 153, 154, 157, 221
  quasinormal, 38, 43
  two-dimensional, 29
  VLS, 43
Growth pyramids, 31, 298
Growth rate, 138
  and amethyst coloring, 316
  anisotropy of, 35-43, 44, 46, 47, 49, 137, 201, 203
  critical, 222
  effect of impurities on, 141
  normal, 46, 128, 132, 221
  tangential, 46, 299
  temperature dependence of, 140-1
Growth steps, elementary, 25, 38, 127
Growth surface
  macroscopic structure of, 298
  step-layer structure, 43
  step-like structure, 35, 38, 43

Habit, faceted, 137
Heteroepitaxy, liquid-phase, 309-10
Heterojunction, 307
Homoepitaxy, liquid-phase, 307
Hydrothermal growth, 255-8, 358, 361
  phase composition of products of, 270

Impurities (dopant), 122
  distribution of, 145, 294-5,
    coefficient of, 142, 159, 160
Impurity bands, 294-5

Interface (see also Growth surface; surface)
  atomically rough, 152-3, 168
  atomically smooth, 137, 138
  control of temperature gradients, 365
  curvilinear, 291, 293, 298, 356-7
  diffuse, 137, 138, 291
  laser detection of, 353
  shape of, 138-40, 224
Interface transition, rough to smooth, 137, 158
Interfacial free energy
  solid-liquid, 28, 96, 178, 307
  vapor-crystal, 28-30,
    anisotropy in, 210
    for Si in contact with Cl-H system, 30
Ion diffusion, 351

Jackson's criterion, 148

Kinematic theory of crystal growth, 54
Kinetic coefficients, 60, 129, 131, 133, 141, 175, 224
Kinetic parameters, 176, 247
Kinetics, atomic, 175
Kinetics-limited processes, 157
Kyropoulos growth technique, 356

Langmuir isotherm, 42
Laplace's capillary equation, 187, 188, 190
Lattice parameter misfit, 308
Legendre elliptic function, 191
Local epitaxial structures, 46-56

Macrosteps, 127-35, 296, 299
Mass transfer, 263-6, 353, 356
  gradient of, 240
Mechanisms of growth, see Growth mechanism
Melted rod growth technique, 366, 367
Melting, incongruent, 234
Meniscus, shape of, 191, 354
Metastable crystals
  carbon, 61
  elemental, 60-1
  gallium, 61
  growth of, 58-64, 121, 157
  heterogeneous nucleation of, 59
  homogeneous nucleation of, 59-60
  phosphorus, 61
  polymorphic modifications of, 64
  structure of, 58, 121, 122
  sulfur, 61
Microcrystals, 89
Mixed crystals, growth rate of, 152-60
Morphological instability, 128

Navier-Stokes equation, 192
Nucleation
  activation barrier for, 36-8
  activation energy of, 71, 307
  theory of, 97
Nucleation rate, 71, 96
Nuclei
  critical, 96-7,
    work of formation, 97, 178
  generation of, 42
  three-dimensional, 60
  two-dimensional, 60, 137, 140, 157, 175, 176, 181

Oxidation of SiC, 117
  rate of, 117

Paratellurite, growth rate of, 264-5
Phase diagrams, 31, 168
  of binary systems, 229-35
  crystallization, 269, 270
  kinetic, 167, 168-71
Phase equilibrium, 131, 132, 133, 229, 305
Phase transition(s), 59, 96, 100, 121
  in AgI, 343, 344
  of the first kind, 110, 111
  in SiC, 119, 121, 122
Pleochroism, anomalous, 319-20
Polymerization, degree of, 270, 272
Polymorphism
  of protein crystals, 11-14
  surface, 60
Polytypism, effect of impurities, 119, 122
Profile curves of melt-grown crystals, 191
Protein
  crystallization of, 9-11
  unit cell dimensions, 3
Protein molecule, globular, 3
Protein structure
  primary, 5
  quaternary, 5
  secondary, 5
  tertiary, 5

Quartz reactor, 261
Quasiliquid layer, 30-1

Replication, 83-4, 113
Ribbon crystals, 204
  shape stability of, 217
Roughness, kinetic, 29
Rounded shapes of crystals, 137, 138, 144, 147

Screw dislocations, 121, 176
Sectorial structure of crystals, 298, 312, 314
Shaped single crystals, 187-93, 198-206
SiC polytypes, 117-22
Singular planes, 40
Singular surfaces, 40
Solid solutions
   continuous, 231-4
   liquidation of, 246
   quaternary, 301-3
   rhombic, 234
   ternary, 305, 307
   volatility of, 246
Solid state electrolytes, 343-51
Stability of growth
   caillary, 193-4, 203
   thermal, 194
Stefan problem, 222, 224, 280
Step generation sites, 42
Step height, 39, 63
Stepanov growth technique, 185-6, 191, 194, 201, 209, 219
Steps
   echelon (train) of, 35, 36, 39, 54, 55
   kinematic density waves, 38
   motion of, 54
   specific energy of, 30
   tangential displacement velocity of, 39, 128
Strain, thermoelastic, 280
Stress
   residual, 279, 284, 292, 303
   thermoelastic, 204, 226, 279, 280, 282, 286, 295,
     profile of, 280-1, 284
   thermoplastic, 284, 285
Striation inhomogeneity, 294
Sublimation, 118
Submonolayer films, 108-15
Substrate defects, 72
Supercooling
   constitutional, 143
   critical, 138, 140, 143, 149, 195
   optimal, 309
   relative, 101, 146-8
Superheating, critical, 177
   critical, 177
Superlattice, two-dimensional, 25
Supersaturation
   chemical vapor deposition, 27, 28, 29
   gallium arsenide, 40, 41
   jumps, 121
   macrosteps and, 128
   maximization of, 95

Supersaturation (continued)
   oscillations of, 122
   protein solutions, 9, 10, 11
   relative, 130
   in vapor, 28
Surface
   atomically rough, 80, 113, 138
   microscopically rough, 42, 43, 137, 244, 246
     step layer structure of, 43
   microscopically smooth, 244
   vicinal, 38
Surface defects, 72
Surface diffusion, 24
   coefficients of, 25, 39
Surface energy of face covered with adsorption layer, 28
Surface melting, 98, 99, 103
Surface tension at solid-melt interface, 138

Tobacco mosaic virus, structure of, 12
Train (echelon) of steps, 35, 36, 39, 54, 55
Tubular crystals of enzymes, 12-13
Twinning, 72, 202, 203, 312, 316
Twins
   Brazil, 312, 316
   Dauphiné, 312, 316
Tysonite, 234

Undersaturation, 130
   critical, 177
Uniaxial crystallization
   horizontal, 353
   vertical, 356

Vapor-liquid-solid (VLS) growth mechanism, 43, 67, 68, 71, 72, 91
   whiskers, 31, 67, 180
Verneuil flame-fusion, 356, 361
   with laser heating, 366
Volmer-Becker-Döring theory, 27

Whiskers
   growth by VLS mechanism, 31, 67, 80
   growth rate, 30
   nucleation of
     at defects on substrate surface, 67
     in liquid phase, 31
     kinetic range of, 67
     role of impurities in, 67-72

Zone melting, 222, 364
   floating, 222, 224, 225, 356, 361
   with laser heating, 368
   temperature-gradient, 174

MIX
Papier aus verantwortungsvollen Quellen
Paper from responsible sources
FSC® C105338

If you have any concerns about our products,
you can contact us on
ProductSafety@springernature.com

In case Publisher is established outside the EU,
the EU authorized representative is:
**Springer Nature Customer Service Center GmbH
Europaplatz 3, 69115 Heidelberg, Germany**

Printed by Libri Plureos GmbH
in Hamburg, Germany